Understanding Linear Algebra

Understanding Linear Algebra

David Austin
Grand Valley State University

December 11, 2023

Edition: 2023 Update

Website: http://gvsu.edu/s/0Ck

©2023   David Austin

This work is licensed under a Creative Commons Attribution 4.0 International License[1].

---

[1]creativecommons.org

For Sam and Henry

Acknowledgements

Many people have supported and shaped this project. First is my colleague Matt Boelkins, whose *Active Calculus*[2] is a model for how an open textbook can encourage and facilitate good pedagogy. The debt that this book owes to Matt's cannot be overstated. In addition, he has provided a great deal of editorial feedback on this text and improved it in countless ways. Over many, many years, I have valued Matt's friendship and wise counsel.

I could not imagine a more supportive environment than the mathematics department at Grand Valley State University. The influence of my colleagues and their deep commitment to student growth is embedded in every page of this book. Conversations about the teaching of linear algebra with Paul Fishback have been especially helpful as has editorial feedback from Lauren Keough and Lora Bailey. I am also grateful for a sabbatical leave in 2017 during which I began this project.

In addition to my colleagues, I am grateful for the many students who have helped me grow as a teacher. Thank you for your willingness to engage in this very human art of learning and for sharing your experiences, frustrations, and successes with me.

The open textbook community that has grown around the PreTeXt authoring and publishing system is a continual source of support and inspiration. The goal of providing all students with high-quality, affordable textbooks is ambitious, but the commitment of this passionate and dedicated group makes clear that it is possible. As part of that community, Mitch Keller and Kathy Yoshiwara have read much of this book and provided detailed and insightful editorial feedback.

There coud be no better partners than Candice Price, Miloš Savić, and their team at 619 Wreath Publishing.[3] Thank you for your support of this project and for everything you do to further your mission to "foster creativity, equity, and scholarship."

Finally, a book is nothing without readers, and I am so thankful for all the instructors, students, and self-learners who have reached out with suggestions, comments, and questions. Hearing from those who are using the book gives meaning to this project, so please know that your voice is always welcomed.

---

[2] activecalculus.org
[3] www.619wreath.com/

Our goals

This is a textbook for a first-year course in linear algebra. Of course, there are already many fine linear algebra textbooks available. Even if you are reading this one online for free, you should know that there are other free linear algebra textbooks available online. You have choices! So why would you choose this one?

This book arises from my belief that linear algebra, as presented in a traditional undergraduate curriculum, has for too long lived in the shadow of calculus. Many mathematics programs currently require their students to complete at least three semesters of calculus, but only one semester of linear algebra, which often has two semesters of calculus as a prerequisite.

In addition, what linear algebra students encounter is frequently presented in an overly formal way that does not fully represent the range of linear algebraic thinking. Indeed, many programs use a first course in linear algebra as an "introduction to proofs" course. While linear algebra provides an excellent introduction to mathematical reasoning, to only emphasize this aspect of the subject neglects some important student needs.

Of course, linear algebra is based on a set of abstract principles. However, these principles underlie an astonishingly wide range of technology that shapes our society in profound ways. The interplay between these principles and their applications provides a unique opportunity for working with students. First, the consideration of significant real-world problems grounds abstract mathematical thinking in a way that deepens students' understanding. At the same time, the variety of ways in which these abstract principles may be applied clearly demonstrates for students the power of mathematical abstraction. Linear algebra empowers students to experience what the physicist Eugene Wigner called "the unreasonable effectiveness of mathematics in the natural sciences," an aspect of mathematics that is both fundamental and mysterious.

Neglecting this experience does not serve our students well. For instance, only about 15% of current mathematics majors will go on to attend graduate school. The remainder are headed for careers that will ask them to use their mathematical training in business, industry, and government. What do these careers look like? Right now, data analytics and data mining, computer graphics, software development, finance, and operations research. These careers depend much more on linear algebra than calculus. In addition to helping students appreciate the profound changes that mathematics has brought to our society, more training in linear algebra will help our students participate in the inevitable developments yet to come.

These thoughts are not uniquely mine nor are they particularly new. The Linear Algebra

Curriculum Study Group, a broadly-based group of mathematicians and mathematics educators funded by the National Science Foundation, formed to improve the teaching of linear algebra. In their final report, they wrote

> There is a growing concern that the linear algebra curriculum at many schools does not adequately address the needs of the students it attempts to serve. In recent years, demand for linear algebra training has risen in client disciplines such as engineering, computer science, operations research, economics, and statistics. At the same time, hardware and software improvements in computer science have raised the power of linear algebra to solve problems that are orders of magnitude greater than dreamed possible a few decades ago. Yet in many courses, the importance of linear algebra in applied fields is not communicated to students, and the influence of the computer is not felt in the classroom, in the selection of topics covered or in the mode of presentation. Furthermore, an overemphasis on abstraction may overwhelm beginning students to the point where they leave the course with little understanding or mastery of the basic concepts they may need in later courses and their careers.

Furthermore, among their recommendations is this:

> We believe that a first course in linear algebra should be taught in a way that reflects its new role as a scientific tool. This implies less emphasis on abstraction and more emphasis on problem solving and motivating applications.

What may be surprising is that this was written in 1993; that is, before the introduction of Google's PageRank algorithm, before Pixar's *Toy Story*, and before the ascendence of what is often called data science or machine learning made these statements only more relevant.

With these thoughts in mind, the aim of this book is to facilitate a fuller, richer experience of linear algebra for all students, which informs the following decisions.

- *This book is written without the assumption that students have taken a calculus course.* In making this decision, I hope that students will gain a more authentic experience of mathematics through linear algebra at an earlier stage of their academic careers.

  Indeed, a common barrier to student success in calculus is its relatively high prerequisite tower culminating in a course often called "Precalculus". By contrast, linear algebra begins with much simpler assumptions about our students' preparation: the expressions studied are linear so that may be manipulated using only the four basic arithmetic operations.

  The most common explanation I hear for requiring calculus as a prerequisite for linear algebra is that calculus develops in students a beneficial "mathematical maturity." Given persistent student struggles with calculus, however, it seems just as reasonable to develop students' abilities to reason mathematically through linear algebra.

- *The text includes a number of significant applications of important linear algebraic concepts,* such as computer animation, the JPEG compression algorithm, and Google's PageRank algorithm. In my experience, students find these applications more authentic and compelling than typical applications presented in a calculus class. These applications also provide a strong justification for mathematical abstraction, which can seem unnecessary to beginning students, and demonstrate how mathematics is currently shaping our world.

- *Each section begins with a preview activity and includes a number of activities that can be used to facilitate active learning in a classroom.* By now, active learning's effectiveness in helping students develop a deep understanding of important mathematical concepts is beyond dispute. The activities here are designed to reinforce ideas already encountered, motivate the need for upcoming ideas, and help students recognize various manifestations of simple underlying themes. As much as possible, students are asked to develop new ideas and take ownership of them.

- *The activities emphasize a broad range of mathematical thinking.* Rather than providing the traditional cycle of Definition-Theorem-Proof, *Understanding Linear Algebra* aims to develop an appreciation of ideas as arising in response to a need that students perceive. Working much as research mathematicians do, students are asked to consider examples that illustrate the importance of key concepts so that definitions arise as natural labels used to identify these concepts. Again using examples as motivation, students are asked to reason mathematically and explain general phenomena they observe, which are then recorded as theorems and propositions. It is not, however, the intention of this book to develop students' formal proof-writing abilities.

- *There are frequent embedded Sage cells that help develop students' computational proficiency.* The impact that linear algebra is having on our society is inextricably tied to the phenomenal increase in computing power witnessed in the last half-century. Indeed, Carl Cowen, former president of the Mathematical Association of America, has said, "No serious application of linear algebra happens without a computer." This means that an understanding of linear algebra is not complete without an understanding of how linear algebraic ideas are deployed in a computational environment.

- *The text aims to leverage geometric intuition to enhance algebraic thinking.* In spite of the fact that it may be difficult to visualize phenomena in thousands of dimensions, many linear algebraic concepts may be effectively illustrated in two or three dimensions and the resulting intuition applied more generally. Indeed, this useful interplay between geometry and algebra illustrates another mysterious mathematical connection between seemingly disparate areas.

I hope that *Understanding Linear Algebra* is useful for you, whether you are a student taking a linear algebra class, someone just interested in self-study, or an instructor seeking out some ideas to use with your students. I would be more than happy to hear your feedback.

A note on the print version

This book aims to develop readers' ability to reason about linear algebraic concepts and to apply that reasoning in a computational environment. In particular, Sage is introduced as a platform for performing many linear algebraic computations since it is freely available and its syntax mirrors common mathematical notation.

Print readers may access Sage online using either the Sage cell server[4] or a provided page of Sage cells.[5]

Throughout the book, Sage cells appear in various places to encourage readers to use Sage to complete some relevant computation. In the print version, these may appear with some pre-populated code, such as the one below, that you will want to copy into an online Sage cell.

```
A = matrix([[1,2], [2,1]])
```

Empty cells appear as shown below and are included to indicate part of an exercise or activity that is meant to be completed in Sage.

---

[4] sagecell.sagemath.org/
[5] https:gvsu.edu/s/0Ng

Contents

Acknowledgements   vii

Our goals   ix

A note on the print version   xiii

1 Systems of equations   1

    1.1   What can we expect 1

    1.2   Finding solutions to linear systems 10

    1.3   Computation with Sage 22

    1.4   Pivots and their influence on solution spaces 34

2 Vectors, matrices, and linear combinations   45

    2.1   Vectors and linear combinations 45

    2.2   Matrix multiplication and linear combinations 63

    2.3   The span of a set of vectors 80

    2.4   Linear independence 96

    2.5   Matrix transformations 108

    2.6   The geometry of matrix transformations 123

3 Invertibility, bases, and coordinate systems   145

    3.1   Invertibility 145

3.2	Bases and coordinate systems	159
3.3	Image compression	178
3.4	Determinants	201
3.5	Subspaces	217

4 Eigenvalues and eigenvectors — 229

4.1	An introduction to eigenvalues and eigenvectors	229
4.2	Finding eigenvalues and eigenvectors	243
4.3	Diagonalization, similarity, and powers of a matrix	256
4.4	Dynamical systems	271
4.5	Markov chains and Google's PageRank algorithm	291

5 Linear algebra and computing — 311

5.1	Gaussian elimination revisited	311
5.2	Finding eigenvectors numerically	324

6 Orthogonality and Least Squares — 335

6.1	The dot product	335
6.2	Orthogonal complements and the matrix transpose	355
6.3	Orthogonal bases and projections	367
6.4	Finding orthogonal bases	383
6.5	Orthogonal least squares	394

7 Singular value decompositions — 415

7.1	Symmetric matrices and variance	415
7.2	Quadratic forms	432
7.3	Principal Component Analysis	446
7.4	Singular Value Decompositions	458
7.5	Using Singular Value Decompositions	474

Appendices

A Sage Reference   491

Back Matter

Index   497

CHAPTER 1

Systems of equations

1.1 What can we expect

At its heart, the subject of linear algebra is about linear equations and, more specifically, sets of two or more linear equations. Google routinely deals with a set of trillions of equations each of which has trillions of unknowns. We will eventually understand how to deal with that kind of complexity. To begin, however, we will look at a more familiar situation in which there are a small number of equations and a small number of unknowns. In spite of its relative simplicity, this situation is rich enough to demonstrate some fundamental concepts that will motivate much of our exploration.

1.1.1 Some simple examples

Activity 1.1.1. In this activity, we consider sets of linear equations having just two unknowns. In this case, we can graph the solutions sets for the equations, which allows us to visualize different types of behavior.

  a. On the grid below, graph the lines

  $$y = x + 1$$
  $$y = 2x - 1.$$

  At what point or points (x, y), do the lines intersect? How many points (x, y) satisfy both equations?

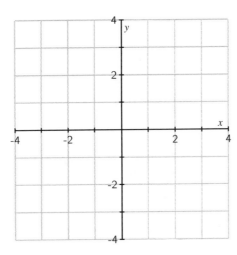

b. On the grid below, graph the lines

$$y = x + 1$$
$$y = x - 1.$$

At what point or points (x, y), do the lines intersect? How many points (x, y) satisfy both equations?

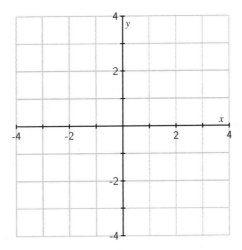

c. On the grid below, graph the line

$$y = x + 1.$$

How many points (x, y) satisfy this equation?

1.1. WHAT CAN WE EXPECT 3

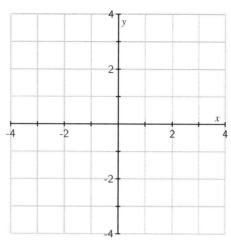

d. On the grid below, graph the lines

$$y = x + 1$$
$$y = 2x - 1$$
$$y = -x.$$

At what point or points (x, y), do the lines intersect? How many points (x, y) satisfy all three equations?

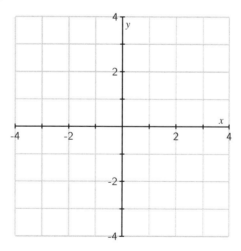

The examples in this introductory activity demonstrate several possible outcomes for the solutions to a set of linear equations. Notice that we are interested in points that satisfy each equation in the set and that these are seen as intersection points of the lines. Similar to the examples considered in the activity, three types of outcomes are seen in Figure 1.1.1.

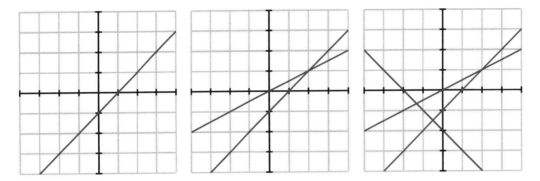

Figure 1.1.1 Three possible graphs for sets of linear equations in two unknowns.

In this figure, we see that

- With a single equation, there are infinitely many points (x, y) satisfying that equation.
- Adding a second equation adds another condition we place on the points (x, y) resulting in a single point that satisfies both equations.
- Adding a third equation adds a third condition on the points (x, y), and there is no point that satisfies all three equations.

Generally speaking, a single equation will have many solutions, in fact, infinitely many. As we add equations, we add conditions which lead, in a sense we will make precise later, to a smaller number of solutions. Eventually, we have too many equations and find there are no points that satisfy all of them.

This example illustrates a general principle to which we will frequently return.

> **Solutions to sets of linear equations.**
>
> Given a set of linear equations, there are either:
> - infinitely many points,
> - exactly one point, or
> - no points
>
> that satisfy every equation in the set.

Notice that we can see a bit more. In Figure 1.1.1, we are looking at equations in two unknowns. Here we see that

- One equation has infinitely many solutions.
- Two equations have exactly one solution.
- Three equations have no solutions.

1.1. WHAT CAN WE EXPECT

It seems reasonable to wonder if the number of solutions depends on whether the number of equations is less than, equal to, or greater than the number of unknowns. Of course, one of the examples in the activity shows that there are exceptions to this simple rule, as seen in Figure 1.1.2. For instance, two equations in two unknowns may correspond to parallel lines so that the set of equations has no solutions. It may also happen that a set of three equations in two unknowns has a single solution. However, it seems safe to think that the more equations we have, the smaller the set of solutions will be.

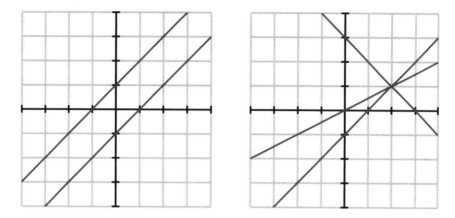

Figure 1.1.2 A set of two equations in two unknowns can have no solutions, and a set of three equations can have one solution.

Let's also consider some examples of equations having three unknowns, which we call x, y, and z. Just as solutions to linear equations in two unknowns formed straight lines, solutions to linear equations in three unknowns form planes.

When we consider an equation in three unknowns graphically, we need to add a third coordinate axis, as shown in Figure 1.1.3.

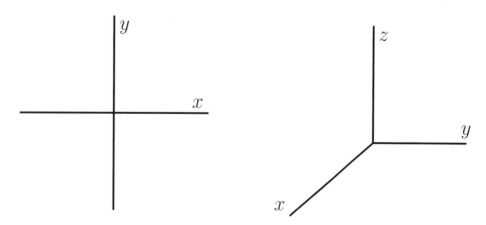

Figure 1.1.3 Coordinate systems in two and three dimensions.

As shown in Figure 1.1.4, a linear equation in two unknowns, such as $y = 0$, is a line while a linear equation in three unknowns, such as $z = 0$, is a plane.

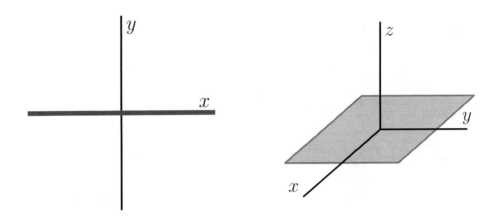

Figure 1.1.4 The solutions to the equation $y = 0$ in two dimensions and $z = 0$ in three.

In three unknowns, the set of solutions to one linear equation forms a plane. The set of solutions to a pair of linear equations is seen graphically as the intersection of the two planes. As in Figure 1.1.5, we typically expect this intersection to be a line.

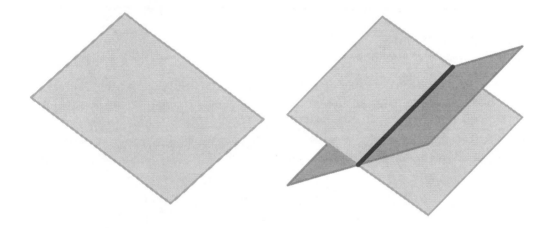

Figure 1.1.5 A single plane and the intersection of two planes.

When we add a third equation, we are looking for the intersection of three planes, which we expect to form a point, as in the left of Figure 1.1.6. However, in certain special cases, it may happen that there are no solutions, as seen on the right.

1.1. WHAT CAN WE EXPECT

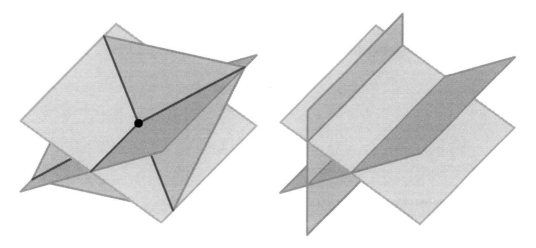

Figure 1.1.6 Two examples showing the intersections of three planes.

> **Activity 1.1.2.** This activity considers sets of equations having three unknowns. In this case, we know that the solutions of a single equation form a plane. If it helps with visualization, consider using 3 × 5-inch index cards to represent planes.
>
> a. Is it possible that there are no solutions to two linear equations in three unknowns? Either sketch an example or state a reason why it can't happen.
>
> b. Is it possible that there is exactly one solution to two linear equations in three unknowns? Either sketch an example or state a reason why it can't happen.
>
> c. Is it possible that the solutions to four equations in three unknowns form a line? Either sketch an example or state a reason why it can't happen.
>
> d. What would you usually expect for the set of solutions to four equations in three unknowns?
>
> e. Suppose we have a set of 500 linear equations in 10 unknowns. Which of the three possibilities would you expect to hold?
>
> f. Suppose we have a set of 10 linear equations in 500 unknowns. Which of the three possibilities would you expect to hold?

1.1.2 Systems of linear equations

Now that we have seen some simple examples, let's agree on some terminology to help us think more precisely about sets of equations.

First, we considered a linear equation having the form

$$y = 2x - 1.$$

It will be convenient for us to rewrite this so that all the unknowns are on one side of the

equation:
$$-2x + y = -1.$$

More generally, the equation of a line can always be expressed in the form
$$ax + by = c,$$
which gives us the flexibility to describe all lines. For instance, vertical lines, such as $x = 3$, may be represented in this form.

Notice that each term on the left is the product of a constant and the first power of an unknown. In the future, we will want to consider equations having many more unknowns, which we will sometimes denote as $x_1, x_2, \ldots, x_n$. This leads to the following definition:

Definition 1.1.7 A linear equation in the unknowns $x_1, x_2, \ldots, x_n$ may be written in the form
$$a_1 x_1 + a_2 x_2 + \ldots + a_n x_n = b,$$
where $a_1, a_2, \ldots, a_n$ are real numbers known as *coefficients*. We also say that $x_1, x_2, \ldots, x_n$ are the *variables* in the equation.

By a **system of linear equations** or a **linear system**, we mean a set of linear equations written in a common set of unknowns.

For instance,
$$\begin{aligned} 2x_1 + 1.2x_2 - 4x_3 &= 3.7 \\ -0.1x_1 + x_3 &= 2 \\ x_1 + x_2 - x_3 &= 1.4 \end{aligned}$$
is an example of a linear system.

Definition 1.1.8 A **solution** to a linear system is simply a set of numbers $x_1 = s_1, x_2 = s_2, \ldots, x_n = s_n$ that satisfy all the equations in the system.

For instance, we earlier considered the linear system
$$\begin{aligned} -x + y &= 1 \\ -2x + y &= -1. \end{aligned}$$

To check that $(x, y) = (2, 3)$ is a solution, we verify that the following equations are true.
$$\begin{aligned} -2 + 3 &= 1 \\ -2(2) + 3 &= -1. \end{aligned}$$

Definition 1.1.9 We call the set of all solutions the **solution space** of the linear system.

Activity 1.1.3 Linear equations and their solutions..

a. Which of the following equations are linear? Please provide a justification for your response.

1.
$$2x + xy - 3y^2 = 2.$$

2.
$$-2x_1 + 3x_2 + 4x_3 - x_5 = 0.$$

3.
$$x = 3z - 4y.$$

b. Consider the system of linear equations:

$$\begin{aligned} x + y &= 3 \\ y - z &= 2 \\ 2x + y + z &= 4. \end{aligned}$$

1. Is $(x, y, z) = (1, 2, 0)$ a solution?
2. Is $(x, y, z) = (-2, 1, 0)$ a solution?
3. Is $(x, y, z) = (0, -3, 1)$ a solution?
4. Can you find a solution in which $y = 0$?
5. Do you think there are other solutions? Please explain your response.

1.1.3 Summary

The point of this section is to build some intuition about the behavior of solutions to linear systems through consideration of some simple examples. We will develop a deeper and more precise understanding of these phenomena in our future explorations.

- A linear equation is one that may be written in the form

$$a_1 x_1 + a_2 x_2 + \ldots + a_n x_n = b.$$

- A linear system is a set of linear equations and a solution is a set of values assigned to the unknowns that make each equation true.

- We came to expect that a linear system has either infinitely many solutions, exactly one solution, or no solutions.

- When we add more equations to a system, the solution space usually seems to become smaller.

1.2 Finding solutions to linear systems

In the previous section, we looked at systems of linear equations from a graphical perspective. Since the equations had only two or three variables, we could study the solution spaces as the intersections of lines and planes.

Because we will eventually consider systems with many equations and many variables, this graphical approach will not generally be a useful strategy. Instead, we will approach this problem algebraically and develop a technique to describe the solution spaces of general linear systems.

1.2.1 Gaussian elimination

We will develop an algorithm, which is usually called *Gaussian elimination*, that allows us to describe the solution space of a linear system. This algorithm plays a central role in much of what is to come.

> **Preview Activity 1.2.1.** In this activity, we will consider some simple examples that will guide us in finding a more general approach.
>
> a. Give a description of the solution space to the linear system:
> $$\begin{aligned} x &= 2 \\ y &= -1. \end{aligned}$$
>
> b. Give a description of the solution space to the linear system:
> $$\begin{aligned} -x + 2y - z &= -3 \\ 3y + z &= -1 \\ 2z &= 4. \end{aligned}$$
>
> c. Give a description of the solution space to the linear system:
> $$\begin{aligned} x + 3y &= -1 \\ 2x + y &= 3. \end{aligned}$$
>
> d. Describe the solution space to the linear equation $0x = 0$.
>
> e. Describe the solution space to the linear equation $0x = 5$.

These examples lead to a few observations that motivate a general approach to finding solutions of linear systems.

Observation 1.2.1 First, finding the solution space to some systems is simple. For example,

1.2. FINDING SOLUTIONS TO LINEAR SYSTEMS

because each equation in the following system

$$x = -4$$
$$y = 2.$$

has only one variable, it prescribes a specific value for that variable. We therefore see that there is exactly one solution, which is $(x, y) = (-4, 2)$. We call such a system *decoupled*.

Observation 1.2.2 Second, there is a process that can be used to find solutions to certain types of linear systems. For instance, let's consider the system

$$x + 2y - 2z = -4$$
$$-y + z = 3$$
$$3z = 3.$$

Multiplying both sides of the last equation by 1/3 gives us

$$x + 2y - 2z = -4$$
$$-y + z = 3$$
$$z = 1.$$

Any solution to this linear system must then have $z = 1$.

Once we know that, we can substitute $z = 1$ into the first and second equations and simplify to obtain a new system of equations having the same solutions:

$$x + 2y = -2$$
$$-y = 2.$$

The second equation, after multiplying both sides by -1, tells us that $y = -2$. We can then substitute this value into the first equation to determine that $x = 2$.

In this way, we arrive at a decoupled system, which shows that there is exactly one solution, namely $(x, y, z) = (2, -2, 1)$.

Our original system,

$$x + 2y - 2z = -4$$
$$-y + z = 3$$
$$3z = 3,$$

is called a *triangular* system due to the shape formed by the coefficients. As this example demonstrates, triangular systems are easily solved by this process, which is called *back substitution*.

Observation 1.2.3 We can use substitution in a more general way to solve linear systems. For example, a natural approach to the system

$$x + 2y = 1$$
$$2x + 3y = 3.$$

is to use the first equation to express x in terms of y:

$$x = 1 - 2y$$

and then substitute this into the second equation and simplify:

$$\begin{aligned} 2x + 3y &= 3 \\ 2(1-2y) + 3y &= 3 \\ 2 - 4y + 3y &= 3 \\ -y &= 1 \\ y &= -1 \end{aligned}$$

From here, we can substitute $y = -1$ into the first equation to arrive at the solution $(x, y) = (3, -1)$.

However, the two-step process of solving for x in terms of y and substituting into the second equation may be performed more efficiently by adding a multiple of the first equation to the second. In this case, we will multiply the first equation by -2 and add to the second equation

$$\begin{array}{r} -2(\text{equation 1}) \\ + \quad \text{equation 2} \\ \hline \end{array}$$

to obtain

$$\begin{array}{r} -2(x + 2y = 1) \\ + \quad 2x + 3y = 3 \\ \hline \end{array} \quad \text{which gives us} \quad \begin{array}{r} -2x - 4y = -2 \\ + \quad 2x + 3y = 3 \\ \hline -y = 1. \end{array}$$

In this way, the system

$$\begin{aligned} x + 2y &= 1 \\ 2x + 3y &= 3 \end{aligned}$$

is transformed into the new triangular system

$$\begin{aligned} x + 2y &= 1 \\ -y &= 1. \end{aligned}$$

Notice that this process can be reversed. Beginning with the triangular system, we can recover the original system by multiplying the first equation by 2 and adding it to the second. Because of this, the two systems have the same solution space. We will revisit this point later and give what may be a more convincing explanation.

Of course, the choice to multiply the first equation by -2 was made so that the terms involving x in the two equations will cancel leading to a triangular system that can be solved using back substitution.

Based on these observations, we take note of three operations that transform a system of linear equations into a new system of equations having the same solution space. Our goal is to create a new system whose solution space is the same as the original system's and may

1.2. FINDING SOLUTIONS TO LINEAR SYSTEMS

be easily described.

Scaling We can multiply one equation by a nonzero number. For instance,
$$2x - 4y = 6$$
has the same set of solutions as
$$\frac{1}{2}(2x - 4y = 6)$$
or
$$x - 2y = 3.$$

Interchange Interchanging equations will not change the set of solutions. For instance,
$$2x + 4y = 1$$
$$x - 3y = 0$$
has the same set of solutions as
$$x - 3y = 0$$
$$2x + 4y = 1.$$

Replacement As we saw above, we may multiply one equation by a real number and add it to another equation. We call this process *replacement*.

Example 1.2.4 Let's illustrate the use of these operations to find the solution space to the system of equations:
$$x + 2y = 4$$
$$2x + y - 3z = 11$$
$$-3x - 2y + z = -10$$

We will first transform the system into a triangular system so we start by eliminating x from the second and third equations.

We begin with a replacement operation where we multiply the first equation by -2 and add the result to the second equation.
$$x + 2y = 4$$
$$-3y - 3z = 3$$
$$-3x - 2y + z = -10$$

Another replacement operation eliminates x from the third equation. We multiply the first equation by 3 and add to the third.
$$x + 2y = 4$$
$$-3y - 3z = 3$$
$$4y + z = 2$$

Scale the second equation by multiplying it by $-1/3$.
$$x + 2y = 4$$
$$y + z = -1$$
$$4y + z = 2$$

Eliminate y from the third equation by multiplying the second equation by -4 and adding it to the third. Notice that we now have a triangular system that can be solved using back substitution.

$$\begin{aligned} x + 2y &= 4 \\ y + z &= -1 \\ -3z &= 6 \end{aligned}$$

After scaling the third equation by $-1/3$, we have found the value for z.

$$\begin{aligned} x + 2y &= 4 \\ y + z &= -1 \\ z &= -2 \end{aligned}$$

We eliminate z from the second equation by multiplying the third equation by -1 and adding to the second.

$$\begin{aligned} x + 2y &= 4 \\ y &= 1 \\ z &= -2 \end{aligned}$$

Finally, multiply the second equation by -2 and add to the first to obtain:

$$\begin{aligned} x &= 2 \\ y &= 1 \\ z &= -2. \end{aligned}$$

Now that we have arrived at a decoupled system, we know that there is exactly one solution to our original system of equations, which is $(x, y, z) = (2, 1, -2)$.

One could find the same result by applying a different sequence of replacement and scaling operations. However, we chose this particular sequence guided by our desire to first transform the system into a triangular one. To do this, we eliminated the first variable x from all but one equation and then proceeded to the next variables working left to right. Once we had a triangular system, we used back substitution moving through the variables right to left.

We call this process *Gaussian elimination* and note that it is our primary tool for solving systems of linear equations.

Activity 1.2.2 Gaussian Elimination.. For each of the following linear systems, use Gaussian elimination to describe the solutions to the following systems of linear equations. In particular, determine whether each linear system has exactly one solution, infinitely many solutions, or no solutions.

a. $\begin{aligned} x + y + 2z &= 1 \\ 2x - y - 2z &= 2 \\ -x + y + z &= 0 \end{aligned}$

b. $\begin{aligned} -x - 2y + 2z &= -1 \\ 2x + 4y - z &= 5 \\ x + 2y &= 3 \end{aligned}$

c. $\begin{aligned} -x - 2y + 2z &= -1 \\ 2x + 4y - z &= 5 \\ x + 2y &= 2 \end{aligned}$

1.2. FINDING SOLUTIONS TO LINEAR SYSTEMS

1.2.2 Augmented matrices

After performing Gaussian elimination a few times, you probably noticed that you spent most of the time concentrating on the coefficients and simply recorded the variables as place holders. Based on this observation, we will introduce a shorthand description of linear systems.

When writing a linear system, we always write the variables in the same order in each equation. We then construct an *augmented matrix* by simply forgetting about the variables and recording the numerical data in a rectangular array. For instance, the system of equations below has the following augmented matrix

$$\begin{aligned} -x - 2y + 2z &= -1 \\ 2x + 4y - z &= 5 \\ x + 2y &= 3 \end{aligned} \qquad \left[\begin{array}{rrr|r} -1 & -2 & 2 & -1 \\ 2 & 4 & -1 & 5 \\ 1 & 2 & 0 & 3 \end{array}\right].$$

The vertical line reminds us where the equals signs appear in the equations. Entries in the matrix to the left of the vertical line correspond to coefficients of the equations. We sometimes choose to focus only on the coefficients of the system in which case we write the *coefficient matrix* as

$$\left[\begin{array}{rrr} -1 & -2 & 2 \\ 2 & 4 & -1 \\ 1 & 2 & 0 \end{array}\right].$$

The three operations we perform on systems of equations translate naturally into operations on matrices. For instance, the replacement operation that multiplies the first equation by 2 and adds it to the second may be performed by multiplying the first row of the augmented matrix by 2 and adding it to the second row:

$$\left[\begin{array}{rrr|r} -1 & -2 & 2 & -1 \\ 2 & 4 & -1 & 5 \\ 1 & 2 & 0 & 3 \end{array}\right] \sim \left[\begin{array}{rrr|r} -1 & -2 & 2 & -1 \\ 0 & 0 & 3 & 3 \\ 1 & 2 & 0 & 3 \end{array}\right].$$

The symbol $\sim$ between the matrices indicates that the two matrices are related by a sequence of scaling, interchange, and replacement operations. Since these operations act on the rows of the matrices, we say that the matrices are *row equivalent*. Notice that the linear systems corresponding to two row equivalent augmented matrices have the same solution space.

> **Activity 1.2.3 Augmented matrices and solution spaces..**
>
> a. Write the augmented matrix for the linear system
>
> $$\begin{aligned} x + 2y - z &= 1 \\ 3x + 2y + 2z &= 7 \\ -x + 4z &= -3 \end{aligned}$$
>
> and perform Gaussian elimination to describe the solution space in as much detail as you can.
>
> b. Suppose that you have a linear system in the variables x and y whose aug-

mented matrix is row equivalent to

$$\begin{bmatrix} 1 & 0 & | & 3 \\ 0 & 1 & | & 0 \\ 0 & 0 & | & 0 \end{bmatrix}.$$

Write the linear system corresponding to this augmented matrix and describe its solution set in as much detail as you can.

c. Suppose that you have a linear system in the variables x and y whose augmented matrix is row equivalent to

$$\begin{bmatrix} 1 & 0 & | & 3 \\ 0 & 1 & | & 0 \\ 0 & 0 & | & 1 \end{bmatrix}.$$

Write the linear system corresponding to this augmented matrix and describe its solution set in as much detail as you can.

d. Suppose that the augmented matrix of a linear system has the following shape where * could be any real number.

$$\begin{bmatrix} * & * & * & * & * & | & * \\ * & * & * & * & * & | & * \\ * & * & * & * & * & | & * \end{bmatrix}.$$

1. How many equations are there in this system and how many variables?
2. Based on our earlier discussion in Section 1.1, do you think it's possible that this system has exactly one solution, infinitely many solutions, or no solutions?
3. Suppose that this augmented matrix is row equivalent to

$$\begin{bmatrix} 1 & 2 & 0 & 0 & 3 & | & 2 \\ 0 & 0 & 1 & 2 & -1 & | & -1 \\ 0 & 0 & 0 & 0 & 0 & | & 0 \end{bmatrix}.$$

Make a choice for the names of the variables and write the corresponding linear system. Does the system have exactly one solution, infinitely many solutions, or no solutions?

1.2.3 Reduced row echelon form

There is a special class of matrices whose form makes it especially easy to describe the solution space of the corresponding linear system. As we describe the properties of this class of

1.2. FINDING SOLUTIONS TO LINEAR SYSTEMS

matrices, it may be helpful to consider an example, such as the following matrix.

$$\begin{bmatrix} 1 & * & 0 & * & 0 & * \\ 0 & 0 & 1 & * & 0 & * \\ 0 & 0 & 0 & 0 & 1 & * \\ 0 & 0 & 0 & 0 & 0 & 0 \\ 0 & 0 & 0 & 0 & 0 & 0 \end{bmatrix}.$$

Definition 1.2.5 We say that a matrix is in *reduced row echelon form* if the following properties are satisfied.

- If the entries in a row are all zero, then the same is true of any row below it.

- If we move across a row from left to right, the first nonzero entry we encounter is 1. We call this entry the *leading entry* in the row.

- The leading entry in any row is to the right of the leading entries in all the rows above it.

- A leading entry is the only nonzero entry in its column.

We call a matrix in reduced row echelon form a *reduced row echelon matrix*.

We have been intentionally vague about whether the matrix we are considering is an augmented matrix corresponding to a linear system or a coefficient matrix since we will consider both possibilities in the future.

> **Activity 1.2.4 Identifying reduced row echelon matrices..** Consider each of the following augmented matrices. Determine if the matrix is in reduced row echelon form. If it is not, perform a sequence of scaling, interchange, and replacement operations to obtain a row equivalent matrix that is in reduced row echelon form. Then use the reduced row echelon matrix to describe the solution space.
>
> a. $\begin{bmatrix} 2 & 0 & 4 & -8 \\ 0 & 1 & 3 & 2 \end{bmatrix}.$
>
> b. $\begin{bmatrix} 1 & 0 & 0 & -1 \\ 0 & 1 & 0 & 3 \\ 0 & 0 & 1 & 1 \end{bmatrix}.$
>
> c. $\begin{bmatrix} 1 & 0 & 4 & 2 \\ 0 & 1 & 3 & 2 \\ 0 & 0 & 0 & 1 \end{bmatrix}.$
>
> d. $\begin{bmatrix} 0 & 1 & 3 & 2 \\ 0 & 0 & 0 & 0 \\ 1 & 0 & 4 & 2 \end{bmatrix}.$
>
> e. $\begin{bmatrix} 1 & 2 & -1 & 2 \\ 0 & 1 & -2 & 0 \\ 0 & 0 & 1 & 1 \end{bmatrix}.$

The examples in the previous activity indicate that there is a sequence of row operations that transforms any matrix into one in reduced row echelon form. Moreover, the conditions that define reduced row echelon matrices guarantee that this matrix is unique.

Theorem 1.2.6 *For any given matrix, there is exactly one reduced row echelon matrix to which it is row equivalent.*

Once we have this reduced row echelon matrix, we may describe the set of solutions to the corresponding linear system with relative ease.

Example 1.2.7 Describing the solution space from a reduced row echelon matrix.

a. Consider the reduced row echelon matrix

$$\left[\begin{array}{ccc|c} 1 & 0 & 2 & -1 \\ 0 & 1 & 1 & 2 \end{array}\right]$$

and its corresponding linear system as

$$\begin{aligned} x + 2z &= -1 \\ y + z &= 2. \end{aligned}$$

Let's rewrite the equations as

$$\begin{aligned} x &= -1 - 2z \\ y &= 2 - z. \end{aligned}$$

From this description, it is clear that we obtain a solution for any value of the variable z. For instance, if $z = 2$, then $x = -5$ and $y = 0$ so that $(x, y, z) = (-5, 0, 2)$ is a solution. Similarly, if $z = 0$, we see that $(x, y, z) = (-1, 2, 0)$ is also a solution.

Because there is no restriction on the value of z, we call it a *free variable*, and note that the linear system has infinitely many solutions. The variables x and y are called *basic variables* as they are determined once we make a choice of the free variable.

We will call this description of the solution space, in which the basic variables are written in terms of the free variables, a *parametric description* of the solution space.

b. Consider the matrix

$$\left[\begin{array}{ccc|c} 1 & 0 & 0 & 4 \\ 0 & 1 & 0 & -3 \\ 0 & 0 & 1 & 1 \\ 0 & 0 & 0 & 0 \end{array}\right].$$

The last equation gives

$$0x + 0y + 0z = 0,$$

which is true for any (x, y, z). We may safely ignore this equation since it does not impose a restriction on (x, y, z). We then see that there is a unique solution $(x, y, z) = (4, -3, 1)$.

c. Consider the matrix

$$\left[\begin{array}{ccc|c} 1 & 0 & 2 & 0 \\ 0 & 1 & -1 & 0 \\ 0 & 0 & 0 & 1 \end{array}\right].$$

1.2. FINDING SOLUTIONS TO LINEAR SYSTEMS

Beginning with the last equation, we see that

$$0x + 0y + 0z = 0 = 1,$$

which is not true for any (x, y, z). There is no solution to this particular equation and therefore no solution to the system of equations.

1.2.4 Summary

We saw several important concepts in this section.

- We can describe the solution space to a linear system by transforming it into a new linear system having the same solution space through a sequence of scaling, interchange, and replacement operations.

- We can represent a linear system by an augmented matrix. Using scaling, interchange, and replacement operations, the augmented matrix is row equivalent to exactly one reduced row echelon matrix. The process of constructing this reduced row echelon matrix is called Gaussian elimination.

- The reduced row echelon matrix allows us to easily describe the solution space of a linear system.

1.2.5 Exercises

1. For each of the linear systems below, write the associated augmented matrix and find the reduced row echelon matrix that is row equivalent to it. Identify the basic and free variables and then describe the solution space of the original linear system using a parametric description, if appropriate.

    a.
    $$\begin{aligned} 2x + y &= 0 \\ x + 2y &= 3 \\ -2x + 2y &= 6 \end{aligned}$$

    b.
    $$\begin{aligned} -x_1 + 2x_2 + x_3 &= 2 \\ 3x_1 + 2x_3 &= -1 \\ -x_1 - x_2 + x_3 &= 2 \end{aligned}$$

    c.
    $$\begin{aligned} x_1 + 2x_2 - 5x_3 - x_4 &= -3 \\ -2x_1 - 2x_2 + 6x_3 - 2x_4 &= 4 \\ x_1 - x_3 + 9x_4 &= -7 \\ -x_2 + 2x_3 - x_4 &= 4 \end{aligned}$$

2. Consider each matrix below and determine if it is in reduced row echelon form. If not, indicate the reason and apply a sequence of row operations to find its reduced row

echelon matrix. For each matrix, indicate whether the corresponding linear system has infinitely many solutions, exactly one solution, or no solutions.

a.
$$\begin{bmatrix} 1 & 1 & 0 & 3 & | & 3 \\ 0 & 1 & 0 & -2 & | & 1 \\ 0 & 0 & 1 & 3 & | & 4 \end{bmatrix}$$

b.
$$\begin{bmatrix} 1 & 0 & 0 & 0 & | & 0 \\ 0 & 2 & 0 & 0 & | & 0 \\ 0 & 0 & -3 & 0 & | & 0 \\ 0 & 0 & 0 & 1 & | & 0 \\ 0 & 0 & 0 & 0 & | & 1 \end{bmatrix}$$

c.
$$\begin{bmatrix} 1 & 0 & 0 & 3 & | & 3 \\ 0 & 1 & 0 & -2 & | & 1 \\ 0 & 0 & 1 & 3 & | & 4 \\ 0 & 0 & 0 & 3 & | & 3 \end{bmatrix}$$

d.
$$\begin{bmatrix} 0 & 0 & 1 & 0 & | & -1 \\ 0 & 1 & 0 & 0 & | & 3 \\ 1 & 1 & 1 & 1 & | & 2 \end{bmatrix}$$

3. Give an example of a reduced row echelon matrix that describes a linear system having the stated properties. If it is not possible to find such an example, explain why not.

   a. Write a reduced row echelon matrix for a linear system having five equations and three variables and having exactly one solution.

   b. Write a reduced row echelon matrix for a linear system having three equations and three variables and having no solution.

   c. Write a reduced row echelon matrix for a linear system having three equations and five variables and having infinitely many solutions.

   d. Write a reduced row echelon matrix for a linear system having three equations and four variables and having exactly one solution.

   e. Write a reduced row echelon matrix for a linear system having four equations and four variables and having exactly one solution.

4. For any given matrix, Theorem 1.2.6 tells us that there is a reduced row echelon matrix that is row equivalent to it. This exercise demonstrates why this is the case. Each of the following matrices satisfies three of the four conditions required of a reduced row echelon matrix as prescribed by Definition 1.2.5. For each, indicate how a sequence of

1.2. FINDING SOLUTIONS TO LINEAR SYSTEMS

row operations can be applied to form a row equivalent reduced row echelon matrix.

a.
$$\begin{bmatrix} 1 & 0 & 2 & 3 \\ 0 & 0 & 0 & 0 \\ 0 & 1 & -2 & 1 \end{bmatrix}$$

b.
$$\begin{bmatrix} 1 & 0 & 0 & 2 \\ 0 & -2 & 0 & -4 \\ 0 & 0 & 1 & 1 \end{bmatrix}$$

c.
$$\begin{bmatrix} 0 & 1 & 0 & -2 \\ 0 & 0 & 1 & 4 \\ 1 & 0 & 0 & 3 \\ 0 & 0 & 0 & 0 \\ 0 & 0 & 0 & 0 \end{bmatrix}$$

d.
$$\begin{bmatrix} 1 & 0 & 2 & 3 \\ 0 & 1 & 3 & 0 \\ 0 & 0 & 1 & -1 \\ 0 & 0 & 0 & 0 \end{bmatrix}$$

5. For each of the questions below, provide a justification for your response.

   a. What does the presence of a row whose entries are all zero in an augmented matrix tell us about the solution space of the linear system?

   b. How can you determine if a linear system has no solutions directly from its reduced row echelon matrix?

   c. How can you determine if a linear system has infinitely many solutions directly from its reduced row echelon matrix?

   d. What can you say about the solution space of a linear system if there are more variables than equations and at least one solution exists?

6. Determine whether the following statements are true or false and explain your reasoning.

   a. If every variable is basic, then the linear system has exactly one solution.

   b. If two augmented matrices are row equivalent to one another, then they describe two linear systems having the same solution spaces.

   c. The presence of a free variable indicates that there are no solutions to the linear system.

   d. If a linear system has exactly one solution, then it must have the same number of equations as variables.

   e. If a linear system has the same number of equations as variables, then it has exactly one solution.

1.3 Computation with Sage

Linear algebra owes its prominence as a powerful scientific tool to the ever-growing power of computers. Carl Cowen, a former president of the Mathematical Association of America, has said, "No serious application of linear algebra happens without a computer." Indeed, Cowen notes that, in the 1950s, working with a system of 100 equations in 100 variables was difficult. Today, scientists and mathematicians routinely work on problems that are vastly larger. This is only possible because of today's computing power.

It is therefore important for any student of linear algebra to become comfortable solving linear algebraic problems on a computer. This section will introduce you to a program called Sage that can help. While you may be able to do much of this work on a graphing calculator, you are encouraged to become comfortable with Sage as we will use increasingly powerful features as we encounter their need.

1.3.1 Introduction to Sage

There are several ways to access Sage.

- If you are reading this book online, there will be embedded "Sage cells" at appropriate places in the text. You have the opportunity to type Sage commands into these cells and execute them, provided you are connected to the Internet. Please be aware that your work will be lost if you reload the page.

    Here is a Sage cell containing a command that asks Sage to multiply 5 and 3. You may execute the command by pressing the *Evaluate* button.

    ```
    5 * 3
    ```

- You may also go to cocalc.com, sign up for an account, open a new project, and create a "Sage worksheet." Once inside the worksheet, you may enter commands as shown here, and evaluate them by pressing *Enter* on your keyboard while holding down the *Shift* key.

- There is a page of Sage cells at gsvu.edu/s/0Ng. Any results obtained by evaluating one cell are available in other cells. However, your work will be lost when the page is reloaded.

Throughout the text, we will introduce new Sage commands that allow us to explore linear algebraic concepts. These commands are collected and summarized in the reference found in Appendix A.

Activity 1.3.1 Basic Sage commands..

    a. Sage uses the standard operators +, -, *, /, and ^ for the usual arithmetic operations. By entering text in the cell below, ask Sage to evaluate

$$3 + 4(2^4 - 1)$$

1.3. COMPUTATION WITH SAGE

b. Notice that we can create new lines by pressing *Enter* and entering additional commands on them. What happens when you evaluate this Sage cell?

```
5 * 3
10 - 4
```

Notice that we only see the result from the last command. With the `print` command, we may see earlier results, if we wish.

```
print(5 * 3)
print(10 - 4)
```

c. We may give a name to the result of one command and refer to it in a later command.

```
income = 1500 * 12
taxes = income * 0.15
print(taxes)
```

Suppose you have three tests in your linear algebra class and your scores are 90, 100, and 98. In the Sage cell below, add your scores together and call the result `total`. On the next line, find the average of your test scores and print it.

d. If you are not a programmer, you may ignore this part. If you are an experienced programmer, however, you should know that Sage is written in the Python programming language and that you may enter Python code into a Sage cell.

```
for i in range(10):
    print(i)
```

1.3.2 Sage and matrices

When we encounter a matrix, Theorem 1.2.6 tells us that there is exactly one reduced row echelon matrix that is row equivalent to it.

In fact, the uniqueness of this reduced row echelon matrix is what motivates us to define this particular form. When solving a system of linear equations using Gaussian elimination, there are other row equivalent matrices that reveal the structure of the solution space. The reduced row echelon matrix is simply a convenience as it is an agreement we make with one another to seek the same matrix.

An added benefit is that we can ask a computer program, like Sage, to find reduced row echelon matrices for us. We will learn how to do this now that we have a little familiarity with Sage.

First, notice that a matrix has a certain number of rows and columns. For instance, the matrix

$$\begin{bmatrix} * & * & * & * & * \\ * & * & * & * & * \\ * & * & * & * & * \end{bmatrix}$$

has three rows and five columns. We consequently refer to this as a 3×5 matrix.

We may ask Sage to create the 2×4 matrix

$$\begin{bmatrix} -1 & 0 & 2 & 7 \\ 2 & 1 & -3 & -1 \end{bmatrix}$$

by entering

```
matrix(2, 4, [-1, 0, 2, 7, 2, 1, -3, -1])
```

When evaluated, Sage will confirm the matrix by writing out the rows of the matrix, each inside square brackets.

Notice that there are three separate things (we call them *arguments*) inside the parentheses: the number of rows, the number of columns, and the entries of the matrix listed by row inside square brackets. These three arguments are separated by commas. Notice that there is no way of specifying whether this is an augmented or coefficient matrix so it will be up to us to interpret our results appropriately.

> **Sage syntax.**
>
> Some common mistakes are
> - to forget the square brackets around the list of entries,
> - to omit an entry from the list or to add an extra one,
> - to forget to separate the rows, columns, and entries by commas, and
> - to omit the parentheses around the arguments after `matrix`.
>
> If you see an error message, carefully proofread your input and try again.

Alternatively, you can create a matrix by simply listing its rows, like this

```
matrix([ [-1, 0, 2, 7],
         [ 2, 1,-3,-1] ])
```

1.3. COMPUTATION WITH SAGE

Activity 1.3.2 Using Sage to find row reduced echelon matrices..

a. Enter the following matrix into Sage.

$$\begin{bmatrix} -1 & -2 & 2 & -1 \\ 2 & 4 & -1 & 5 \\ 1 & 2 & 0 & 3 \end{bmatrix}$$

b. Give the matrix the name A by entering

```
A = matrix( ..., ..., [ ... ])
```

We may then find its reduced row echelon form by entering

```
A = matrix( ..., ..., [ ... ])
A.rref()
```

A common mistake is to forget the parentheses after rref.

Use Sage to find the reduced row echelon form of the matrix from Item a of this activity.

c. Use Sage to describe the solution space of the system of linear equations

$$\begin{aligned} -x_1 \phantom{{}+3x_2+x_3} + 2x_4 &= 4 \\ 3x_2 + x_3 + 2x_4 &= 3 \\ 4x_1 - 3x_2 \phantom{{}+x_3} + x_4 &= 14 \\ 2x_2 + 2x_3 + x_4 &= 1 \end{aligned}$$

d. Consider the two matrices:

$$A = \begin{bmatrix} 1 & -2 & 1 & -3 \\ -2 & 4 & 1 & 1 \\ -4 & 8 & -1 & 7 \end{bmatrix}$$

$$B = \begin{bmatrix} 1 & -2 & 1 & -3 & 0 & 3 \\ -2 & 4 & 1 & 1 & 1 & -1 \\ -4 & 8 & -1 & 7 & 3 & 2 \end{bmatrix}$$

We say that B is an *augmentation* of A because it is obtained from A by adding some more columns.

Using Sage, define the matrices and compare their reduced row echelon forms. What do you notice about the relationship between the two reduced row echelon forms?

e. Using the system of equations in Item c, write the augmented matrix corresponding to the system of equations. What did you find for the reduced row echelon form of the augmented matrix?

Now write the coefficient matrix of this system of equations. What does Item d of this activity tell you about its reduced row echelon form?

> **Sage practices.**
>
> Here are some practices that you may find helpful when working with matrices in Sage.
>
> - Break the matrix entries across lines, one for each row, for better readability by pressing *Enter* between rows.
>
> ```
> A = matrix(2, 4, [ 1, 2, -1, 0,
>                   -3, 0,  4, 3 ])
> ```
>
> - Print your original matrix to check that you have entered it correctly. You may want to also print a dividing line to separate matrices.
>
> ```
> A = matrix(2, 2, [ 1, 2,
>                    2, 2])
> print (A)
> print ("---------")
> A.rref()
> ```

The last part of the previous activity, Item d, demonstrates something that will be helpful for us in the future. In that activity, we started with a matrix A, which we augmented by adding some columns to obtain a matrix B. We then noticed that the reduced row echelon form of B is itself an augmentation of the reduced row echelon form of A.

To illustrate, we can consider the reduced row echelon form of the augmented matrix:

$$\left[\begin{array}{ccc|c} -2 & 3 & 0 & 2 \\ -1 & 4 & 1 & 3 \\ 3 & 0 & 2 & 2 \\ 1 & 5 & 3 & 7 \end{array}\right] \sim \left[\begin{array}{ccc|c} 1 & 0 & 0 & -4 \\ 0 & 1 & 0 & -2 \\ 0 & 0 & 1 & 7 \\ 0 & 0 & 0 & 0 \end{array}\right]$$

We can then determine the reduced row echelon form of the coefficient matrix by looking inside the augmented matrix.

$$\begin{bmatrix} -2 & 3 & 0 \\ -1 & 4 & 1 \\ 3 & 0 & 2 \\ 1 & 5 & 3 \end{bmatrix} \sim \begin{bmatrix} 1 & 0 & 0 \\ 0 & 1 & 0 \\ 0 & 0 & 1 \\ 0 & 0 & 0 \end{bmatrix}$$

If we trace through the steps in the Gaussian elimination algorithm carefully, we see that this is a general principle, which we now state.

Proposition 1.3.1 Augmentation Principle. *If matrix B is an augmentation of matrix A, then the reduced row echelon form of B is an augmentation of the reduced row echelon form of A.*

1.3.3 Computational effort

At the beginning of this section, we indicated that linear algebra has become more prominent as computers have grown more powerful. Computers, however, still have limits. Let's consider how much effort is expended when we ask to find the reduced row echelon form of a matrix. We will measure, very roughly, the effort by the number of times the algorithm requires us to multiply or add two numbers.

We will assume that our matrix has the same number of rows as columns, which we call n. We are mainly interested in the case when n is very large, which is when we need to worry about how much effort is required.

Let's first consider the effort required for each of our row operations.

- Scaling a row multiplies each of the n entries in a row by some number, which requires n operations.

- Interchanging two rows requires no multiplications or additions so we won't worry about the effort required by an interchange.

- A replacement requires us to multiply each entry in a row by some number, which takes n operations, and then add the resulting entries to another row, which requires another n operations. The total number of operations is $2n$.

Our goal is to transform a matrix to its reduced row echelon form, which looks something like this:

$$\begin{bmatrix} 1 & 0 & \ldots & 0 \\ 0 & 1 & \ldots & 0 \\ \vdots & \vdots & \ddots & 0 \\ 0 & 0 & \ldots & 1 \end{bmatrix}.$$

We roughly perform one replacement operation for every 0 entry in the reduced row echelon matrix. When n is very large, most of the n^2 entries in the reduced row echelon form are 0 so we need roughly n^2 replacements. Since each replacement operation requires $2n$ operations, the number of operations resulting from the needed replacements is roughly $n^2(2n) = 2n^3$.

Each row is scaled roughly one time so there are roughly n scaling operations, each of which requires n operations. The number of operations due to scaling is roughly n^2.

Therefore, the total number of operations is roughly

$$2n^3 + n^2.$$

When n is very large, the n^2 term is much smaller than the n^3 term. We therefore state that

Observation 1.3.2 The number of operations required to find the reduced row echelon form of an $n \times n$ matrix is roughly proportional to n^3.

This is a very rough measure of the effort required to find the reduced row echelon form; a more careful accounting shows that the number of arithmetic operations is roughly $\frac{2}{3}n^3$. As we have seen, some matrices require more effort than others, but the upshot of this observation is that the effort is proportional to n^3. We can think of this in the following way: If the size of the matrix grows by a factor of 10, then the effort required grows by a factor of $10^3 = 1000$.

While today's computers are powerful, they cannot handle every problem we might ask of them. Eventually, we would like to be able to consider matrices that have $n = 10^{12}$ (a trillion) rows and columns. In very broad terms, the effort required to find the reduced row echelon matrix will require roughly $(10^{12})^3 = 10^{36}$ operations.

To put this into context, imagine we need to solve a linear system with a trillion equations and a trillion variables and that we have a computer that can perform a trillion, 10^{12}, operations every second. Finding the reduced row echelon form would take about 10^{16} years. At this time, the universe is estimated to be approximately 10^{10} years old. If we started the calculation when the universe was born, we'd be about one-millionth of the way through.

This may seem like an absurd situation, but we'll see in Subsection 4.5.3 how we use the results of such a computation every day. Clearly, we will need some better tools to deal with *really* big problems like this one.

1.3.4 Summary

We learned some basic features of Sage with an emphasis on finding the reduced row echelon form of a matrix.

- Sage can perform basic arithmetic using standard operators. Sage can also save results from one command to be reused in a later command.

- We may define matrices in Sage and find the reduced row echelon form using the rref command.

- We saw an example of the Augmentation Principle, which we then stated as a general principle.

- We saw that the computational effort required to find the reduced row echelon form of an $n \times n$ matrix is proportional to n^3.

Appendix A contains a reference outlining the Sage commands that we have encountered.

1.3. COMPUTATION WITH SAGE

1.3.5 Exercises

1. Consider the linear system
    $$\begin{aligned} x + 2y - z &= 1 \\ 3x + 2y + 2z &= 7 \\ -x + 4z &= -3 \end{aligned}$$

    Write this system as an augmented matrix and use Sage to find a description of the solution space.

2. Shown below are some traffic patterns in the downtown area of a large city. The figures give the number of cars per hour traveling along each road. Any car that drives into an intersection must also leave the intersection. This means that the number of cars entering an intersection in an hour is equal to the number of cars leaving the intersection.

    a. Let's begin with the following traffic pattern.

       i. How many cars per hour enter the upper left intersection? How many cars per hour leave this intersection? Use this to form a linear equation in the variables x, y, z, and w.

    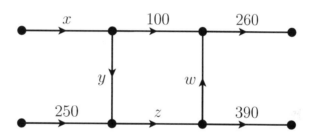

       ii. Form three more linear equations from the other three intersections to form a linear system having four equations in four variables. Then use Sage to find the solution space to this system.

       iii. Is there exactly one solution or infinitely many solutions? Explain why you would expect this given the information provided.

    b. Another traffic pattern is shown below.

    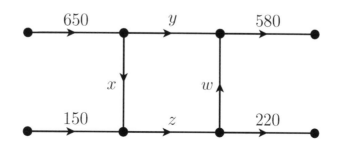

i. Once again, write a linear system for the quantities x, y, z, and w and solve the system using the Sage cell below.

  ii. What can you say about the solution of this linear system? Is there exactly one solution or infinitely many solutions? Explain why you would expect this given the information provided.

  iii. What is the smallest possible amount of traffic flowing through x?

3. A typical problem in thermodynamics is to find the steady-state temperature distribution inside a thin plate if we know the temperature around the boundary. Let $T_1, T_2, \ldots, T_6$ be the temperatures at the six nodes inside the plate as shown below.

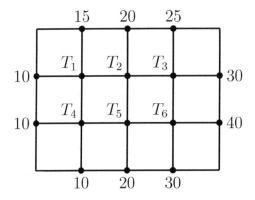

The temperature at a node is approximately the average of the four nearest nodes: for instance,
$$T_1 = (10 + 15 + T_2 + T_4)/4,$$
which we may rewrite as
$$4T_1 - T_2 - T_4 = 25.$$

Set up a linear system to find the temperature at these six points inside the plate. Then use Sage to solve the linear system. If all the entries of the matrix are integers, Sage will compute the reduced row echelon form using rational numbers. To view a decimal approximation of the results, you may use

```
A.rref().numerical_approx(digits=4)
```

In the real world, the approximation becomes better as we add more and more points into the grid. This is a situation where we may want to solve a linear system having millions of equations and millions of variables.

4. The fuel inside model rocket motors is a black powder mixture that ideally consists of 60% charcoal, 30% potassium nitrate, and 10% sulfur by weight.

Suppose you work at a company that makes model rocket motors. When you come into work one morning, you learn that yesterday's first shift made a perfect batch of fuel. The second shift, however, misread the recipe and used 50% charcoal, 20% potassium nitrate and 30% sulfur. Then the two batches were mixed together. A chemical analysis shows that there are 100.3 pounds of charcoal in the mixture and 46.9 pounds of potassium nitrate.

   a. Assuming the first shift produced x pounds of fuel and the second y pounds, set up a linear system in terms of x and y. How many pounds of fuel did the first shift produce and how many did the second shift produce?

   b. How much sulfur would you expect to find in the mixture?

5. This exercise is about balancing chemical reactions.

   a. Chemists denote a molecule of water as H_2O, which means it is composed of two atoms of hydrogen (H) and one atom of oxygen (O). The process by which hydrogen burns is described by the chemical reaction

   $$x\, H_2 + y\, O_2 \to z\, H_2O$$

   This means that x molecules of hydrogen H_2 combine with y molecules of oxygen O_2 to produce z water molecules. The number of hydrogen atoms is the same before and after the reaction; the same is true of the oxygen atoms.

   1. In terms of x, y, and z, how many hydrogen atoms are there before the reaction? How many hydrogen atoms are there after the reaction? Find a linear equation in x, y, and z by equating these quantities.

   2. Find a second linear equation in x, y, and z by equating the number of oxygen atoms before and after the reaction.

   3. Find the solutions of this linear system. Why are there infinitely many solutions?

   4. In this chemical setting, x, y, and z should be positive integers. Find the solution where x, y, and z are the smallest possible positive integers.

   b. Now consider the reaction where potassium permanganate and manganese sulfate combine with water to produce manganese dioxide, potassium sulfate, and sulfuric acid:

   $$x_1\, KMnO_4 + x_2\, MnSO_4 + x_3\, H_2O \to x_4\, MnO_2 + x_5\, K_2SO_4 + x_6\, H_2SO_4.$$

   As in the previous exercise, find the appropriate values for $x_1, x_2, \ldots, x_6$ to balance the chemical reaction.

6. We began this section by stating that increasing computational power has helped linear algebra assume a prominent role as a scientific tool. Later, we looked at one computa-

tional limitation: once a matrix gets to be too big, it is not reasonable to apply Gaussian elimination to find its reduced row echelon form.

In this exercise, we will see another limitation: computer arithmetic with real numbers is only an approximation because computers represent real numbers with only a finite number of bits. For instance, the number pi

$$\pi = 3.1415926535897932384626433832795028841971693399\ldots$$

would be approximated inside a computer by, say,

$$\pi \approx 3.141592653589793$$

Most of the time, this is not a problem. However, when we perform millions or even billions of arithmetic operations, the error in these approximations starts to accumulate and can lead to results that are wildly inaccurate. Here are two examples demonstrating this.

a. Let's first see an example showing that computer arithmetic really is an approximation. First, consider the linear system

$$x + \frac{1}{2}y + \frac{1}{3}z = 1$$
$$\frac{1}{2}x + \frac{1}{3}y + \frac{1}{4}z = 0$$
$$\frac{1}{3}x + \frac{1}{4}y + \frac{1}{5}z = 0$$

If the coefficients are entered into Sage as fractions, Sage will find the exact reduced row echelon form. Find the exact solution to this linear system.

Now let's ask Sage to compute with real numbers. We can do this by representing one of the coefficients as a decimal. For instance, the same linear system can be represented as

$$x + 0.5y + \frac{1}{3}z = 1$$
$$\frac{1}{2}x + \frac{1}{3}y + \frac{1}{4}z = 0$$
$$\frac{1}{3}x + \frac{1}{4}y + \frac{1}{5}z = 0$$

Most computers do arithmetic using either 32 or 64 bits. To magnify the problem so that we can see it better, we will ask Sage to do arithmetic using only 10 bits as follows.

```
R = RealNumber
RealNumber = RealField(10)

# enter the matrix below
A = matrix( ..., ..., [ ... ] )

print (A.rref())
RealNumber = R
```

1.3. COMPUTATION WITH SAGE

What does Sage give for the solution now? Compare this to the exact solution that you found previously.

b. Some types of linear systems are particularly sensitive to errors resulting from computers' approximate arithmetic. For instance, suppose we are interested in the linear system

$$x + y = 2$$
$$x + 1.001y = 2$$

Find the solution to this linear system.

Suppose now that the computer has accumulated some error in one of the entries of this system so that it incorrectly stores the system as

$$x + y = 2$$
$$x + 1.001y = 2.001$$

Find the solution to this linear system.

Notice how a small error in one of the entries in the linear system leads to a solution that has a dramatically large error. Fortunately, this is an issue that has been well studied, and there are techniques that mitigate this type of behavior.

1.4 Pivots and their influence on solution spaces

By now, we have seen several examples illustrating how the reduced row echelon matrix leads to a convenient description of the solution space to a linear system. In this section, we will use this understanding to make some general observations about how certain features of the reduced row echelon matrix reflect the nature of the solution space.

Remember that a leading entry in a reduced row echelon matrix is the leftmost nonzero entry in a row of the matrix. As we'll see, the positions of these leading entries encode a lot of information about the solution space of the corresponding linear system. For this reason, we make the following definition.

Definition 1.4.1 A **pivot position** in a matrix A is the position of a leading entry in the reduced row echelon matrix of A.

For instance, in this reduced row echelon matrix, the pivot positions are indicated in bold:

$$\begin{bmatrix} \mathbf{1} & 0 & * & 0 \\ 0 & \mathbf{1} & * & 0 \\ 0 & 0 & 0 & \mathbf{1} \\ 0 & 0 & 0 & 0 \end{bmatrix}.$$

We can refer to pivot positions by their row and column number saying, for instance, that there is a pivot position in the second row and fourth column.

> **Preview Activity 1.4.1 Some basic observations about pivots..**
>
> a. Shown below is a matrix and its reduced row echelon form. Indicate the pivot positions.
>
> $$\begin{bmatrix} 2 & 4 & 6 & -1 \\ -3 & 1 & 5 & 0 \\ 1 & 3 & 5 & 1 \end{bmatrix} \sim \begin{bmatrix} 1 & 0 & -1 & 0 \\ 0 & 1 & 2 & 0 \\ 0 & 0 & 0 & 1 \end{bmatrix}.$$
>
> b. How many pivot positions can there be in one row? In a 3×5 matrix, what is the largest possible number of pivot positions? Give an example of a 3×5 matrix that has the largest possible number of pivot positions.
>
> c. How many pivots can there be in one column? In a 5×3 matrix, what is the largest possible number of pivot positions? Give an example of a 5×3 matrix that has the largest possible number of pivot positions.
>
> d. Give an example of a matrix with a pivot position in every row and every column. What is special about such a matrix?

When we have looked at solution spaces of linear systems, we have frequently asked whether there are infinitely many solutions, exactly one solution, or no solutions. We will now break this question into two separate questions.

Question 1.4.2 Two Fundamental Questions. When we encounter a linear system, we often ask

1.4. PIVOTS AND THEIR INFLUENCE ON SOLUTION SPACES

Existence Is there a solution to the linear system? If so, we say that the system is *consistent*; if not, we say it is *inconsistent*.

Uniqueness If the linear system is consistent, is the solution unique or are there infinitely many solutions?

These two questions represent two sides of a coin that appear in many variations throughout our explorations. In this section, we will study how the location of the pivots influence the answers to these two questions. We begin by considering the first question concerning the existence of solutions.

1.4.1 The existence of solutions

Activity 1.4.2.

a. Shown below are three augmented matrices in reduced row echelon form.

$$\begin{bmatrix} 1 & 0 & 0 & | & 3 \\ 0 & 1 & 0 & | & 0 \\ 0 & 0 & 1 & | & -2 \\ 0 & 0 & 0 & | & 0 \end{bmatrix} \quad \begin{bmatrix} 1 & 0 & 2 & | & 3 \\ 0 & 1 & -1 & | & 0 \\ 0 & 0 & 0 & | & 0 \\ 0 & 0 & 0 & | & 0 \end{bmatrix} \quad \begin{bmatrix} 1 & 0 & 2 & | & 0 \\ 0 & 1 & -1 & | & 0 \\ 0 & 0 & 0 & | & 1 \\ 0 & 0 & 0 & | & 0 \end{bmatrix}$$

For each matrix, identify the pivot positions and determine if the corresponding linear system is consistent. Explain how the location of the pivots determines whether the system is consistent or inconsistent.

b. Each of the augmented matrices above has a row in which each entry is zero. What, if anything, does the presence of such a row tell us about the consistency of the corresponding linear system?

c. Give an example of a 3×5 augmented matrix in reduced row echelon form that represents a consistent system. Indicate the pivot positions in your matrix and explain why these pivot positions guarantee a consistent system.

d. Give an example of a 3×5 augmented matrix in reduced row echelon form that represents an inconsistent system. Indicate the pivot positions in your matrix and explain why these pivot positions guarantee an inconsistent system.

e. Write the reduced row echelon form of the coefficient matrix of the corresponding linear system in Item d? (Remember that the Augmentation Principle says that the reduced row echelon form of the coefficient matrix simply consists of the first four columns of the augmented matrix.) What do you notice about the pivot positions in this coefficient matrix?

f. Suppose we have a linear system for which the *coefficient* matrix has the following reduced row echelon form.

$$\begin{bmatrix} 1 & 0 & 0 & 0 & -1 \\ 0 & 1 & 0 & 0 & 2 \\ 0 & 0 & 1 & 0 & 0 \\ 0 & 0 & 0 & 1 & -3 \end{bmatrix}$$

What can you say about the consistency of the linear system?

Let's summarize the results of this activity by considering the following reduced row echelon matrix:

$$\left[\begin{array}{ccc|c} 1 & * & 0 & 0 \\ 0 & 0 & 1 & 0 \\ 0 & 0 & 0 & 1 \\ 0 & 0 & 0 & 0 \end{array}\right].$$

In terms of variables x, y, and z, the final equation says

$$0x + 0y + 0z = 0.$$

If we evaluate the left-hand side with any values of x, y, and z, we get 0, which means that the equation always holds. Therefore, its presence has no effect on the solution space defined by the other three equations.

The third equation, however, says that

$$0x + 0y + 0z = 1.$$

Again, if we evaluate the left-hand side with any values of x, y, and z, we get 0 so this equation cannot be satisfied for any (x, y, z). This means that the entire linear system has no solution and is therefore inconsistent.

An equation like this appears in the reduced row echelon matrix as

$$\left[\begin{array}{ccccc|c} \vdots & \vdots & \vdots & \vdots & \vdots \\ 0 & 0 & \cdots & 0 & 1 \\ \vdots & \vdots & \vdots & \vdots & \vdots \end{array}\right].$$

The pivot positions make this condition clear: *the system is inconsistent if there is a pivot position in the rightmost column of the corresponding augmented matrix.*

In fact, we will soon see that the system is consistent if there is not a pivot in the rightmost column of the corresponding augmented matrix. This leaves us with the following

Proposition 1.4.3 *A linear system is inconsistent if and only if there is a pivot position in the rightmost column of the corresponding augmented matrix.*

This also says something about the pivot positions of the coefficient matrix. Consider an example of an inconsistent system corresponding to the reduced row echelon form of the

following augmented matrix
$$\begin{bmatrix} 1 & 0 & * & 0 \\ 0 & 1 & * & 0 \\ 0 & 0 & 0 & 1 \end{bmatrix}.$$

The Augmentation Principle says that that the reduced row echelon form of the coefficient matrix is
$$\begin{bmatrix} 1 & 0 & * \\ 0 & 1 & * \\ 0 & 0 & 0 \end{bmatrix},$$

which shows that the coefficient matrix has a row without a pivot position. To turn this around, we see that *if every row of the coefficient matrix has a pivot position, then the system must be consistent*. For instance, if our linear system has a coefficient matrix whose reduced row echelon form is
$$\begin{bmatrix} 1 & 0 & 0 \\ 0 & 1 & 0 \\ 0 & 0 & 1 \end{bmatrix},$$

then we can guarantee that the linear system is consistent because there is no way to obtain a pivot in the rightmost column of the augmented matrix.

Proposition 1.4.4 *If every row of the coefficient matrix has a pivot position, then the corresponding system of linear equations is consistent.*

1.4.2 The uniqueness of solutions

Now that we have studied the role that pivot positions play in the existence of solutions, let's turn to the question of uniqueness.

Activity 1.4.3.

a. Here are the three augmented matrices in reduced row echelon form that we considered in the previous section.

$$\begin{bmatrix} 1 & 0 & 0 & 3 \\ 0 & 1 & 0 & 0 \\ 0 & 0 & 1 & -2 \\ 0 & 0 & 0 & 0 \end{bmatrix} \quad \begin{bmatrix} 1 & 0 & 2 & 3 \\ 0 & 1 & -1 & 0 \\ 0 & 0 & 0 & 0 \\ 0 & 0 & 0 & 0 \end{bmatrix} \quad \begin{bmatrix} 1 & 0 & 2 & 0 \\ 0 & 1 & -1 & 0 \\ 0 & 0 & 0 & 1 \\ 0 & 0 & 0 & 0 \end{bmatrix}$$

For each matrix, identify the pivot positions and state whether the corresponding linear system is consistent. If the system is consistent, explain whether the solution is unique or whether there are infinitely many solutions.

b. If possible, give an example of a 3×5 augmented matrix that corresponds to a linear system having a unique solution. If it is not possible, explain why.

c. If possible, give an example of a 5×3 augmented matrix that corresponds to a linear system having a unique solution. If it is not possible, explain why.

d. What condition on the pivot positions guarantees that a linear system has a unique solution?

e. If a linear system has a unique solution, what can we say about the relationship between the number of equations and the number of variables?

Let's consider what we've learned in this activity. Since we are interested in the question of whether a consistent linear system has a unique solution or infinitely many, we will only consider consistent systems. By the results of the previous section, this means that there is not a pivot in the rightmost column of the augmented matrix. Here are two possible examples:

$$\left[\begin{array}{ccc|c} 1 & 0 & 0 & 4 \\ 0 & 1 & 0 & -1 \\ 0 & 0 & 1 & 2 \end{array}\right] \qquad \left[\begin{array}{ccc|c} 1 & 0 & 2 & -2 \\ 0 & 1 & 1 & 4 \\ 0 & 0 & 0 & 0 \end{array}\right]$$

In the first example, we have the equations

$$\begin{aligned} x_1 &= 4 \\ x_2 &= -1 \\ x_3 &= 2 \end{aligned}$$

demonstrating the fact that there is a unique solution $(x_1, x_2, x_3) = (4, -1, 2)$.

In the second example, we have the equations

$$\begin{aligned} x_1 \phantom{{}+x_2} + 2x_3 &= -2 \\ x_2 + x_3 &= 4 \end{aligned}$$

that we may rewrite in parametric form as

$$\begin{aligned} x_1 &= -2 - 2x_3 \\ x_2 &= 4 - x_3 \end{aligned}.$$

Here we see that x_1 and x_2 are basic variables that may be expressed in terms of the free variable x_3. In this case, the presence of the free variable leads to infinitely many solutions.

Remember that every column of the coefficient matrix corresponds to a variable in our linear system. In the first example, we see that every column of the coefficient contains a pivot position, which means that every variable is uniquely determined. In the second example, the column of the coefficient matrix corresponding to x_3 does not contain a pivot position, which results in x_3 appearing as a free variable. This illustrates the following principle.

Principle 1.4.5 *Suppose that we consider a consistent linear system.*

- *If every column of the coefficient matrix contains a pivot position, then the system has a unique solution.*

- *If there is a column in the coefficient matrix that contains no pivot position, then the system has infinitely many solutions.*

- *Columns that contain a pivot position correspond to basic variables while columns that do not correspond to free variables.*

When a linear system has a unique solution, every column of the coefficient matrix has a pivot position. Since every row contains at most one pivot position, there must be at least as many rows as columns in the coefficient matrix. Therefore, the linear system has at least as many equations as variables, which is something we intuitively suspected in Section 1.1.

It is reasonable to ask how we choose the free variables. For instance, if we have a single equation
$$x + 2y = 4,$$
then we may write
$$x = 4 - 2y$$
or, equivalently,
$$y = 2 - \frac{1}{2}x.$$
Clearly, either variable may be considered as a free variable in this case.

As we'll see in the future, we are more interested in the *number* of free variables rather than in their choice. For convenience, we will adopt the convention that free variables correspond to columns without a pivot position, which allows us to quickly identify them. For example, the variables x_2 and x_4 appear as free variables in the following linear system:

$$\left[\begin{array}{cccc|c} 1 & 0 & 0 & 2 & 3 \\ 0 & 0 & 1 & -1 & 0 \end{array}\right].$$

1.4.3 Summary

We have seen how the locations of pivot positions, in both the augmented and coefficient matrices, give vital information about the existence and uniqueness of solutions to linear systems. More specifically,

- A linear system is inconsistent exactly when a pivot position appears in the rightmost column of the *augmented* matrix.

- If a linear system is consistent, the solution is unique when every column of the *coefficient* matrix contains a pivot position. There are infinitely many solutions when there is a column of the *coefficient* matrix without a pivot position.

- If a linear system is consistent, the columns of the coefficient matrix containing pivot positions correspond to basic variables and the columns without pivot positions correspond to free variables.

1.4.4 Exercises

1. For each of the augmented matrices in reduced row echelon form given below, determine whether the corresponding linear system is consistent and, if so, determine whether the solution is unique. If the system is consistent, identify the free variables and the basic variables and give a description of the solution space in parametric form.

    a.
    $$\begin{bmatrix} 0 & 1 & 0 & 0 & | & 2 \\ 0 & 0 & 1 & 0 & | & 3 \\ 0 & 0 & 0 & 1 & | & -2 \end{bmatrix}.$$

    b.
    $$\begin{bmatrix} 1 & 0 & 0 & 0 & | & 0 \\ 0 & 1 & 0 & 0 & | & 0 \\ 0 & 0 & 1 & 1 & | & 0 \\ 0 & 0 & 0 & 0 & | & 0 \end{bmatrix}.$$

    c.
    $$\begin{bmatrix} 1 & 0 & 0 & 0 & | & 0 \\ 0 & 1 & 0 & 0 & | & 0 \\ 0 & 0 & 1 & 0 & | & 0 \\ 0 & 0 & 0 & 1 & | & 0 \\ 0 & 0 & 0 & 0 & | & 1 \end{bmatrix}.$$

    d.
    $$\begin{bmatrix} 1 & 0 & 0 & | & -3 \\ 0 & 1 & 0 & | & -1 \\ 0 & 0 & 1 & | & -2 \end{bmatrix}.$$

2. For each of the following linear systems, determine whether the system is consistent, and, if so, determine whether there are infinitely many solutions.

    a.
    $$\begin{aligned} 2x_1 - x_2 + 3x_3 &= 10 \\ -x_1 + x_2 + 3x_4 &= 8 \\ 2x_2 + 2x_3 - x_4 &= -4 \\ 3x_1 + 2x_2 - x_3 + x_4 &= 10 \end{aligned}$$

    b.
    $$\begin{aligned} 2x_1 - x_2 + 3x_3 + 3x_4 &= 8 \\ -x_1 + x_2 + 2x_4 &= -1 \\ 2x_2 + 2x_3 + 6x_4 &= 4 \\ 3x_1 + 2x_2 - x_3 - 3x_4 &= 1 \end{aligned}$$

c.
$$2x_1 - x_2 + 3x_3 + 3x_4 = 8$$
$$-x_1 + x_2 + 2x_4 = -1$$
$$2x_2 + 2x_3 + 6x_4 = 4$$
$$3x_1 + 2x_2 - x_3 - 3x_4 = 0$$

3. Include an example of an appropriate matrix as you justify your responses to the following questions.

   a. Suppose a linear system having six equations and three variables is consistent. Can you guarantee that the solution is unique? Can you guarantee that there are infinitely many solutions?

   b. Suppose that a linear system having three equations and six variables is consistent. Can you guarantee that the solution is unique? Can you guarantee that there are infinitely many solutions?

   c. Suppose that a linear system is consistent and has a unique solution. What can you guarantee about the pivot positions in the augmented matrix?

4. Determine whether the following statements are true or false and provide a justification for your response.

   a. If the coefficient matrix of a linear system has a pivot in the rightmost column, then the system is inconsistent.

   b. If a linear system has two equations and four variables, then it must be consistent.

   c. If a linear system having four equations and three variables is consistent, then the solution is unique.

   d. Suppose that a linear system has four equations and four variables and that the coefficient matrix has four pivots. Then the linear system is consistent and has a unique solution.

   e. Suppose that a linear system has five equations and three variables and that the coefficient matrix has a pivot position in every column. Then the linear system is consistent and has a unique solution.

5. We began our explorations in Section 1.1 by noticing that the solution spaces of linear systems with more equations seem to be smaller. Let's reexamine this idea using what we know about pivot positions.

   a. Remember that the solution space of a single linear equation in three variables is a plane. Can two planes ever intersect in a single point? What are the possible ways in which two planes can intersect? How can our understanding of pivot positions help answer these questions?

   b. Suppose that a consistent linear system has more variables than equations. By considering the possible pivot positions, what can you say with certainty about the solution space?

   c. If a linear system has many more equations than variables, why is it reasonable to expect the system to be inconsistent?

6. The following linear systems contain either one or two parameters.

   a. For what values of the parameter k is the following system consistent? For which of those values is the solution unique?
   $$-x_1 + 2x_2 = 3$$
   $$2x_1 - 4x_2 = k.$$

   b. For what values of the parameters k and l is the following system consistent? For which of those values is the solution unique?
   $$2x_1 + 4x_2 = 3$$
   $$-x_1 + kx_2 = l.$$

7. Consider the linear system described by the following augmented matrix.
   $$\begin{bmatrix} 1 & 2 & 3 & | & 1 \\ 4 & 5 & 6 & | & 4 \\ a & b & c & | & 9 \end{bmatrix}.$$

   a. Find a choice for the parameters a, b, and c that causes the linear system to be inconsistent. Explain why your choice has this property.

   b. Find a choice for the parameters a, b, and c that causes the linear system to have a unique solution. Explain why your choice has this property.

   c. Find a choice for the parameters a, b, and c that causes the linear system to have infinitely many solutions. Explain why your choice has this property.

8. A linear system where the right hand side of every equation is 0 is called *homogeneous*. The augmented matrix of a homogeneous system, for instance, has the following form:
   $$\begin{bmatrix} * & * & * & * & | & 0 \\ * & * & * & * & | & 0 \\ * & * & * & * & | & 0 \end{bmatrix}.$$

   a. Using the concepts we've seen in this section, explain why a homogeneous linear system must be consistent.

   b. What values for the variables are guaranteed to give a solution? Use this to offer another explanation for why a homogeneous linear system is consistent.

   c. Suppose that a homogeneous linear system has a unique solution.

      1. Give an example of such a system by writing its augmented matrix in reduced row echelon form.
      2. Write just the coefficient matrix for the example you gave in the previous part. What can you say about the pivot positions in the coefficient matrix? Explain why your observation must hold for any homogeneous system having a unique solution.
      3. If a homogeneous system of equations has a unique solution, what can you say about the number of equations compared to the number of variables?

9. In a previous math class, you have probably seen the fact that, if we are given two points in the plane, then there is a unique line passing through both of them. In this problem, we will begin with the four points on the left below and ask to find a polynomial that passes through these four points as shown on the right.

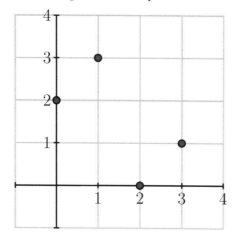 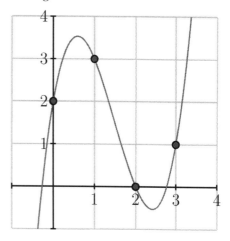

A degree three polynomial can be written as

$$p(x) = a + bx + cx^2 + dx^3$$

where $a, b, c,$ and d are coefficients that we would like to determine. Since we want the polynomial to pass through the point $(3, 1)$, we should require that

$$p(3) = a + 3b + 9c + 27d = 1.$$

In this way, we obtain a linear equation for the coefficients $a, b, c,$ and d.

(a) Write the four linear equations for the coefficients obtained by requiring that the graph of the polynomial $p(x)$ passes through the four points above.

(b) Write the augmented matrix corresponding to this system of equations and use the Sage cell below to solve for the coefficients.

(c) Write the polynomial $p(x)$ that you found and check your work by graphing it in the Sage cell below and verifying that it passes through the four points. To plot a function over a range, you may use a command like `plot(1 + x- 2*x^2, xmin = -1, xmax = 4)`.

(d) Rather than looking for a degree three polynomial, suppose we wanted to find a polynomial that passes through the four points and that has degree two, such as

$$p(x) = a + bx + cx^2.$$

Solve the linear system for the coefficients. What can you say about the existence and uniqueness of a degree two polynomial passing through these four points?

(e) Rather than looking for a degree three polynomial, suppose we wanted to find a polynomial that passes through the four points and that has degree four, such as

$$p(x) = a + bx + cx^2 + dx^3 + ex^4.$$

Solve the linear system for the coefficients. What can you say about the existence and uniqueness of a degree four polynomial passing through these four points?

(f) Suppose you had 10 points and you wanted to find a polynomial passing through each of them. What should the degree of the polynomial be to guarantee that there is exactly one such polynomial? Explain your response.

CHAPTER 2

Vectors, matrices, and linear combinations

We began our study of linear systems in Chapter 1 where we described linear systems in terms of augmented matrices, such as

$$\left[\begin{array}{rrr|r} 1 & 2 & -1 & 3 \\ -3 & 3 & -1 & 2 \\ 2 & 3 & 2 & -1 \end{array}\right]$$

In this chapter, we will uncover geometric information in a matrix like this, which will lead to an intuitive understanding of the insights we previously gained into the solutions of linear systems.

2.1 Vectors and linear combinations

It is a remarkable fact that algebra, which is about symbolic equations and their solutions, and geometry are intimately connected. For instance, the solution set of a linear equation in two unknowns, such as $2x + y = 1$, can be represented graphically by a straight line. The aim of this section is to further this connection by introducing vectors, which will help us to apply geometric intuition to our thinking about linear systems.

2.1.1 Vectors

A *vector* is most simply thought of as a matrix with a single column. For instance, $\mathbf{v} = \begin{bmatrix} 2 \\ 1 \end{bmatrix}$ and $\mathbf{w} = \begin{bmatrix} -3 \\ 1 \\ 0 \\ 2 \end{bmatrix}$ are both vectors. The entries in a vector are called its components. Since the vector $\mathbf{v}$ has two components, we say that it is a two-dimensional vector; in the same way, the vector $\mathbf{w}$ is a four-dimensional vector.

We denote the set of all m-dimensional vectors by $\mathbb{R}^m$. Consequently, if **u** is a 3-dimensional vector, we say that **u** is in $\mathbb{R}^3$.

While it can be difficult to visualize a four-dimensional vector, we can draw a simple picture describing the two-dimensional vector **v**, as shown in Figure 2.1.1.

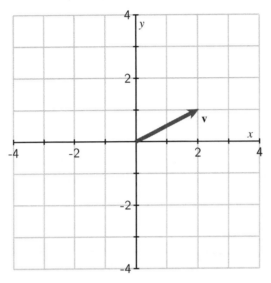

Figure 2.1.1 A graphical representation of the vector $\mathbf{v} = \begin{bmatrix} 2 \\ 1 \end{bmatrix}$.

We can think of **v** as describing a walk in the plane where we move two units horizontally and one unit vertically. Though we allow ourselves to begin walking from any point in the plane, we will most frequently begin at the origin in which case we arrive at the the point (2, 1), as shown in the figure.

There are two simple algebraic operations we often perform on vectors.

Scalar Multiplication We multiply a vector **v** by a real number c by multiplying each of the components of **v** by c. For instance,

$$-3 \begin{bmatrix} 2 \\ -4 \\ 1 \end{bmatrix} = \begin{bmatrix} -6 \\ 12 \\ -3 \end{bmatrix}.$$

We will frequently refer to real numbers, such as -3 in this example, as *scalars* to distinguish them from vectors.

Vector Addition We add two vectors of the same dimension by adding their components. For instance,

$$\begin{bmatrix} 2 \\ -4 \\ 3 \end{bmatrix} + \begin{bmatrix} -5 \\ 6 \\ -3 \end{bmatrix} = \begin{bmatrix} -3 \\ 2 \\ 0 \end{bmatrix}.$$

2.1. VECTORS AND LINEAR COMBINATIONS

Preview Activity 2.1.1 Scalar Multiplication and Vector Addition.. Suppose that

$$\mathbf{v} = \begin{bmatrix} 3 \\ 1 \end{bmatrix}, \mathbf{w} = \begin{bmatrix} -1 \\ 2 \end{bmatrix}.$$

a. Find expressions for the vectors

$$\mathbf{v}, \quad 2\mathbf{v}, \quad -\mathbf{v}, \quad -2\mathbf{v},$$
$$\mathbf{w}, \quad 2\mathbf{w}, \quad -\mathbf{w}, \quad -2\mathbf{w}.$$

and sketch them using Figure 2.1.2.

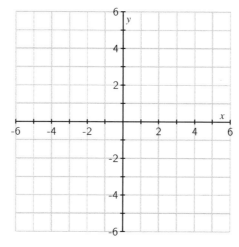

Figure 2.1.2 Sketch the vectors on this grid.

b. What geometric effect does scalar multiplication have on a vector? Also, describe the effect that multiplying by a negative scalar has.

c. Sketch the vectors $\mathbf{v}, \mathbf{w}, \mathbf{v} + \mathbf{w}$ using Figure 2.1.3.

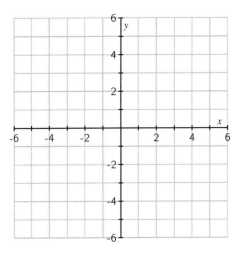

Figure 2.1.3 Sketch the vectors on this grid.

d. Consider vectors that have the form $\mathbf{v} + c\mathbf{w}$ where c is any scalar. Sketch a few of these vectors when, say, $c = -2, -1, 0, 1$, and 2. Give a geometric description of this set of vectors.

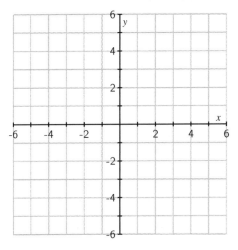

Figure 2.1.4 Sketch the vectors on this grid.

e. If c and d are two scalars, then the vector

$$c\mathbf{v} + d\mathbf{w}$$

is called a *linear combination* of the vectors $\mathbf{v}$ and $\mathbf{w}$. Find the vector that is the linear combination when $c = -2$ and $d = 1$.

f. Can the vector $\begin{bmatrix} -31 \\ 37 \end{bmatrix}$ be represented as a linear combination of $\mathbf{v}$ and $\mathbf{w}$?

Asked differently, can we find scalars c and d such that $c\mathbf{v} + d\mathbf{w} = \begin{bmatrix} -31 \\ 37 \end{bmatrix}$.

2.1. VECTORS AND LINEAR COMBINATIONS

The preview activity demonstrates how we may interpret scalar multiplication and vector addition geometrically.

First, we see that scalar multiplication has the effect of stretching or compressing a vector. Multiplying by a negative scalar changes the direction of the vector. In either case, Figure 2.1.5 shows that a scalar multiple of a vector **v** lies on the same line defined by **v**.

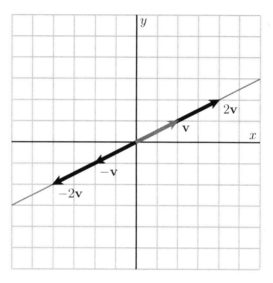

Figure 2.1.5 Scalar multiples of the vector $\mathbf{v} = \begin{bmatrix} 2 \\ 1 \end{bmatrix}$.

To represent the sum **v** + **w**, we imagine walking from the origin with the appropriate horizontal and vertical changes given by **v**. From there, we continue our walk using the horizontal and vertical changes prescribed by **w**, after which we arrive at the sum **v** + **w**. This is illustrated on the left of Figure 2.1.6 where the tail of **w** is placed on the tip of **v**.

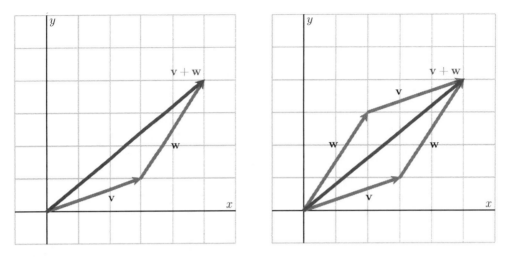

Figure 2.1.6 Vector addition as a simple walk in the plane is illustrated on the left. The vector sum is represented as the diagonal of a parallelogram on the right.

Alternatively, we may construct the parallelogram with **v** and **w** as two sides. The sum is then the diagonal of the parallelogram, as illustrated on the right of Figure 2.1.6.

We have now seen that the set of vectors having the form $c\mathbf{v}$ is a line. To form the set of vectors $c\mathbf{v} + \mathbf{w}$, we can begin with the vector **w** and add multiples of **v**. Geometrically, this means that we begin from the tip of **w** and move in a direction parallel to **v**. The effect is to translate the line $c\mathbf{v}$ by the vector **w**, as shown in Figure 2.1.7.

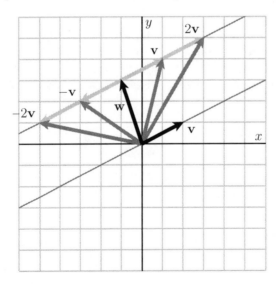

Figure 2.1.7 The set of vectors $c\mathbf{v} + \mathbf{w}$ form a line.

At times, it will be useful for us to think of vectors and points interchangeably. That is, we may wish to think of the vector $\begin{bmatrix} 2 \\ 1 \end{bmatrix}$ as describing the point (2, 1) and vice-versa. When we say that the vectors having the form $c\mathbf{v} + \mathbf{w}$ form a line, we really mean that the tips of the

2.1. VECTORS AND LINEAR COMBINATIONS

vectors all lie on the line passing through **w** and parallel to **v**.

Observation 2.1.8 Even though these vector operations are new, it is straightforward to check that some familiar properties hold.

Commutativity $\mathbf{v} + \mathbf{w} = \mathbf{w} + \mathbf{v}.$

Distributivity $a(\mathbf{v} + \mathbf{w}) = a\mathbf{v} + a\mathbf{w}.$

Sage can perform scalar multiplication and vector addition. We define a vector using the vector command; then * and + denote scalar multiplication and vector addition.

```
v = vector([3,1])
w = vector([-1,2])
print (2*v)
print (v + w)
```

2.1.2 Linear combinations

Linear combinations, which we encountered in the preview activity, provide the link between vectors and linear systems. In particular, they will help us apply geometric intuition to problems involving linear systems.

Definition 2.1.9 The *linear combination* of the vectors $\mathbf{v}_1, \mathbf{v}_2, \ldots, \mathbf{v}_n$ with scalars $c_1, c_2, \ldots, c_n$ is the vector

$$c_1 \mathbf{v}_1 + c_2 \mathbf{v}_2 + \ldots + c_n \mathbf{v}_n.$$

The scalars $c_1, c_2, \ldots, c_n$ are called the *weights* of the linear combination.

Activity 2.1.2. In this activity, we will look at linear combinations of a pair of vectors,
$\mathbf{v} = \begin{bmatrix} 2 \\ 1 \end{bmatrix}$ and $\mathbf{w} = \begin{bmatrix} 1 \\ 2 \end{bmatrix}$.

There is an interactive diagram, available at gvsu.edu/s/0Je, that accompanies this activity.

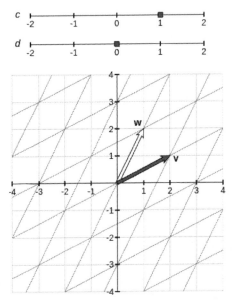

Figure 2.1.10 Linear combinations of vectors **v** and **w**.

a. The weight d is initially set to 0. Explain what happens as you vary c while keeping $d = 0$. How is this related to scalar multiplication?

b. What is the linear combination of **v** and **w** when $c = 1$ and $d = -2$? You may find this result using the diagram, but you should also verify it by computing the linear combination.

c. Describe the vectors that arise when the weight d is set to 1 and c is varied. How is this related to our investigations in the preview activity?

d. Can the vector $\begin{bmatrix} 0 \\ 0 \end{bmatrix}$ be expressed as a linear combination of **v** and **w**? If so, what are the weights c and d?

e. Can the vector $\begin{bmatrix} 3 \\ 0 \end{bmatrix}$ be expressed as a linear combination of **v** and **w**? If so, what are the weights c and d?

f. Verify the result from the previous part by algebraically finding the weights c and d that form the linear combination $\begin{bmatrix} 3 \\ 0 \end{bmatrix}$.

g. Can the vector $\begin{bmatrix} 1.3 \\ -1.7 \end{bmatrix}$ be expressed as a linear combination of **v** and **w**? What

2.1. VECTORS AND LINEAR COMBINATIONS

about the vector $\begin{bmatrix} 15.2 \\ 7.1 \end{bmatrix}$?

h. Are there any two-dimensional vectors that cannot be expressed as linear combinations of **v** and **w**?

This activity illustrates how linear combinations are constructed geometrically: the linear combination $c\mathbf{v} + d\mathbf{w}$ is found by walking along **v** a total of c times followed by walking along **w** a total of d times. When one of the weights is held constant while the other varies, the vector moves along a line.

Example 2.1.11 The previous activity also shows that questions about linear combinations lead naturally to linear systems. Suppose we have vectors $\mathbf{v} = \begin{bmatrix} 3 \\ -1 \end{bmatrix}$ and $\mathbf{w} = \begin{bmatrix} 4 \\ 3 \end{bmatrix}$. Let's determine whether we can describe the vector $\mathbf{b} = \begin{bmatrix} -11 \\ -18 \end{bmatrix}$ as a linear combination of **v** and **w**. In other words, we would like to know whether there are weights c and d such that

$$c\mathbf{v} + d\mathbf{w} = \mathbf{b}.$$

This leads to the equations

$$c \begin{bmatrix} 3 \\ -1 \end{bmatrix} + d \begin{bmatrix} 4 \\ 3 \end{bmatrix} = \begin{bmatrix} -11 \\ -18 \end{bmatrix}$$

$$\begin{bmatrix} 3c \\ -c \end{bmatrix} + \begin{bmatrix} 4d \\ 3d \end{bmatrix} = \begin{bmatrix} -11 \\ -18 \end{bmatrix}$$

$$\begin{bmatrix} 3c + 4d \\ -c + 3d \end{bmatrix} = \begin{bmatrix} -11 \\ -18 \end{bmatrix}$$

Equating the components of the vectors on each side of the equation, we arrive at the linear system

$$3a + 4b = -11$$
$$-a + 3b = -18$$

This means that **b** is a linear combination of **v** and **w** if this linear system is consistent.

To solve this linear system, we construct its corresponding augmented matrix and find its reduced row echelon form,

$$\left[\begin{array}{cc|c} 3 & 4 & -11 \\ -1 & 3 & -18 \end{array}\right] \sim \left[\begin{array}{cc|c} 1 & 0 & 3 \\ 0 & 1 & -5 \end{array}\right],$$

giving us the weights $c = 3$ and $d = -5$; that is,

$$3\mathbf{v} - 5\mathbf{w} = \mathbf{b}.$$

In fact, we know more because the reduced row echelon matrix tells us that these are the only possible weights. Therefore, **b** may be expressed as a linear combination of **v** and **w** in exactly one way.

This example demonstrates the connection between linear combinations and linear systems. Asking whether a vector **b** is a linear combination of vectors $\mathbf{v}_1, \mathbf{v}_2, \ldots, \mathbf{v}_n$ is equivalent to asking whether an associated linear system is consistent.

In fact, we may easily describe the associated linear system in terms of the vectors **v**, **w**, and **b**. Notice that the augmented matrix we found in our example was $\left[\begin{array}{cc|c} 3 & 4 & -11 \\ -1 & 3 & -18 \end{array}\right]$. The first two columns of this matrix are **v** and **w** and the rightmost column is **b**. As shorthand, we will write this augmented matrix replacing the columns with their vector representation:

$$\left[\begin{array}{cc|c} \mathbf{v} & \mathbf{w} & \mathbf{b} \end{array}\right].$$

This fact is generally true so we record it in the following proposition.

Proposition 2.1.12 *The vector* **b** *is a linear combination of the vectors* $\mathbf{v}_1, \mathbf{v}_2, \ldots, \mathbf{v}_n$ *if and only if the linear system corresponding to the augmented matrix*

$$\left[\begin{array}{cccc|c} \mathbf{v}_1 & \mathbf{v}_2 & \ldots & \mathbf{v}_n & \mathbf{b} \end{array}\right]$$

is consistent. A solution to this linear system gives weights $c_1, c_2, \ldots, c_n$ *such that*

$$c_1 \mathbf{v}_1 + c_2 \mathbf{v}_2 + \ldots + c_n \mathbf{v}_n = \mathbf{b}.$$

The next activity puts this proposition to use.

Activity 2.1.3 Linear combinations and linear systems..

a. Given the vectors

$$\mathbf{v}_1 = \begin{bmatrix} 4 \\ 0 \\ 2 \\ 1 \end{bmatrix}, \mathbf{v}_2 = \begin{bmatrix} 1 \\ -3 \\ 3 \\ 1 \end{bmatrix}, \mathbf{v}_3 = \begin{bmatrix} -2 \\ 1 \\ 1 \\ 0 \end{bmatrix}, \mathbf{b} = \begin{bmatrix} 0 \\ 1 \\ 2 \\ -2 \end{bmatrix},$$

can **b** be expressed as a linear combination of $\mathbf{v}_1$, $\mathbf{v}_2$, and $\mathbf{v}_3$? Rephrase this question by writing a linear system for the weights c_1, c_2, and c_3 and use the Sage cell below to answer this question.

b. Consider the following linear system.

$$\begin{aligned} 3x_1 + 2x_2 - x_3 &= 4 \\ x_1 \phantom{{}+ 2x_2} + 2x_3 &= 0 \\ -x_1 - x_2 + 3x_3 &= 1 \end{aligned}$$

Identify vectors $\mathbf{v}_1$, $\mathbf{v}_2$, $\mathbf{v}_3$, and **b** such that the question "Is this linear system consistent?" is equivalent to the question "Can **b** be expressed as a linear combination of $\mathbf{v}_1$, $\mathbf{v}_2$, and $\mathbf{v}_3$?"

2.1. VECTORS AND LINEAR COMBINATIONS

c. Consider the vectors

$$\mathbf{v}_1 = \begin{bmatrix} 0 \\ -2 \\ 1 \end{bmatrix}, \mathbf{v}_2 = \begin{bmatrix} 1 \\ 1 \\ -1 \end{bmatrix}, \mathbf{v}_3 = \begin{bmatrix} 2 \\ 0 \\ -1 \end{bmatrix}, \mathbf{b} = \begin{bmatrix} -1 \\ 3 \\ -1 \end{bmatrix}.$$

Can $\mathbf{b}$ be expressed as a linear combination of $\mathbf{v}_1$, $\mathbf{v}_2$, and $\mathbf{v}_3$? If so, can $\mathbf{b}$ be written as a linear combination of these vectors in more than one way?

d. Considering the vectors $\mathbf{v}_1$, $\mathbf{v}_2$, and $\mathbf{v}_3$ from the previous part, can we write every three-dimensional vector $\mathbf{b}$ as a linear combination of these vectors? Explain how the pivot positions of the matrix $\begin{bmatrix} \mathbf{v}_1 & \mathbf{v}_2 & \mathbf{v}_3 \end{bmatrix}$ help answer this question.

e. Now consider the vectors

$$\mathbf{v}_1 = \begin{bmatrix} 0 \\ -2 \\ 1 \end{bmatrix}, \mathbf{v}_2 = \begin{bmatrix} 1 \\ 1 \\ -1 \end{bmatrix}, \mathbf{v}_3 = \begin{bmatrix} 1 \\ -1 \\ -2 \end{bmatrix}, \mathbf{b} = \begin{bmatrix} 0 \\ 8 \\ -4 \end{bmatrix}.$$

Can $\mathbf{b}$ be expressed as a linear combination of $\mathbf{v}_1$, $\mathbf{v}_2$, and $\mathbf{v}_3$? If so, can $\mathbf{b}$ be written as a linear combination of these vectors in more than one way?

f. Considering the vectors $\mathbf{v}_1$, $\mathbf{v}_2$, and $\mathbf{v}_3$ from the previous part, can we write every three-dimensional vector $\mathbf{b}$ as a linear combination of these vectors? Explain how the pivot positions of the matrix $\begin{bmatrix} \mathbf{v}_1 & \mathbf{v}_2 & \mathbf{v}_3 \end{bmatrix}$ help answer this question.

Example 2.1.13 Consider the vectors $\mathbf{v} = \begin{bmatrix} -1 \\ 1 \end{bmatrix}$ and $\mathbf{w} = \begin{bmatrix} 2 \\ -2 \end{bmatrix}$, as shown in Figure 2.1.14.

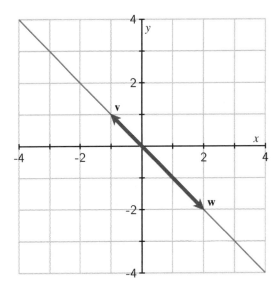

Figure 2.1.14 Vectors **v** and **w**.

These vectors appear to lie on the same line, a fact that becomes apparent once we notice that **w** = −2**v**. Intuitively, we think of the linear combination

$$c\mathbf{v} + d\mathbf{w}$$

as the result of walking c times in the **v** direction and d times in the **w** direction. With these vectors, we are always walking along the same line so it would seem that any linear combination of these vectors should lie on the same line. In addition, a vector that is not on the line, say $\mathbf{b} = \begin{bmatrix} 3 \\ 0 \end{bmatrix}$, should be not be expressible as a linear combination of **v** and **w**.

We can verify this by checking

$$\begin{bmatrix} -1 & 2 & | & 3 \\ 1 & -2 & | & 0 \end{bmatrix} \sim \begin{bmatrix} 1 & -2 & | & 0 \\ 0 & 0 & | & 1 \end{bmatrix}.$$

This shows that the associated linear system is inconsistent, which means that the vector $\mathbf{b} = \begin{bmatrix} 3 \\ 0 \end{bmatrix}$ cannot be written as a linear combination of **v** and **w**.

Notice that the reduced row echelon form of the coefficient matrix

$$\begin{bmatrix} \mathbf{v} & \mathbf{w} \end{bmatrix} = \begin{bmatrix} -1 & 2 \\ 1 & -2 \end{bmatrix} \sim \begin{bmatrix} 1 & -2 \\ 0 & 0 \end{bmatrix}$$

tells us to expect this. Since there is not a pivot position in the second row of the coefficient matrix $\begin{bmatrix} \mathbf{v} & \mathbf{w} \end{bmatrix}$, it is possible for a pivot position to appear in the rightmost column of the augmented matrix

$$\begin{bmatrix} \mathbf{v} & \mathbf{w} & | & \mathbf{b} \end{bmatrix}$$

for some choice of **b**.

2.1. VECTORS AND LINEAR COMBINATIONS

2.1.3 Summary

This section has introduced vectors, linear combinations, and their connection to linear systems.

- There are two operations we can perform with vectors: scalar multiplication and vector addition. Both of these operations have geometric meaning.
- Given a set of vectors and a set of scalars we call weights, we can create a linear combination using scalar multiplication and vector addition.
- A solution to the linear system whose augmented matrix is

$$\left[\begin{array}{cccc|c} \mathbf{v}_1 & \mathbf{v}_2 & \ldots & \mathbf{v}_n & \mathbf{b} \end{array}\right]$$

is a set of weights that expresses $\mathbf{b}$ as a linear combination of $\mathbf{v}_1, \mathbf{v}_2, \ldots, \mathbf{v}_n$.

2.1.4 Exercises

1. Consider the vectors

$$\mathbf{v} = \begin{bmatrix} 1 \\ -1 \end{bmatrix}, \mathbf{w} = \begin{bmatrix} 3 \\ 1 \end{bmatrix}$$

   a. Sketch these vectors below.

   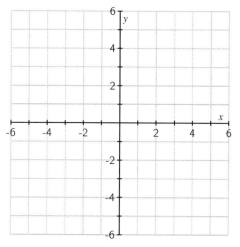

   b. Compute the vectors $-3\mathbf{v}$, $2\mathbf{w}$, $\mathbf{v} + \mathbf{w}$, and $\mathbf{v} - \mathbf{w}$ and add them into the sketch above.

   c. Sketch below the set of vectors having the form $2\mathbf{v} + c\mathbf{w}$ where c is any scalar.

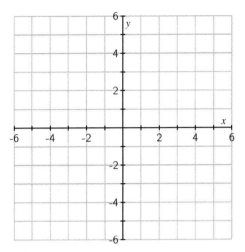

d. Sketch below the line $y = 3x - 2$. Then identify two vectors **v** and **w** so that this line is described by $\mathbf{v} + c\mathbf{w}$. Are there other choices for the vectors **v** and **w**?

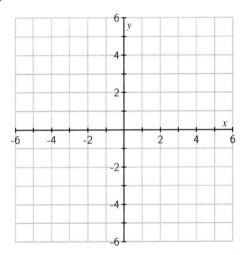

2. Shown below are two vectors **v** and **w**

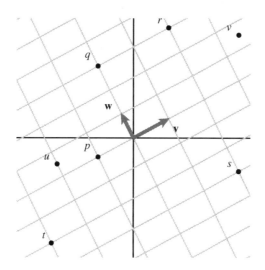

a. Express the labeled points as linear combinations of **v** and **w**.

b. Sketch the line described parametrically as $-2\mathbf{v} + c\mathbf{w}$.

3. Consider the vectors

$$\mathbf{v}_1 = \begin{bmatrix} 2 \\ 1 \end{bmatrix}, \mathbf{v}_2 = \begin{bmatrix} -1 \\ 1 \end{bmatrix}, \mathbf{v}_3 = \begin{bmatrix} -2 \\ 0 \end{bmatrix}$$

a. Find the linear combination with weights $c_1 = 2$, $c_2 = -3$, and $c_3 = 1$.

b. Can you write the vector $\mathbf{0} = \begin{bmatrix} 0 \\ 0 \end{bmatrix}$ as a linear combination of $\mathbf{v}_1$, $\mathbf{v}_2$, and $\mathbf{v}_3$? If so, describe all the ways in which you can do so.

c. Can you write the vector $\mathbf{0} = \begin{bmatrix} 0 \\ 0 \end{bmatrix}$ as a linear combination using just the first two vectors $\mathbf{v}_1$ $\mathbf{v}_2$? If so, describe all the ways in which you can do so.

d. Can you write $\mathbf{v}_3$ as a linear combination of $\mathbf{v}_1$ and $\mathbf{v}_2$? If so, in how many ways?

4. Nutritional information about a breakfast cereal is printed on the box. For instance, one serving of Frosted Flakes has 111 calories, 140 milligrams of sodium, and 1.2 grams of protein. We may represent this as a vector

$$\begin{bmatrix} 111 \\ 140 \\ 1.2 \end{bmatrix}.$$

One serving of Cocoa Puffs has 120 calories, 105 milligrams of sodium, and 1.0 grams of protein.

a. Write the vector describing the nutritional content of Cocoa Puffs.

b. Suppose you eat c servings of Frosted Flakes and d servings of Cocoa Puffs. Use the language of vectors and linear combinations to express the quantities of calories, sodium, and protein you have consumed.

c. How many servings of each cereal have you eaten if you have consumed 342 calories, 385 milligrams of sodium, and 3.4 grams of protein.

d. Suppose your sister consumed 250 calories, 200 milligrams of sodium, and 4 grams of protein. What can you conclude about her breakfast?

5. Consider the vectors
$$\mathbf{v}_1 = \begin{bmatrix} 2 \\ -1 \\ -2 \end{bmatrix}, \mathbf{v}_2 = \begin{bmatrix} 0 \\ 3 \\ 1 \end{bmatrix}, \mathbf{v}_3 = \begin{bmatrix} 4 \\ 4 \\ -2 \end{bmatrix}.$$

a. Can you express the vector $\mathbf{b} = \begin{bmatrix} 10 \\ 1 \\ -8 \end{bmatrix}$ as a linear combination of $\mathbf{v}_1, \mathbf{v}_2$, and $\mathbf{v}_3$? If so, describe all the ways in which you can do so.

b. Can you express the vector $\mathbf{b} = \begin{bmatrix} 3 \\ 7 \\ 1 \end{bmatrix}$ as a linear combination of $\mathbf{v}_1, \mathbf{v}_2$, and $\mathbf{v}_3$? If so, describe all the ways in which you can do so.

c. Show that $\mathbf{v}_3$ can be written as a linear combination of $\mathbf{v}_1$ and $\mathbf{v}_2$.

d. Explain why any linear combination of $\mathbf{v}_1, \mathbf{v}_2$, and $\mathbf{v}_3$,
$$a\mathbf{v}_1 + b\mathbf{v}_2 + c\mathbf{v}_3,$$
can be rewritten as a linear combination of just $\mathbf{v}_1$ and $\mathbf{v}_2$.

6. Consider the vectors
$$\mathbf{v}_1 = \begin{bmatrix} 3 \\ -1 \\ 1 \end{bmatrix}, \mathbf{v}_2 = \begin{bmatrix} 1 \\ 1 \\ 2 \end{bmatrix}.$$

For what value(s) of k, if any, can the vector $\begin{bmatrix} k \\ -2 \\ 5 \end{bmatrix}$ be written as a linear combination of $\mathbf{v}_1$ and $\mathbf{v}_2$?

7. Determine whether the following statements are true or false and provide a justification

2.1. VECTORS AND LINEAR COMBINATIONS

for your response.

a. Given two vectors **v** and **w**, the vector 2**v** is a linear combination of **v** and **w**.

b. Suppose $\mathbf{v}_1, \mathbf{v}_2, \ldots, \mathbf{v}_n$ is a collection of m-dimensional vectors and that the matrix $\begin{bmatrix} \mathbf{v}_1 & \mathbf{v}_2 & \ldots & \mathbf{v}_n \end{bmatrix}$ has a pivot position in every row. If **b** is any m-dimensional vector, then **b** can be written as a linear combination of $\mathbf{v}_1, \mathbf{v}_2, \ldots, \mathbf{v}_n$.

c. Suppose $\mathbf{v}_1, \mathbf{v}_2, \ldots, \mathbf{v}_n$ is a collection of m-dimensional vectors and that the matrix $\begin{bmatrix} \mathbf{v}_1 & \mathbf{v}_2 & \ldots & \mathbf{v}_n \end{bmatrix}$ has a pivot position in every row and every column. If **b** is any m-dimensional vector, then **b** can be written as a linear combination of $\mathbf{v}_1, \mathbf{v}_2, \ldots, \mathbf{v}_n$ in exactly one way.

d. It is possible to find two 3-dimensional vectors $\mathbf{v}_1$ and $\mathbf{v}_2$ such that every 3-dimensional vector can be written as a linear combination of $\mathbf{v}_1$ and $\mathbf{v}_2$.

8. A theme that will later unfold concerns the use of coordinate systems. We can identify the point (x, y) with the tip of the vector $\begin{bmatrix} x \\ y \end{bmatrix}$, drawn emanating from the origin. We can then think of the usual Cartesian coordinate system in terms of linear combinations of the vectors

$$\mathbf{e}_1 = \begin{bmatrix} 1 \\ 0 \end{bmatrix}, \mathbf{e}_2 = \begin{bmatrix} 0 \\ 1 \end{bmatrix}.$$

For instance, the point $(2, -3)$ is identified with the vector

$$\begin{bmatrix} 2 \\ -3 \end{bmatrix} = 2\mathbf{e}_1 - 3\mathbf{e}_2,$$

as shown on the left in Figure 2.1.15.

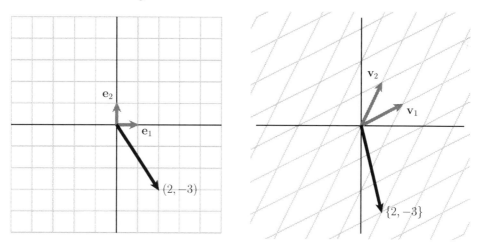

Figure 2.1.15 The usual Cartesian coordinate system, defined by the vectors $\mathbf{e}_1$ and $\mathbf{e}_2$, is shown on the left along with the representation of the point $(2, -3)$. The right shows a nonstandard coordinate system defined by vectors $\mathbf{v}_1$ and $\mathbf{v}_2$.

If instead we have vectors

$$\mathbf{v}_1 = \begin{bmatrix} 2 \\ 1 \end{bmatrix}, \mathbf{v}_2 = \begin{bmatrix} 1 \\ 2 \end{bmatrix},$$

we may define a new coordinate system in which a point $\{c, d\}$ will correspond to the vector
$$c\mathbf{v}_1 + d\mathbf{v}_2.$$
For instance, the point $\{2, -3\}$ is shown on the right side of Figure 2.1.15.

a. Write the point $\{2, -3\}$ in standard coordinates; that is, find x and y such that
$$(x, y) = \{2, -3\}.$$

b. Write the point $(2, -3)$ in the new coordinate system; that is, find c and d such that
$$\{c, d\} = (2, -3).$$

c. Convert a general point $\{c, d\}$, expressed in the new coordinate system, into standard Cartesian coordinates (x, y).

d. What is the general strategy for converting a point from standard Cartesian coordinates (x, y) to the new coordinates $\{c, d\}$? Actually implementing this strategy in general may take a bit of work so just describe the strategy. We will study this in more detail later.

2.2 Matrix multiplication and linear combinations

The previous section introduced vectors and linear combinations and demonstrated how they provide a way to think about linear systems geometrically. In particular, we saw that the vector $\mathbf{b}$ is a linear combination of the vectors $\mathbf{v}_1, \mathbf{v}_2, \ldots, \mathbf{v}_n$ precisely when the linear system corresponding to the augmented matrix

$$\left[\begin{array}{cccc|c} \mathbf{v}_1 & \mathbf{v}_2 & \cdots & \mathbf{v}_n & \mathbf{b} \end{array} \right]$$

is consistent.

Our goal in this section is to introduce matrix multiplication, another algebraic operation that deepens the connection between linear systems and linear combinations.

2.2.1 Scalar multiplication and addition of matrices

We first thought of a matrix as a rectangular array of numbers. If we say that the *shape* of a matrix is $m \times n$, we mean that it has m rows and n columns. For instance, the shape of the matrix below is 3×4:

$$\begin{bmatrix} 0 & 4 & -3 & 1 \\ 3 & -1 & 2 & 0 \\ 2 & 0 & -1 & 1 \end{bmatrix}.$$

We may also think of the columns of a matrix as a set of vectors. For instance, the matrix above may be represented as

$$\begin{bmatrix} \mathbf{v}_1 & \mathbf{v}_2 & \mathbf{v}_3 & \mathbf{v}_4 \end{bmatrix}$$

where

$$\mathbf{v}_1 = \begin{bmatrix} 0 \\ 3 \\ 2 \end{bmatrix}, \mathbf{v}_2 = \begin{bmatrix} 4 \\ -1 \\ 0 \end{bmatrix}, \mathbf{v}_3 = \begin{bmatrix} -3 \\ 2 \\ -1 \end{bmatrix}, \mathbf{v}_4 = \begin{bmatrix} 1 \\ 0 \\ 1 \end{bmatrix}.$$

In this way, we see that the 3×4 matrix is equivalent to an ordered set of 4 vectors in $\mathbb{R}^3$.

This means that we may define scalar multiplication and matrix addition operations using the corresponding column-wise vector operations. For instance,

$$c \begin{bmatrix} \mathbf{v}_1 & \mathbf{v}_2 & \cdots & \mathbf{v}_n \end{bmatrix} = \begin{bmatrix} c\mathbf{v}_1 & c\mathbf{v}_2 & \cdots & c\mathbf{v}_n \end{bmatrix}$$

$$\begin{bmatrix} \mathbf{v}_1 & \mathbf{v}_2 & \cdots & \mathbf{v}_n \end{bmatrix} + \begin{bmatrix} \mathbf{w}_1 & \mathbf{w}_2 & \cdots & \mathbf{w}_n \end{bmatrix}$$
$$= \begin{bmatrix} \mathbf{v}_1+\mathbf{w}_1 & \mathbf{v}_2+\mathbf{w}_2 & \cdots & \mathbf{v}_n+\mathbf{w}_n \end{bmatrix}.$$

Preview Activity 2.2.1 Matrix operations..

a. Compute the scalar multiple

$$-3 \begin{bmatrix} 3 & 1 & 0 \\ -4 & 3 & -1 \end{bmatrix}.$$

b. Find the sum
$$\begin{bmatrix} 0 & -3 \\ 1 & -2 \\ 3 & 4 \end{bmatrix} + \begin{bmatrix} 4 & -1 \\ -2 & 2 \\ 1 & 1 \end{bmatrix}.$$

c. Suppose that A and B are two matrices. What do we need to know about their shapes before we can form the sum $A + B$?

d. The matrix I_n, which we call the *identity* matrix, is the $n \times n$ matrix whose entries are zero except for the diagonal entries, all of which are 1. For instance,
$$I_3 = \begin{bmatrix} 1 & 0 & 0 \\ 0 & 1 & 0 \\ 0 & 0 & 1 \end{bmatrix}.$$

If we can form the sum $A + I_n$, what must be true about the matrix A?

e. Find the matrix $A - 2I_3$ where
$$A = \begin{bmatrix} 1 & 2 & -2 \\ 2 & -3 & 3 \\ -2 & 3 & 4 \end{bmatrix}.$$

As this preview activity shows, the operations of scalar multiplication and addition of matrices are natural extensions of their vector counterparts. Some care, however, is required when adding matrices. Since we need the same number of vectors to add and since those vectors must be of the same dimension, two matrices must have the same shape if we wish to form their sum.

2.2.2 Matrix-vector multiplication and linear combinations

A more important operation will be matrix multiplication as it allows us to compactly express linear systems. We now introduce the product of a matrix and a vector with an example.

Example 2.2.1 Matrix-vector multiplication. Suppose we have the matrix A and vector $\mathbf{x}$:
$$A = \begin{bmatrix} -2 & 3 \\ 0 & 2 \\ 3 & 1 \end{bmatrix}, \quad \mathbf{x} = \begin{bmatrix} 2 \\ 3 \end{bmatrix}.$$

Their product will be defined to be the linear combination of the columns of A using the

2.2. MATRIX MULTIPLICATION AND LINEAR COMBINATIONS

components of **x** as weights. This means that

$$A\mathbf{x} = \begin{bmatrix} -2 & 3 \\ 0 & 2 \\ 3 & 1 \end{bmatrix} \begin{bmatrix} 2 \\ 3 \end{bmatrix} = 2 \begin{bmatrix} -2 \\ 0 \\ 3 \end{bmatrix} + 3 \begin{bmatrix} 3 \\ 2 \\ 1 \end{bmatrix}$$

$$= \begin{bmatrix} -4 \\ 0 \\ 6 \end{bmatrix} + \begin{bmatrix} 9 \\ 6 \\ 3 \end{bmatrix}$$

$$= \begin{bmatrix} 5 \\ 6 \\ 9 \end{bmatrix}.$$

Because A has two columns, we need two weights to form a linear combination of those columns, which means that **x** must have two components. In other words, the number of columns of A must equal the dimension of the vector **x**.

Similarly, the columns of A are 3-dimensional so any linear combination of them is 3-dimensional as well. Therefore, $A\mathbf{x}$ will be 3-dimensional.

We then see that if A is a 3×2 matrix, **x** must be a 2-dimensional vector and $A\mathbf{x}$ will be 3-dimensional.

More generally, we have the following definition.

Definition 2.2.2 Matrix-vector multiplication. The product of a matrix A by a vector **x** will be the linear combination of the columns of A using the components of **x** as weights. More specifically, if

$$A = \begin{bmatrix} \mathbf{v}_1 & \mathbf{v}_2 & \ldots & \mathbf{v}_n \end{bmatrix}, \quad \mathbf{x} = \begin{bmatrix} c_1 \\ c_2 \\ \vdots \\ c_n \end{bmatrix},$$

then

$$A\mathbf{x} = c_1 \mathbf{v}_1 + c_2 \mathbf{v}_2 + \ldots + c_n \mathbf{v}_n.$$

If A is an $m \times n$ matrix, then **x** must be an n-dimensional vector, and the product $A\mathbf{x}$ will be an m-dimensional vector.

The next activity explores some properties of matrix multiplication.

Activity 2.2.2 Matrix-vector multiplication..

a. Find the matrix product

$$\begin{bmatrix} 1 & 2 & 0 & -1 \\ 2 & 4 & -3 & -2 \\ -1 & -2 & 6 & 1 \end{bmatrix} \begin{bmatrix} 3 \\ 1 \\ -1 \\ 1 \end{bmatrix}.$$

b. Suppose that A is the matrix

$$\begin{bmatrix} 3 & -1 & 0 \\ 0 & -2 & 4 \\ 2 & 1 & 5 \\ 1 & 0 & 3 \end{bmatrix}.$$

If $A\mathbf{x}$ is defined, what is the dimension of the vector $\mathbf{x}$ and what is the dimension of $A\mathbf{x}$?

c. A vector whose entries are all zero is denoted by $\mathbf{0}$. If A is a matrix, what is the product $A\mathbf{0}$?

d. Suppose that $I = \begin{bmatrix} 1 & 0 & 0 \\ 0 & 1 & 0 \\ 0 & 0 & 1 \end{bmatrix}$ is the identity matrix and $\mathbf{x} = \begin{bmatrix} x_1 \\ x_2 \\ x_3 \end{bmatrix}$. Find the product $I\mathbf{x}$ and explain why I is called the identity matrix.

e. Suppose we write the matrix A in terms of its columns as

$$A = \begin{bmatrix} \mathbf{v}_1 & \mathbf{v}_2 & \cdots & \mathbf{v}_n \end{bmatrix}.$$

If the vector $\mathbf{e}_1 = \begin{bmatrix} 1 \\ 0 \\ \vdots \\ 0 \end{bmatrix}$, what is the product $A\mathbf{e}_1$?

f. Suppose that

$$A = \begin{bmatrix} 1 & 2 \\ -1 & 1 \end{bmatrix}, \mathbf{b} = \begin{bmatrix} 6 \\ 0 \end{bmatrix}.$$

Is there a vector $\mathbf{x}$ such that $A\mathbf{x} = \mathbf{b}$?

Multiplication of a matrix A and a vector is defined as a linear combination of the columns of A. However, there is a shortcut for computing such a product. Let's look at our previous example and focus on the first row of the product.

$$\begin{bmatrix} -2 & 3 \\ 0 & 2 \\ 3 & 1 \end{bmatrix} \begin{bmatrix} 2 \\ 3 \end{bmatrix} = 2 \begin{bmatrix} -2 \\ * \\ * \end{bmatrix} + 3 \begin{bmatrix} 3 \\ * \\ * \end{bmatrix} = \begin{bmatrix} 2(-2) + 3(3) \\ * \\ * \end{bmatrix} = \begin{bmatrix} 5 \\ * \\ * \end{bmatrix}.$$

2.2. MATRIX MULTIPLICATION AND LINEAR COMBINATIONS

To find the first component of the product, we consider the first row of the matrix. We then multiply the first entry in that row by the first component of the vector, the second entry by the second component of the vector, and so on, and add the results. In this way, we see that the third component of the product would be obtained from the third row of the matrix by computing $2(3) + 3(1) = 9$.

You are encouraged to evaluate the product Item a of the previous activity using this shortcut and compare the result to what you found while completing that activity.

Activity 2.2.3. Sage can find the product of a matrix and vector using the * operator. For example,

```
A = matrix(2,2,[1,2,2,1])
v = vector([3,-1])
A*v
```

a. Use Sage to evaluate the product

$$\begin{bmatrix} 1 & 2 & 0 & -1 \\ 2 & 4 & -3 & -2 \\ -1 & -2 & 6 & 1 \end{bmatrix} \begin{bmatrix} 3 \\ 1 \\ -1 \\ 1 \end{bmatrix}$$

from Item a of the previous activity.

b. In Sage, define the matrix and vectors

$$A = \begin{bmatrix} -2 & 0 \\ 3 & 1 \\ 4 & 2 \end{bmatrix}, \mathbf{0} = \begin{bmatrix} 0 \\ 0 \end{bmatrix}, \mathbf{v} = \begin{bmatrix} -2 \\ 3 \end{bmatrix}, \mathbf{w} = \begin{bmatrix} 1 \\ 2 \end{bmatrix}.$$

c. What do you find when you evaluate $A\mathbf{0}$?

d. What do you find when you evaluate $A(3\mathbf{v})$ and $3(A\mathbf{v})$ and compare your results?

e. What do you find when you evaluate $A(\mathbf{v} + \mathbf{w})$ and $A\mathbf{v} + A\mathbf{w}$ and compare your results?

This activity demonstrates several general properties satisfied by matrix multiplication that we record here.

Proposition 2.2.3 Linearity of matrix multiplication. *If A is a matrix, $\mathbf{v}$ and $\mathbf{w}$ vectors of the appropriate dimensions, and c a scalar, then*

- $A\mathbf{0} = \mathbf{0}$.

- $A(c\mathbf{v}) = cA\mathbf{v}$.
- $A(\mathbf{v} + \mathbf{w}) = A\mathbf{v} + A\mathbf{w}$.

2.2.3 Matrix-vector multiplication and linear systems

So far, we have begun with a matrix A and a vector $\mathbf{x}$ and formed their product $A\mathbf{x} = \mathbf{b}$. We would now like to turn this around: beginning with a matrix A and a vector $\mathbf{b}$, we will ask if we can find a vector $\mathbf{x}$ such that $A\mathbf{x} = \mathbf{b}$. This will naturally lead back to linear systems.

To see the connection between the matrix equation $A\mathbf{x} = \mathbf{b}$ and linear systems, let's write the matrix A in terms of its columns $\mathbf{v}_i$ and $\mathbf{x}$ in terms of its components.

$$A = \begin{bmatrix} \mathbf{v}_1 & \mathbf{v}_2 & \ldots & \mathbf{v}_n \end{bmatrix}, \mathbf{x} = \begin{bmatrix} c_1 \\ c_2 \\ \vdots \\ c_n \end{bmatrix}.$$

We know that the matrix product $A\mathbf{x}$ forms a linear combination of the columns of A. Therefore, the equation $A\mathbf{x} = \mathbf{b}$ is merely a compact way of writing the equation for the weights c_i:

$$c_1 \mathbf{v}_1 + c_2 \mathbf{v}_2 + \ldots + c_n \mathbf{v}_n = \mathbf{b}.$$

We have seen this equation before: Remember that Proposition 2.1.12 says that the solutions of this equation are the same as the solutions to the linear system whose augmented matrix is

$$\begin{bmatrix} \mathbf{v}_1 & \mathbf{v}_2 & \ldots & \mathbf{v}_n & | & \mathbf{b} \end{bmatrix}.$$

This gives us three different ways of looking at the same solution space.

Proposition 2.2.4 *If* $A = \begin{bmatrix} \mathbf{v}_1 & \mathbf{v}_2 & \ldots \mathbf{v}_n \end{bmatrix}$ *and* $\mathbf{x} = \begin{bmatrix} x_1 \\ x_2 \\ \vdots \\ x_n \end{bmatrix}$, *then the following statements are equivalent.*

- *The vector* $\mathbf{x}$ *satisfies the equation* $A\mathbf{x} = \mathbf{b}$.
- *The vector* $\mathbf{b}$ *is a linear combination of the columns of* A *with weights* x_j:

$$x_1 \mathbf{v}_1 + x_2 \mathbf{v}_2 + \ldots + x_n \mathbf{v}_n = \mathbf{b}.$$

- *The components of* $\mathbf{x}$ *form a solution to the linear system corresponding to the augmented matrix*

$$\begin{bmatrix} \mathbf{v}_1 & \mathbf{v}_2 & \ldots & \mathbf{v}_n & | & \mathbf{b} \end{bmatrix}.$$

When the matrix $A = \begin{bmatrix} \mathbf{v}_1 & \mathbf{v}_2 & \ldots & \mathbf{v}_n \end{bmatrix}$, we will frequently write

$$\begin{bmatrix} \mathbf{v}_1 & \mathbf{v}_2 & \ldots & \mathbf{v}_n & | & \mathbf{b} \end{bmatrix} = \begin{bmatrix} A & | & \mathbf{b} \end{bmatrix}$$

2.2. MATRIX MULTIPLICATION AND LINEAR COMBINATIONS

and say that the matrix A is augmented by the vector **b**.

The equation $A\mathbf{x} = \mathbf{b}$ gives a notationally compact way to write a linear system. Moreover, this notation will allow us to focus on important features of the system that determine its solution space.

Example 2.2.5 We will describe the solution space of the equation

$$\begin{bmatrix} 2 & 0 & 2 \\ 4 & -1 & 6 \\ 1 & 3 & -5 \end{bmatrix} \mathbf{x} = \begin{bmatrix} 0 \\ -5 \\ 15 \end{bmatrix}.$$

By Proposition 2.2.4, this equation may be equivalently expressed as

$$x_1 \begin{bmatrix} 2 \\ 4 \\ 1 \end{bmatrix} + x_2 \begin{bmatrix} 0 \\ -1 \\ 3 \end{bmatrix} + x_3 \begin{bmatrix} 2 \\ 6 \\ -5 \end{bmatrix} = \begin{bmatrix} 0 \\ -5 \\ 15 \end{bmatrix},$$

which is the linear system corresponding to the augmented matrix

$$\left[\begin{array}{ccc|c} 2 & 0 & 2 & 0 \\ 4 & -1 & 6 & -5 \\ 1 & 3 & -5 & 15 \end{array} \right].$$

The reduced row echelon form of the augmented matrix is

$$\left[\begin{array}{ccc|c} 2 & 0 & 2 & 0 \\ 4 & -1 & 6 & -5 \\ 1 & 3 & -5 & 15 \end{array} \right] \sim \left[\begin{array}{ccc|c} 1 & 0 & 1 & 0 \\ 0 & 1 & -2 & 5 \\ 0 & 0 & 0 & 0 \end{array} \right],$$

which corresponds to the linear system

$$\begin{aligned} x_1 + x_3 &= 0 \\ x_2 - 2x_3 &= 5. \end{aligned}$$

The variable x_3 is free so we may write the solution space parametrically as

$$\begin{aligned} x_1 &= -x_3 \\ x_2 &= 5 + 2x_3. \end{aligned}$$

Since we originally asked to describe the solutions to the equation $A\mathbf{x} = \mathbf{b}$, we will express the solution in terms of the vector **x**:

$$\mathbf{x} = \begin{bmatrix} x_1 \\ x_2 \\ x_3 \end{bmatrix} = \begin{bmatrix} -x_3 \\ 5 + 2x_3 \\ x_3 \end{bmatrix} = \begin{bmatrix} 0 \\ 5 \\ 0 \end{bmatrix} + x_3 \begin{bmatrix} -1 \\ 2 \\ 1 \end{bmatrix}.$$

As before, we call this a parametric description of the solution space.

This shows that the solutions **x** may be written in the form $\mathbf{v} + x_3\mathbf{w}$, for appropriate vectors **v** and **w**. Geometrically, the solution space is a line in $\mathbb{R}^3$ through **v** moving parallel to **w**.

Activity 2.2.4 The equation $A\mathbf{x} = \mathbf{b}$..

a. Consider the linear system
$$\begin{aligned} 2x + y - 3z &= 4 \\ -x + 2y + z &= 3 \\ 3x - y &= -4. \end{aligned}$$

Identify the matrix A and vector $\mathbf{b}$ to express this system in the form $A\mathbf{x} = \mathbf{b}$.

b. If A and $\mathbf{b}$ are as below, write the linear system corresponding to the equation $A\mathbf{x} = \mathbf{b}$ and describe its solution space, using a parametric description if appropriate:

$$A = \begin{bmatrix} 3 & -1 & 0 \\ -2 & 0 & 6 \end{bmatrix}, \quad \mathbf{b} = \begin{bmatrix} -6 \\ 2 \end{bmatrix}.$$

c. Describe the solution space of the equation
$$\begin{bmatrix} 1 & 2 & 0 & -1 \\ 2 & 4 & -3 & -2 \\ -1 & -2 & 6 & 1 \end{bmatrix} \mathbf{x} = \begin{bmatrix} -1 \\ 1 \\ 5 \end{bmatrix}.$$

d. Suppose A is an $m \times n$ matrix. What can you guarantee about the solution space of the equation $A\mathbf{x} = \mathbf{0}$?

2.2.4 Matrix-matrix products

In this section, we have developed some algebraic operations on matrices with the aim of simplifying our description of linear systems. We now introduce a final operation, the product of two matrices, that will become important when we study linear transformations in Section 2.5.

Definition 2.2.6 Matrix-matrix multiplication. Given matrices A and B, we form their product AB by first writing B in terms of its columns

$$B = \begin{bmatrix} \mathbf{v}_1 & \mathbf{v}_2 & \cdots & \mathbf{v}_p \end{bmatrix}$$

and then defining
$$AB = \begin{bmatrix} A\mathbf{v}_1 & A\mathbf{v}_2 & \cdots & A\mathbf{v}_p \end{bmatrix}.$$

2.2. MATRIX MULTIPLICATION AND LINEAR COMBINATIONS

Example 2.2.7 Given the matrices

$$A = \begin{bmatrix} 4 & 2 \\ 0 & 1 \\ -3 & 4 \\ 2 & 0 \end{bmatrix}, \quad B = \begin{bmatrix} -2 & 3 & 0 \\ 1 & 2 & -2 \end{bmatrix},$$

we have

$$AB = \begin{bmatrix} A\begin{bmatrix} -2 \\ 1 \end{bmatrix} & A\begin{bmatrix} 3 \\ 2 \end{bmatrix} & A\begin{bmatrix} 0 \\ -2 \end{bmatrix} \end{bmatrix} = \begin{bmatrix} -6 & 16 & -4 \\ 1 & 2 & -2 \\ 10 & -1 & -8 \\ -4 & 6 & 0 \end{bmatrix}.$$

Observation 2.2.8 It is important to note that we can only multiply matrices if the shapes of the matrices are compatible. More specifically, when constructing the product AB, the matrix A multiplies the columns of B. Therefore, the number of columns of A must equal the number of rows of B. When this condition is met, the number of rows of AB is the number of rows of A, and the number of columns of AB is the number of columns of B.

Activity 2.2.5. Consider the matrices

$$A = \begin{bmatrix} 1 & 3 & 2 \\ -3 & 4 & -1 \end{bmatrix}, \quad B = \begin{bmatrix} 3 & 0 \\ 1 & 2 \\ -2 & -1 \end{bmatrix}.$$

a. Before computing, first explain why the shapes of A and B enable us to form the product AB. Then describe the shape of AB.

b. Compute the product AB.

c. Sage can multiply matrices using the * operator. Define the matrices A and B in the Sage cell below and check your work by computing AB.

d. Are we able to form the matrix product BA? If so, use the Sage cell above to find BA. Is it generally true that $AB = BA$?

e. Suppose we form the three matrices.

$$A = \begin{bmatrix} 1 & 2 \\ 3 & -2 \end{bmatrix}, B = \begin{bmatrix} 0 & 4 \\ 2 & -1 \end{bmatrix}, C = \begin{bmatrix} -1 & 3 \\ 4 & 3 \end{bmatrix}.$$

Compare what happens when you compute $A(B+C)$ and $AB + AC$. State your finding as a general principle.

f. Compare the results of evaluating $A(BC)$ and $(AB)C$ and state your finding as a general principle.

g. When we are dealing with real numbers, we know if $a \neq 0$ and $ab = ac$, then $b = c$. Define matrices

$$A = \begin{bmatrix} 1 & 2 \\ -2 & -4 \end{bmatrix}, B = \begin{bmatrix} 3 & 0 \\ 1 & 3 \end{bmatrix}, C = \begin{bmatrix} 1 & 2 \\ 2 & 2 \end{bmatrix}$$

and compute AB and AC.

If $AB = AC$, is it necessarily true that $B = C$?

h. Again, with real numbers, we know that if $ab = 0$, then either $a = 0$ or $b = 0$. Define

$$A = \begin{bmatrix} 1 & 2 \\ -2 & -4 \end{bmatrix}, B = \begin{bmatrix} 2 & -4 \\ -1 & 2 \end{bmatrix}$$

and compute AB.

If $AB = 0$, is it necessarily true that either $A = 0$ or $B = 0$?

This activity demonstrated some general properties about products of matrices, which mirror some properties about operations with real numbers.

> **Properties of Matrix-matrix Multiplication.**
>
> If A, B, and C are matrices such that the following operations are defined, it follows that
>
> **Associativity:** $A(BC) = (AB)C$.
> **Distributivity:** $A(B + C) = AB + AC$.
> $(A + B)C = AC + BC$.

At the same time, there are a few properties that hold for real numbers that do not hold for matrices.

> **Caution.**
>
> The following properties hold for real numbers but not for matrices.
>
> **Commutativity:** It is *not* generally true that $AB = BA$.
> **Cancellation:** It is *not* generally true that $AB = AC$ implies that $B = C$.
> **Zero divisors:** It is *not* generally true that $AB = 0$ implies that either $A = 0$ or $B = 0$.

2.2.5 Summary

In this section, we have found an especially simple way to express linear systems using matrix multiplication.

- If A is an $m \times n$ matrix and $\mathbf{x}$ an n-dimensional vector, then $A\mathbf{x}$ is the linear combination of the columns of A using the components of $\mathbf{x}$ as weights. The vector $A\mathbf{x}$ is m-dimensional.

- The solution space to the equation $A\mathbf{x} = \mathbf{b}$ is the same as the solution space to the linear system corresponding to the augmented matrix $[\, A \mid \mathbf{b} \,]$.

- If A is an $m \times n$ matrix and B is an $n \times p$ matrix, we can form the product AB, which is an $m \times p$ matrix whose columns are the products of A and the columns of B.

2.2.6 Exercises

1. Consider the system of linear equations

$$\begin{aligned} x + 2y - z &= 1 \\ 3x + 2y + 2z &= 7. \\ -x + 4z &= -3 \end{aligned}$$

   a. Find the matrix A and vector $\mathbf{b}$ that expresses this linear system in the form $A\mathbf{x} = \mathbf{b}$.

   b. Give a description of the solution space to the equation $A\mathbf{x} = \mathbf{b}$.

2. Suppose that A is a 135×2201 matrix, and that $\mathbf{x}$ is a vector. If $A\mathbf{x}$ is defined, what is the dimension of $\mathbf{x}$? What is the dimension of $A\mathbf{x}$?

3. Suppose that A is a 3×2 matrix whose columns are $\mathbf{v}_1$ and $\mathbf{v}_2$; that is,

$$A = [\, \mathbf{v}_1 \quad \mathbf{v}_2 \,].$$

   a. What is the dimension of the vectors $\mathbf{v}_1$ and $\mathbf{v}_2$?

   b. What is the product $A \begin{bmatrix} 1 \\ 0 \end{bmatrix}$ in terms of $\mathbf{v}_1$ and $\mathbf{v}_2$? What is the product $A \begin{bmatrix} 0 \\ 1 \end{bmatrix}$? What is the product $A \begin{bmatrix} 2 \\ 3 \end{bmatrix}$?

   c. If we know that

$$A \begin{bmatrix} 1 \\ 0 \end{bmatrix} = \begin{bmatrix} 3 \\ -2 \\ 1 \end{bmatrix}, \quad A \begin{bmatrix} 0 \\ 1 \end{bmatrix} = \begin{bmatrix} 0 \\ 3 \\ 2 \end{bmatrix},$$

   what is the matrix A?

4. Suppose that the matrix $A = \begin{bmatrix} \mathbf{v}_1 & \mathbf{v}_2 \end{bmatrix}$ where $\mathbf{v}_1$ and $\mathbf{v}_2$ are shown in Figure 2.2.9.

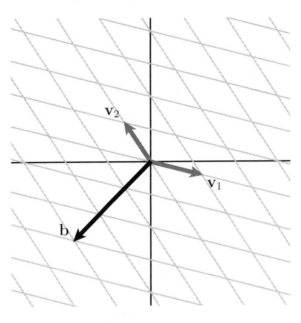

Figure 2.2.9 Two vectors $\mathbf{v}_1$ and $\mathbf{v}_2$ that form the columns of the matrix A.

   a. What is the shape of the matrix A?

   b. On Figure 2.2.9, indicate the vectors
   $$A\begin{bmatrix} 1 \\ 0 \end{bmatrix}, \quad A\begin{bmatrix} 2 \\ 3 \end{bmatrix}, \quad A\begin{bmatrix} 0 \\ -3 \end{bmatrix}.$$

   c. Find all vectors $\mathbf{x}$ such that $A\mathbf{x} = \mathbf{b}$.

   d. Find all vectors $\mathbf{x}$ such that $A\mathbf{x} = \mathbf{0}$.

5. Suppose that
$$A = \begin{bmatrix} 1 & 0 & 2 \\ 2 & 2 & 2 \\ -1 & -3 & 1 \end{bmatrix}.$$

   a. Describe the solution space to the equation $A\mathbf{x} = \mathbf{0}$.

   b. Find a 3×2 matrix B with no zero entries such that $AB = 0$.

6. Consider the matrix
$$A = \begin{bmatrix} 1 & 2 & -4 & -4 \\ 2 & 3 & 0 & 1 \\ 1 & 0 & 4 & 6 \end{bmatrix}.$$

a. Find the product $A\mathbf{x}$ where

$$\mathbf{x} = \begin{bmatrix} 1 \\ -2 \\ 0 \\ 2 \end{bmatrix}.$$

b. Give a description of the vectors $\mathbf{x}$ such that

$$A\mathbf{x} = \begin{bmatrix} -1 \\ 15 \\ 17 \end{bmatrix}.$$

c. Find the reduced row echelon form of A and identify the pivot positions.

d. Can you find a vector $\mathbf{b}$ such that $A\mathbf{x} = \mathbf{b}$ is inconsistent?

e. For a general 3-dimensional vector $\mathbf{b}$, what can you say about the solution space of the equation $A\mathbf{x} = \mathbf{b}$?

7. The operations that we perform in Gaussian elimination can be accomplished using matrix multiplication. This observation is the basis of an important technique that we will investigate in a subsequent chapter.

Let's consider the matrix

$$A = \begin{bmatrix} 1 & 2 & -1 \\ 2 & 0 & 2 \\ -3 & 2 & 3 \end{bmatrix}.$$

a. Suppose that

$$S = \begin{bmatrix} 1 & 0 & 0 \\ 0 & 7 & 0 \\ 0 & 0 & 1 \end{bmatrix}.$$

Verify that SA is the matrix that results when the second row of A is scaled by a factor of 7. What matrix S would scale the third row by -3?

b. Suppose that

$$P = \begin{bmatrix} 0 & 1 & 0 \\ 1 & 0 & 0 \\ 0 & 0 & 1 \end{bmatrix}.$$

Verify that PA is the matrix that results from interchanging the first and second rows. What matrix P would interchange the first and third rows?

c. Suppose that

$$L_1 = \begin{bmatrix} 1 & 0 & 0 \\ -2 & 1 & 0 \\ 0 & 0 & 1 \end{bmatrix}.$$

Verify that $L_1 A$ is the matrix that results from multiplying the first row of A by -2 and adding it to the second row. What matrix L_2 would multiply the first row by 3 and add it to the third row?

d. When we performed Gaussian elimination, our first goal was to perform row operations that brought the matrix into a triangular form. For our matrix A, find the row operations needed to find a row equivalent matrix U in triangular form. By expressing these row operations in terms of matrix multiplication, find a matrix L such that $LA = U$.

8. In this exercise, you will construct the *inverse* of a matrix, a subject that we will investigate more fully in the next chapter. Suppose that A is the 2×2 matrix:

$$A = \begin{bmatrix} 3 & -2 \\ -2 & 1 \end{bmatrix}.$$

a. Find the vectors $\mathbf{b}_1$ and $\mathbf{b}_2$ such that the matrix $B = \begin{bmatrix} \mathbf{b}_1 & \mathbf{b}_2 \end{bmatrix}$ satisfies

$$AB = I = \begin{bmatrix} 1 & 0 \\ 0 & 1 \end{bmatrix}.$$

b. In general, it is not true that $AB = BA$. Check that it is true, however, for the specific A and B that appear in this problem.

c. Suppose that $\mathbf{x} = \begin{bmatrix} x_1 \\ x_2 \end{bmatrix}$. What do you find when you evaluate $I\mathbf{x}$?

d. Suppose that we want to solve the equation $A\mathbf{x} = \mathbf{b}$. We know how to do this using Gaussian elimination; let's use our matrix B to find a different way:

$$A\mathbf{x} = \mathbf{b}$$
$$B(A\mathbf{x}) = B\mathbf{b}$$
$$(BA)\mathbf{x} = B\mathbf{b}.$$
$$I\mathbf{x} = B\mathbf{b}$$
$$\mathbf{x} = B\mathbf{b}$$

In other words, the solution to the equation $A\mathbf{x} = \mathbf{b}$ is $\mathbf{x} = B\mathbf{b}$.

Consider the equation $A\mathbf{x} = \begin{bmatrix} 5 \\ -2 \end{bmatrix}$. Find the solution in two different ways, first using Gaussian elimination and then as $\mathbf{x} = B\mathbf{b}$, and verify that you have found the same result.

9. Determine whether the following statements are true or false and provide a justification

2.2. MATRIX MULTIPLICATION AND LINEAR COMBINATIONS

for your response.

a. If $A\mathbf{x}$ is defined, then the number of components of $\mathbf{x}$ equals the number of rows of A.

b. The solution space to the equation $A\mathbf{x} = \mathbf{b}$ is equivalent to the solution space to the linear system whose augmented matrix is $\begin{bmatrix} A & | & \mathbf{b} \end{bmatrix}$.

c. If a linear system of equations has 8 equations and 5 unknowns, then the shape of the matrix A in the corresponding equation $A\mathbf{x} = \mathbf{b}$ is 5×8.

d. If A has a pivot position in every row, then every equation $A\mathbf{x} = \mathbf{b}$ is consistent.

e. If A is a 9×5 matrix, then $A\mathbf{x} = \mathbf{b}$ is inconsistent for some vector $\mathbf{b}$.

10. Suppose that A is a 4×4 matrix and that the equation $A\mathbf{x} = \mathbf{b}$ has a unique solution for some vector $\mathbf{b}$.

    a. What does this say about the pivot positions of the matrix A? Write the reduced row echelon form of A.

    b. Can you find another vector $\mathbf{c}$ such that $A\mathbf{x} = \mathbf{c}$ is inconsistent?

    c. What can you say about the solution space to the equation $A\mathbf{x} = \mathbf{0}$?

    d. Suppose $A = \begin{bmatrix} \mathbf{v}_1 & \mathbf{v}_2 & \mathbf{v}_3 & \mathbf{v}_4 \end{bmatrix}$. Explain why every four-dimensional vector can be written as a linear combination of the vectors $\mathbf{v}_1, \mathbf{v}_2, \mathbf{v}_3$, and $\mathbf{v}_4$ in exactly one way.

11. Define the matrix
$$A = \begin{bmatrix} 1 & 2 & 4 \\ -2 & 1 & -3 \\ 3 & 1 & 7 \end{bmatrix}.$$

    a. Describe the solution space to the homogeneous equation $A\mathbf{x} = \mathbf{0}$ using a parametric description, if appropriate. What does this solution space represent geometrically?

    b. Describe the solution space to the equation $A\mathbf{x} = \mathbf{b}$ where $\mathbf{b} = \begin{bmatrix} -3 \\ -4 \\ 1 \end{bmatrix}$. What does this solution space represent geometrically and how does it compare to the previous solution space?

    c. We will now explain the relationship between the previous two solution spaces. Suppose that $\mathbf{x}_h$ is a solution to the homogeneous equation; that is $A\mathbf{x}_h = \mathbf{0}$. Suppose also that $\mathbf{x}_p$ is a solution to the equation $A\mathbf{x} = \mathbf{b}$; that is, $A\mathbf{x}_p = \mathbf{b}$.

    Use the Linearity Principle expressed in Proposition 2.2.3 to explain why $\mathbf{x}_h + \mathbf{x}_p$ is a solution to the equation $A\mathbf{x} = \mathbf{b}$. You may do this by evaluating $A(\mathbf{x}_h + \mathbf{x}_p)$.

    That is, if we find one solution $\mathbf{x}_p$ to an equation $A\mathbf{x} = \mathbf{b}$, we may add any solution to the homogeneous equation to $\mathbf{x}_p$ and still have a solution to the equation $A\mathbf{x} = \mathbf{b}$.

In other words, the solution space to the equation $A\mathbf{x} = \mathbf{b}$ is given by translating the solution space to the homogeneous equation by the vector $\mathbf{x}_p$.

12. Suppose that a city is starting a bicycle sharing program with bicycles at locations B and C. Bicycles that are rented at one location may be returned to either location at the end of the day. Over time, the city finds that 80% of bicycles rented at location B are returned to B with the other 20% returned to C. Similarly, 50% of bicycles rented at location C are returned to B and 50% to C.

    To keep track of the bicycles, we form a vector

    $$\mathbf{x}_k = \begin{bmatrix} B_k \\ C_k \end{bmatrix}$$

    where B_k is the number of bicycles at location B at the beginning of day k and C_k is the number of bicycles at C. The information above tells us how to determine the distribution of bicycles the following day:

    $$B_{k+1} = 0.8 B_k + 0.5 C_k$$
    $$C_{k+1} = 0.2 B_k + 0.5 C_k.$$

    Expressed in matrix-vector form, these expressions give

    $$\mathbf{x}_{k+1} = A \mathbf{x}_k$$

    where

    $$A = \begin{bmatrix} 0.8 & 0.5 \\ 0.2 & 0.5 \end{bmatrix}.$$

    a. Let's check that this makes sense.

        1. Suppose that there are 1000 bicycles at location B and none at C on day 1. This means we have $\mathbf{x}_1 = \begin{bmatrix} 1000 \\ 0 \end{bmatrix}$. Find the number of bicycles at both locations on day 2 by evaluating $\mathbf{x}_2 = A\mathbf{x}_1$.

        2. Suppose that there are 1000 bicycles at location C and none at B on day 1. Form the vector $\mathbf{x}_1$ and determine the number of bicycles at the two locations the next day by finding $\mathbf{x}_2 = A\mathbf{x}_1$.

    b. Suppose that one day there are 1050 bicycles at location B and 450 at location C. How many bicycles were there at each location the previous day?

    c. Suppose that there are 500 bicycles at location B and 500 at location C on Monday. How many bicycles are there at the two locations on Tuesday? on Wednesday? on Thursday?

2.2. MATRIX MULTIPLICATION AND LINEAR COMBINATIONS

13. This problem is a continuation of the previous problem.

    a. Let us define vectors
    $$\mathbf{v}_1 = \begin{bmatrix} 5 \\ 2 \end{bmatrix}, \quad \mathbf{v}_2 = \begin{bmatrix} -1 \\ 1 \end{bmatrix}.$$
    Show that
    $$A\mathbf{v}_1 = \mathbf{v}_1, \quad A\mathbf{v}_2 = 0.3\mathbf{v}_2.$$

    b. Suppose that $\mathbf{x}_1 = c_1\mathbf{v}_1 + c_2\mathbf{v}_2$ where c_1 and c_2 are scalars. Use the Linearity Principle expressed in Proposition 2.2.3 to explain why
    $$\mathbf{x}_2 = A\mathbf{x}_1 = c_1\mathbf{v}_1 + 0.3c_2\mathbf{v}_2.$$

    c. Continuing in this way, explain why
    $$\mathbf{x}_3 = A\mathbf{x}_2 = c_1\mathbf{v}_1 + 0.3^2 c_2\mathbf{v}_2$$
    $$\mathbf{x}_4 = A\mathbf{x}_3 = c_1\mathbf{v}_1 + 0.3^3 c_2\mathbf{v}_2.$$
    $$\mathbf{x}_5 = A\mathbf{x}_4 = c_1\mathbf{v}_1 + 0.3^4 c_2\mathbf{v}_2$$

    d. Suppose that there are initially 500 bicycles at location B and 500 at location C. Write the vector $\mathbf{x}_1$ and find the scalars c_1 and c_2 such that $\mathbf{x}_1 = c_1\mathbf{v}_1 + c_2\mathbf{v}_2$.

    e. Use the previous part of this problem to determine $\mathbf{x}_2$, $\mathbf{x}_3$ and $\mathbf{x}_4$.

    f. After a very long time, how are all the bicycles distributed?

2.3 The span of a set of vectors

Matrix multiplication allows us to rewrite a linear system in the form $A\mathbf{x} = \mathbf{b}$. Besides being a more compact way of expressing a linear system, this form allows us to think about linear systems geometrically since matrix multiplication is defined in terms of linear combinations of vectors.

We now return to our two fundamental questions, rephrased here in terms of matrix multiplication.

- *Existence:* Is there a solution to the equation $A\mathbf{x} = \mathbf{b}$?
- *Uniqueness:* If there is a solution to the equation $A\mathbf{x} = \mathbf{b}$, is it unique?

In this section, we focus on the existence question and see how it leads to the concept of the *span* of a set of vectors.

Preview Activity 2.3.1 The existence of solutions..

a. If the equation $A\mathbf{x} = \mathbf{b}$ is inconsistent, what can we say about the pivot positions of the augmented matrix $\left[\begin{array}{c|c} A & \mathbf{b} \end{array}\right]$?

b. Consider the matrix A
$$A = \begin{bmatrix} 1 & 0 & -2 \\ -2 & 2 & 2 \\ 1 & 1 & -3 \end{bmatrix}.$$

If $\mathbf{b} = \begin{bmatrix} 2 \\ 2 \\ 5 \end{bmatrix}$, is the equation $A\mathbf{x} = \mathbf{b}$ consistent? If so, find a solution.

c. If $\mathbf{b} = \begin{bmatrix} 2 \\ 2 \\ 6 \end{bmatrix}$, is the equation $A\mathbf{x} = \mathbf{b}$ consistent? If so, find a solution.

d. Identify the pivot positions of A.

e. For our two choices of the vector $\mathbf{b}$, one equation $A\mathbf{x} = \mathbf{b}$ has a solution and the other does not. What feature of the pivot positions of the matrix A tells us to expect this?

2.3.1 The span of a set of vectors

In the preview activity, we considered a 3 × 3 matrix A and found that the equation $A\mathbf{x} = \mathbf{b}$ has a solution for some vectors $\mathbf{b}$ in $\mathbb{R}^3$ and has no solution for others. We will introduce a

2.3. THE SPAN OF A SET OF VECTORS

concept called *span* that describes the vectors **b** for which there is a solution.

We can write an $m \times n$ matrix A in terms of its columns

$$A = \begin{bmatrix} \mathbf{v}_1 & \mathbf{v}_2 & \cdots & \mathbf{v}_n \end{bmatrix}.$$

Remember that Proposition 2.2.4 says that the equation $A\mathbf{x} = \mathbf{b}$ is consistent if and only if we can express **b** as a linear combination of $\mathbf{v}_1, \mathbf{v}_2, \ldots, \mathbf{v}_n$.

Definition 2.3.1 The **span** of a set of vectors $\mathbf{v}_1, \mathbf{v}_2, \ldots, \mathbf{v}_n$ is the set of all linear combinations that can be formed from the vectors.

Alternatively, if $A = \begin{bmatrix} \mathbf{v}_1 & \mathbf{v}_2 & \cdots & \mathbf{v}_n \end{bmatrix}$, then the span of the vectors consists of all vectors **b** for which the equation $A\mathbf{x} = \mathbf{b}$ is consistent.

Example 2.3.2 Considering the set of vectors $\mathbf{v} = \begin{bmatrix} -2 \\ 1 \end{bmatrix}$ and $\mathbf{w} = \begin{bmatrix} 8 \\ -4 \end{bmatrix}$, we see that the vector

$$\mathbf{b} = 3\mathbf{v} + \mathbf{w} = \begin{bmatrix} 2 \\ -1 \end{bmatrix}$$

is one vector in the span of the vectors **v** and **w** because it is a linear combination of **v** and **w**.

To determine whether the vector $\mathbf{b} = \begin{bmatrix} 5 \\ 2 \end{bmatrix}$ is in the span of **v** and **w**, we form the matrix

$$A = \begin{bmatrix} \mathbf{v} & \mathbf{w} \end{bmatrix} = \begin{bmatrix} -2 & 8 \\ 1 & -4 \end{bmatrix}$$

and consider the equation $A\mathbf{x} = \mathbf{b}$. We have

$$\left[\begin{array}{cc|c} -2 & 8 & 5 \\ 1 & -4 & 2 \end{array}\right] \sim \left[\begin{array}{cc|c} 1 & -4 & 0 \\ 0 & 0 & 1 \end{array}\right],$$

which shows that the equation $A\mathbf{x} = \mathbf{b}$ is inconsistent. Therefore, $\mathbf{b} = \begin{bmatrix} 5 \\ 2 \end{bmatrix}$ is one vector that is not in the span of **v** and **w**.

Activity 2.3.2. Let's look at two examples to develop some intuition for the concept of span.

a. First, we will consider the set of vectors

$$\mathbf{v} = \begin{bmatrix} 1 \\ 2 \end{bmatrix}, \quad \mathbf{w} = \begin{bmatrix} -2 \\ -4 \end{bmatrix}.$$

There is an interactive diagram, available at gvsu.edu/s/0Jg, that accompanies this activity. The diagram at the top of that page accompanies part a of this activity.

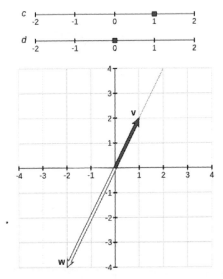

Figure 2.3.3 An interactive diagram for constructing linear combinations of the vectors **v** and **w**.

1. What vector is the linear combination of **v** and **w** with weights:
   - $c = 2$ and $d = 0$?
   - $c = 1$ and $d = 1$?
   - $c = 0$ and $d = -1$?

2. Can the vector $\begin{bmatrix} 2 \\ 4 \end{bmatrix}$ be expressed as a linear combination of **v** and **w**? Is the vector $\begin{bmatrix} 2 \\ 4 \end{bmatrix}$ in the span of **v** and **w**?

3. Can the vector $\begin{bmatrix} 3 \\ 0 \end{bmatrix}$ be expressed as a linear combination of **v** and **w**? Is the vector $\begin{bmatrix} 3 \\ 0 \end{bmatrix}$ in the span of **v** and **w**?

4. Describe the set of vectors in the span of **v** and **w**.

5. For what vectors **b** does the equation
$$\begin{bmatrix} 1 & -2 \\ 2 & -4 \end{bmatrix} \mathbf{x} = \mathbf{b}$$
have a solution?

2.3. THE SPAN OF A SET OF VECTORS

b. We will now look at an example where

$$\mathbf{v} = \begin{bmatrix} 2 \\ 1 \end{bmatrix}, \quad \mathbf{w} = \begin{bmatrix} 1 \\ 2 \end{bmatrix}.$$

The diagram at the bottom of the page at gvsu.edu/s/0Jg accompanies part b of this activity.

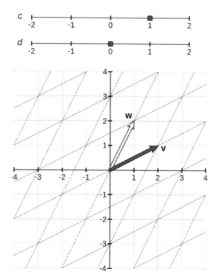

Figure 2.3.4 An interactive diagram for constructing linear combinations of the vectors **v** and **w**.

1. What vector is the linear combination of **v** and **w** with weights:
   - $c = 2$ and $d = 0$?
   - $c = 1$ and $d = 1$?
   - $c = 0$ and $d = -1$?

2. Can the vector $\begin{bmatrix} -2 \\ 2 \end{bmatrix}$ be expressed as a linear combination of **v** and **w**? Is the vector $\begin{bmatrix} -2 \\ 2 \end{bmatrix}$ in the span of **v** and **w**?

3. Can the vector $\begin{bmatrix} 3 \\ 0 \end{bmatrix}$ be expressed as a linear combination of **v** and **w**? Is the vector $\begin{bmatrix} 3 \\ 0 \end{bmatrix}$ in the span of **v** and **w**?

4. Describe the set of vectors in the span of **v** and **w**.

5. For what vectors **b** does the equation

$$\begin{bmatrix} 2 & 1 \\ 1 & 2 \end{bmatrix} \mathbf{x} = \mathbf{b}$$

have a solution?

This activity aims to convey the geometric meaning of span. Remember that we can think of a linear combination of the two vectors **v** and **w** as a recipe for walking in the plane $\mathbb{R}^2$. We first move a prescribed amount in the direction of **v** and then a prescribed amount in the direction of **w**. The span consists of all the places we can walk to.

Example 2.3.5 Let's consider the vectors $\mathbf{v} = \begin{bmatrix} 2 \\ 0 \end{bmatrix}$ and $\mathbf{w} = \begin{bmatrix} -1 \\ 1 \end{bmatrix}$ as shown in Figure 2.3.6.

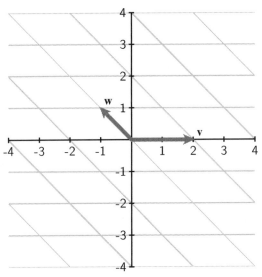

Figure 2.3.6 The vectors **v** and **w** and some linear combinations they create.

The figure shows us that $\mathbf{b} = \mathbf{v} + 2\mathbf{w} = \begin{bmatrix} 0 \\ 2 \end{bmatrix}$ is a linear combination of **v** and **w**. Indeed, we can verify this algebraically by constructing the linear system

$$\begin{bmatrix} \mathbf{v} & \mathbf{w} \end{bmatrix} \mathbf{x} = \begin{bmatrix} 0 \\ 2 \end{bmatrix},$$

whose corresponding augmented matrix has the reduced row echelon form

$$\left[\begin{array}{cc|c} 2 & -1 & 0 \\ 0 & 1 & 2 \end{array}\right] \sim \left[\begin{array}{cc|c} 1 & 0 & 1 \\ 0 & 1 & 2 \end{array}\right].$$

Because this system is consistent, we know that $\mathbf{b} = \begin{bmatrix} 0 \\ 2 \end{bmatrix}$ is in the span of **v** and **w**.

In fact, we can say more. Notice that the coefficient matrix

$$\begin{bmatrix} 2 & -1 \\ 0 & 1 \end{bmatrix} \sim \begin{bmatrix} 1 & 0 \\ 0 & 1 \end{bmatrix}$$

has a pivot position in every row. This means that for any other vector **b**, the augmented matrix corresponding to the equation $\begin{bmatrix} \mathbf{v} & \mathbf{w} \end{bmatrix} \mathbf{x} = \mathbf{b}$ cannot have a pivot position in its rightmost

2.3. THE SPAN OF A SET OF VECTORS

column:
$$\begin{bmatrix} 2 & -1 & * \\ 0 & 1 & * \end{bmatrix} \sim \begin{bmatrix} 1 & 0 & * \\ 0 & 1 & * \end{bmatrix}.$$

Therefore, the equation $\begin{bmatrix} \mathbf{v} & \mathbf{w} \end{bmatrix} \mathbf{x} = \mathbf{b}$ is consistent for every two-dimensional vector $\mathbf{b}$, which tells us that every two-dimensional vector is in the span of $\mathbf{v}$ and $\mathbf{w}$. In this case, we say that the span of $\mathbf{v}$ and $\mathbf{w}$ is $\mathbb{R}^2$.

The intuitive meaning is that we can walk to any point in the plane by moving an appropriate distance in the $\mathbf{v}$ and $\mathbf{w}$ directions.

Example 2.3.7 Now let's consider the vectors $\mathbf{v} = \begin{bmatrix} -1 \\ 1 \end{bmatrix}$ and $\mathbf{w} = \begin{bmatrix} 2 \\ -2 \end{bmatrix}$ as shown in Figure 2.3.8.

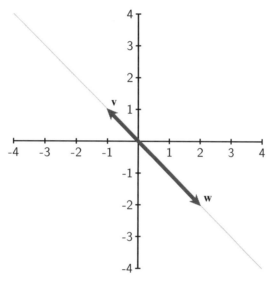

Figure 2.3.8 The vectors $\mathbf{v}$ and $\mathbf{w}$ and some linear combinations they create.

From the figure, we expect that $\mathbf{b} = \begin{bmatrix} 0 \\ 2 \end{bmatrix}$ is not a linear combination of $\mathbf{v}$ and $\mathbf{w}$. Once again, we can verify this algebraically by constructing the linear system

$$\begin{bmatrix} \mathbf{v} & \mathbf{w} \end{bmatrix} \mathbf{x} = \begin{bmatrix} 0 \\ 2 \end{bmatrix}.$$

The augmented matrix has the reduced row echelon form

$$\begin{bmatrix} -1 & 2 & 0 \\ 1 & -2 & 2 \end{bmatrix} \sim \begin{bmatrix} 1 & -2 & 0 \\ 0 & 0 & 1 \end{bmatrix},$$

from which we see that the system is inconsistent. Therefore, $\mathbf{b} = \begin{bmatrix} 0 \\ 2 \end{bmatrix}$ is not in the span of $\mathbf{v}$ and $\mathbf{w}$.

We should expect this behavior from the coefficient matrix

$$\begin{bmatrix} -1 & 2 \\ 1 & -2 \end{bmatrix} \sim \begin{bmatrix} 1 & -2 \\ 0 & 0 \end{bmatrix}.$$

Because the second row of the coefficient matrix does not have a pivot position, it is possible for a linear system $\begin{bmatrix} \mathbf{v} & \mathbf{w} \end{bmatrix} \mathbf{x} = \mathbf{b}$ to have a pivot position in its rightmost column:

$$\left[\begin{array}{cc|c} -1 & 2 & * \\ 1 & -2 & * \end{array}\right] \sim \left[\begin{array}{cc|c} 1 & -2 & 0 \\ 0 & 0 & 1 \end{array}\right].$$

If we notice that $\mathbf{w} = -2\mathbf{v}$, we see that any linear combination of $\mathbf{v}$ and $\mathbf{w}$,

$$c\mathbf{v} + d\mathbf{w} = c\mathbf{v} - 2d\mathbf{v} = (c - 2d)\mathbf{v},$$

is actually a scalar multiple of $\mathbf{v}$. Therefore, the span of $\mathbf{v}$ and $\mathbf{w}$ is the line defined by the vector $\mathbf{v}$. Intuitively, this means that we can only walk to points on this line using these two vectors.

Notation 2.3.9 We will denote the span of the set of vectors $\mathbf{v}_1, \mathbf{v}_2, \ldots, \mathbf{v}_n$ by $\text{Span}\{\mathbf{v}_1, \mathbf{v}_2, \ldots, \mathbf{v}_n\}$.

In Example 2.3.5, we saw that $\text{Span}\{\mathbf{v}, \mathbf{w}\} = \mathbb{R}^2$. However, for the vectors in Example 2.3.7, we saw that $\text{Span}\{\mathbf{v}, \mathbf{w}\}$ is simply a line.

2.3.2 Pivot positions and span

A set of vectors $\mathbf{v}_1, \mathbf{v}_2, \ldots, \mathbf{v}_n$ naturally defines a matrix $A = \begin{bmatrix} \mathbf{v}_1 & \mathbf{v}_2 & \cdots & \mathbf{v}_n \end{bmatrix}$ whose columns are the given vectors. As we've seen, a vector $\mathbf{b}$ is in $\text{Span}\{\mathbf{v}_1, \mathbf{v}_2, \ldots, \mathbf{v}_n\}$ precisely when the linear system $A\mathbf{x} = \mathbf{b}$ is consistent.

The previous examples point to the fact that the span is related to the pivot positions of A. While Section 2.4 and Section 3.5 develop this idea more fully, we will now examine the possibilities in $\mathbb{R}^3$.

Activity 2.3.3. In this activity, we will look at the span of sets of vectors in $\mathbb{R}^3$.

a. Suppose $\mathbf{v} = \begin{bmatrix} 1 \\ 2 \\ 1 \end{bmatrix}$. Give a geometric description of $\text{Span}\{\mathbf{v}\}$ and a rough sketch of $\mathbf{v}$ and its span in Figure 2.3.10.

2.3. THE SPAN OF A SET OF VECTORS

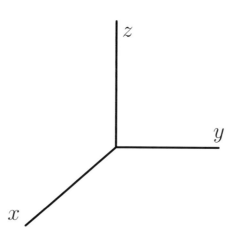

Figure 2.3.10 A three-dimensional coordinate system for sketching **v** and its span.

b. Now consider the two vectors

$$\mathbf{e}_1 = \begin{bmatrix} 1 \\ 0 \\ 0 \end{bmatrix}, \quad \mathbf{e}_2 = \begin{bmatrix} 0 \\ 1 \\ 0 \end{bmatrix}.$$

Sketch the vectors below. Then give a geometric description of Span$\{\mathbf{e}_1, \mathbf{e}_2\}$ and a rough sketch of the span in Figure 2.3.11.

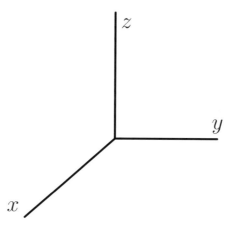

Figure 2.3.11 A coordinate system for sketching $\mathbf{e}_1$, $\mathbf{e}_2$, and Span$\{\mathbf{e}_1, \mathbf{e}_2\}$.

c. Let's now look at this situation algebraically by writing write $\mathbf{b} = \begin{bmatrix} b_1 \\ b_2 \\ b_3 \end{bmatrix}$. Determine the conditions on b_1, b_2, and b_3 so that **b** is in Span$\{\mathbf{e}_1, \mathbf{e}_2\}$ by considering

the linear system
$$[\,\mathbf{e}_1\ \mathbf{e}_2\,]\,\mathbf{x} = \mathbf{b}$$
or
$$\begin{bmatrix} 1 & 0 \\ 0 & 1 \\ 0 & 0 \end{bmatrix}\mathbf{x} = \begin{bmatrix} b_1 \\ b_2 \\ b_3 \end{bmatrix}.$$

Explain how this relates to your sketch of Span$\{\mathbf{e}_1, \mathbf{e}_2\}$.

d. Consider the vectors
$$\mathbf{v}_1 = \begin{bmatrix} 1 \\ 1 \\ -1 \end{bmatrix},\ \mathbf{v}_2 = \begin{bmatrix} 0 \\ 2 \\ 1 \end{bmatrix}.$$

1. Is the vector $\mathbf{b} = \begin{bmatrix} 1 \\ -2 \\ 4 \end{bmatrix}$ in Span$\{\mathbf{v}_1, \mathbf{v}_2\}$?

2. Is the vector $\mathbf{b} = \begin{bmatrix} -2 \\ 0 \\ 3 \end{bmatrix}$ in Span$\{\mathbf{v}_1, \mathbf{v}_2\}$?

3. Give a geometric description of Span$\{\mathbf{v}_1, \mathbf{v}_2\}$.

e. Consider the vectors
$$\mathbf{v}_1 = \begin{bmatrix} 1 \\ 1 \\ -1 \end{bmatrix},\ \mathbf{v}_2 = \begin{bmatrix} 0 \\ 2 \\ 1 \end{bmatrix},\ \mathbf{v}_3 = \begin{bmatrix} 1 \\ -2 \\ 4 \end{bmatrix}.$$

Form the matrix $[\,\mathbf{v}_1\ \mathbf{v}_2\ \mathbf{v}_3\,]$ and find its reduced row echelon form.

What does this tell you about Span$\{\mathbf{v}_1, \mathbf{v}_2, \mathbf{v}_3\}$?

f. If the span of a set of vectors $\mathbf{v}_1, \mathbf{v}_2, \ldots, \mathbf{v}_n$ is $\mathbb{R}^3$, what can you say about the pivot positions of the matrix $[\,\mathbf{v}_1\ \mathbf{v}_2\ \cdots\ \mathbf{v}_n\,]$?

g. What is the smallest number of vectors such that Span$\{\mathbf{v}_1, \mathbf{v}_2, \ldots, \mathbf{v}_n\} = \mathbb{R}^3$?

The types of sets that appear as the span of a set of vectors in $\mathbb{R}^3$ are relatively simple.

- First, with a single nonzero vector, all linear combinations are simply scalar multiples of that vector so that the span of this vector is a line, as shown in Figure 2.3.12.

2.3. THE SPAN OF A SET OF VECTORS

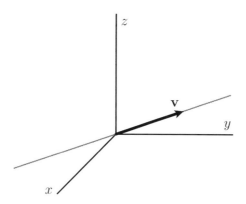

Figure 2.3.12 The span of a single nonzero vector is a line.

Notice that the matrix formed by this vector has one pivot position. For example,

$$\begin{bmatrix} -2 \\ 3 \\ 1 \end{bmatrix} \sim \begin{bmatrix} 1 \\ 0 \\ 0 \end{bmatrix}.$$

- The span of two vectors in $\mathbb{R}^3$ that do not lie on the same line will be a plane, as seen in Figure 2.3.13.

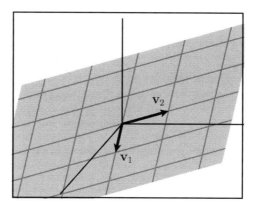

Figure 2.3.13 The span of these two vectors in $\mathbb{R}^3$ is a plane.

For example, the vectors

$$\mathbf{v}_1 = \begin{bmatrix} -2 \\ 3 \\ 1 \end{bmatrix}, \quad \mathbf{v}_2 = \begin{bmatrix} 1 \\ -1 \\ 3 \end{bmatrix}$$

lead to the matrix

$$\begin{bmatrix} -2 & 1 \\ 3 & -1 \\ 1 & 3 \end{bmatrix} \sim \begin{bmatrix} 1 & 0 \\ 0 & 1 \\ 0 & 0 \end{bmatrix}$$

with two pivot positions.

- Finally, a set of three vectors, such as

$$\mathbf{v}_1 = \begin{bmatrix} 1 \\ 2 \\ -1 \end{bmatrix}, \quad \mathbf{v}_2 = \begin{bmatrix} 2 \\ 0 \\ 1 \end{bmatrix}, \quad \mathbf{v}_3 = \begin{bmatrix} -2 \\ 2 \\ 0 \end{bmatrix}$$

may form a matrix having three pivot positions

$$\begin{bmatrix} \mathbf{v}_1 & \mathbf{v}_2 & \mathbf{v}_3 \end{bmatrix} = \begin{bmatrix} 1 & 2 & -2 \\ 2 & 0 & 2 \\ -1 & 1 & 0 \end{bmatrix} \sim \begin{bmatrix} 1 & 0 & 0 \\ 0 & 1 & 0 \\ 0 & 0 & 1 \end{bmatrix},$$

one in every row. When this happens, no matter how we augment this matrix, it is impossible to obtain a pivot position in the rightmost column:

$$\left[\begin{array}{rrr|r} 1 & 2 & -2 & * \\ 2 & 0 & 2 & * \\ -1 & 1 & 0 & * \end{array}\right] \sim \left[\begin{array}{rrr|r} 1 & 0 & 0 & * \\ 0 & 1 & 0 & * \\ 0 & 0 & 1 & * \end{array}\right].$$

Therefore, any linear system $\begin{bmatrix} \mathbf{v}_1 & \mathbf{v}_2 & \mathbf{v}_3 \end{bmatrix} \mathbf{x} = \mathbf{b}$ is consistent, which tells us that Span$\{\mathbf{v}_1, \mathbf{v}_2, \mathbf{v}_3\} = \mathbb{R}^3$.

To summarize, we looked at the pivot positions in a matrix whose columns are the three-dimensional vectors $\mathbf{v}_1, \mathbf{v}_2, \ldots, \mathbf{v}_n$. We found that with

- one pivot position, the span was a line.
- two pivot positions, the span was a plane.
- three pivot positions, the span was $\mathbb{R}^3$.

Though we will return to these ideas later, for now take note of the fact that the span of a set of vectors in $\mathbb{R}^3$ is a relatively simple, familiar geometric object.

The reasoning that led us to conclude that the span of a set of vectors is $\mathbb{R}^3$ when the associated matrix has a pivot position in every row applies more generally.

Proposition 2.3.14 *Suppose we have vectors* $\mathbf{v}_1, \mathbf{v}_2, \ldots, \mathbf{v}_n$ *in* $\mathbb{R}^m$. *Then* Span$\{\mathbf{v}_1, \mathbf{v}_2, \ldots, \mathbf{v}_n\} = \mathbb{R}^m$ *if and only if the matrix* $\begin{bmatrix} \mathbf{v}_1 & \mathbf{v}_2 & \cdots & \mathbf{v}_n \end{bmatrix}$ *has a pivot position in every row.*

This tells us something important about the number of vectors needed to span $\mathbb{R}^m$. Suppose we have n vectors $\mathbf{v}_1, \mathbf{v}_2, \ldots, \mathbf{v}_n$ that span $\mathbb{R}^m$. The proposition tells us that the matrix $A = \begin{bmatrix} \mathbf{v}_1 & \mathbf{v}_2 & \cdots & \mathbf{v}_n \end{bmatrix}$ has a pivot position in every row, such as in this reduced row echelon matrix.

$$\begin{bmatrix} 1 & 0 & * & 0 & * & 0 \\ 0 & 1 & * & 0 & * & 0 \\ 0 & 0 & 0 & 1 & * & 0 \\ 0 & 0 & 0 & 0 & 0 & 1 \end{bmatrix}.$$

Since a matrix can have at most one pivot position in a column, there must be at least as many columns as there are rows, which implies that $n \geq m$. For instance, if we have a set of vectors that span $\mathbb{R}^{632}$, there must be at least 632 vectors in the set.

2.3. THE SPAN OF A SET OF VECTORS

Proposition 2.3.15 *A set of vectors whose span is $\mathbb{R}^m$ contains at least m vectors.*

We have thought about a linear combination of a set of vectors $v_1, v_2, \ldots, v_n$ as the result of walking a certain distance in the direction of v_1, followed by walking a certain distance in the direction of v_2, and so on. If $\text{Span}\{v_1, v_2, \ldots, v_n\} = \mathbb{R}^m$, this means that we can walk to every point in $\mathbb{R}^m$ using the directions $v_1, v_2, \ldots, v_n$. Intuitively, this proposition is telling us that we need at least m directions to have the flexibility needed to reach every point in $\mathbb{R}^m$.

> **Terminology.**
>
> Because span is a concept that is connected to a set of vectors, we say, "The span of the set of vectors $v_1, v_2, \ldots, v_n$ is ...." While it may be tempting to say, "The span of the matrix A is ...," we should instead say "The span of the columns of the matrix A is ...."

2.3.3 Summary

We defined the span of a set of vectors and developed some intuition for this concept through a series of examples.

- The span of a set of vectors $v_1, v_2, \ldots, v_n$ is the set of linear combinations of the vectors. We denote the span by $\text{Span}\{v_1, v_2, \ldots, v_n\}$.

- A vector b is in $\text{Span}\{v_1, v_2, \ldots, v_n\}$ if and only if the linear system

$$\begin{bmatrix} v_1 & v_2 & \ldots v_n \end{bmatrix} x = b$$

  is consistent.

- If the $m \times n$ matrix

$$\begin{bmatrix} v_1 & v_2 & \ldots v_n \end{bmatrix}$$

  has a pivot position in every row, then the span of these vectors is $\mathbb{R}^m$; that is,

$$\text{Span}\{v_1, v_2, \ldots, v_n\} = \mathbb{R}^m.$$

- Any set of vectors that spans $\mathbb{R}^m$ must have at least m vectors.

2.3.4 Exercises

1. In this exercise, we will consider the span of some sets of two- and three-dimensional vectors.

   a. Consider the vectors

   $$v_1 = \begin{bmatrix} 1 \\ -2 \end{bmatrix}, v_2 = \begin{bmatrix} 4 \\ 3 \end{bmatrix}.$$

1. Is $\mathbf{b} = \begin{bmatrix} 2 \\ 1 \end{bmatrix}$ in Span$\{\mathbf{v}_1, \mathbf{v}_2\}$?

2. Give a geometric description of Span$\{\mathbf{v}_1, \mathbf{v}_2\}$.

b. Consider the vectors

$$\mathbf{v}_1 = \begin{bmatrix} 2 \\ 1 \\ 3 \end{bmatrix}, \mathbf{v}_2 = \begin{bmatrix} -2 \\ 0 \\ 2 \end{bmatrix}, \mathbf{v}_3 = \begin{bmatrix} 6 \\ 1 \\ -1 \end{bmatrix}.$$

1. Is the vector $\mathbf{b} = \begin{bmatrix} -10 \\ -1 \\ 5 \end{bmatrix}$ in Span$\{\mathbf{v}_1, \mathbf{v}_2, \mathbf{v}_3\}$?

2. Is the vector $\mathbf{v}_3$ in Span$\{\mathbf{v}_1, \mathbf{v}_2, \mathbf{v}_3\}$?

3. Is the vector $\mathbf{b} = \begin{bmatrix} 3 \\ 3 \\ -1 \end{bmatrix}$ in Span$\{\mathbf{v}_1, \mathbf{v}_2, \mathbf{v}_3\}$?

4. Give a geometric description of Span$\{\mathbf{v}_1, \mathbf{v}_2, \mathbf{v}_3\}$.

2. Provide a justification for your response to the following questions.

   a. Suppose you have a set of vectors $\mathbf{v}_1, \mathbf{v}_2, \ldots, \mathbf{v}_n$. Can you guarantee that $\mathbf{0}$ is in Span$\{\mathbf{v}_1 \mathbf{v}_2, \ldots, \mathbf{v}_n\}$?

   b. Suppose that A is an $m \times n$ matrix. Can you guarantee that the equation $A\mathbf{x} = \mathbf{0}$ is consistent?

   c. What is Span$\{\mathbf{0}, \mathbf{0}, \ldots, \mathbf{0}\}$?

3. For both parts of this exercise, give a geometric description of sets of the vectors $\mathbf{b}$ and include a sketch.

   a. For which vectors $\mathbf{b}$ in $\mathbb{R}^2$ is the equation
   $$\begin{bmatrix} 3 & -6 \\ -2 & 4 \end{bmatrix} \mathbf{x} = \mathbf{b}$$
   consistent?

   b. For which vectors $\mathbf{b}$ in $\mathbb{R}^2$ is the equation
   $$\begin{bmatrix} 3 & -6 \\ -2 & 2 \end{bmatrix} \mathbf{x} = \mathbf{b}$$
   consistent?

4. Consider the following matrices:

$$A = \begin{bmatrix} 3 & 0 & -1 & 1 \\ 1 & -1 & 3 & 7 \\ 3 & -2 & 1 & 5 \\ -1 & 2 & 2 & 3 \end{bmatrix}, B = \begin{bmatrix} 3 & 0 & -1 & 4 \\ 1 & -1 & 3 & -1 \\ 3 & -2 & 1 & 3 \\ -1 & 2 & 2 & 1 \end{bmatrix}.$$

2.3. THE SPAN OF A SET OF VECTORS

Do the columns of A span $\mathbb{R}^4$? Do the columns of B span $\mathbb{R}^4$?

5. Determine whether the following statements are true or false and provide a justification for your response. Throughout, we will assume that the matrix A has columns $\mathbf{v}_1, \mathbf{v}_2, \ldots, \mathbf{v}_n$; that is,
$$A = \begin{bmatrix} \mathbf{v}_1 & \mathbf{v}_2 & \cdots & \mathbf{v}_n \end{bmatrix}.$$

   a. If the equation $A\mathbf{x} = \mathbf{b}$ is consistent, then $\mathbf{b}$ is in $\text{Span}\{\mathbf{v}_1, \mathbf{v}_2, \ldots, \mathbf{v}_n\}$.

   b. The equation $A\mathbf{x} = \mathbf{v}_1$ is consistent.

   c. If $\mathbf{v}_1, \mathbf{v}_2, \mathbf{v}_3$, and $\mathbf{v}_4$ are vectors in $\mathbb{R}^3$, then their span is $\mathbb{R}^3$.

   d. If $\mathbf{b}$ is a linear combination of $\mathbf{v}_1, \mathbf{v}_2, \ldots, \mathbf{v}_n$, then $\mathbf{b}$ is in $\text{Span}\{\mathbf{v}_1, \mathbf{v}_2, \ldots, \mathbf{v}_n\}$.

   e. If A is an 8032×427 matrix, then the span of the columns of A is a set of vectors in $\mathbb{R}^{427}$.

6. This exercise asks you to construct some matrices whose columns span a given set.

   a. Construct a 3×3 matrix whose columns span $\mathbb{R}^3$.

   b. Construct a 3×3 matrix whose columns span a plane in $\mathbb{R}^3$.

   c. Construct a 3×3 matrix whose columns span a line in $\mathbb{R}^3$.

7. Provide a justification for your response to the following questions.

   a. Suppose that we have vectors in $\mathbb{R}^8$, $\mathbf{v}_1, \mathbf{v}_2, \ldots, \mathbf{v}_{10}$, whose span is $\mathbb{R}^8$. Can every vector $\mathbf{b}$ in $\mathbb{R}^8$ be written as a linear combination of $\mathbf{v}_1, \mathbf{v}_2, \ldots, \mathbf{v}_{10}$?

   b. Suppose that we have vectors in $\mathbb{R}^8$, $\mathbf{v}_1, \mathbf{v}_2, \ldots, \mathbf{v}_{10}$, whose span is $\mathbb{R}^8$. Can every vector $\mathbf{b}$ in $\mathbb{R}^8$ be written *uniquely* as a linear combination of $\mathbf{v}_1, \mathbf{v}_2, \ldots, \mathbf{v}_{10}$?

   c. Do the vectors
   $$\mathbf{e}_1 = \begin{bmatrix} 1 \\ 0 \\ 0 \end{bmatrix}, \mathbf{e}_2 = \begin{bmatrix} 0 \\ 1 \\ 0 \end{bmatrix}, \mathbf{e}_3 = \begin{bmatrix} 0 \\ 0 \\ 1 \end{bmatrix}$$
   span $\mathbb{R}^3$?

   d. Suppose that $\mathbf{v}_1, \mathbf{v}_2, \ldots, \mathbf{v}_n$ span $\mathbb{R}^{438}$. What can you guarantee about the value of n?

   e. Can 17 vectors in $\mathbb{R}^{20}$ span $\mathbb{R}^{20}$?

8. The following observation will be helpful in this exercise. If we want to find a solution to the equation $AB\mathbf{x} = \mathbf{b}$, we could first find a solution to the equation $A\mathbf{y} = \mathbf{b}$ and then find a solution to the equation $B\mathbf{x} = \mathbf{y}$.

   Suppose that A is a 3×4 matrix whose columns span $\mathbb{R}^3$ and B is a 4×5 matrix. In this case, we can form the product AB.

   a. What is the shape of the product AB?

   b. Can you guarantee that the columns of AB span $\mathbb{R}^3$?

c. If you know additionally that the span of the columns of B is $\mathbb{R}^4$, can you guarantee that the columns of AB span $\mathbb{R}^3$?

9. Suppose that A is a 12×12 matrix and that, for some vector $\mathbf{b}$, the equation $A\mathbf{x} = \mathbf{b}$ has a unique solution.

   a. What can you say about the pivot positions of A?

   b. What can you say about the span of the columns of A?

   c. If $\mathbf{c}$ is some other vector in $\mathbb{R}^{12}$, what can you conclude about the equation $A\mathbf{x} = \mathbf{c}$?

   d. What can you about the solution space to the equation $A\mathbf{x} = \mathbf{0}$?

10. Suppose that

$$\mathbf{v}_1 = \begin{bmatrix} 3 \\ 1 \\ 3 \\ -1 \end{bmatrix}, \mathbf{v}_2 = \begin{bmatrix} 0 \\ -1 \\ -2 \\ 2 \end{bmatrix}, \mathbf{v}_3 = \begin{bmatrix} -3 \\ -3 \\ -7 \\ 5 \end{bmatrix}.$$

   a. Is $\mathbf{v}_3$ a linear combination of $\mathbf{v}_1$ and $\mathbf{v}_2$? If so, find weights such that $\mathbf{v}_3 = a\mathbf{v}_1 + b\mathbf{v}_2$.

   b. Show that a linear combination

$$a\mathbf{v}_1 + b\mathbf{v}_2 + c\mathbf{v}_3$$

   can be rewritten as a linear combination of $\mathbf{v}_1$ and $\mathbf{v}_2$.

   c. Explain why $\text{Span}\{\mathbf{v}_1, \mathbf{v}_2, \mathbf{v}_3\} = \text{Span}\{\mathbf{v}_1, \mathbf{v}_2\}$.

11. As defined in this section, the span of a set of vectors is generated by taking all possible linear combinations of those vectors. This exercise will demonstrate the fact that the span can also be realized as the solution space to a linear system.

   We will consider the vectors

$$\mathbf{v}_1 = \begin{bmatrix} 1 \\ 0 \\ -2 \end{bmatrix}, \mathbf{v}_2 = \begin{bmatrix} 2 \\ 1 \\ 0 \end{bmatrix}, \mathbf{v}_3 = \begin{bmatrix} 1 \\ 1 \\ 2 \end{bmatrix}.$$

   a. Is every vector in $\mathbb{R}^3$ in $\text{Span}\{\mathbf{v}_1, \mathbf{v}_2, \mathbf{v}_3\}$? If not, describe the span.

   b. To describe $\text{Span}\{\mathbf{v}_1, \mathbf{v}_2, \mathbf{v}_3\}$ as the solution space of a linear system, we will write

$$\mathbf{b} = \begin{bmatrix} a \\ b \\ c \end{bmatrix}.$$

   If $\mathbf{b}$ is in $\text{Span}\{\mathbf{v}_1, \mathbf{v}_2, \mathbf{v}_3\}$, then the linear system corresponding to the augmented matrix

$$\begin{bmatrix} 1 & 2 & 1 & a \\ 0 & 1 & 1 & b \\ -2 & 0 & 2 & c \end{bmatrix}$$

2.3. THE SPAN OF A SET OF VECTORS

must be consistent. This means that a pivot cannot occur in the rightmost column. Perform row operations to put this augmented matrix into a triangular form. Now identify an equation in a, b, and c that tells us when there is no pivot in the rightmost column. The solution space to this equation describes Span$\{\mathbf{v}_1, \mathbf{v}_2, \mathbf{v}_3\}$.

c. In this example, the matrix formed by the vectors $\begin{bmatrix} \mathbf{v}_1 & \mathbf{v}_2 & \mathbf{v}_2 \end{bmatrix}$ has two pivot positions. Suppose we were to consider another example in which this matrix had had only one pivot position. How would this have changed the linear system describing Span$\{\mathbf{v}_1, \mathbf{v}_2, \mathbf{v}_3\}$?

2.4 Linear independence

In the previous section, questions about the existence of solutions of a linear system led to the concept of the span of a set of vectors. In particular, the span of a set of vectors $\mathbf{v}_1, \mathbf{v}_2, \ldots, \mathbf{v}_n$ is the set of vectors $\mathbf{b}$ for which a solution to the linear system $\begin{bmatrix} \mathbf{v}_1 & \mathbf{v}_2 & \cdots & \mathbf{v}_n \end{bmatrix} \mathbf{x} = \mathbf{b}$ exists.

In this section, we turn to the uniqueness of solutions of a linear system, the second of our two fundamental questions. This will lead us to the concept of linear independence.

Preview Activity 2.4.1. Let's begin by looking at some sets of vectors in $\mathbb{R}^3$. As we saw in the previous section, the span of a set of vectors in $\mathbb{R}^3$ will be either a line, a plane, or $\mathbb{R}^3$ itself.

a. Consider the following vectors in $\mathbb{R}^3$:

$$\mathbf{v}_1 = \begin{bmatrix} 0 \\ -1 \\ 2 \end{bmatrix}, \mathbf{v}_2 = \begin{bmatrix} 3 \\ 1 \\ -1 \end{bmatrix}, \mathbf{v}_3 = \begin{bmatrix} 2 \\ 0 \\ 1 \end{bmatrix}.$$

Describe the span of these vectors, Span$\{\mathbf{v}_1, \mathbf{v}_2, \mathbf{v}_3\}$, as a line, a plane, or $\mathbb{R}^3$.

b. Now consider the set of vectors:

$$\mathbf{w}_1 = \begin{bmatrix} 0 \\ -1 \\ 2 \end{bmatrix}, \mathbf{w}_2 = \begin{bmatrix} 3 \\ 1 \\ -1 \end{bmatrix}, \mathbf{w}_3 = \begin{bmatrix} 3 \\ 0 \\ 1 \end{bmatrix}.$$

Describe the span of these vectors, Span$\{\mathbf{w}_1, \mathbf{w}_2, \mathbf{w}_3\}$, as a line, a plane, or $\mathbb{R}^3$.

c. Show that the vector $\mathbf{w}_3$ is a linear combination of $\mathbf{w}_1$ and $\mathbf{w}_2$ by finding weights such that
$$\mathbf{w}_3 = c\mathbf{w}_1 + d\mathbf{w}_2.$$

d. Explain why any linear combination of $\mathbf{w}_1$, $\mathbf{w}_2$, and $\mathbf{w}_3$,

$$c_1\mathbf{w}_1 + c_2\mathbf{w}_2 + c_3\mathbf{w}_3$$

can be written as a linear combination of $\mathbf{w}_1$ and $\mathbf{w}_2$.

e. Explain why
$$\text{Span}\{\mathbf{w}_1, \mathbf{w}_2, \mathbf{w}_3\} = \text{Span}\{\mathbf{w}_1, \mathbf{w}_2\}.$$

2.4. LINEAR INDEPENDENCE

2.4.1 Linear dependence

We have seen examples where the span of a set of three vectors in $\mathbb{R}^3$ is $\mathbb{R}^3$ and other examples where the span of three vectors is a plane. We would like to understand the difference between these two situations.

Example 2.4.1 Let's consider the set of three vectors in $\mathbb{R}^3$:

$$\mathbf{v}_1 = \begin{bmatrix} 2 \\ 2 \\ 0 \end{bmatrix}, \quad \mathbf{v}_2 = \begin{bmatrix} 1 \\ 1 \\ -1 \end{bmatrix}, \quad \mathbf{v}_3 = \begin{bmatrix} -1 \\ 0 \\ 1 \end{bmatrix}.$$

Forming the associated matrix gives

$$\begin{bmatrix} \mathbf{v}_1 & \mathbf{v}_2 & \mathbf{v}_3 \end{bmatrix} = \begin{bmatrix} 2 & 1 & -1 \\ 2 & 1 & 0 \\ 0 & -1 & 1 \end{bmatrix} \sim \begin{bmatrix} 1 & 0 & 0 \\ 0 & 1 & 0 \\ 0 & 0 & 1 \end{bmatrix}.$$

Because there is a pivot position in every row, Proposition 2.3.14 tells us that $\text{Span}\{\mathbf{v}_1, \mathbf{v}_2, \mathbf{v}_3\} = \mathbb{R}^3$.

Example 2.4.2 Now let's consider the set of three vectors:

$$\mathbf{w}_1 = \begin{bmatrix} 2 \\ 2 \\ 0 \end{bmatrix}, \quad \mathbf{w}_2 = \begin{bmatrix} 1 \\ 1 \\ -1 \end{bmatrix}, \quad \mathbf{w}_3 = \begin{bmatrix} -5 \\ -5 \\ 1 \end{bmatrix}.$$

Forming the associated matrix gives

$$\begin{bmatrix} \mathbf{w}_1 & \mathbf{w}_2 & \mathbf{w}_3 \end{bmatrix} = \begin{bmatrix} 2 & 1 & -5 \\ 2 & 1 & -5 \\ 0 & -1 & 1 \end{bmatrix} \sim \begin{bmatrix} 1 & 0 & -2 \\ 0 & 1 & -1 \\ 0 & 0 & 0 \end{bmatrix}.$$

Since the last row does not have a pivot position, we know that the span of these vectors is not $\mathbb{R}^3$ but is instead a plane.

In fact, we can say more if we shift our perspective slightly and view this as an augmented matrix:

$$\begin{bmatrix} \mathbf{w}_1 & \mathbf{w}_2 & | & \mathbf{w}_3 \end{bmatrix} = \left[\begin{array}{cc|c} 2 & 1 & -5 \\ 2 & 1 & -5 \\ 0 & -1 & 1 \end{array}\right] \sim \left[\begin{array}{cc|c} 1 & 0 & -2 \\ 0 & 1 & -1 \\ 0 & 0 & 0 \end{array}\right].$$

In this way, we see that $\mathbf{w}_3 = -2\mathbf{w}_1 - \mathbf{w}_2$, which enables us to rewrite any linear combination of these three vectors:

$$c_1\mathbf{w}_1 + c_2\mathbf{w}_2 + c_3\mathbf{w}_3 = c_1\mathbf{w}_1 + c_2\mathbf{w}_2 + c_3(-2\mathbf{w}_1 - \mathbf{w}_2)$$
$$= (c_1 - 2c_3)\mathbf{w}_1 + (c_2 - c_3)\mathbf{w}_2.$$

In other words, any linear combination of $\mathbf{w}_1$, $\mathbf{w}_2$, and $\mathbf{w}_3$ may be written as a linear combination using only the vectors $\mathbf{w}_1$ and $\mathbf{w}_2$. Since the span of a set of vectors is simply the set of their linear combinations, this shows that

$$\text{Span}\{\mathbf{w}_1, \mathbf{w}_2, \mathbf{w}_3\} = \text{Span}\{\mathbf{w}_1, \mathbf{w}_2\}.$$

As a result, adding the vector $\mathbf{w}_3$ to the set of vectors $\mathbf{w}_1$ and $\mathbf{w}_2$ does not change the span.

Before exploring this type of behavior more generally, let's think about it from a geometric point of view. Suppose that we begin with the two vectors $\mathbf{v}_1$ and $\mathbf{v}_2$ in Example 2.4.1. The span of these two vectors is a plane in $\mathbb{R}^3$, as seen on the left of Figure 2.4.3.

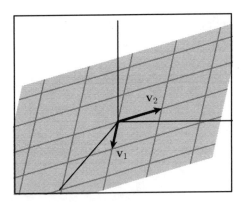 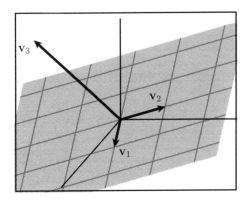

Figure 2.4.3 The span of the vectors $\mathbf{v}_1$, $\mathbf{v}_2$, and $\mathbf{v}_3$.

Because the vector $\mathbf{v}_3$ is not a linear combination of $\mathbf{v}_1$ and $\mathbf{v}_2$, it provides a direction to move that is independent of $\mathbf{v}_1$ and $\mathbf{v}_2$. Adding this third vector $\mathbf{v}_3$ therefore forms a set whose span is $\mathbb{R}^3$, as seen on the right of Figure 2.4.3.

Similarly, the span of the vectors $\mathbf{w}_1$ and $\mathbf{w}_2$ in Example 2.4.2 is also a plane. However, the third vector $\mathbf{w}_3$ is a linear combination of $\mathbf{w}_1$ and $\mathbf{w}_2$, which means that it already lies in the plane formed by $\mathbf{w}_1$ and $\mathbf{w}_2$, as seen in Figure 2.4.4. Since we can already move in this direction using just $\mathbf{w}_1$ and $\mathbf{w}_2$, adding $\mathbf{w}_3$ to the set does not change the span. As a result, it remains a plane.

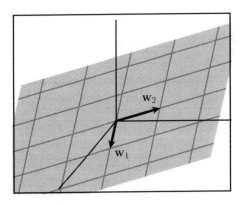 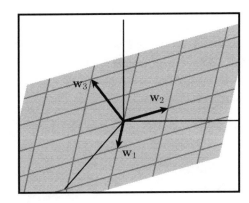

Figure 2.4.4 The span of the vectors $\mathbf{w}_1$, $\mathbf{w}_2$, and $\mathbf{w}_3$.

What distinguishes these two examples is whether one of the vectors is a linear combination of the others, an observation that leads to the following definition.

Definition 2.4.5 A set of vectors is called *linearly dependent* if one of the vectors is a linear combination of the others. Otherwise, the set of vectors is called *linearly independent*.

2.4. LINEAR INDEPENDENCE

For the sake of completeness, we say that a set of vectors containing only one nonzero vector is linearly independent.

2.4.2 How to recognize linear dependence

Activity 2.4.2. We would like to develop a means to detect when a set of vectors is linearly dependent. This activity will point the way.

a. Suppose we have five vectors in $\mathbb{R}^4$ that form the columns of a matrix having reduced row echelon form

$$\begin{bmatrix} \mathbf{v}_1 & \mathbf{v}_2 & \mathbf{v}_3 & \mathbf{v}_4 & \mathbf{v}_5 \end{bmatrix} \sim \begin{bmatrix} 1 & 0 & -1 & 0 & 2 \\ 0 & 1 & 2 & 0 & 3 \\ 0 & 0 & 0 & 1 & -1 \\ 0 & 0 & 0 & 0 & 0 \end{bmatrix}.$$

Is it possible to write one of the vectors $\mathbf{v}_1, \mathbf{v}_2, \ldots, \mathbf{v}_5$ as a linear combination of the others? If so, show explicitly how one vector appears as a linear combination of some of the other vectors. Is this set of vectors linearly dependent or independent?

b. Suppose we have another set of three vectors in $\mathbb{R}^4$ that form the columns of a matrix having reduced row echelon form

$$\begin{bmatrix} \mathbf{w}_1 & \mathbf{w}_2 & \mathbf{w}_3 \end{bmatrix} \sim \begin{bmatrix} 1 & 0 & 0 \\ 0 & 1 & 0 \\ 0 & 0 & 1 \\ 0 & 0 & 0 \end{bmatrix}.$$

Is it possible to write one of these vectors $\mathbf{w}_1, \mathbf{w}_2, \mathbf{w}_3$ as a linear combination of the others? If so, show explicitly how one vector appears as a linear combination of some of the other vectors. Is this set of vectors linearly dependent or independent?

c. By looking at the pivot positions, how can you determine whether the columns of a matrix are linearly dependent or independent?

d. If one vector in a set is the zero vector $\mathbf{0}$, can the set of vectors be linearly independent?

e. Suppose a set of vectors in $\mathbb{R}^{10}$ has twelve vectors. Is it possible for this set to be linearly independent?

By now, we should expect that the pivot positions play an important role in determining whether the columns of a matrix are linearly dependent. For instance, suppose we have four vectors and their associated matrix

$$\begin{bmatrix} \mathbf{v}_1 & \mathbf{v}_2 & \mathbf{v}_3 & \mathbf{v}_4 \end{bmatrix} \sim \begin{bmatrix} 1 & 0 & 2 & 0 \\ 0 & 1 & -3 & 0 \\ 0 & 0 & 0 & 1 \\ 0 & 0 & 0 & 0 \end{bmatrix}.$$

Since the third column does not contain a pivot position, let's just focus on the first three columns and view them as an augmented matrix:

$$\begin{bmatrix} \mathbf{v}_1 & \mathbf{v}_2 & | & \mathbf{v}_3 \end{bmatrix} \sim \left[\begin{array}{cc|c} 1 & 0 & 2 \\ 0 & 1 & -3 \\ 0 & 0 & 0 \\ 0 & 0 & 0 \end{array}\right].$$

This says that
$$\mathbf{v}_3 = 2\mathbf{v}_1 - 3\mathbf{v}_2,$$
which tells us that the set of vectors $\mathbf{v}_1, \mathbf{v}_2, \mathbf{v}_3, \mathbf{v}_4$ is linearly dependent. Moreover, we see that
$$\text{Span}\{\mathbf{v}_1, \mathbf{v}_2, \mathbf{v}_3, \mathbf{v}_4\} = \text{Span}\{\mathbf{v}_1, \mathbf{v}_2, \mathbf{v}_4\}.$$

More generally, the same reasoning implies that a set of vectors is linearly dependent if the associated matrix has a column without a pivot position. Indeed, as illustrated here, a vector corresponding to a column without a pivot position can be expressed as a linear combination of the vectors whose columns do contain pivot positions.

Suppose instead that the matrix associated to a set of vectors has a pivot position in every column.

$$\begin{bmatrix} \mathbf{w}_1 & \mathbf{w}_2 & \mathbf{w}_3 & \mathbf{w}_4 \end{bmatrix} \sim \begin{bmatrix} 1 & 0 & 0 & 0 \\ 0 & 1 & 0 & 0 \\ 0 & 0 & 1 & 0 \\ 0 & 0 & 0 & 1 \\ 0 & 0 & 0 & 0 \end{bmatrix}.$$

Viewing this as an augmented matrix again, we see that the linear system is inconsistent since there is a pivot in the rightmost column, which means that $\mathbf{w}_4$ cannot be expressed as a linear combination of the other vectors. Similarly, $\mathbf{w}_3$ cannot be expressed as a linear combination of $\mathbf{w}_1$ and $\mathbf{w}_2$. In fact, none of the vectors can be written as a linear combination of the others so this set of vectors is linearly independent.

The following proposition summarizes these findings.

Proposition 2.4.6 *The columns of a matrix are linearly independent if and only if every column contains a pivot position.*

This condition imposes a constraint on how many vectors we can have in a linearly independent set. Here is an example of the reduced row echelon form of a matrix whose columns

2.4. LINEAR INDEPENDENCE

form a set of three linearly independent vectors in $\mathbb{R}^5$:

$$\begin{bmatrix} 1 & 0 & 0 \\ 0 & 1 & 0 \\ 0 & 0 & 1 \\ 0 & 0 & 0 \\ 0 & 0 & 0 \end{bmatrix}.$$

Notice that there are at least as many rows as columns, which must be the case if every column is to have a pivot position.

More generally, if $\mathbf{v}_1, \mathbf{v}_2, \ldots, \mathbf{v}_n$ is a linearly independent set of vectors in $\mathbb{R}^m$, the associated matrix must have a pivot position in every column. Since every row contains at most one pivot position, the number of columns can be no greater than the number of rows. This means that the number of vectors in a linearly independent set can be no greater than the number of dimensions.

Proposition 2.4.7 *A linearly independent set of vectors in $\mathbb{R}^m$ contains at most m vectors.*

This says, for instance, that any linearly independent set of vectors in $\mathbb{R}^3$ can contain no more three vectors. We usually imagine three independent directions, such as up/down, front/back, left/right, in our three-dimensional world. This proposition tells us that there can be no more independent directions.

The proposition above says that a set of vectors in $\mathbb{R}^m$ that is linear independent has at most m vectors. By comparison, Proposition 2.3.15 says that a set of vectors whose span is $\mathbb{R}^m$ has at least m vectors.

2.4.3 Homogeneous equations

If A is a matrix, we call the equation $A\mathbf{x} = \mathbf{0}$ a *homogeneous* equation. As we'll see, the uniqueness of solutions to this equation reflects on the linear independence of the columns of A.

Activity 2.4.3 Linear independence and homogeneous equations..

a. Explain why the homogeneous equation $A\mathbf{x} = \mathbf{0}$ is consistent no matter the matrix A.

b. Consider the matrix

$$A = \begin{bmatrix} 3 & 2 & 0 \\ -1 & 0 & -2 \\ 2 & 1 & 1 \end{bmatrix}$$

whose columns we denote by $\mathbf{v}_1$, $\mathbf{v}_2$, and $\mathbf{v}_3$. Describe the solution space of the homogeneous equation $A\mathbf{x} = \mathbf{0}$ using a parametric description, if appropriate.

c. Find a nonzero solution to the homogeneous equation and use it to find weights c_1, c_2, and c_3 such that

$$c_1\mathbf{v}_1 + c_2\mathbf{v}_2 + c_3\mathbf{v}_3 = \mathbf{0}.$$

d. Use the equation you found in the previous part to write one of the vectors as a linear combination of the others.

e. Are the vectors $\mathbf{v}_1$, $\mathbf{v}_2$, and $\mathbf{v}_3$ linearly dependent or independent?

This activity shows how the solution space of the homogeneous equation $A\mathbf{x} = \mathbf{0}$ indicates whether the columns of A are linearly dependent or independent. First, we know that the equation $A\mathbf{x} = \mathbf{0}$ always has at least one solution, the vector $\mathbf{x} = \mathbf{0}$. Any other solution is a nonzero solution.

Example 2.4.8 Let's consider the vectors

$$\mathbf{v}_1 = \begin{bmatrix} 2 \\ -4 \\ 1 \\ 0 \end{bmatrix}, \quad \mathbf{v}_2 = \begin{bmatrix} 1 \\ 1 \\ 3 \\ -2 \end{bmatrix}, \quad \mathbf{v}_3 = \begin{bmatrix} 3 \\ -3 \\ 4 \\ -2 \end{bmatrix}$$

and their associated matrix $A = \begin{bmatrix} \mathbf{v}_1 & \mathbf{v}_2 & \mathbf{v}_3 \end{bmatrix}$.

The homogeneous equation $A\mathbf{x} = \mathbf{0}$ has the associated augmented matrix

$$\left[\begin{array}{ccc|c} 2 & 1 & 3 & 0 \\ -4 & 1 & -3 & 0 \\ 1 & 3 & 4 & 0 \\ 0 & -2 & -2 & 0 \end{array}\right] \sim \left[\begin{array}{ccc|c} 1 & 0 & 1 & 0 \\ 0 & 1 & 1 & 0 \\ 0 & 0 & 0 & 0 \\ 0 & 0 & 0 & 0 \end{array}\right].$$

Therefore, A has a column without a pivot position, which tells us that the vectors $\mathbf{v}_1$, $\mathbf{v}_2$, and $\mathbf{v}_3$ are linearly dependent. However, we can also see this fact in another way.

The reduced row echelon matrix tells us that the homogeneous equation has a free variable so that there must be infinitely many solutions. In particular, we have

$$x_1 = -x_3$$
$$x_2 = -x_3$$

so the solutions have the form

$$\mathbf{x} = \begin{bmatrix} x_1 \\ x_2 \\ x_3 \end{bmatrix} = \begin{bmatrix} -x_3 \\ -x_3 \\ x_3 \end{bmatrix} = x_3 \begin{bmatrix} -1 \\ -1 \\ 1 \end{bmatrix}.$$

If we choose $x_3 = 1$, then we obtain the nonzero solution to the homogeneous equation $\mathbf{x} = \begin{bmatrix} -1 \\ -1 \\ 1 \end{bmatrix}$, which implies that

$$A\begin{bmatrix} -1 \\ -1 \\ 1 \end{bmatrix} = \begin{bmatrix} \mathbf{v}_1 & \mathbf{v}_2 & \mathbf{v}_3 \end{bmatrix}\begin{bmatrix} -1 \\ -1 \\ 1 \end{bmatrix} = -\mathbf{v}_1 - \mathbf{v}_2 + \mathbf{v}_3 = \mathbf{0}.$$

2.4. LINEAR INDEPENDENCE

In other words,
$$-\mathbf{v}_1 - \mathbf{v}_2 + \mathbf{v}_3 = \mathbf{0}$$
$$\mathbf{v}_3 = \mathbf{v}_1 + \mathbf{v}_2.$$

Because $\mathbf{v}_3$ is a linear combination of $\mathbf{v}_1$ and $\mathbf{v}_2$, we know that this set of vectors is linearly dependent.

As this example demonstrates, there are many ways we can view the question of linear independence, some of which are recorded in the following proposition.

Proposition 2.4.9 *For a matrix* $A = \begin{bmatrix} \mathbf{v}_1 & \mathbf{v}_2 & \ldots & \mathbf{v}_n \end{bmatrix}$, *the following statements are equivalent:*

- *The columns of A are linearly dependent.*

- *One of the vectors in the set* $\mathbf{v}_1, \mathbf{v}_2, \ldots, \mathbf{v}_n$ *is a linear combination of the others.*

- *The matrix A has a column without a pivot position.*

- *The homogeneous equation* $A\mathbf{x} = \mathbf{0}$ *has infinitely many solutions and hence a nonzero solution.*

- *There are weights* $c_1, c_2, \ldots, c_n$, *not all of which are zero, such that*

$$c_1 \mathbf{v}_1 + c_2 \mathbf{v}_2 + \ldots + c_n \mathbf{v}_n = \mathbf{0}.$$

2.4.4 Summary

This section developed the concept of linear dependence of a set of vectors. More specifically, we saw that:

- A set of vectors is linearly dependent if one of the vectors is a linear combination of the others.

- A set of vectors is linearly independent if and only if the vectors form a matrix that has a pivot position in every column.

- A set of linearly independent vectors in $\mathbb{R}^m$ contains no more than m vectors.

- The columns of the matrix A are linearly dependent if the homogeneous equation $A\mathbf{x} = \mathbf{0}$ has a nonzero solution.

- A set of vectors $\mathbf{v}_1, \mathbf{v}_2, \ldots, \mathbf{v}_n$ is linearly dependent if there are weights $c_1, c_2, \ldots, c_n$, not all of which are zero, such that

$$c_1 \mathbf{v}_1 + c_2 \mathbf{v}_2 + \ldots + c_n \mathbf{v}_n = \mathbf{0}.$$

At the beginning of the section, we said that this concept addressed the second of our two fundamental questions concerning the uniqueness of solutions to a linear system. It is worth comparing the results of this section with those of the previous one so that the parallels between them become clear.

As usual, we will write a matrix as a collection of vectors,

$$A = \begin{bmatrix} \mathbf{v}_1 & \mathbf{v}_2 & \ldots & \mathbf{v}_n \end{bmatrix}.$$

Table 2.4.10 Span and Linear Independence

Span	Linear independence
A vector **b** is in the span of a set of vectors if it is a linear combination of those vectors.	A set of vectors is linearly dependent if one of the vectors is a linear combination of the others.
A vector **b** is in the span of $\mathbf{v}_1, \mathbf{v}_2, \ldots, \mathbf{v}_n$ if there exists a solution to $A\mathbf{x} = \mathbf{b}$.	The vectors $\mathbf{v}_1, \mathbf{v}_2, \ldots, \mathbf{v}_n$ are linearly independent if $\mathbf{x} = \mathbf{0}$ is the unique solution to $A\mathbf{x} = \mathbf{0}$.
The columns of an $m \times n$ matrix span $\mathbb{R}^m$ if the matrix has a pivot position in every row.	The columns of a matrix are linearly independent if the matrix has a pivot position in every column.
A set of vectors that span $\mathbb{R}^m$ has at least m vectors.	A set of linearly independent vectors in $\mathbb{R}^m$ has at most m vectors.

2.4.5 Exercises

1. Consider the set of vectors

$$\mathbf{v}_1 = \begin{bmatrix} 1 \\ 2 \\ 1 \end{bmatrix}, \mathbf{v}_2 = \begin{bmatrix} 0 \\ 1 \\ 3 \end{bmatrix}, \mathbf{v}_3 = \begin{bmatrix} 2 \\ 3 \\ -1 \end{bmatrix}, \mathbf{v}_4 = \begin{bmatrix} -2 \\ 4 \\ -1 \end{bmatrix}.$$

    a. Explain why this set of vectors is linearly dependent.

    b. Write one of the vectors as a linear combination of the others.

    c. Find weights $c_1, c_2, c_3,$ and c_4, not all of which are zero, such that

    $$c_1\mathbf{v}_1 + c_2\mathbf{v}_2 + c_3\mathbf{v}_3 + c_4\mathbf{v}_4 = \mathbf{0}.$$

    d. Suppose $A = \begin{bmatrix} \mathbf{v}_1 & \mathbf{v}_2 & \mathbf{v}_3 & \mathbf{v}_4 \end{bmatrix}$. Find a nonzero solution to the homogenous equation $A\mathbf{x} = \mathbf{0}$.

2. Consider the vectors

$$\mathbf{v}_1 = \begin{bmatrix} 2 \\ -1 \\ 0 \end{bmatrix}, \mathbf{v}_2 = \begin{bmatrix} 1 \\ 2 \\ 1 \end{bmatrix}, \mathbf{v}_3 = \begin{bmatrix} 2 \\ -2 \\ 3 \end{bmatrix}.$$

    a. Are these vectors linearly independent or linearly dependent?

    b. Describe the Span$\{\mathbf{v}_1, \mathbf{v}_2, \mathbf{v}_3\}$.

    c. Suppose that **b** is a vector in $\mathbb{R}^3$. Explain why we can guarantee that **b** may be written as a linear combination of $\mathbf{v}_1, \mathbf{v}_2,$ and $\mathbf{v}_3$.

2.4. LINEAR INDEPENDENCE

d. Suppose that $\mathbf{b}$ is a vector in $\mathbb{R}^3$. In how many ways can $\mathbf{b}$ be written as a linear combination of $\mathbf{v}_1$, $\mathbf{v}_2$, and $\mathbf{v}_3$?

3. Respond to the following questions and provide a justification for your responses.

   a. If the vectors $\mathbf{v}_1$ and $\mathbf{v}_2$ form a linearly dependent set, must one vector be a scalar multiple of the other?

   b. Suppose that $\mathbf{v}_1, \mathbf{v}_2, \ldots, \mathbf{v}_n$ is a linearly independent set of vectors. What can you say about the linear independence or dependence of a subset of these vectors?

   c. Suppose $\mathbf{v}_1, \mathbf{v}_2, \ldots, \mathbf{v}_n$ is a linearly independent set of vectors that form the columns of a matrix A. If the equation $A\mathbf{x} = \mathbf{b}$ is inconsistent, what can you say about the linear independence or dependence of the set of vectors $\mathbf{v}_1, \mathbf{v}_2, \ldots, \mathbf{v}_n, \mathbf{b}$?

4. Determine whether the following statements are true or false and provide a justification for your response.

   a. If $\mathbf{v}_1, \mathbf{v}_2, \ldots, \mathbf{v}_n$ are linearly dependent, then one vector is a scalar multiple of one of the others.

   b. If $\mathbf{v}_1, \mathbf{v}_2, \ldots, \mathbf{v}_{10}$ are vectors in $\mathbb{R}^5$, then the set of vectors is linearly dependent.

   c. If $\mathbf{v}_1, \mathbf{v}_2, \ldots, \mathbf{v}_5$ are vectors in $\mathbb{R}^{10}$, then the set of vectors is linearly independent.

   d. Suppose we have a set of vectors $\mathbf{v}_1, \mathbf{v}_2, \ldots, \mathbf{v}_n$ and that $\mathbf{v}_2$ is a scalar multiple of $\mathbf{v}_1$. Then the set is linearly dependent.

   e. Suppose that $\mathbf{v}_1, \mathbf{v}_2, \ldots, \mathbf{v}_n$ are linearly independent and form the columns of a matrix A. If $A\mathbf{x} = \mathbf{b}$ is consistent, then there is exactly one solution.

5. Suppose we have a set of vectors $\mathbf{v}_1, \mathbf{v}_2, \mathbf{v}_3, \mathbf{v}_4$ in $\mathbb{R}^5$ that satisfy the relationship:

$$2\mathbf{v}_1 - \mathbf{v}_2 + 3\mathbf{v}_3 + \mathbf{v}_4 = 0$$

   and suppose that A is the matrix $A = \begin{bmatrix} \mathbf{v}_1 & \mathbf{v}_2 & \mathbf{v}_3 & \mathbf{v}_4 \end{bmatrix}$.

   a. Find a nonzero solution to the equation $A\mathbf{x} = \mathbf{0}$.

   b. Explain why the matrix A has a column without a pivot position.

   c. Write one of the vectors as a linear combination of the others.

   d. Explain why the set of vectors is linearly dependent.

6. Suppose that $\mathbf{v}_1, \mathbf{v}_2, \ldots, \mathbf{v}_n$ is a set of vectors in $\mathbb{R}^{27}$ that form the columns of a matrix A.

   a. Suppose that the vectors span $\mathbb{R}^{27}$. What can you say about the number of vectors n in this set?

   b. Suppose instead that the vectors are linearly independent. What can you say about the number of vectors n in this set?

   c. Suppose that the vectors are both linearly independent and span $\mathbb{R}^{27}$. What can you say about the number of vectors in the set?

   d. Assume that the vectors are both linearly independent and span $\mathbb{R}^{27}$. Given a vector **b** in $\mathbb{R}^{27}$, what can you say about the solution space to the equation $A\mathbf{x} = \mathbf{b}$?

7. Given below are some descriptions of sets of vectors that form the columns of a matrix A. For each description, give a possible reduced row echelon form for A or indicate why there is no set of vectors satisfying the description by stating why the required reduced row echelon matrix cannot exist.

    a. A set of 4 linearly independent vectors in $\mathbb{R}^5$.

    b. A set of 4 linearly independent vectors in $\mathbb{R}^4$.

    c. A set of 3 vectors whose span is $\mathbb{R}^4$.

    d. A set of 5 linearly independent vectors in $\mathbb{R}^3$.

    e. A set of 5 vectors whose span is $\mathbb{R}^4$.

8. When we explored matrix multiplication in Section 2.2, we saw that some properties that are true for real numbers are not true for matrices. This exercise will investigate that in some more depth.

    a. Suppose that A and B are two matrices and that $AB = 0$. If $B \neq 0$, what can you say about the linear independence of the columns of A?

    b. Suppose that we have matrices A, B and C such that $AB = AC$. We have seen that we cannot generally conclude that $B = C$. If we assume additionally that A is a matrix whose columns are linearly independent, explain why $B = C$. You may wish to begin by rewriting the equation $AB = AC$ as $AB - AC = A(B - C) = 0$.

9. Suppose that k is an unknown parameter and consider the set of vectors

    $$\mathbf{v}_1 = \begin{bmatrix} 2 \\ 0 \\ 1 \end{bmatrix}, \mathbf{v}_2 = \begin{bmatrix} 4 \\ -2 \\ -1 \end{bmatrix}, \mathbf{v}_3 = \begin{bmatrix} 0 \\ 2 \\ k \end{bmatrix}.$$

    a. For what values of k is the set of vectors linearly dependent?

    b. For what values of k does the set of vectors span $\mathbb{R}^3$?

10. Given a set of linearly dependent vectors, we can eliminate some of the vectors to create a smaller, linearly independent set of vectors.

    a. Suppose that **w** is a linear combination of the vectors $\mathbf{v}_1$ and $\mathbf{v}_2$. Explain why Span$\{\mathbf{v}_1, \mathbf{v}_2, \mathbf{w}\}$ = Span$\{\mathbf{v}_1, \mathbf{v}_2\}$.

    b. Consider the vectors

    $$\mathbf{v}_1 = \begin{bmatrix} 2 \\ -1 \\ 0 \end{bmatrix}, \mathbf{v}_2 = \begin{bmatrix} 1 \\ 2 \\ 1 \end{bmatrix}, \mathbf{v}_3 = \begin{bmatrix} -2 \\ 6 \\ 2 \end{bmatrix}, \mathbf{v}_4 = \begin{bmatrix} 7 \\ -1 \\ 1 \end{bmatrix}.$$

    Write one of the vectors as a linear combination of the others. Find a set of three vectors whose span is the same as Span$\{\mathbf{v}_1, \mathbf{v}_2, \mathbf{v}_3, \mathbf{v}_4\}$.

    c. Are the three vectors you are left with linearly independent? If not, express one of the vectors as a linear combination of the others and find a set of two vectors

2.4. LINEAR INDEPENDENCE

whose span is the same as $\text{Span}\{\mathbf{v}_1, \mathbf{v}_2, \mathbf{v}_3, \mathbf{v}_4\}$.

d. Give a geometric description of $\text{Span}\{\mathbf{v}_1, \mathbf{v}_2, \mathbf{v}_3, \mathbf{v}_4\}$ in $\mathbb{R}^3$ as we did in Section 2.3.

2.5 Matrix transformations

The past few sections introduced us to matrix-vector multiplication as a means of thinking geometrically about the solutions to a linear system. In particular, we rewrote a linear system as a matrix equation $A\mathbf{x} = \mathbf{b}$ and developed the concepts of span and linear independence in response to our two fundamental questions.

In this section, we will explore how matrix-vector multiplication defines certain types of functions, which we call *matrix transformations*, similar to those encountered in previous algebra courses. In particular, we will develop some algebraic tools for thinking about matrix transformations and look at some motivating examples. In the next section, we will see how matrix transformations describe important geometric operations and how they are used in computer animation.

Preview Activity 2.5.1. We will begin by considering a more familiar situation; namely, the function $f(x) = x^2$, which takes a real number x as an input and produces its square x^2 as its output.

a. What is the value of $f(3)$?

b. Can we solve the equation $f(x) = 4$? If so, is the solution unique?

c. Can we solve the equation $f(x) = -10$? If so, is the solution unique?

d. Sketch a graph of the function $f(x) = x^2$ in Figure 2.5.1

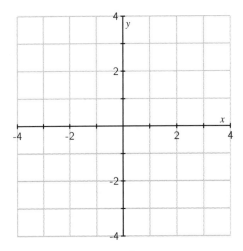

Figure 2.5.1 Graph the function $f(x) = x^2$ above.

e. We will now consider functions having the form $g(x) = mx$. Draw a graph of the function $g(x) = 2x$ on the left in Figure 2.5.2.

2.5. MATRIX TRANSFORMATIONS

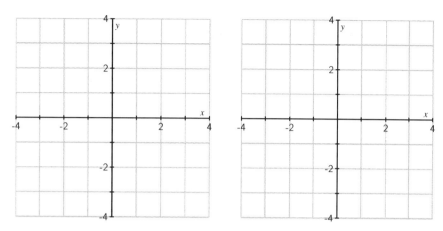

Figure 2.5.2 Graphs of the function $g(x) = 2x$ and $h(x) = -\frac{1}{3}x$.

f. Draw a graph of the function $h(x) = -\frac{1}{3}x$ on the right of Figure 2.5.2.

g. Remember that composing two functions means we use the output from one function as the input into the other; that is, $(g \circ h)(x) = g(h(x))$. What function results from composing $(g \circ h)(x)$?

2.5.1 Matrix transformations

In the preview activity, we considered familiar linear functions of a single variable, such as $g(x) = 2x$. We construct a function like this by choosing a number m; when given an input x, the output $g(x) = mx$ is formed by multiplying x by m.

In this section, we will consider functions whose inputs are vectors and whose outputs are vectors defined through matrix-vector multiplication. That is, if A is a matrix and $\mathbf{x}$ is a vector, the function $T(\mathbf{x}) = A\mathbf{x}$ forms the product $A\mathbf{x}$ as its output. Such a function is called a matrix transformation.

Definition 2.5.3 The **matrix transformation** associated to the matrix A is the function that assigns to the vector $\mathbf{x}$ the vector $A\mathbf{x}$; that is, $T(\mathbf{x}) = A\mathbf{x}$.

Example 2.5.4 The matrix $A = \begin{bmatrix} 3 & -2 \\ 1 & 2 \\ 0 & 3 \end{bmatrix}$ defines a matrix transformation $T(\mathbf{x}) = A\mathbf{x}$ in the following way:

$$T\left(\begin{bmatrix} x_1 \\ x_2 \end{bmatrix}\right) = \begin{bmatrix} 3 & -2 \\ 1 & 2 \\ 0 & 3 \end{bmatrix} \begin{bmatrix} x_1 \\ x_2 \end{bmatrix} = \begin{bmatrix} 3x_1 - 2x_2 \\ x_1 + 2x_2 \\ 3x_2 \end{bmatrix}.$$

Notice that the input to T is a two-dimensional vector $\begin{bmatrix} x_1 \\ x_2 \end{bmatrix}$ and the output is a three-dimensional vector $\begin{bmatrix} 3x_1 - 2x_2 \\ x_1 + 2x_2 \\ 3x_2 \end{bmatrix}$. As a shorthand, we will write

$$T : \mathbb{R}^2 \to \mathbb{R}^3$$

to indicate that the inputs are two-dimensional vectors and the outputs are three-dimensional vectors.

Example 2.5.5 Suppose we have a function $T : \mathbb{R}^3 \to \mathbb{R}^2$ that has the form

$$T\left(\begin{bmatrix} x_1 \\ x_2 \\ x_3 \end{bmatrix}\right) = \begin{bmatrix} -4x_1 - x_2 + 2x_3 \\ x_1 + 2x_2 - x_3 \end{bmatrix}.$$

We may write

$$\begin{aligned}
T\left(\begin{bmatrix} x_1 \\ x_2 \\ x_3 \end{bmatrix}\right) &= \begin{bmatrix} -4x_1 - x_2 + 2x_3 \\ x_1 + 2x_2 - x_3 \end{bmatrix} \\
&= \begin{bmatrix} -4x_1 \\ x_1 \end{bmatrix} + \begin{bmatrix} -x_2 \\ 2x_2 \end{bmatrix} + \begin{bmatrix} 2x_3 \\ -x_3 \end{bmatrix} \\
&= x_1 \begin{bmatrix} -4 \\ 1 \end{bmatrix} + x_2 \begin{bmatrix} -1 \\ 2 \end{bmatrix} + x_3 \begin{bmatrix} 2 \\ -1 \end{bmatrix} \\
&= \begin{bmatrix} -4 & -1 & 2 \\ 1 & 2 & -1 \end{bmatrix} \begin{bmatrix} x_1 \\ x_2 \\ x_3 \end{bmatrix}.
\end{aligned}$$

This shows that T is a matrix transformation $T(\mathbf{x}) = A\mathbf{x}$ associated to the matrix $A = \begin{bmatrix} -4 & -1 & 2 \\ 1 & 2 & -1 \end{bmatrix}$.

Activity 2.5.2. In this activity, we will look at some examples of matrix transformations.

a. To begin, suppose that A is the matrix

$$A = \begin{bmatrix} 2 & 1 \\ 1 & 2 \end{bmatrix}.$$

with associated matrix transformation $T(\mathbf{x}) = A\mathbf{x}$.

1. What is $T\left(\begin{bmatrix} 1 \\ -2 \end{bmatrix}\right)$?

2. What is $T\left(\begin{bmatrix} 1 \\ 0 \end{bmatrix}\right)$?

3. What is $T\left(\begin{bmatrix} 0 \\ 1 \end{bmatrix}\right)$?

2.5. MATRIX TRANSFORMATIONS

4. Is there a vector **x** such that $T(\mathbf{x}) = \begin{bmatrix} 3 \\ 0 \end{bmatrix}$?

5. Write $T\left(\begin{bmatrix} x \\ y \end{bmatrix}\right)$ as a two-dimensional vector.

b. Suppose that $T(\mathbf{x}) = A\mathbf{x}$ where

$$A = \begin{bmatrix} 3 & 3 & -2 & 1 \\ 0 & 2 & 1 & -3 \\ -2 & 1 & 4 & -4 \end{bmatrix}.$$

1. What is the dimension of the vectors **x** that are inputs for T?
2. What is the dimension of the vectors $T(\mathbf{x}) = A\mathbf{x}$ that are outputs?
3. If we describe this transformation as $T : \mathbb{R}^n \to \mathbb{R}^m$, what are the values of n and m and how do they relate to the shape of A?
4. Describe the vectors **x** for which $T(\mathbf{x}) = \mathbf{0}$.

c. If A is the matrix $A = \begin{bmatrix} \mathbf{v}_1 & \mathbf{v}_2 \end{bmatrix}$, what is $T\left(\begin{bmatrix} 1 \\ 0 \end{bmatrix}\right)$ in terms of the vectors $\mathbf{v}_1$ and $\mathbf{v}_2$? What about $T\left(\begin{bmatrix} 0 \\ 1 \end{bmatrix}\right)$?

d. Suppose that A is a 3×2 matrix and that $T(\mathbf{x}) = A\mathbf{x}$. If

$$T\left(\begin{bmatrix} 1 \\ 0 \end{bmatrix}\right) = \begin{bmatrix} 3 \\ -1 \\ 1 \end{bmatrix}, T\left(\begin{bmatrix} 0 \\ 1 \end{bmatrix}\right) = \begin{bmatrix} 2 \\ 2 \\ -1 \end{bmatrix},$$

what is the matrix A?

Let's discuss a few of the issues that appear in this activity. First, notice that the shape of the matrix A and the dimension of the input vector **x** must be compatible if the product $A\mathbf{x}$ is to be defined. In particular, if A is an $m \times n$ matrix, **x** needs to be an n-dimensional vector, and the resulting product $A\mathbf{x}$ will be an m-dimensional vector. For the associated matrix transformation, we therefore write $T : \mathbb{R}^n \to \mathbb{R}^m$ meaning T takes vectors in $\mathbb{R}^n$ as inputs and produces vectors in $\mathbb{R}^m$ as outputs. For instance, if

$$A = \begin{bmatrix} 4 & 0 & -3 & 2 \\ 0 & 1 & 3 & 1 \end{bmatrix},$$

then $T : \mathbb{R}^4 \to \mathbb{R}^2$.

Second, we can often reconstruct the matrix A if we only know some output values from its associated linear transformation T by remembering that matrix-vector multiplication constructs linear combinations. For instance, if A is an $m \times 2$ matrix $A = \begin{bmatrix} \mathbf{v}_1 & \mathbf{v}_2 \end{bmatrix}$, then

$$T\left(\begin{bmatrix} 1 \\ 0 \end{bmatrix}\right) = \begin{bmatrix} \mathbf{v}_1 & \mathbf{v}_2 \end{bmatrix} \begin{bmatrix} 1 \\ 0 \end{bmatrix} = 1\mathbf{v}_1 + 0\mathbf{v}_2 = \mathbf{v}_1.$$

That is, we can find the first column of A by evaluating $T\left(\begin{bmatrix} 1 \\ 0 \end{bmatrix}\right)$. Similarly, the second column of A is found by evaluating $T\left(\begin{bmatrix} 0 \\ 1 \end{bmatrix}\right)$.

More generally, we will write the columns of the $n \times n$ identity matrix as

$$\mathbf{e}_1 = \begin{bmatrix} 1 \\ 0 \\ \vdots \\ 0 \end{bmatrix}, \quad \mathbf{e}_2 = \begin{bmatrix} 0 \\ 1 \\ \vdots \\ 0 \end{bmatrix}, \quad \ldots, \quad \mathbf{e}_n = \begin{bmatrix} 0 \\ 0 \\ \vdots \\ 1 \end{bmatrix}$$

so that

$$T(\mathbf{e}_j) = \begin{bmatrix} \mathbf{v}_1 & \mathbf{v}_2 & \cdots & \mathbf{v}_n \end{bmatrix} \mathbf{e}_j = \mathbf{v}_j.$$

This means that the jth column of A is found by evaluating $T(\mathbf{e}_j)$. We record this fact in the following proposition.

Proposition 2.5.6 *If $T : \mathbb{R}^n \to \mathbb{R}^m$ is a matrix transformation given by $T(\mathbf{x}) = A\mathbf{x}$, then the matrix A has columns $T(\mathbf{e}_j)$; that is,*

$$A = \begin{bmatrix} T(\mathbf{e}_1) & T(\mathbf{e}_2) & \cdots & T(\mathbf{e}_n) \end{bmatrix}.$$

Activity 2.5.3. Let's look at some examples and apply these observations.

a. To begin, suppose that T is the matrix transformation that takes a two-dimensional vector $\mathbf{x}$ as an input and outputs $T(\mathbf{x})$, the two-dimensional vector obtained by rotating $\mathbf{x}$ counterclockwise by $90°$, as shown in Figure 2.5.7.

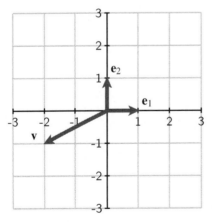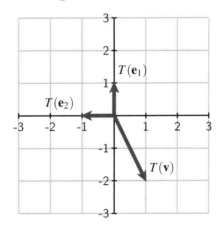

Figure 2.5.7 The matrix transformation T takes two-dimensional vectors on the left and rotates them by $90°$ counterclockwise into the vectors on the right.

We will see in the next section that many geometric operations like this one can be performed by matrix transformations.

1. If we write $T : \mathbb{R}^n \to \mathbb{R}^m$, what are the values of m and n, and what is the shape of the associated matrix A?

2.5. MATRIX TRANSFORMATIONS

2. Determine the matrix A by applying Proposition 2.5.6.

3. If $\mathbf{v} = \begin{bmatrix} -2 \\ -1 \end{bmatrix}$ as shown on the left in Figure 2.5.7, use your matrix to determine $T(\mathbf{v})$ and verify that it agrees with that shown on the right of Figure 2.5.7.

4. If $\mathbf{x} = \begin{bmatrix} x \\ y \end{bmatrix}$, determine the vector $T(\mathbf{x})$ obtained by rotating $\mathbf{x}$ counterclockwise by $90°$.

b. Suppose that we work for a company that makes baked goods, including cakes, doughnuts, and eclairs. The company operates two bakeries, Bakery 1 and Bakery 2. In one hour of operation,

- Bakery 1 produces 10 cakes, 50 doughnuts, and 30 eclairs.
- Bakery 2 produces 20 cakes, 30 doughnuts, and 30 eclairs.

If Bakery 1 operates for x_1 hours and Bakery 2 for x_2 hours, we will use the vector $\mathbf{x} = \begin{bmatrix} x_1 \\ x_2 \end{bmatrix}$ to describe the operation of the two bakeries.

We would like to describe a matrix transformation T where $\mathbf{x}$ describes the number of hours the bakeries operate and $T(\mathbf{x})$ describes the total number of cakes, doughnuts, and eclairs produced. That is, $T(\mathbf{x}) = \begin{bmatrix} y_1 \\ y_2 \\ y_3 \end{bmatrix}$ where y_1 is the number of cakes, y_2 is the number of doughnuts, and y_3 is the number of eclairs produced.

1. If $T : \mathbb{R}^n \to \mathbb{R}^m$, what are the values of m and n, and what is the shape of the associated matrix A?

2. We can determine the matrix A using Proposition 2.5.6. For instance, $T\left(\begin{bmatrix} 1 \\ 0 \end{bmatrix}\right)$ will describe the number of cakes, doughnuts, and eclairs produced when Bakery 1 operates for one hour and Bakery 2 sits idle. What is this vector?

3. In the same way, determine $T\left(\begin{bmatrix} 0 \\ 1 \end{bmatrix}\right)$. What is the matrix A?

4. If Bakery 1 operates for 120 hours and Bakery 2 for 180 hours, what is the total number of cakes, doughnuts, and eclairs produced?

5. Suppose that in one period of time, the company produces 5060 cakes, 14310 doughnuts, and 10470 eclairs. How long did each bakery operate?

6. Suppose that the company receives an order for a certain number of cakes, doughnuts, and eclairs. Can you guarantee that you can fill the order without having leftovers?

In these examples, we glided over an important point: how do we know these functions

$T: \mathbb{R}^n \to \mathbb{R}^m$ can be expressed as matrix transformations? We will take up this question in detail in the next section and not worry about it for now.

2.5.2 Composing matrix transformations

It sometimes happens that we want to combine matrix transformations by performing one and then another. In the last activity, for instance, we considered the matrix transformation where $T(\mathbf{x})$ is the result of rotating the two-dimensional vector $\mathbf{x}$ by $90°$. Now suppose we are interested in rotating that vector twice; that is, we take a vector $\mathbf{x}$, rotate it by $90°$ to obtain $T(\mathbf{x})$, and then rotate the result by $90°$ again to obtain $T(T(\mathbf{x}))$.

This process is called function *composition* and likely appeared in an earlier algebra course. For instance, if $g(x) = 2x + 1$ and $h(x) = x^2$, the composition of these functions obtained by first performing g and then performing h is denoted by

$$(h \circ g)(x) = h(g(x)) = h(2x + 1) = (2x + 1)^2.$$

Composing matrix transformations is similar. Suppose that we have two matrix transformations, $T: \mathbb{R}^n \to \mathbb{R}^m$ and $S: \mathbb{R}^m \to \mathbb{R}^p$. Their associated matrices will be denoted by A and B so that $T(\mathbf{x}) = A\mathbf{x}$ and $S(\mathbf{x}) = B\mathbf{x}$. If we apply T to a vector $\mathbf{x}$ to obtain $T(\mathbf{x})$ and then apply S to the result, we have

$$(S \circ T)(\mathbf{x}) = S(T(\mathbf{x})) = S(A\mathbf{x}) = BA\mathbf{x} = (BA)\mathbf{x}.$$

Notice that this implies that the composition $(S \circ T)$ is itself a matrix transformation and that the associated matrix is the product BA.

Proposition 2.5.8 *If $T: \mathbb{R}^n \to \mathbb{R}^m$ and $S: \mathbb{R}^m \to \mathbb{R}^p$ are matrix transformations with associated matrices A and B respectively, then the composition $(S \circ T)$ is also a matrix transformation whose associated matrix is the product BA.*

Notice that the matrix transformations must be compatible if they are to be composed. In particular, the vector $T(\mathbf{x})$, an m-dimensional vector, must be a suitable input vector for S, which means that the inputs to S must be m-dimensional. In fact, this is the same condition we need to form the product BA of their associated matrices, namely, that the number of columns of B is the same as the number of rows of A.

Activity 2.5.4. We will explore the composition of matrix transformations by revisiting the matrix transformations from Activity 2.5.3.

a. Let's begin with the matrix transformation $T: \mathbb{R}^2 \to \mathbb{R}^2$ that rotates a two-dimensional vector $\mathbf{x}$ by $90°$ to produce $T(\mathbf{x})$. We saw in the earlier activity that the associated matrix is $A = \begin{bmatrix} 0 & -1 \\ 1 & 0 \end{bmatrix}$. Suppose that we compose this matrix transformation with itself to obtain $(T \circ T)(\mathbf{x}) = T(T(\mathbf{x}))$, which is the result of rotating $\mathbf{x}$ by $90°$ twice.

1. What is the matrix associated to the composition $(T \circ T)$?

2.5. MATRIX TRANSFORMATIONS

2. What is the result of rotating $\mathbf{v} = \begin{bmatrix} -2 \\ -1 \end{bmatrix}$ twice?

3. Suppose that $R : \mathbb{R}^2 \to \mathbb{R}^2$ is the matrix transformation that rotates vectors by $180°$, as shown in Figure 2.5.9.

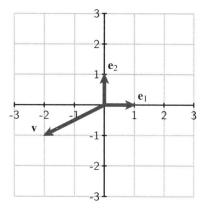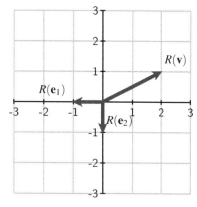

Figure 2.5.9 The matrix transformation R takes two-dimensional vectors on the left and rotates them by $180°$ into the vectors on the right.

Use Proposition 2.5.6 to find the matrix associated to R and explain why it is the same matrix associated to $(T \circ T)$.

4. Write the two-dimensional vector $(T \circ T)\left(\begin{bmatrix} x \\ y \end{bmatrix}\right)$. How might this vector be expressed in terms of scalar multiplication and why does this make sense geometrically?

b. In the previous activity, we imagined a company that operates two bakeries. We found the matrix transformation $T : \mathbb{R}^2 \to \mathbb{R}^3$ where $T\left(\begin{bmatrix} x_1 \\ x_2 \end{bmatrix}\right)$ describes the number of cakes, doughnuts, and eclairs when Bakery1 runs for x_1 hours and Bakery 2 runs for x_2 hours. The associated matrix is $A = \begin{bmatrix} 10 & 20 \\ 50 & 30 \\ 30 & 30 \end{bmatrix}$.

Suppose now that

- Each cake requires 4 cups of flour and and 2 cups of sugar.
- Each doughnut requires 1 cup of flour and 1 cup of sugar.
- Each eclair requires 1 cup of flour and 2 cups of sugar.

We will describe a matrix transformation $S : \mathbb{R}^3 \to \mathbb{R}^2$ where $S\left(\begin{bmatrix} y_1 \\ y_2 \\ y_3 \end{bmatrix}\right)$ is a two-dimensional vector describing the number of cups of flour and sugar required to make y_1 cakes, y_2 doughnuts, and y_3 eclairs.

1. Use Proposition 2.5.6 to write the matrix B associated to the transformation S.

2. If we make 1200 cakes, 2850 doughnuts, and 2250 eclairs, how many cups of flour and sugar are required?

3. Suppose that Bakery 1 operates for 75 hours and Bakery 2 operates for 53 hours. How many cakes, doughnuts, and eclairs are produced? How many cups of flour and sugar are required?

4. What is the meaning of the composition $(S \circ T)$ and what is its associated matrix?

5. In a certain time interval, both bakeries use a total of 5800 cups of flour and 5980 cups of sugar. How long have the two bakeries been operating?

2.5.3 Discrete Dynamical Systems

In Chapter 4, we will give considerable attention to a specific type of matrix transformation, which is illustrated in the next activity.

Activity 2.5.5. Suppose we run a company that has two warehouses, which we will call P and Q, and a fleet of 1000 delivery trucks. Every morning, a delivery truck goes out from one of the warehouses and returns in the evening to one of the warehouses. It is observed that

- 70% of the trucks that leave P return to P. The other 30% return to Q.
- 50% of the trucks that leave Q return to Q and 50% return to P.

The distribution of trucks is represented by the vector $\mathbf{x} = \begin{bmatrix} x_1 \\ x_2 \end{bmatrix}$ when there are x_1 trucks at location P and x_2 trucks at Q. If $\mathbf{x}$ describes the distribution of trucks in the morning, then the matrix transformation $T(\mathbf{x})$ will describe the distribution in the evening.

a. Suppose that all 1000 trucks begin the day at location P and none at Q. How many trucks are at each location that evening? Using our vector representation, what is $T\left(\begin{bmatrix} 1000 \\ 0 \end{bmatrix}\right)$?

So that we can find the matrix A associated to T, what does this tell us about $T\left(\begin{bmatrix} 1 \\ 0 \end{bmatrix}\right)$?

b. In the same way, suppose that all 1000 trucks begin the day at location Q and

2.5. MATRIX TRANSFORMATIONS

none at P. How many trucks are at each location that evening? What is the result $T\left(\begin{bmatrix} 0 \\ 1000 \end{bmatrix}\right)$ and what is $T\left(\begin{bmatrix} 0 \\ 1 \end{bmatrix}\right)$?

c. Find the matrix A such that $T(\mathbf{x}) = A\mathbf{x}$.

d. Suppose that there are 100 trucks at P and 900 at Q in the morning. How many are there at the two locations in the evening?

e. Suppose that there are 550 trucks at P and 450 at Q in the evening. How many trucks were there at the two locations that morning?

f. Suppose that all of the trucks are at location Q on Monday morning.

   1. How many trucks are at each location Monday evening?
   2. How many trucks are at each location Tuesday evening?
   3. How many trucks are at each location Wednesday evening?

g. Suppose that S is the matrix transformation that transforms the distribution of trucks $\mathbf{x}$ one morning into the distribution of trucks in the morning one week (seven days) later. What is the matrix that defines the transformation S?

As we will see later, this type of situation occurs frequently. We have a vector $\mathbf{x}$ that describes the state of some system; in this case, $\mathbf{x}$ describes the distribution of trucks between the two locations at a particular time. Then there is a matrix transformation $T(\mathbf{x}) = A\mathbf{x}$ that describes the state at some later time. We call $\mathbf{x}$ the *state* vector and T the *transition* function, as it describes the transition of the state vector from one time to the next.

Beginning with an initial state $\mathbf{x}_0$, we would like to know how the state evolves over time. For instance,

$$\mathbf{x}_1 = T(\mathbf{x}_0) = A\mathbf{x}_0$$
$$\mathbf{x}_2 = T(\mathbf{x}_1) = (T \circ T)(\mathbf{x}_0) = A^2\mathbf{x}_0$$
$$\mathbf{x}_3 = T(\mathbf{x}_2) = A^3\mathbf{x}_0$$

and so on.

We call this situation where the state of a system evolves from one time to the next according to the rule $\mathbf{x}_{k+1} = A\mathbf{x}_k$ a *discrete dynamical system*. In Chapter 4, we will develop a theory that enables us to make long-term predictions about the evolution of the state vector.

2.5.4 Summary

This section introduced matrix transformations, functions that are defined by matrix-vector multiplication, such as $T(\mathbf{x}) = A\mathbf{x}$ for some matrix A.

- If A is an $m \times n$ matrix, then $T : \mathbb{R}^n \to \mathbb{R}^m$.

- The columns of the matrix A are given by evaluating the transformation T on the vec-

tors e_j; that is,
$$A = \begin{bmatrix} T(e_1) & T(e_2) & \cdots & T(e_n) \end{bmatrix}.$$

- The composition of matrix transformations corresponds to matrix multiplication.
- A discrete dynamical system consists of a state vector x along with a transition function $T(x) = Ax$ that describes how the state vector evolves from one time to the next. Powers of the matrix A determine the long-term behavior of the state vector.

2.5.5 Exercises

1. Suppose that T is the matrix transformation defined by the matrix A and S is the matrix transformation defined by B where
$$A = \begin{bmatrix} 3 & -1 & 0 \\ 1 & 2 & 2 \\ -1 & 3 & 2 \end{bmatrix}, \quad B = \begin{bmatrix} 1 & -1 & 0 \\ 2 & 1 & 2 \end{bmatrix}.$$

    a. If $T : \mathbb{R}^n \to \mathbb{R}^m$, what are the values of m and n? What values of m and n are appropriate for the transformation S?

    b. Evaluate $T\left(\begin{bmatrix} 1 \\ -3 \\ 2 \end{bmatrix}\right)$.

    c. Evaluate $S\left(\begin{bmatrix} -2 \\ 2 \\ 1 \end{bmatrix}\right)$.

    d. Evaluate $S \circ T\left(\begin{bmatrix} 1 \\ -3 \\ 2 \end{bmatrix}\right)$.

    e. Find the matrix C that defines the matrix transformation $S \circ T$.

2. This problem concerns the identification of matrix transformations, about which more will be said in the next section.

    a. Check that the following function $T : \mathbb{R}^3 \to \mathbb{R}^2$ is a matrix transformation by finding a matrix A such that $T(x) = Ax$.
    $$T\left(\begin{bmatrix} x_1 \\ x_2 \\ x_3 \end{bmatrix}\right) = \begin{bmatrix} 3x_1 - x_2 + 4x_3 \\ 5x_2 - x_3 \end{bmatrix}.$$

    b. Explain why
    $$T\left(\begin{bmatrix} x_1 \\ x_2 \\ x_3 \end{bmatrix}\right) = \begin{bmatrix} 3x_1^4 - x_2 + 4x_3 \\ 5x_2 - x_3 \end{bmatrix}$$
    is not a matrix transformation.

2.5. MATRIX TRANSFORMATIONS

3. Suppose that the matrix
$$A = \begin{bmatrix} 1 & 3 & 1 \\ -2 & 1 & 5 \\ 0 & 2 & 2 \end{bmatrix}$$
defines the matrix transformation $T : \mathbb{R}^3 \to \mathbb{R}^3$.

   a. Describe the vectors $\mathbf{x}$ that satisfy $T(\mathbf{x}) = \mathbf{0}$.

   b. Describe the vectors $\mathbf{x}$ that satisfy $T(\mathbf{x}) = \begin{bmatrix} -8 \\ 9 \\ 2 \end{bmatrix}$.

   c. Describe the vectors $\mathbf{x}$ that satisfy $T(\mathbf{x}) = \begin{bmatrix} -8 \\ 2 \\ -4 \end{bmatrix}$.

4. Suppose $T : \mathbb{R}^3 \to \mathbb{R}^2$ is a matrix transformation with $T(\mathbf{e}_j) = \mathbf{v}_j$ where $\mathbf{v}_1$, $\mathbf{v}_2$, and $\mathbf{v}_3$ are as shown in Figure 2.5.10.

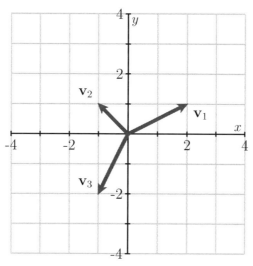

Figure 2.5.10 The vectors $T(\mathbf{e}_j) = \mathbf{v}_j$.

   a. Sketch the vector $T\left(\begin{bmatrix} 2 \\ 1 \\ 2 \end{bmatrix}\right)$.

   b. What is the vector $T\left(\begin{bmatrix} 0 \\ 1 \\ 0 \end{bmatrix}\right)$?

   c. Find all the vectors $\mathbf{x}$ such that $T(\mathbf{x}) = \mathbf{0}$.

5. In Example 2.5.5 and Example 2.5.4, we wrote matrix transformations in terms of the

components of $T(\mathbf{x})$. This exercise makes use of that form.

   a. Let's return to the example in Activity 2.5.3 concerning the company that operates two bakeries. We used a matrix transformation with input $\mathbf{x}$, which recorded the amount of time the two bakeries operated, and output $T(\mathbf{x})$, the number of cakes, doughnuts, and eclairs produced. The associated matrix is $A = \begin{bmatrix} 10 & 20 \\ 50 & 30 \\ 30 & 30 \end{bmatrix}$.

   1. If $\mathbf{x} = \begin{bmatrix} x_1 \\ x_2 \end{bmatrix}$, write the output $T(\mathbf{x})$ as a three-dimensional vector in terms of x_1 and x_2.
   2. If Bakery 1 operates for x_1 hours and Bakery 2 for x_2 hours, how many cakes are produced?
   3. Explain how you may have discovered this expression by considering the rates at which the two locations make cakes.

   b. Suppose that a bicycle sharing program has two locations P and Q. Bicycles are rented from some location in the morning and returned to a location in the evening. Suppose that

   - 60% of bicycles that begin at P in the morning are returned to P in the evening while the other 40% are returned to Q.
   - 30% of bicycles that begin at Q are returned to Q and the other 70% are returned to P.

   1. If x_1 is the number of bicycles at location P and x_2 the number at Q in the morning, write an expression for the number of bicycles at P in the evening.
   2. Write an expression for the number of bicycles at Q in the evening.
   3. Write an expression for $T\left(\begin{bmatrix} x_1 \\ x_2 \end{bmatrix}\right)$, the vector that describs the distribution of bicycles in the evening.
   4. Use this expression to identify the matrix A associated to the matrix transformation T.

6. Determine whether the following statements are true or false and provide a justification for your response.

   a. A matrix transformation $T : \mathbb{R}^4 \to \mathbb{R}^5$ is defined by $T(\mathbf{x}) = A\mathbf{x}$ where A is a 4×5 matrix.

   b. If $T : \mathbb{R}^3 \to \mathbb{R}^2$ is a matrix transformation, then there are infinitely many vectors $\mathbf{x}$ such that $T(\mathbf{x}) = \mathbf{0}$.

   c. If $T : \mathbb{R}^2 \to \mathbb{R}^3$ is a matrix transformation, then it is possible that every equation $T(\mathbf{x}) = \mathbf{b}$ has a solution for every vector $\mathbf{b}$.

   d. If $T : \mathbb{R}^n \to \mathbb{R}^m$ is a matrix transformation, then the equation $T(\mathbf{x}) = \mathbf{0}$ always has a solution.

2.5. MATRIX TRANSFORMATIONS

7. Suppose that a company has three plants, called Plants 1, 2, and 3, that produce milk M and yogurt Y. For every hour of operation,

    - Plant 1 produces 20 units of milk and 15 units of yogurt.
    - Plant 2 produces 30 units of milk and 5 units of yogurt.
    - Plant 3 produces 0 units of milk and 40 units of yogurt.

    a. Suppose that x_1, x_2, and x_3 record the amounts of time that the three plants are operated and that M and Y record the amount of milk and yogurt produced. If we write $\mathbf{x} = \begin{bmatrix} x_1 \\ x_2 \\ x_3 \end{bmatrix}$ and $\mathbf{y} = \begin{bmatrix} M \\ Y \end{bmatrix}$, find the matrix A that defines the matrix transformation $T(\mathbf{x}) = \mathbf{y}$.

    b. Furthermore, suppose that producing each unit of
        - milk requires 5 units of electricity and 8 units of labor.
        - yogurt requires 6 units of electricity and 10 units of labor.

    If we write the vector $\mathbf{z} = \begin{bmatrix} E \\ L \end{bmatrix}$ to record the required amounts of electricity E and labor L, find the matrix B that defines the matrix transformation $S(\mathbf{y}) = \mathbf{z}$.

    c. If $\mathbf{x} = \begin{bmatrix} 30 \\ 20 \\ 10 \end{bmatrix}$ describes the amounts of time that the three plants are operated, how much milk and yogurt is produced? How much electricity and labor are required?

    d. Find the matrix C that describes the matrix transformation $R(\mathbf{x}) = \mathbf{z}$ that gives the required amounts of electricity and labor when the each plants is operated an amount of time given by the vector $\mathbf{x}$.

8. Suppose that $T : \mathbb{R}^2 \to \mathbb{R}^2$ is a matrix transformation and that

    $$T\left(\begin{bmatrix} 1 \\ 1 \end{bmatrix}\right) = \begin{bmatrix} 3 \\ -2 \end{bmatrix}, \quad T\left(\begin{bmatrix} -1 \\ 1 \end{bmatrix}\right) = \begin{bmatrix} 1 \\ 2 \end{bmatrix}.$$

    a. Find the vector $T\left(\begin{bmatrix} 1 \\ 0 \end{bmatrix}\right)$.

    b. Find the matrix A that defines T.

    c. Find the vector $T\left(\begin{bmatrix} 4 \\ -5 \end{bmatrix}\right)$.

9. Suppose that two species P and Q interact with one another and that we measure their populations every month. We record their populations in a state vector $\mathbf{x} = \begin{bmatrix} p \\ q \end{bmatrix}$,

where p and q are the populations of P and Q, respectively. We observe that there is a matrix

$$A = \begin{bmatrix} 0.8 & 0.3 \\ 0.7 & 1.2 \end{bmatrix}$$

such that the matrix transformation $T(\mathbf{x}) = A\mathbf{x}$ is the transition function describing how the state vector evolves from month to month. We also observe that, at the beginning of July, the populations are described by the state vector $\mathbf{x} = \begin{bmatrix} 1 \\ 2 \end{bmatrix}$.

a. What will the populations be at the beginning of August?

b. What were the populations at the beginning of June?

c. What will the populations be at the beginning of December?

d. What will the populations be at the beginning of July in the following year?

10. Students in a school are sometimes absent due to an illness. Suppose that

- 95% of the students who attend school will attend school the next day.
- 50% of the students who are absent one day will be absent the next day.

We will record the number of present students p and the number of absent students a in a state vector $\mathbf{x} = \begin{bmatrix} p \\ a \end{bmatrix}$ and note that that state vector evolves from one day to the next according to the transition function $T : \mathbb{R}^2 \to \mathbb{R}^2$. On Tuesday, the state vector is $\mathbf{x} = \begin{bmatrix} 1700 \\ 100 \end{bmatrix}$.

a. Suppose we initially have 1000 students who are present and none absent. Find $T\left(\begin{bmatrix} 1000 \\ 0 \end{bmatrix}\right)$.

b. Suppose we initially have 1000 students who are absent and none present. Find $T\left(\begin{bmatrix} 0 \\ 1000 \end{bmatrix}\right)$.

c. Use the results of parts a and b to find the matrix A that defines the matrix transformation T.

d. If $\mathbf{x} = \begin{bmatrix} 1700 \\ 100 \end{bmatrix}$ on Tuesday, how are the students distributed on Wednesday?

e. How many students were present on Monday?

f. How many students are present on the following Tuesday?

g. What happens to the number of students who are present after a very long time?

2.6 The geometry of matrix transformations

Matrix transformations, which we explored in the last section, allow us to describe certain functions $T : \mathbb{R}^n \to \mathbb{R}^m$. In this section, we will demonstrate how matrix transformations provide a convenient way to describe geometric operations, such as rotations, reflections, and scalings. We will then explore how matrix transformations are used in computer animation.

> **Preview Activity 2.6.1.** We will describe the matrix transformation T that reflects 2-dimensional vectors across the horizontal axis. For instance, Figure 2.6.1 illustrates how a vector **x** is reflected onto the vector $T(\mathbf{x})$.
>
>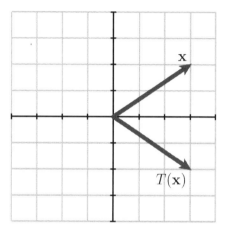
>
> **Figure 2.6.1** A vector **x** and its reflection $T(\mathbf{x})$ across the horizontal axis.
>
> a. If $\mathbf{x} = \begin{bmatrix} 2 \\ 4 \end{bmatrix}$, what is the vector $T(\mathbf{x})$? Sketch the vectors **x** and $T(\mathbf{x})$.
>
> b. More generally, if $\mathbf{x} = \begin{bmatrix} x \\ y \end{bmatrix}$, what is $T(\mathbf{x})$?
>
> c. Find the vectors $T\left(\begin{bmatrix} 1 \\ 0 \end{bmatrix}\right)$ and $T\left(\begin{bmatrix} 0 \\ 1 \end{bmatrix}\right)$.
>
> d. Use your results to write the matrix A so that $T(\mathbf{x}) = A\mathbf{x}$. Then verify that $T\left(\begin{bmatrix} x \\ y \end{bmatrix}\right)$ agrees with what you found in part b.
>
> e. Describe the transformation that results from composing T with itself; that is, what is the transformation $T \circ T$? Explain how matrix multiplication can be used to justify your response.

2.6.1 The geometry of 2×2 matrix transformations

We have now seen how a few geometric operations, such as rotations and reflections, can be described using matrix transformations. The following activity shows, more generally, that matrix transformations can perform a variety of important geometric operations.

Activity 2.6.2 Using matrix transformations to describe geometric operations..

This activity uses an interactive diagram that is available at gvsu.edu/s/0Jf.

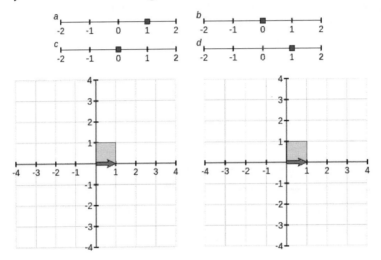

Figure 2.6.2 The matrix transformation T transforms features shown on the left into features shown on the right.

For the following 2×2 matrices A, use the diagram to study the effect of the corresponding matrix transformation $T(\mathbf{x}) = A\mathbf{x}$. For each transformation, describe the geometric effect the transformation has on the plane.

a. $A = \begin{bmatrix} 2 & 0 \\ 0 & 1 \end{bmatrix}$.

b. $A = \begin{bmatrix} 2 & 0 \\ 0 & 2 \end{bmatrix}$.

c. $A = \begin{bmatrix} 0 & 1 \\ -1 & 0 \end{bmatrix}$.

d. $A = \begin{bmatrix} 1 & 1 \\ 0 & 1 \end{bmatrix}$.

e. $A = \begin{bmatrix} -1 & 0 \\ 0 & 1 \end{bmatrix}$.

2.6. THE GEOMETRY OF MATRIX TRANSFORMATIONS

f. $A = \begin{bmatrix} 1 & 0 \\ 0 & 0 \end{bmatrix}$.

g. $A = \begin{bmatrix} 1 & 0 \\ 0 & 1 \end{bmatrix}$.

h. $A = \begin{bmatrix} 1 & -1 \\ -2 & 2 \end{bmatrix}$.

The previous activity presented some examples showing that matrix transformations can perform interesting geometric operations, such as rotations, scalings, and reflections. Before we go any further, we should explain why it is possible to represent these operations by matrix transformations. In fact, we ask more generally: what types of functions $T : \mathbb{R}^n \to \mathbb{R}^m$ are represented as matrix transformations?

The linearity of matrix-vector multiplication provides the key to answering this question. Remember that if A is a matrix, $\mathbf{v}$ and $\mathbf{w}$ vectors, and c a scalar, then

$$A(c\mathbf{v}) = cA\mathbf{v}$$
$$A(\mathbf{v} + \mathbf{w}) = A\mathbf{v} + A\mathbf{w}.$$

This means that a matrix transformation $T(\mathbf{x}) = A\mathbf{x}$ satisfies the corresponding linearity property:

Linearity of Matrix Transformations.

$$T(c\mathbf{v}) = cT(\mathbf{v})$$
$$T(\mathbf{v} + \mathbf{w}) = T(\mathbf{v}) + T(\mathbf{w}).$$

It turns out that, if $T : \mathbb{R}^n \to \mathbb{R}^m$ satisfies these two linearity properties, then we can find a matrix A such that $T(\mathbf{x}) = A\mathbf{x}$. In fact, Proposition 2.5.6 tells us how to form A; we simply write

$$A = \begin{bmatrix} T(\mathbf{e}_1) & T(\mathbf{e}_2) & \dots T(\mathbf{e}_n) \end{bmatrix}.$$

We will now check that $T(\mathbf{x}) = A\mathbf{x}$ using the linearity of T:

$$T(\mathbf{x}) = T\left(\begin{bmatrix} x_1 \\ x_2 \\ \vdots \\ x_n \end{bmatrix}\right) = T(x_1\mathbf{e}_1 + x_2\mathbf{e}_2 + \ldots + x_n\mathbf{e}_n)$$

$$= x_1 T(\mathbf{e}_1) + x_2 T(\mathbf{e}_2) + \ldots + x_n T(\mathbf{e}_n)$$

$$= x_1 A\mathbf{e}_1 + x_2 A\mathbf{e}_2 + \ldots + x_n A\mathbf{e}_n$$

$$= A(x_1\mathbf{e}_1 + x_2\mathbf{e}_2 + \ldots + x_n\mathbf{e}_n)$$

$$= A\begin{bmatrix} x_1 \\ x_2 \\ \vdots \\ x_n \end{bmatrix}$$

$$= A\mathbf{x}$$

The result is the following proposition.

Proposition 2.6.3 *The function $T : \mathbb{R}^n \to \mathbb{R}^m$ is a matrix transformation where $T(\mathbf{x}) = A\mathbf{x}$ for some $m \times n$ matrix A if and only if*

$$T(c\mathbf{v}) = cT(\mathbf{v})$$
$$T(\mathbf{v} + \mathbf{w}) = T(\mathbf{v}) + T(\mathbf{w}).$$

In this case, A is the matrix whose columns are $T(\mathbf{e}_j)$; that is,

$$A = \begin{bmatrix} T(\mathbf{e}_1) & T(\mathbf{e}_2) & \ldots & T(\mathbf{e}_n) \end{bmatrix}.$$

Said simply, this proposition means says that if have a function $T : \mathbb{R}^n \to \mathbb{R}^m$ and can verify the two linearity properties stated in the proposition, then we know that T is a matrix transformation. Let's see how this works in practice.

Example 2.6.4 We will consider the function $T : \mathbb{R}^2 \to \mathbb{R}^2$ that rotates a vector $\mathbf{x}$ by 45° in the counterclockwise direction to obtain $T(\mathbf{x})$ as seen in Figure 2.6.5.

2.6. THE GEOMETRY OF MATRIX TRANSFORMATIONS

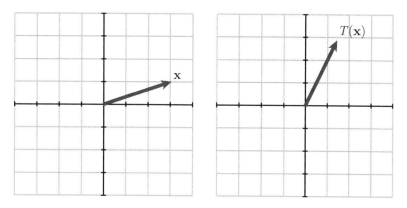

Figure 2.6.5 The function T rotates a vector counterclockwise by $45°$.

We first need to know that T can be represented by a matrix transformation, which means, by Proposition 2.6.3, that we need to verify the linearity properties:

$$T(c\mathbf{v}) = cT(\mathbf{v})$$
$$T(\mathbf{v} + \mathbf{w}) = T(\mathbf{v}) + T(\mathbf{w}).$$

The next two figures illustrate why these properties hold. For instance, Figure 2.6.6 shows the relationship between $T(\mathbf{v})$ and $T(c\mathbf{v})$ when c is a scalar. In particular, scaling a vector and then rotating it is the same as rotating and then scaling it, which means that $T(c\mathbf{v}) = cT(\mathbf{v})$.

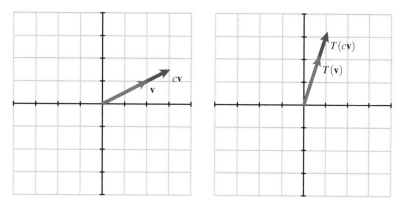

Figure 2.6.6 We see that the vector $T(c\mathbf{v})$ is a scalar multiple to $T(\mathbf{v})$ so that $T(c\mathbf{v}) = cT(\mathbf{v})$.

Similarly, Figure 2.6.7 shows the relationship between $T(\mathbf{v} + \mathbf{w})$, $T(\mathbf{v})$, and $T(\mathbf{w})$. Remember that the sum of two vectors is represented by the diagonal of the parallelogram defined by the two vectors. The rotation T has the effect of rotating the parallelogram defined by $\mathbf{v}$ and $\mathbf{w}$ into the parallelogram defined by $T(\mathbf{v})$ and $T(\mathbf{w})$, explaining why $T(\mathbf{v} + \mathbf{w}) = T(\mathbf{v}) + T(\mathbf{w})$.

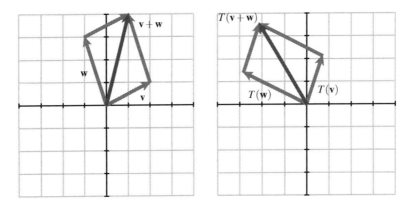

Figure 2.6.7 We see that the vector $T(\mathbf{v} + \mathbf{w})$ is the sum of $T(\mathbf{v})$ and $T(\mathbf{w})$ so that $T(\mathbf{v} + \mathbf{w}) = T(\mathbf{v}) + T(\mathbf{w})$.

Having verified these two properties, we now know that the function T that rotates vectors by $45°$ is a matrix transformation. We may therefore write it as $T(\mathbf{x}) = A\mathbf{x}$ where A is the 2×2 matrix $A = \begin{bmatrix} T(\mathbf{e}_1) & T(\mathbf{e}_2) \end{bmatrix}$. The columns of this matrix, $T(\mathbf{e}_1)$ and $T(\mathbf{e}_2)$, are shown on the right of Figure 2.6.8.

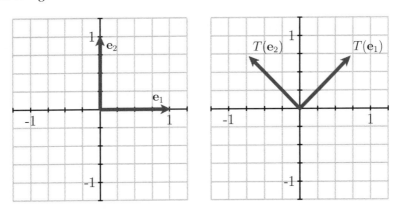

Figure 2.6.8 The matrix transformation T rotates $\mathbf{e}_1$ and $\mathbf{e}_2$ by $45°$.

Notice that $T(\mathbf{e}_1)$ forms an isosceles right triangle, as shown in Figure 2.6.9. Since the length of $\mathbf{e}_1$ is 1, the length of $T(\mathbf{e}_1)$, the hypotenuse of the triangle, is also 1, and by Pythagoras' theorem, the lengths of its legs are $1/\sqrt{2}$.

2.6. THE GEOMETRY OF MATRIX TRANSFORMATIONS

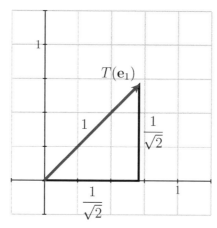

Figure 2.6.9 The vector $T(\mathbf{e}_1)$ has length 1 and is the hypotenuse of a right isosceles triangle.

This leads to $T(\mathbf{e}_1) = \begin{bmatrix} \frac{1}{\sqrt{2}} \\ \frac{1}{\sqrt{2}} \end{bmatrix}$. In the same way, we find that $T(\mathbf{e}_2) = \begin{bmatrix} -\frac{1}{\sqrt{2}} \\ \frac{1}{\sqrt{2}} \end{bmatrix}$ so that the matrix A is

$$A = \begin{bmatrix} \frac{1}{\sqrt{2}} & -\frac{1}{\sqrt{2}} \\ \frac{1}{\sqrt{2}} & \frac{1}{\sqrt{2}} \end{bmatrix}.$$

You may wish to check this using the interactive diagram in the previous activity using the approximation $1/\sqrt{2} \approx 0.7$.

In this example, we found that T, a function describing a rotation in the plane, was in fact a matrix transformation by checking that

$$T(c\mathbf{v}) = cT(\mathbf{v})$$
$$T(\mathbf{v} + \mathbf{w}) = T(\mathbf{v}) + T(\mathbf{w}).$$

The same kind of thinking applies more generally to show that rotations, reflections, and scalings are matrix transformations. Similarly, we could revisit the functions in Activity 2.5.3 and verify that they are matrix transformations.

> **Activity 2.6.3.** In this activity, we seek to describe various matrix transformations by finding the matrix that gives the desired transformation. All of the transformations that we study here have the form $T : \mathbb{R}^2 \to \mathbb{R}^2$.
>
> a. Find the matrix of the transformation that has no effect on vectors; that is, $T(\mathbf{x}) = \mathbf{x}$.
>
> b. Find the matrix of the transformation that reflects vectors in $\mathbb{R}^2$ across the line $y = x$.
>
> c. What is the result of composing the reflection you found in the previous part with itself; that is, what is the effect of reflecting across the line $y = x$ and then reflecting across this line again? Provide a geometric explanation for your result as well as an algebraic one obtained by multiplying matrices.

d. Find the matrix that rotates vectors counterclockwise in the plane by 90°.

e. Compare the result of rotating by 90° and then reflecting in the line $y = x$ to the result of first reflecting in $y = x$ and then rotating 90°.

f. Find the matrix that results from composing a 90° rotation with itself four times; that is, if T is the matrix transformation that rotates vectors by 90°, find the matrix for $T \circ T \circ T \circ T$. Explain why your result makes sense geometrically.

g. Explain why the matrix that rotates vectors counterclockwise by an angle θ is

$$\begin{bmatrix} \cos\theta & -\sin\theta \\ \sin\theta & \cos\theta \end{bmatrix}.$$

2.6.2 Matrix transformations and computer animation

Linear algebra plays a significant role in computer animation. We will now illustrate how matrix transformations and some of the ideas we have developed in this section are used by computer animators to create the illusion of motion in their characters.

Figure 2.6.10 shows a test character used by Pixar animators. On the left is the original definition of the character; on the right, we see that the character has been moved into a different pose. To make it appear that the character is moving, animators create a sequence of frames in which the character's pose is modified slightly from one frame to the next often using matrix transformations.

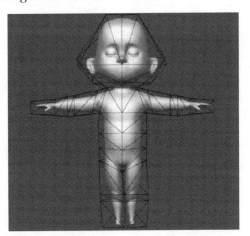 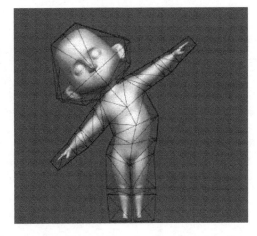

Figure 2.6.10 Computer animators define a character and create motion by drawing it in a sequence of poses. © Disney/Pixar

2.6. THE GEOMETRY OF MATRIX TRANSFORMATIONS

Of course, realistic characters will be drawn in three-dimensions. To keep things a little more simple, however, we will look at this two-dimensional character and devise matrix transformations that move them into different poses.

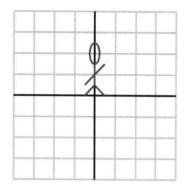

Of course, the first thing we may wish to do is simply move them to a different position in the plane, such as that shown in Figure 2.6.11. Motions like this are called *translations*.

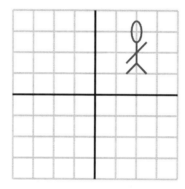

Figure 2.6.11 Translating our character to a new position in the plane.

This presents a problem because a matrix transformation $T : \mathbb{R}^2 \to \mathbb{R}^2$ has the property that $T(0) = A0 = 0$. This means that a matrix transformation cannot move the origin of the coordinate plane. To address this restriction, animators use *homogeneous coordinates*, which are formed by placing the two-dimensional coordinate plane inside $\mathbb{R}^3$ as the plane $z = 1$, as shown in Figure 2.6.12.

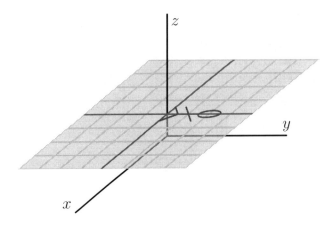

Figure 2.6.12 Include the two-dimensional coordinate plane in $\mathbb{R}^3$ as the plane $z = 1$ so that we can translate the character.

As a result, rather than describing points in the plane as vectors $\begin{bmatrix} x \\ y \end{bmatrix}$, we describe them as three-dimensional vectors $\begin{bmatrix} x \\ y \\ 1 \end{bmatrix}$. As we see in the next activity, this allows us to translate our character in the plane.

> **Activity 2.6.4.** In this activity, we will use homogeneous coordinates and matrix transformations to move our character into a variety of poses.
>
> a. Since we regard our character as living in $\mathbb{R}^3$, we will consider matrix transformations defined by matrices
>
> $$\begin{bmatrix} a & b & c \\ d & e & f \\ 0 & 0 & 1 \end{bmatrix}.$$
>
> Verify that such a matrix transformation transforms points in the plane $z = 1$ into points in the same plane; that is, verify that
>
> $$\begin{bmatrix} a & b & c \\ d & e & f \\ 0 & 0 & 1 \end{bmatrix} \begin{bmatrix} x \\ y \\ 1 \end{bmatrix} = \begin{bmatrix} x' \\ y' \\ 1 \end{bmatrix}.$$
>
> Express the coordinates of the resulting point x' and y' in terms of the coordinates of the original point x and y.

2.6. THE GEOMETRY OF MATRIX TRANSFORMATIONS   133

This activity uses an interactive diagram that is available at gvsu.edu/s/0Jb. Using the six sliders, you may choose the matrix $\begin{bmatrix} a & b & c \\ d & e & f \\ 0 & 0 & 1 \end{bmatrix}$ that will move our character in the plane.

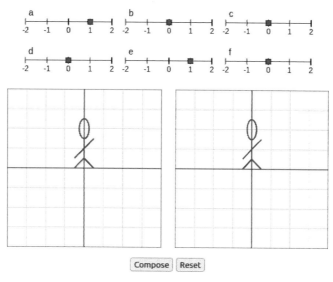

Figure 2.6.13 An interactive diagram that allows us to move the character using homogeneous coordinates.

b. Find the matrix transformation that translates our character to a new position in the plane, as shown in Figure 2.6.14

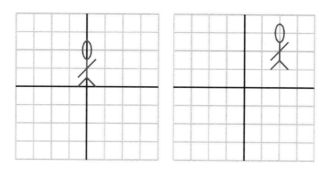

Figure 2.6.14 Translating to a new position.

c. As originally drawn, our character is waving with one of their hands. In one of the movie's scenes, we would like them to wave with their other hand, as shown in Figure 2.6.15. Find the matrix transformation that moves them into this pose.

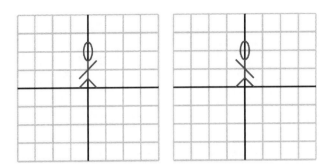

Figure 2.6.15 Waving with the other hand.

d. Later, our character performs a cartwheel by moving through the sequence of poses shown in Figure 2.6.16. Find the matrix transformations that create these poses.

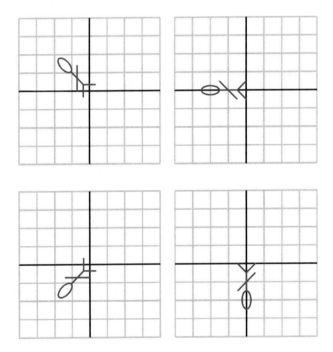

Figure 2.6.16 Performing a cartwheel.

e. Next, we would like to find the transformations that zoom in on our character's face, as shown in Figure 2.6.17. To do this, you should think about composing matrix transformations. This can be accomplished in the diagram by using the *Compose* button, which makes the current pose, displayed on the right, the new beginning pose, displayed on the left. What is the matrix transformation that moves the character from the original pose, shown in the upper left, to the final pose, shown in the lower right?

2.6. THE GEOMETRY OF MATRIX TRANSFORMATIONS 135

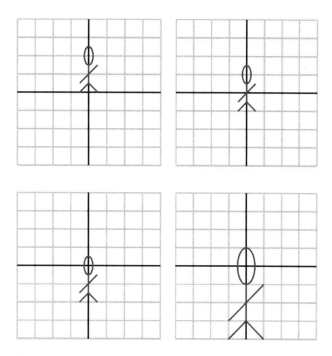

Figure 2.6.17 Zooming in on our characters' face.

f. We would also like to create our character's shadow, shown in the sequence of poses in Figure 2.6.18. Find the sequence of matrix transformations that achieves this. In particular, find the matrix transformation that takes our character from their original pose to their shadow in the lower right.

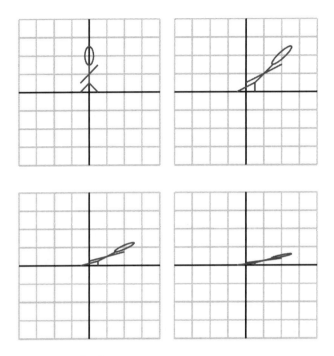

Figure 2.6.18 Casting a shadow.

g. Write a final scene to the movie and describe how to construct a sequence of matrix transformations that create your scene.

2.6.3 Summary

This section explored how geometric operations are performed by matrix transformations.

- A function $T : \mathbb{R}^n \to \mathbb{R}^m$ is a matrix transformation if and only if these properties are satisfied:
$$T(c\mathbf{v}) = cT(\mathbf{v})$$
$$T(\mathbf{v} + \mathbf{w}) = T(\mathbf{v}) + T(\mathbf{w}).$$

- Geometric operations, such as rotations, reflections, and scalings, can be represented as matrix transformations.

- Composing geometric operations corresponds to matrix multiplication.

- Computer animators use homogeneous coordinates and matrix transformations to create the illusion of motion.

2.6. THE GEOMETRY OF MATRIX TRANSFORMATIONS

2.6.4 Exercises

1. For each of the following geometric operations in the plane, find a 2×2 matrix that defines the matrix transformation performing the operation.

   a. Rotates vectors by $180°$.

   b. Reflects vectors across the vertical axis.

   c. Reflects vectors across the line $y = -x$.

   d. Rotates vectors counterclockwise by $60°$.

   e. First rotates vectors counterclockwise by $60°$ and then reflects in the line $y = x$.

2. This exercise investigates the composition of reflections in the plane.

   a. Find the result of first reflecting across the line $y = 0$ and then $y = x$. What familiar operation is the cumulative effect of this composition?

   b. What happens if you compose the operations in the opposite order; that is, what happens if you first reflect across $y = x$ and then $y = 0$? What familiar operation results?

   c. What familiar geometric operation results if you first reflect across the line $y = x$ and then $y = -x$?

   d. What familiar geometric operation results if you first rotate by $90°$ and then reflect across the line $y = x$?

   It is a general fact that the composition of two reflections results in a rotation through twice the angle from the first line of reflection to the second. We will investigate this more generally in Exercise 2.6.4.8

3. Shown below in Figure 2.6.19 are the vectors e_1, e_2, and e_3 in $\mathbb{R}^3$.

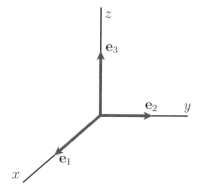

Figure 2.6.19 The vectors e_1, e_2, and e_3 in $\mathbb{R}^3$.

   a. Imagine that the thumb of your right hand points in the direction of e_1. A positive rotation about the x axis corresponds to a rotation in the direction in which your fingers point. Find the matrix definining the matrix transformation T that rotates vectors by 90° around the x-axis.

b. In the same way, find the matrix that rotates vectors by 90° around the y-axis.

c. Find the matrix that rotates vectors by 90° around the z-axis.

d. What is the cumulative effect of rotating by 90° about the x-axis, followed by a 90° rotation about the y-axis, followed by a −90° rotation about the x-axis.

4. If a matrix transformation performs a geometric operation, we would like to find a matrix transformation that undoes that operation.

   a. Suppose that $T : \mathbb{R}^2 \to \mathbb{R}^2$ is the matrix transformation that rotates vectors by 90°. Find a matrix transformation $S : \mathbb{R}^2 \to \mathbb{R}^2$ that undoes the rotation; that is, S takes $T(\mathbf{x})$ back into $\mathbf{x}$ so that $(S \circ T)(\mathbf{x}) = \mathbf{x}$. Think geometrically about what the transformation S should be and then verify it algebraically.

   We say that S is the *inverse* of T and we will write it as T^{-1}.

   b. Verify algebraically that the reflection $R : \mathbb{R}^2 \to \mathbb{R}^2$ across the line $y = x$ is its own inverse; that is, $R^{-1} = R$.

   c. The matrix transformation $T : \mathbb{R}^2 \to \mathbb{R}^2$ defined by the matrix

   $$A = \begin{bmatrix} 1 & 1 \\ 0 & 1 \end{bmatrix}$$

   is called a *shear*. Find the inverse of T.

   d. Describe the geometric effect of the matrix transformation defined by

   $$A = \begin{bmatrix} \frac{1}{2} & 0 \\ 0 & 3 \end{bmatrix}$$

   and then find its inverse.

5. We have seen that the matrix

   $$\begin{bmatrix} \cos\theta & -\sin\theta \\ \sin\theta & \cos\theta \end{bmatrix}$$

   performs a rotation through an angle θ about the origin. Suppose instead that we would like to rotate by 90° about the point $(1,2)$. Using homogeneous coordinates, we will develop a matrix that performs this operation.

   Our strategy is to

   - begin with a vector whose tail is at the point $(1,2)$,
   - translate the vector so that its tail is at the origin,
   - rotate by 90°, and
   - translate the vector so that its tail is back at $(1,2)$.

2.6. THE GEOMETRY OF MATRIX TRANSFORMATIONS

This is shown in Figure 2.6.20.

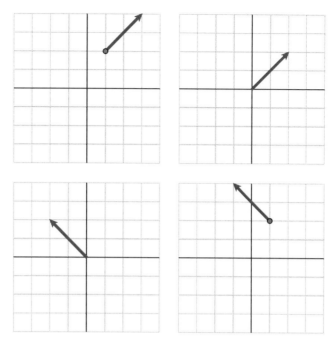

Figure 2.6.20 A sequence of matrix transformations that, when read right to left and top to bottom, rotate a vector about the point $(1,2)$.

Remember that, when working with homogeneous coordinates, we consider matrices of the form
$$\begin{bmatrix} a & b & c \\ d & e & f \\ 0 & 0 & 1 \end{bmatrix}.$$

a. The first operation is a translation by $(-1, -2)$. Find the matrix that performs this translation.

b. The second operation is a 90° rotation about the origin. Find the matrix that performs this rotation.

c. The third operation is a translation by $(1, 2)$. Find the matrix that performs this translation.

d. Use these matrices to find the matrix that performs a 90° rotation about $(1, 2)$.

e. Use your matrix to determine where the point $(-10, 5)$ ends up if rotated by 90° about the $(1, 2)$.

6. Consider the matrix transformation $T : \mathbb{R}^2 \to \mathbb{R}^2$ that assigns to a vector $\mathbf{x}$ the closest vector on horizontal axis as illustrated in Figure 2.6.21. This transformation is called the *projection* onto the horizontal axis. You may imagine $T(\mathbf{x})$ as the shadow cast by $\mathbf{x}$ from a flashlight far up on the positive y-axis.

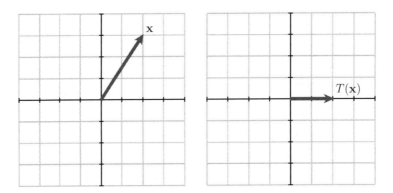

Figure 2.6.21 Projection onto the x-axis.

  a. Find the matrix that defines this matrix transformation T.

  b. Find the matrix that defines projection on the vertical axis.

  c. What is the result of composing the projection onto the horizontal axis with the projection onto the vertical axis?

  d. Find the matrix that defines projection onto the line $y = x$.

7. This exericse concerns the matrix transformations defined by matrices of the form

$$A = \begin{bmatrix} a & -b \\ b & a \end{bmatrix}.$$

Let's begin by looking at two special types of these matrices.

  a. First, consider the matrix where $a = 2$ and $b = 0$ so that

$$A = \begin{bmatrix} 2 & 0 \\ 0 & 2 \end{bmatrix}.$$

Describe the geometric effect of this matrix. More generally, suppose we have

$$A = \begin{bmatrix} r & 0 \\ 0 & r \end{bmatrix},$$

where r is a positive number. What is the geometric effect of A on vectors in the plane?

  b. Suppose now that $a = 0$ and $b = 1$ so that

$$A = \begin{bmatrix} 0 & -1 \\ 1 & 0 \end{bmatrix}.$$

What is the geometric effect of A on vectors in the plane? More generally, suppose we have

$$A = \begin{bmatrix} \cos\theta & -\sin\theta \\ \sin\theta & \cos\theta \end{bmatrix}.$$

What is the geometric effect of A on vectors in the plane?

2.6. THE GEOMETRY OF MATRIX TRANSFORMATIONS

c. In general, the composition of matrix transformation depends on the order in which we compose them. For these transformations, however, it is not the case. Check this by verifying that

$$\begin{bmatrix} r & 0 \\ 0 & r \end{bmatrix} \begin{bmatrix} \cos\theta & -\sin\theta \\ \sin\theta & \cos\theta \end{bmatrix} = \begin{bmatrix} \cos\theta & -\sin\theta \\ \sin\theta & \cos\theta \end{bmatrix} \begin{bmatrix} r & 0 \\ 0 & r \end{bmatrix}.$$

d. Let's now look at the general case where

$$A = \begin{bmatrix} a & -b \\ b & a \end{bmatrix}.$$

We will draw the vector $\begin{bmatrix} a \\ b \end{bmatrix}$ in the plane and express it using polar coordinates r and θ as shown in Figure 2.6.22.

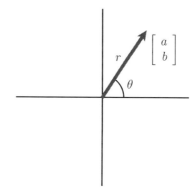

Figure 2.6.22 A vector may be expressed in polar coordinates.

We then have

$$\begin{bmatrix} a \\ b \end{bmatrix} = \begin{bmatrix} r\cos\theta \\ r\sin\theta \end{bmatrix}.$$

Show that the matrix

$$\begin{bmatrix} a & -b \\ b & a \end{bmatrix} = \begin{bmatrix} r & 0 \\ 0 & r \end{bmatrix} \begin{bmatrix} \cos\theta & -\sin\theta \\ \sin\theta & \cos\theta \end{bmatrix}.$$

e. Using this description, describe the geometric effect on vectors in the plane of the matrix transformation defined by

$$A = \begin{bmatrix} a & -b \\ b & a \end{bmatrix}.$$

f. Suppose we have a matrix transformation T defined by a matrix A and another transformation S defined by B where

$$A = \begin{bmatrix} a & -b \\ b & a \end{bmatrix}, \quad B = \begin{bmatrix} c & -d \\ d & c \end{bmatrix}.$$

Describe the geometric effect of the composition $S \circ T$ in terms of the a, b, c, and d.

The matrices of this form give a model for the complex numbers and will play an important role in Section 4.4.

8. We saw earlier that the rotation in the plane through an angle θ is given by the matrix:

$$\begin{bmatrix} \cos\theta & -\sin\theta \\ \sin\theta & \cos\theta \end{bmatrix}.$$

We would like to find a similar expression for the matrix that represents the reflection across L_θ, the line passing through the origin and making an angle of θ with the positive x-axis, as shown in Figure 2.6.23.

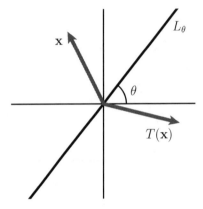

Figure 2.6.23 The reflection across the line L_θ.

a. To do this, notice that this reflection can be obtained by composing three separate transformations as shown in Figure 2.6.24. Beginning with the vector **x**, we apply the transformation R to rotate by $-\theta$ and obtain $R(\mathbf{x})$. Next, we apply S, a reflection in the horizontal axis, followed by T, a rotation by θ. We see that $T(S(R(\mathbf{x})))$ is the same as the reflection of **x** in the original line L_θ.

2.6. THE GEOMETRY OF MATRIX TRANSFORMATIONS

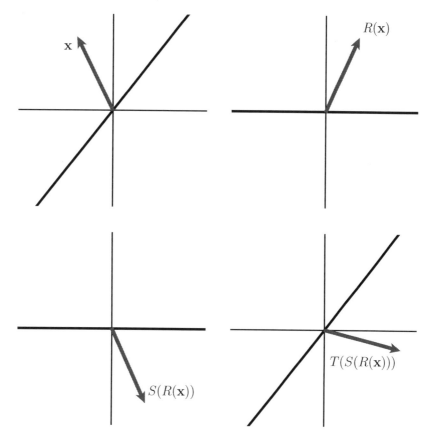

Figure 2.6.24 Reflection in the line L_θ as a composition of three transformations.

Using this decomposition, show that the reflection in the line L_θ is described by the matrix

$$\begin{bmatrix} \cos(2\theta) & \sin(2\theta) \\ \sin(2\theta) & -\cos(2\theta) \end{bmatrix}.$$

You will need to remember the trigonometric identities:

$$\cos(2\theta) = \cos^2 \theta - \sin^2 \theta$$
$$\sin(2\theta) = 2 \sin \theta \cos \theta$$

b. Now that we have a matrix that describes the reflection in the line L_θ, show that the composition of the reflection in the horizontal axis followed by the reflection in L_θ is a counterclockwise rotation by an angle 2θ. We saw some examples of this earlier in Exercise 2.6.4.2.

CHAPTER 3

Invertibility, bases, and coordinate systems

In Chapter 2, we examined the two fundamental questions concerning the existence and uniqueness of solutions to linear systems independently of one another. We found that every equation of the form $A\mathbf{x} = \mathbf{b}$ has a solution when the span of the columns of A is $\mathbb{R}^m$. We also found that the solution $\mathbf{x} = \mathbf{0}$ of the homogeneous equation $A\mathbf{x} = \mathbf{0}$ is unique when the columns of A are linearly independent. In this chapter, we explore the situation in which these two conditions hold simultaneously.

3.1 Invertibility

Up to this point, we have used the Gaussian elimination algorithm to find solutions to linear systems. We now investigate another way to find solutions to the equation $A\mathbf{x} = \mathbf{b}$ when the matrix A has the same number of rows and columns. To get started, let's look at some familiar examples.

Preview Activity 3.1.1.

a. Explain how you would solve the equation $3x = 5$ using multiplication rather than division.

b. Find the 2×2 matrix A that rotates vectors counterclockwise by $90°$.

c. Find the 2×2 matrix B that rotates vectors *clockwise* by $90°$.

d. What do you expect the product AB to be? Explain the reasoning behind your expectation and then compute AB to verify it.

e. Solve the equation $A\mathbf{x} = \begin{bmatrix} 3 \\ -2 \end{bmatrix}$ using Gaussian elimination.

f. Explain why your solution may also be found by computing $\mathbf{x} = B \begin{bmatrix} 3 \\ -2 \end{bmatrix}$.

3.1.1 Invertible matrices

The preview activity began with a familiar type of equation, $3x = 5$, and asked for a strategy to solve it. One possible response is to divide both sides by 3. Instead, let's rephrase this as multiplying by $3^{-1} = \frac{1}{3}$, the multiplicative inverse of 3.

Now that we are interested in solving equations of the form $A\mathbf{x} = \mathbf{b}$, we might try to find a similar approach. Is there a matrix A^{-1} that plays the role of the multiplicative inverse of A? Of course, the real number 0 does not have a multiplicative inverse so we probably shouldn't expect every matrix to have a multiplicative inverse. We will see, however, that many do.

Definition 3.1.1 An $n \times n$ matrix A is called *invertible* if there is a matrix B such that $AB = I_n$, where I_n is the $n \times n$ identity matrix. The matrix B is called the *inverse* of A and denoted A^{-1}.

Notice that we only define invertibility for matrices that have the same number of rows and columns in which case we say that the matrix is *square*.

Example 3.1.2 Suppose that A is the matrix that rotates two-dimensional vectors counter-clockwise by $90°$ and that B rotates vectors by $-90°$. We have

$$A = \begin{bmatrix} 0 & -1 \\ 1 & 0 \end{bmatrix}, \quad B = \begin{bmatrix} 0 & 1 \\ -1 & 0 \end{bmatrix}.$$

We can check that

$$AB = \begin{bmatrix} 0 & -1 \\ 1 & 0 \end{bmatrix} \begin{bmatrix} 0 & 1 \\ -1 & 0 \end{bmatrix} = \begin{bmatrix} 1 & 0 \\ 0 & 1 \end{bmatrix} = I$$

which shows that A is invertible and that $A^{-1} = B$.

Notice that if we multiply the matrices in the opposite order, we find that $BA = I$, which says that B is also invertible and that $B^{-1} = A$. In other words, A and B are inverses of each other.

Activity 3.1.2. This activity demonstrates a procedure for finding the inverse of a matrix A.

a. Suppose that $A = \begin{bmatrix} 3 & -2 \\ 1 & -1 \end{bmatrix}$. To find an inverse B, we write its columns as $B = \begin{bmatrix} \mathbf{b}_1 & \mathbf{b}_2 \end{bmatrix}$ and require that

$$AB = I$$

$$\begin{bmatrix} A\mathbf{b}_1 & A\mathbf{b}_2 \end{bmatrix} = \begin{bmatrix} 1 & 0 \\ 0 & 1 \end{bmatrix}.$$

In other words, we can find the columns of B by solving the equations

$$A\mathbf{b}_1 = \begin{bmatrix} 1 \\ 0 \end{bmatrix}, \quad A\mathbf{b}_2 = \begin{bmatrix} 0 \\ 1 \end{bmatrix}.$$

Solve these equations to find $\mathbf{b}_1$ and $\mathbf{b}_2$. Then write the matrix B and verify that $AB = I$. This is enough for us to conclude that B is the inverse of A.

3.1. INVERTIBILITY

b. Find the product BA and explain why we now know that B is invertible and $B^{-1} = A$.

c. What happens when you try to find the inverse of $C = \begin{bmatrix} -2 & 1 \\ 4 & -2 \end{bmatrix}$?

d. We now develop a condition that must be satisfied by an invertible matrix. Suppose that A is an invertible $n \times n$ matrix with inverse B and suppose that $\mathbf{b}$ is any n-dimensional vector. Since $AB = I$, we have

$$A(B\mathbf{b}) = (AB)\mathbf{b} = I\mathbf{b} = \mathbf{b}.$$

This says that the equation $A\mathbf{x} = \mathbf{b}$ is consistent and that $\mathbf{x} = B\mathbf{b}$ is a solution.

Since we know that $A\mathbf{x} = \mathbf{b}$ is consistent for any vector $\mathbf{b}$, what does this say about the span of the columns of A?

e. Since A is a square matrix, what does this say about the pivot positions of A? What is the reduced row echelon form of A?

f. In this activity, we have studied the matrices

$$A = \begin{bmatrix} 3 & -2 \\ 1 & -1 \end{bmatrix}, \quad C = \begin{bmatrix} -2 & 1 \\ 4 & -2 \end{bmatrix}.$$

Find the reduced row echelon form of each and explain how those forms enable us to conclude that one matrix is invertible and the other is not.

Example 3.1.3 We can reformulate this procedure for finding the inverse of a matrix. For the sake of convenience, suppose that A is a 2×2 invertible matrix with inverse $B = \begin{bmatrix} \mathbf{b}_1 & \mathbf{b}_2 \end{bmatrix}$. Rather than solving the equations

$$A\mathbf{b}_1 = \begin{bmatrix} 1 \\ 0 \end{bmatrix}, \quad A\mathbf{b}_2 = \begin{bmatrix} 0 \\ 1 \end{bmatrix}$$

separately, we can solve them at the same time by augmenting A by both vectors $\begin{bmatrix} 1 \\ 0 \end{bmatrix}$ and $\begin{bmatrix} 0 \\ 1 \end{bmatrix}$ and finding the reduced row echelon form.

For example, if $A = \begin{bmatrix} 1 & 2 \\ 1 & 1 \end{bmatrix}$, we form

$$\left[\begin{array}{cc|cc} 1 & 2 & 1 & 0 \\ 1 & 1 & 0 & 1 \end{array}\right] \sim \left[\begin{array}{cc|cc} 1 & 0 & -1 & 2 \\ 0 & 1 & 1 & -1 \end{array}\right].$$

This shows that the matrix $B = \begin{bmatrix} -1 & 2 \\ 1 & 1 \end{bmatrix}$ is the inverse of A.

In other words, beginning with A, we augment by the identify and find the reduced row echelon form to determine A^{-1}:

$$[\,A\,|\,I\,] \sim [\,I\,|\,A^{-1}\,].$$

In fact, this reformulation will always work. Suppose that A is an invertible $n \times n$ matrix with inverse B. Suppose furthermore that $\mathbf{b}$ is any n-dimensional vector and consider the equation $A\mathbf{x} = \mathbf{b}$. We know that $x = B\mathbf{b}$ is a solution because $A(B\mathbf{b}) = (AB)\mathbf{b} = I\mathbf{b} = \mathbf{b}$.

Proposition 3.1.4 *If A is an invertible matrix with inverse B, then any equation $A\mathbf{x} = \mathbf{b}$ is consistent and $\mathbf{x} = B\mathbf{b}$ is a solution. In other words, the solution to $A\mathbf{x} = \mathbf{b}$ is $\mathbf{x} = A^{-1}\mathbf{b}$.*

Notice that this is similar to saying that the solution to $3x = 5$ is $x = \frac{1}{3} \cdot 5$, as we saw in the preview activity.

Now since $A\mathbf{x} = \mathbf{b}$ is consistent for every vector $\mathbf{b}$, the columns of A must span $\mathbb{R}^n$ so there is a pivot position in every row. Since A is also square, this means that the reduced row echelon form of A is the identity matrix.

Proposition 3.1.5 *The matrix A is invertible if and only if the reduced row echelon form of A is the identity matrix: $A \sim I$. In addition, we can find the inverse by augmenting A by the identity and finding the reduced row echelon form:*

$$[\,A\,|\,I\,] \sim [\,I\,|\,A^{-1}\,].$$

You may have noticed that Proposition 3.1.4 says that *the* solution to the equation $A\mathbf{x} = \mathbf{b}$ is $\mathbf{x} = A^{-1}\mathbf{b}$. Indeed, we know that this equation has a unique solution because A has a pivot position in every column.

It is important to remember that the product of two matrices depends on the order in which they are multiplied. That is, if C and D are matrices, then it sometimes happens that $CD \neq DC$. However, something fortunate happens when we consider invertibility. It turns out that if A is an $n \times n$ matrix and that $AB = I$, then it is also true that $BA = I$. We have verified this in a few examples so far, and Exercise 3.1.5.12 explains why it always happens. This leads to the following proposition.

Proposition 3.1.6 *If A is a $n \times n$ invertible matrix with inverse B, then $BA = I$, which tells us that B is invertible with inverse A. In other words,*

$$(A^{-1})^{-1} = A.$$

3.1.2 Solving equations with an inverse

If A is an invertible matrix, then Proposition 3.1.4 shows us how to use A^{-1} to solve equations involving A. In particular, the solution to $A\mathbf{x} = \mathbf{b}$ is $\mathbf{x} = A^{-1}\mathbf{b}$.

3.1. INVERTIBILITY

Activity 3.1.3. We'll begin by considering the square matrix

$$A = \begin{bmatrix} 1 & 0 & 2 \\ 2 & 2 & 1 \\ 1 & 1 & 1 \end{bmatrix}.$$

a. Describe the solution space to the equation $A\mathbf{x} = \begin{bmatrix} 3 \\ 4 \\ 3 \end{bmatrix}$ by augmenting A and finding the reduced row echelon form.

b. Using Proposition 3.1.5, explain why A is invertible and find its inverse.

c. Now use the inverse to solve the equation $A\mathbf{x} = \begin{bmatrix} 3 \\ 4 \\ 3 \end{bmatrix}$ and verify that your result agrees with what you found in part a.

d. If you have defined a matrix B in Sage, you can find it's inverse as B.inverse() or B^-1. Use Sage to find the inverse of the matrix

$$B = \begin{bmatrix} 1 & -2 & -1 \\ -1 & 5 & 6 \\ 5 & -4 & 6 \end{bmatrix}$$

and use it to solve the equation $B\mathbf{x} = \begin{bmatrix} 8 \\ 3 \\ 36 \end{bmatrix}$.

e. If A and B are the two matrices defined in this activity, find their product AB and verify that it is invertible.

f. Compute the products $A^{-1}B^{-1}$ and $B^{-1}A^{-1}$. Which one agrees with $(AB)^{-1}$?

g. Explain your finding by considering the product

$$(AB)(B^{-1}A^{-1})$$

and using associativity to regroup the products so that the middle two terms are multiplied first.

The next proposition summarizes much of what we have found about invertible matrices.

Proposition 3.1.7 Properties of invertible matrices.
- An $n \times n$ matrix A is invertible if and only if $A \sim I$.

- If A is invertible, then the solution to the equation $A\mathbf{x} = \mathbf{b}$ is given by $\mathbf{x} = A^{-1}\mathbf{b}$.

- We can find A^{-1} by finding the reduced row echelon form of $\begin{bmatrix} A & | & I \end{bmatrix}$; namely,

$$\begin{bmatrix} A & | & I \end{bmatrix} \sim \begin{bmatrix} I & | & A^{-1} \end{bmatrix}.$$

- If A and B are two invertible $n \times n$ matrices, then their product AB is also invertible and $(AB)^{-1} = B^{-1}A^{-1}$.

There is a simple formula for finding the inverse of a 2×2 matrix:

$$\begin{bmatrix} a & b \\ c & d \end{bmatrix}^{-1} = \frac{1}{ad - bc} \begin{bmatrix} d & -b \\ -c & a \end{bmatrix},$$

which can be easily checked. The condition that A be invertible is, in this case, reduced to the condition that $ad - bc \neq 0$. We will understand this condition better once we have explored determinants in Section 3.4. There is a similar formula for the inverse of a 3×3 matrix, but there is not a good reason to write it here.

3.1.3 Triangular matrices and Gaussian elimination

With some of the ideas we've developed, we can recast the Gaussian elimination algorithm in terms of matrix multiplication and invertibility. This will be especially helpful later when we consider determinants and LU factorizations. Triangular matrices will play an important role.

Definition 3.1.8 We say that a matrix A is *lower triangular* if all its entries above the diagonal are zero. Similarly, A is *upper triangular* if all the entries below the diagonal are zero.

For example, the matrix L below is a lower triangular matrix while U is an upper triangular one.

$$L = \begin{bmatrix} * & 0 & 0 & 0 \\ * & * & 0 & 0 \\ * & * & * & 0 \\ * & * & * & * \end{bmatrix}, \quad U = \begin{bmatrix} * & * & * & * \\ 0 & * & * & * \\ 0 & 0 & * & * \\ 0 & 0 & 0 & * \end{bmatrix}.$$

We can develop a simple test to determine whether an $n \times n$ lower triangular matrix is invertible. Let's use Gaussian elimination to find the reduced row echelon form of the lower triangular matrix

$$\begin{bmatrix} 1 & 0 & 0 \\ 2 & -2 & 0 \\ -3 & 4 & -4 \end{bmatrix} \sim \begin{bmatrix} 1 & 0 & 0 \\ 0 & -2 & 0 \\ 0 & 4 & -4 \end{bmatrix}$$

$$\sim \begin{bmatrix} 1 & 0 & 0 \\ 0 & -2 & 0 \\ 0 & 0 & -4 \end{bmatrix} \sim \begin{bmatrix} 1 & 0 & 0 \\ 0 & 1 & 0 \\ 0 & 0 & 1 \end{bmatrix}.$$

Because the entries on the diagonal are nonzero, we find a pivot position in every row, which tells us that the matrix is invertible.

3.1. INVERTIBILITY

If, however, there is a zero entry on the diagonal, the matrix cannot be invertible. Considering the matrix below, we see that having a zero on the diagonal leads to a row without a pivot position.

$$\begin{bmatrix} 1 & 0 & 0 \\ 2 & 0 & 0 \\ -3 & 4 & -4 \end{bmatrix} \sim \begin{bmatrix} 1 & 0 & 0 \\ 0 & 0 & 0 \\ 0 & 4 & -4 \end{bmatrix} \sim \begin{bmatrix} 1 & 0 & 0 \\ 0 & 1 & -1 \\ 0 & 0 & 0 \end{bmatrix}.$$

Proposition 3.1.9 *An $n \times n$ triangular matrix is invertible if and only if the entries on the diagonal are all nonzero.*

Activity 3.1.4 Gaussian elimination and matrix multiplication.. This activity explores how the row operations of scaling, interchange, and replacement can be performed using matrix multiplication.

As an example, we consider the matrix

$$A = \begin{bmatrix} 1 & 2 & 1 \\ 2 & 0 & -2 \\ -1 & 2 & -1 \end{bmatrix}$$

and apply a replacement operation that multiplies the first row by -2 and adds it to the second row. Rather than performing this operation in the usual way, we construct a new matrix by applying the desired replacement operation to the identity matrix. To illustrate, we begin with the identity matrix

$$I = \begin{bmatrix} 1 & 0 & 0 \\ 0 & 1 & 0 \\ 0 & 0 & 1 \end{bmatrix}$$

and form a new matrix by multiplying the first row by -2 and adding it to the second row to obtain

$$R = \begin{bmatrix} 1 & 0 & 0 \\ -2 & 1 & 0 \\ 0 & 0 & 1 \end{bmatrix}.$$

a. Show that the product RA is the result of applying the replacement operation to A.

b. Explain why R is invertible and find its inverse R^{-1}.

c. Describe the relationship between R and R^{-1} and use the connection to replacement operations to explain why it holds.

d. Other row operations can be performed using a similar procedure. For instance, suppose we want to scale the second row of A by 4. Find a matrix S so that SA is the same as that obtained from the scaling operation. Why is S invertible and what is S^{-1}?

e. Finally, suppose we want to interchange the first and third rows of A. Find a matrix P, usually called a *permutation matrix* that performs this operation. What is P^{-1}?

f. The original matrix A is seen to be row equivalent to the upper triangular matrix U by performing three replacement operations on A:

$$A = \begin{bmatrix} 1 & 2 & 1 \\ 2 & 0 & -2 \\ -1 & 2 & -1 \end{bmatrix} \sim \begin{bmatrix} 1 & 2 & 1 \\ 0 & -4 & -4 \\ 0 & 0 & -4 \end{bmatrix} = U.$$

Find the matrices L_1, L_2, and L_3 that perform these row replacement operations so that $L_3 L_2 L_1 A = U$.

g. Explain why the matrix product $L_3 L_2 L_1$ is invertible and use this fact to write $A = LU$. What is the matrix L that you find? Why do you think we denote it by L?

The following are examples of matrices, known as *elementary matrices*, that perform the row operations on a matrix having three rows.

Replacement Multiplying the second row by 3 and adding it to the third row is performed by

$$L = \begin{bmatrix} 1 & 0 & 0 \\ 0 & 1 & 0 \\ 0 & 3 & 1 \end{bmatrix}.$$

We often use L to describe these matrices because they are lower triangular.

Scaling Multiplying the third row by 2 is performed by

$$S = \begin{bmatrix} 1 & 0 & 0 \\ 0 & 1 & 0 \\ 0 & 0 & 2 \end{bmatrix}.$$

Interchange Interchanging the first two rows is performed by

$$P = \begin{bmatrix} 0 & 1 & 0 \\ 1 & 0 & 0 \\ 0 & 0 & 1 \end{bmatrix}.$$

Example 3.1.10 Suppose we have

$$A = \begin{bmatrix} 1 & 3 & -2 \\ -3 & -6 & 3 \\ 2 & 0 & -2 \end{bmatrix}.$$

For the forward substitution phase of Gaussian elimination, we perform a sequence of three

3.1. INVERTIBILITY

replacement operations. The first replacement operation multiplies the first row by 3 and adds the result to the second row. We can perform this operation by multiplying A by the lower triangular matrix L_1 where

$$L_1 A = \begin{bmatrix} 1 & 0 & 0 \\ 3 & 1 & 0 \\ 0 & 0 & 1 \end{bmatrix} \begin{bmatrix} 1 & 3 & -2 \\ -3 & -6 & 3 \\ 2 & 0 & -2 \end{bmatrix} = \begin{bmatrix} 1 & 3 & -2 \\ 0 & 3 & -3 \\ 2 & 0 & -1 \end{bmatrix}.$$

The next two replacement operations are performed by the matrices

$$L_2 = \begin{bmatrix} 1 & 0 & 0 \\ 0 & 1 & 0 \\ -2 & 0 & 1 \end{bmatrix}, \quad L_3 = \begin{bmatrix} 1 & 0 & 0 \\ 0 & 1 & 0 \\ 0 & 2 & 1 \end{bmatrix}$$

so that

$$L_3 L_2 L_1 A = U = \begin{bmatrix} 1 & 3 & -2 \\ 0 & 3 & -3 \\ 0 & 0 & -3 \end{bmatrix}.$$

Notice that the inverse of L_1 has the simple form:

$$L_1 = \begin{bmatrix} 1 & 0 & 0 \\ 3 & 1 & 0 \\ 0 & 0 & 1 \end{bmatrix}, \quad L_1^{-1} = \begin{bmatrix} 1 & 0 & 0 \\ -3 & 1 & 0 \\ 0 & 0 & 1 \end{bmatrix}.$$

This says that if we want to undo the operation of multiplying the first row by 3 and adding to the second row, we should multiply the first row by -3 and add it to the second row. That is the effect of L_1^{-1}.

Notice that we now have $L_3 L_2 L_1 A = U$, which gives

$$(L_3 L_2 L_1) A = U$$
$$(L_3 L_2 L_1)^{-1} (L_3 L_2 L_1) A = (L_3 L_2 L_1)^{-1} U$$
$$A = (L_3 L_2 L_1)^{-1} U = LU$$

where L is the lower triangular matrix

$$L = (L_3 L_2 L_1)^{-1} = \begin{bmatrix} 1 & 0 & 0 \\ -3 & 1 & 0 \\ 2 & -2 & 1 \end{bmatrix}.$$

This way of writing $A = LU$ as the product of a lower and an upper triangular matrix is known as an LU factorization of A, and its usefulness will be explored in Section 5.1.

3.1.4 Summary

In this section, we found conditions guaranteeing that a matrix has an inverse. When these conditions hold, we also found an algorithm for finding the inverse.

- A square matrix is invertible if there is a matrix B, known as the inverse of A, such that $AB = I$. We usually write $A^{-1} = B$.

- The $n \times n$ matrix A is invertible if and only if it is row equivalent to I_n, the $n \times n$ identity matrix.

- If a matrix A is invertible, we can use Gaussian elimination to find its inverse:
$$[\ A\ |\ I\ ] \sim [\ I\ |\ A^{-1}\ ].$$

- If a matrix A is invertible, then the solution to the equation $A\mathbf{x} = \mathbf{b}$ is $\mathbf{x} = A^{-1}\mathbf{b}$.

- The row operations of replacement, scaling, and interchange can be performed by multiplying by elementary matrices.

3.1.5 Exercises

1. Consider the matrix
$$A = \begin{bmatrix} 3 & -1 & 1 & 4 \\ 0 & 2 & 3 & 1 \\ -2 & 1 & 0 & -2 \\ 3 & 0 & 1 & 2 \end{bmatrix}.$$

    a. Explain why A has an inverse.

    b. Find the inverse of A by augmenting by the identity I to form $[\ A\ |\ I\ ]$.

    c. Use your inverse to solve the equation $A\mathbf{x} = \begin{bmatrix} 3 \\ 2 \\ -3 \\ -1 \end{bmatrix}.$

2. In this exercise, we will consider 2×2 matrices as defining matrix transformations.

    a. Write the matrix A that performs a 45° rotation. What geometric operation undoes this rotation? Find the matrix that perform this operation and verify that it is A^{-1}.

    b. Write the matrix A that performs a 180° rotation. Verify that $A^2 = I$ so that $A^{-1} = A$, and explain geometrically why this is the case.

    c. Find three more matrices A that satisfy $A^2 = I$.

3.1. INVERTIBILITY

3. Inverses for certain types of matrices can be found in a relatively straightforward fashion.

   a. The matrix $D = \begin{bmatrix} 2 & 0 & 0 \\ 0 & -1 & 0 \\ 0 & 0 & -4 \end{bmatrix}$ is called *diagonal* since the only nonzero entries are on the diagonal of the matrix.

      1. Find D^{-1} by augmenting D by the identity and finding its reduced row echelon form.
      2. Under what conditions is a diagonal matrix invertible?
      3. Explain why the inverse of a diagonal matrix is also diagonal and explain the relationship between the diagonal entries in D and D^{-1}.

   b. Consider the lower triangular matrix $L = \begin{bmatrix} 1 & 0 & 0 \\ -2 & 1 & 0 \\ 3 & -4 & 1 \end{bmatrix}$.

      1. Find L^{-1} by augmenting L by the identity and finding its reduced row echelon form.
      2. Explain why the inverse of a lower triangular matrix is also lower triangular.

4. Our definition of an invertible matrix requires that A be a square $n \times n$ matrix. Let's examine what happens when A is not square. For instance, suppose that

$$A = \begin{bmatrix} -1 & -1 \\ -2 & -1 \\ 3 & 0 \end{bmatrix}, \quad B = \begin{bmatrix} -2 & 2 & 1 \\ 1 & -2 & -1 \end{bmatrix}.$$

   a. Verify that $BA = I_2$. In this case, we say that B is a *left* inverse of A.

   b. If A has a left inverse B, we can still use it to find solutions to linear equations. If we know there is a solution to the equation $A\mathbf{x} = \mathbf{b}$, we can multiply both sides of the equation by B to find $\mathbf{x} = B\mathbf{b}$.

   Suppose you know there is a solution to the equation $A\mathbf{x} = \begin{bmatrix} -1 \\ -3 \\ 6 \end{bmatrix}$. Use the left inverse B to find $\mathbf{x}$ and verify that it is a solution.

   c. Now consider the matrix
   $$C = \begin{bmatrix} 1 & -1 & 0 \\ -2 & 1 & 0 \end{bmatrix}$$
   and verify that C is also a left inverse of A. This shows that the matrix A may have more than one left inverse.

5. If a matrix A is invertible, there is a sequence of row operations that transforms A into the identity matrix I. We have seen that every row operation can be performed by matrix multiplication. If the jth step in the Gaussian elimination process is performed

by multiplying by E_j, then we have

$$E_p \cdots E_2 E_1 A = I,$$

which means that

$$A^{-1} = E_p \cdots E_2 E_1.$$

For each of the following matrices, find a sequence of row operations that transforms the matrix to the identity I. Write the matrices E_j that perform the steps and use them to find A^{-1}.

a.
$$A = \begin{bmatrix} 0 & 2 & 0 \\ -3 & 0 & 0 \\ 0 & 0 & 1 \end{bmatrix}.$$

b.
$$A = \begin{bmatrix} 1 & 0 & 0 & 0 \\ 2 & 1 & 0 & 0 \\ 0 & -3 & 1 & 0 \\ 0 & 0 & 2 & 1 \end{bmatrix}.$$

c.
$$A = \begin{bmatrix} 1 & 1 & 1 \\ 0 & 1 & 1 \\ 0 & 0 & 2 \end{bmatrix}.$$

6. Suppose that A is an $n \times n$ matrix.

   a. Suppose that $A^2 = AA$ is invertible with inverse B. This means that $A^2 B = AAB = I$. Explain why A must be invertible with inverse AB.

   b. Suppose that A^{100} is invertible with inverse B. Explain why A is invertible. What is A^{-1} in terms of A and B?

7. Determine whether the following statements are true or false and explain your reasoning.

   a. If A is invertible, then the columns of A are linearly independent.

   b. If A is a square matrix whose diagonal entries are all nonzero, then A is invertible.

   c. If A is an invertible $n \times n$ matrix, then span of the columns of A is $\mathbb{R}^n$.

   d. If A is invertible, then there is a nonzero solution to the homogeneous equation $A\mathbf{x} = \mathbf{0}$.

   e. If A is an $n \times n$ matrix and the equation $A\mathbf{x} = \mathbf{b}$ has a solution for every vector $\mathbf{b}$, then A is invertible.

8. Provide a justification for your response to the following questions.

   a. Suppose that A is a square matrix with two identical columns. Can A be invertible?

   b. Suppose that A is a square matrix with two identical rows. Can A be invertible?

3.1. INVERTIBILITY

c. Suppose that A is an invertible matrix and that $AB = AC$. Can you conclude that $B = C$?

d. Suppose that A is an invertible $n \times n$ matrix. What can you say about the span of the columns of A^{-1}?

e. Suppose that A is an invertible matrix and that B is row equivalent to A. Can you guarantee that B is invertible?

9. Suppose that we start with the 3×3 matrix A, perform the following sequence of row operations:

   1. Multiply row 1 by -2 and add to row 2.
   2. Multiply row 1 by 4 and add to row 3.
   3. Scale row 2 by 1/2.
   4. Multiply row 2 by -1 and add to row 3,

   and arrive at the upper triangular matrix

   $$U = \begin{bmatrix} 3 & 2 & -1 \\ 0 & 1 & 3 \\ 0 & 0 & -4 \end{bmatrix}.$$

   a. Write the matrices E_1, E_2, E_3, and E_4 that perform the four row operations.

   b. Find the matrix $E = E_4 E_3 E_2 E_1$.

   c. We then have $E_4 E_3 E_2 E_1 A = EA = U$. Now that we have the matrix E, find the original matrix $A = E^{-1}U$.

10. We say that two square matrices A and B are *similar* if there is an invertible matrix P such that $B = PAP^{-1}$.

    a. If A and B are similar, explain why A^2 and B^2 are similar as well. In particular, if $B = PAP^{-1}$, explain why $B^2 = PA^2P^{-1}$.

    b. If A and B are similar and A is invertible, explain why B is also invertible.

    c. If A and B are similar and both are invertible, explain why A^{-1} and B^{-1} are similar.

    d. If A is similar to B and B is similar to C, explain why A is similar to C. To begin, you may wish to assume that $B = PAP^{-1}$ and $C = QBQ^{-1}$.

11. Suppose that A and B are two $n \times n$ matrices and that AB is invertible. We would like to explain why both A and B are invertible.

    a. We first explain why B is invertible.

       1. Since AB is invertible, explain why any solution to the homogeneous equation $AB\mathbf{x} = \mathbf{0}$ is $\mathbf{x} = \mathbf{0}$.

2. Use this fact to explain why any solution to $B\mathbf{x} = \mathbf{0}$ must be $\mathbf{x} = \mathbf{0}$.

3. Explain why B must be invertible.

b. Now we explain why A is invertible.

1. Since AB is invertible, explain why the equation $AB\mathbf{x} = \mathbf{b}$ is consistent for every vector $\mathbf{b}$.

2. Using the fact that $AB\mathbf{x} = A(B\mathbf{x}) = \mathbf{b}$ is consistent for every $\mathbf{b}$, explain why every equation $A\mathbf{x} = \mathbf{b}$ is consistent.

3. Explain why A must be invertible.

12. We defined an $n \times n$ matrix to be invertible if there is a matrix B such that $AB = I_n$. In this exercise, we will explain why it is also true that $BA = I$, which is the statement of Proposition 3.1.6. This means that, if $B = A^{-1}$, then $A = B^{-1}$.

a. Suppose that $\mathbf{x}$ is an n-dimensional vector. Since $AB = I$, explain why $AB\mathbf{x} = \mathbf{x}$ and use this to explain why the only vector for which $B\mathbf{x} = \mathbf{0}$ is $\mathbf{x} = \mathbf{0}$.

b. Explain why this implies that B must be invertible. We will call the inverse C so that $BC = I$.

c. Beginning with $AB = I$, explain why $B(AB)C = BIC$ and why this tells us that $BA = I$.

3.2 Bases and coordinate systems

Standard Cartesian coordinates are commonly used to describe points in the plane. If we mention the point $(4, 3)$, we know that we arrive at this point from the origin by moving four units to the right and three units up.

Sometimes, however, it is more natural to work in a different coordinate system. Suppose that you live in the city whose map is shown in Figure 3.2.1 and that you would like to give a guest directions for getting from your house to the store. You would probably say something like, "Go four blocks up Maple. Then turn left on Main for three blocks." The grid of streets in the city gives a more natural coordinate system than standard north-south, east-west coordinates.

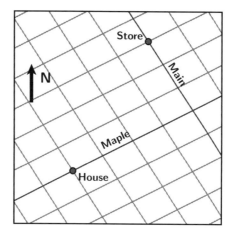

Figure 3.2.1 A city map.

In this section, we will develop the concept of a *basis* through which we will create new coordinate systems in $\mathbb{R}^m$. We will see that the right choice of a coordinate system provides a more natural way to approach some problems.

Preview Activity 3.2.1. Consider the vectors
$$\mathbf{v}_1 = \begin{bmatrix} 2 \\ 1 \end{bmatrix}, \mathbf{v}_2 = \begin{bmatrix} 1 \\ 2 \end{bmatrix}$$
in $\mathbb{R}^2$, which are shown in Figure 3.2.2.

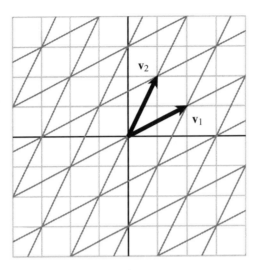

Figure 3.2.2 Linear combinations of $\mathbf{v}_1$ and $\mathbf{v}_2$.

a. Indicate the linear combination $\mathbf{v}_1 - 2\mathbf{v}_2$ on the figure.

b. Express the vector $\begin{bmatrix} -3 \\ 0 \end{bmatrix}$ as a linear combination of $\mathbf{v}_1$ and $\mathbf{v}_2$.

c. Find the linear combination $10\mathbf{v}_1 - 13\mathbf{v}_2$.

d. Express the vector $\begin{bmatrix} 16 \\ -4 \end{bmatrix}$ as a linear combination of $\mathbf{v}_1$ and $\mathbf{v}_2$.

e. Explain why every vector in $\mathbb{R}^2$ can be written as a linear combination of $\mathbf{v}_1$ and $\mathbf{v}_2$ in exactly one way.

In the preview activity, we worked with a set of two vectors in $\mathbb{R}^2$ and found that we could express any vector in $\mathbb{R}^2$ in two different ways: in the usual way where the components of the vector describe horizontal and vertical changes, and in a new way as a linear combination of $\mathbf{v}_1$ and $\mathbf{v}_2$. We could also translate between these two descriptions. This example illustrates the central idea of this section.

3.2.1 Bases

In the preview activity, we created a new coordinate system for $\mathbb{R}^2$ using linear combinations of a set of two vectors. More generally, the following definition will guide us.

Definition 3.2.3 A set of vectors $\mathbf{v}_1, \mathbf{v}_2, \ldots, \mathbf{v}_n$ in $\mathbb{R}^m$ is called a *basis* for $\mathbb{R}^m$ if the set of vectors spans $\mathbb{R}^m$ and is linearly independent.

3.2. BASES AND COORDINATE SYSTEMS

Activity 3.2.2. We will look at some examples of bases in this activity.

a. In the preview activity, we worked with the set of vectors in $\mathbb{R}^2$:

$$\mathbf{v}_1 = \begin{bmatrix} 2 \\ 1 \end{bmatrix}, \mathbf{v}_2 = \begin{bmatrix} 1 \\ 2 \end{bmatrix}.$$

Explain why these vectors form a basis for $\mathbb{R}^2$.

b. Consider the set of vectors in $\mathbb{R}^3$

$$\mathbf{v}_1 = \begin{bmatrix} 1 \\ 1 \\ 1 \end{bmatrix}, \mathbf{v}_2 = \begin{bmatrix} 0 \\ 1 \\ -1 \end{bmatrix}, \mathbf{v}_3 = \begin{bmatrix} 1 \\ 0 \\ -1 \end{bmatrix}$$

and determine whether they form a basis for $\mathbb{R}^3$.

c. Do the vectors

$$\mathbf{v}_1 = \begin{bmatrix} -2 \\ 1 \\ 3 \end{bmatrix}, \mathbf{v}_2 = \begin{bmatrix} 3 \\ 0 \\ -1 \end{bmatrix}, \mathbf{v}_3 = \begin{bmatrix} 1 \\ 1 \\ 0 \end{bmatrix}, \mathbf{v}_4 = \begin{bmatrix} 0 \\ 3 \\ -2 \end{bmatrix}$$

form a basis for $\mathbb{R}^3$?

d. Explain why the vectors $\mathbf{e}_1, \mathbf{e}_2, \mathbf{e}_3$ form a basis for $\mathbb{R}^3$.

e. If a set of vectors $\mathbf{v}_1, \mathbf{v}_2, \ldots, \mathbf{v}_n$ forms a basis for $\mathbb{R}^m$, what can you guarantee about the pivot positions of the matrix

$$\begin{bmatrix} \mathbf{v}_1 & \mathbf{v}_2 & \cdots & \mathbf{v}_n \end{bmatrix}?$$

f. If the set of vectors $\mathbf{v}_1, \mathbf{v}_2, \ldots, \mathbf{v}_n$ is a basis for $\mathbb{R}^{10}$, how many vectors must be in the set?

We can develop a test to determine if a set of vectors $\mathbf{v}_1, \mathbf{v}_2, \ldots, \mathbf{v}_n$ forms a basis for $\mathbb{R}^m$ by considering the matrix

$$A = \begin{bmatrix} \mathbf{v}_1 & \mathbf{v}_2 & \cdots & \mathbf{v}_n \end{bmatrix}.$$

To be a basis, this set of vectors must span $\mathbb{R}^m$ and be linearly independent.

We know that the span of the set of vectors is $\mathbb{R}^m$ if and only if A has a pivot position in every row. We also know that the set of vectors is linearly independent if and only if A has a pivot position in every column. This means that a set of vectors forms a basis if and only if A has a pivot in every row and every column. Therefore, A must be row equivalent to the identity matrix I:

$$A \sim \begin{bmatrix} 1 & 0 & \cdots & 0 \\ 0 & 0 & \cdots & 0 \\ \vdots & \vdots & \ddots & \vdots \\ 0 & 0 & \cdots & 1 \end{bmatrix} = I.$$

In addition to helping identify bases, this fact tells us something important about the number of vectors in a basis. Since the matrix A has a pivot position in every row and every column, it must have the same number of rows as columns. Therefore, the number of vectors in a basis for $\mathbb{R}^m$ must be m. For example, a basis for $\mathbb{R}^{23}$ must have exactly 23 vectors.

Proposition 3.2.4 *A set of vectors forms a basis for $\mathbb{R}^m$ if and only if the matrix*

$$A = \begin{bmatrix} \mathbf{v}_1 & \mathbf{v}_2 & \cdots & \mathbf{v}_n \end{bmatrix} \sim I.$$

This means there must be m vectors in a basis for $\mathbb{R}^m$.

Example 3.2.5 Notice that the vectors

$$\mathbf{e}_1 = \begin{bmatrix} 1 \\ 0 \\ 0 \end{bmatrix}, \mathbf{e}_2 = \begin{bmatrix} 0 \\ 1 \\ 0 \end{bmatrix}, \mathbf{e}_3 = \begin{bmatrix} 0 \\ 0 \\ 1 \end{bmatrix}$$

form the columns of the 3×3 identity matrix, which implies that this set forms a basis for $\mathbb{R}^3$. More generally, the set of vectors $\mathbf{e}_1, \mathbf{e}_2, \ldots, \mathbf{e}_m$ forms a basis for $\mathbb{R}^m$, which we call the *standard* basis for $\mathbb{R}^m$.

3.2.2 Coordinate systems

A basis for $\mathbb{R}^m$ forms a coordinate system for $\mathbb{R}^m$, as we will describe. Rather than continuing to write a list of vectors, we will find it convenient to denote a basis using a single symbol, such as

$$\mathcal{B} = \{\mathbf{v}_1, \mathbf{v}_2, \ldots, \mathbf{v}_m\}$$

Example 3.2.6 In this section's preview activity, we considered the vectors

$$\mathbf{v}_1 = \begin{bmatrix} 2 \\ 1 \end{bmatrix}, \mathbf{v}_2 = \begin{bmatrix} 1 \\ 2 \end{bmatrix},$$

which form a basis $\mathcal{B} = \{\mathbf{v}_1, \mathbf{v}_2\}$ for $\mathbb{R}^2$.

3.2. BASES AND COORDINATE SYSTEMS

In the standard coordinate system, the point $(2, -3)$ is found by moving 2 units to the right and 3 units down. We would like to define a new coordinate system where we interpret $(2, -3)$ to mean we move two times along $\mathbf{v}_1$ and 3 times along $-\mathbf{v}_2$. As we see in the figure, doing so leaves us at the point $(1, -4)$, expressed in the usual coordinate system.

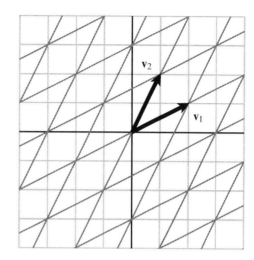

We have seen that

$$\mathbf{x} = \begin{bmatrix} 1 \\ -4 \end{bmatrix} = 2\mathbf{v}_1 - 3\mathbf{v}_2.$$

The coordinates of the vector $\mathbf{x}$ in the new coordinate system are the weights that we use to create $\mathbf{x}$ as a linear combination of $\mathbf{v}_1$ and $\mathbf{v}_2$.

Since we now have two descriptions of the vector $\mathbf{x}$, we need some notation to keep track of which coordinate system we are using. Because $\begin{bmatrix} 1 \\ -4 \end{bmatrix} = 2\mathbf{v}_1 - 3\mathbf{v}_2$, we will write

$$\left\{ \begin{bmatrix} 1 \\ -4 \end{bmatrix} \right\}_{\mathcal{B}} = \begin{bmatrix} 2 \\ -3 \end{bmatrix}.$$

More generally, $\{\mathbf{x}\}_{\mathcal{B}}$ will denote the coordinates of $\mathbf{x}$ in the basis $\mathcal{B}$; that is, $\{\mathbf{x}\}_{\mathcal{B}}$ is the vector $\begin{bmatrix} c_1 \\ c_2 \end{bmatrix}$ of weights such that

$$\mathbf{x} = c_1 \mathbf{v}_1 + c_2 \mathbf{v}_2.$$

For example, if the coordinates of $\mathbf{x}$ in the basis $\mathcal{B}$ are

$$\{\mathbf{x}\}_{\mathcal{B}} = \begin{bmatrix} 5 \\ -2 \end{bmatrix},$$

then

$$\mathbf{x} = 5\mathbf{v}_1 - 2\mathbf{v}_2 = 5 \begin{bmatrix} 2 \\ 1 \end{bmatrix} - 2 \begin{bmatrix} 1 \\ 2 \end{bmatrix} = \begin{bmatrix} 8 \\ 3 \end{bmatrix},$$

and we conclude that

$$\left\{ \begin{bmatrix} 8 \\ 3 \end{bmatrix} \right\}_{\mathcal{B}} = \begin{bmatrix} 5 \\ -2 \end{bmatrix}.$$

This demonstrates how we can translate coordinates in the basis $\mathcal{B}$ into standard coordinates.

Suppose we know the expression of a vector $\mathbf{x}$ in standard coordinates. How can we find its coordinates in the basis $\mathcal{B}$? For instance, suppose $\mathbf{x} = \begin{bmatrix} -8 \\ 2 \end{bmatrix}$ and that we would like to find

$\{\mathbf{x}\}_{\mathcal{B}}$. We can write
$$\left\{\begin{bmatrix} -8 \\ 2 \end{bmatrix}\right\}_{\mathcal{B}} = \begin{bmatrix} c_1 \\ c_2 \end{bmatrix},$$
which means that
$$\begin{bmatrix} -8 \\ 2 \end{bmatrix} = c_1\mathbf{v}_1 + c_2\mathbf{v}_2$$
or
$$c_1 \begin{bmatrix} 2 \\ 1 \end{bmatrix} + c_2 \begin{bmatrix} 1 \\ 2 \end{bmatrix} = \begin{bmatrix} -8 \\ 2 \end{bmatrix}.$$

This linear system for the weights defines an augmented matrix
$$\left[\begin{array}{cc|c} 2 & 1 & -8 \\ 1 & 2 & 2 \end{array}\right] \sim \left[\begin{array}{cc|c} 1 & 0 & -6 \\ 0 & 1 & 4 \end{array}\right],$$
which means that
$$\left\{\begin{bmatrix} -8 \\ 2 \end{bmatrix}\right\}_{\mathcal{B}} = \begin{bmatrix} -6 \\ 4 \end{bmatrix}.$$

This example illustrates how a basis in $\mathbb{R}^2$ provides a new coordinate system for $\mathbb{R}^2$ and shows how we may translate between this coordinate system and the standard one.

More generally, suppose that $\mathcal{B} = \{\mathbf{v}_1, \mathbf{v}_2, \ldots, \mathbf{v}_m\}$ is a basis for $\mathbb{R}^m$. We know that the span of the vectors is $\mathbb{R}^m$, which implies that any vector $\mathbf{x}$ in $\mathbb{R}^m$ can be written as a linear combination of the vectors. In addition, we know that the vectors are linearly independent, which means that we can write $\mathbf{x}$ as a linear combination of the vectors in exactly one way. Therefore, we have
$$\mathbf{x} = c_1\mathbf{v}_1 + c_2\mathbf{v}_2 + \ldots + c_m\mathbf{v}_m$$
where the weights $c_1, c_2, \ldots, c_m$ are unique. In this case, we write the coordinate description of $\mathbf{x}$ in the basis $\mathcal{B}$ as
$$\{\mathbf{x}\}_{\mathcal{B}} = \begin{bmatrix} c_1 \\ c_2 \\ \vdots \\ c_m \end{bmatrix}.$$

Activity 3.2.3. Let's begin with the basis $\mathcal{B} = \{\mathbf{v}_1, \mathbf{v}_2\}$ of $\mathbb{R}^2$ where
$$\mathbf{v}_1 = \begin{bmatrix} 3 \\ -2 \end{bmatrix}, \mathbf{v}_2 = \begin{bmatrix} 2 \\ 1 \end{bmatrix}.$$

a. If the coordinates of $\mathbf{x}$ in the basis $\mathcal{B}$ are $\{\mathbf{x}\}_{\mathcal{B}} = \begin{bmatrix} -2 \\ 4 \end{bmatrix}$, what is the vector $\mathbf{x}$?

b. If $\mathbf{x} = \begin{bmatrix} 3 \\ 5 \end{bmatrix}$, find the coordinates of $\mathbf{x}$ in the basis $\mathcal{B}$; that is, find $\{\mathbf{x}\}_{\mathcal{B}}$.

c. Find a matrix A such that, for any vector $\mathbf{x}$, we have $\mathbf{x} = A\{\mathbf{x}\}_{\mathcal{B}}$. Explain why this matrix is invertible.

3.2. BASES AND COORDINATE SYSTEMS

d. Using what you found in the previous part, find a matrix B such that, for any vector $\mathbf{x}$, we have $\{\mathbf{x}\}_\mathcal{B} = B\mathbf{x}$. What is the relationship between the two matrices A and B? Explain why this relationship holds.

e. Suppose we consider the standard basis

$$\mathcal{E} = \{\mathbf{e}_1, \mathbf{e}_2\}.$$

What is the relationship between $\mathbf{x}$ and $\{\mathbf{x}\}_\mathcal{E}$?

f. Suppose we also consider the basis

$$\mathcal{C} = \left\{ \begin{bmatrix} 1 \\ 2 \end{bmatrix}, \begin{bmatrix} -2 \\ 1 \end{bmatrix} \right\}.$$

Find a matrix C that converts coordinates in the basis $\mathcal{C}$ into coordinates in the basis $\mathcal{B}$; that is,

$$\{\mathbf{x}\}_\mathcal{B} = C\,\{\mathbf{x}\}_\mathcal{C}.$$

You may wish to think about converting coordinates from the basis $\mathcal{C}$ into the standard coordinate system and then into the basis $\mathcal{B}$.

This activity demonstrates how we can efficiently convert between coordinate systems defined by different bases. Let's consider a basis $\mathcal{B} = \{\mathbf{v}_1, \mathbf{v}_2, \ldots, \mathbf{v}_m\}$ and a vector $\mathbf{x}$. We know that

$$\mathbf{x} = c_1\mathbf{v}_1 + c_2\mathbf{v}_2 + \ldots + c_m\mathbf{v}_m$$

$$= \begin{bmatrix} \mathbf{v}_1 & \mathbf{v}_2 & \cdots & \mathbf{v}_m \end{bmatrix} \begin{bmatrix} c_1 \\ c_2 \\ \vdots \\ c_m \end{bmatrix}$$

$$= \begin{bmatrix} \mathbf{v}_1 & \mathbf{v}_2 & \cdots & \mathbf{v}_m \end{bmatrix} \{\mathbf{x}\}_\mathcal{B}.$$

If we use $P_\mathcal{B}$ to denote the matrix whose columns are the basis vectors, then we find that

$$\mathbf{x} = P_\mathcal{B}\,\{\mathbf{x}\}_\mathcal{B}$$

where $P_\mathcal{B} = \begin{bmatrix} \mathbf{v}_1 & \mathbf{v}_2 & \cdots & \mathbf{v}_m \end{bmatrix}$. This means that the matrix $P_\mathcal{B}$ converts coordinates in the basis $\mathcal{B}$ into standard coordinates.

Since the columns of $P_\mathcal{B}$ are the basis vectors $\mathbf{v}_1, \mathbf{v}_2, \ldots, \mathbf{v}_m$, we know that $P_\mathcal{B} \sim I_m$, and $P_\mathcal{B}$ is therefore invertible. Since we have

$$\mathbf{x} = P_\mathcal{B}\,\{\mathbf{x}\}_\mathcal{B},$$

we must also have

$$P_\mathcal{B}^{-1}\mathbf{x} = \{\mathbf{x}\}_\mathcal{B}.$$

Proposition 3.2.7 *If $\mathcal{B}$ is a basis and $P_{\mathcal{B}}$ the matrix whose columns are the basis vectors, then*

$$\mathbf{x} = P_{\mathcal{B}} \{\mathbf{x}\}_{\mathcal{B}},$$

$$\{\mathbf{x}\}_{\mathcal{B}} = P_{\mathcal{B}}^{-1} \mathbf{x}.$$

If we have another basis C, we find, in the same way, that $\mathbf{x} = P_C \{\mathbf{x}\}_C$ for the conversion between coordinates in the basis C into standard coordinates. We then have

$$\{\mathbf{x}\}_{\mathcal{B}} = P_{\mathcal{B}}^{-1} \mathbf{x} = P_{\mathcal{B}}^{-1}(P_C \{\mathbf{x}\}_C) = (P_{\mathcal{B}}^{-1} P_C) \{\mathbf{x}\}_C.$$

Therefore, $P_{\mathcal{B}}^{-1} P_C$ is the matrix that converts C-coordinates into $\mathcal{B}$-coordinates.

3.2.3 Examples of bases

We will now look at some examples of bases that illustrate how it can be useful to study a problem using a different coordinate system.

Example 3.2.8 Let's consider the basis of $\mathbb{R}^3$:

$$\mathcal{B} = \left\{ \begin{bmatrix} 1 \\ 0 \\ -2 \end{bmatrix}, \begin{bmatrix} -2 \\ 1 \\ 0 \end{bmatrix}, \begin{bmatrix} 1 \\ 1 \\ 2 \end{bmatrix} \right\}.$$

It is relatively straightforward to convert a vector's representation in this basis into to the standard basis using the matrix whose columns are the basis vectors:

$$P_{\mathcal{B}} = \begin{bmatrix} 1 & -2 & 1 \\ 0 & 1 & 1 \\ -2 & 0 & 2 \end{bmatrix}.$$

For example, suppose that the vector $\mathbf{x}$ is described in the coordinate system defined by the basis as $\{\mathbf{x}\}_{\mathcal{B}} = \begin{bmatrix} 2 \\ -2 \\ 1 \end{bmatrix}$. We then have

$$\mathbf{x} = P_{\mathcal{B}} \{\mathbf{x}\}_{\mathcal{B}} = \begin{bmatrix} 1 & -2 & 1 \\ 0 & 1 & 1 \\ -2 & 0 & 2 \end{bmatrix} \begin{bmatrix} 2 \\ -2 \\ 1 \end{bmatrix} = \begin{bmatrix} 7 \\ -1 \\ -2 \end{bmatrix}.$$

Consider now the vector $\mathbf{x} = \begin{bmatrix} 3 \\ 1 \\ -2 \end{bmatrix}$. If we would like to express $\mathbf{x}$ in the coordinate system defined by $\mathcal{B}$, then we compute

$$\{\mathbf{x}\}_{\mathcal{B}} = P_{\mathcal{B}}^{-1} \mathbf{x} = \begin{bmatrix} \frac{1}{4} & \frac{1}{2} & -\frac{3}{8} \\ -\frac{1}{4} & \frac{1}{2} & -\frac{1}{8} \\ \frac{1}{4} & \frac{1}{2} & \frac{1}{8} \end{bmatrix} \begin{bmatrix} 3 \\ 1 \\ -2 \end{bmatrix} = \begin{bmatrix} 2 \\ 0 \\ 1 \end{bmatrix}.$$

3.2. BASES AND COORDINATE SYSTEMS

Example 3.2.9 Suppose we work for a company that records its quarterly revenue, in millions of dollars, as:

Table 3.2.10 A company's quarterly revenue

Quarter	Revenue
1	10.3
2	13.1
3	7.5
4	8.2

Rather than using a table to record the data, we could display it in a graph or write it as a vector in $\mathbb{R}^4$:

$$\mathbf{x} = \begin{bmatrix} 10.3 \\ 13.1 \\ 7.5 \\ 8.2 \end{bmatrix}.$$

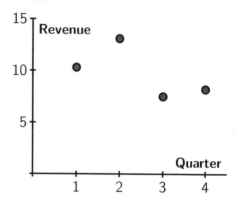

Let's consider a new basis $\mathcal{B}$ for $\mathbb{R}^4$ using vectors

$$\mathbf{v}_1 = \begin{bmatrix} 1 \\ 1 \\ 1 \\ 1 \end{bmatrix}, \quad \mathbf{v}_2 = \begin{bmatrix} 1 \\ 1 \\ -1 \\ -1 \end{bmatrix}, \quad \mathbf{v}_3 = \begin{bmatrix} 1 \\ -1 \\ 0 \\ 0 \end{bmatrix}, \quad \mathbf{v}_4 = \begin{bmatrix} 0 \\ 0 \\ 1 \\ -1 \end{bmatrix}.$$

We may view these basis elements graphically, as in Figure 3.2.11

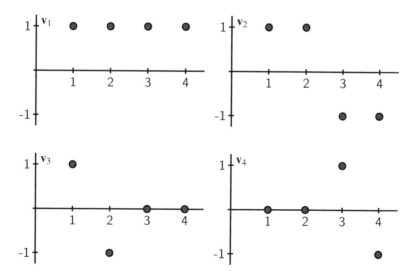

Figure 3.2.11 A representation of the basis elements of $\mathcal{B}$.

To convert our revenue vectors into the coordinates given by $\mathcal{B}$, we form the matrices:

$$P_{\mathcal{B}} = \begin{bmatrix} 1 & 1 & 1 & 0 \\ 1 & 1 & -1 & 0 \\ 1 & -1 & 0 & 1 \\ 1 & -1 & 0 & -1 \end{bmatrix}, \quad P_{\mathcal{B}}^{-1} = \begin{bmatrix} \frac{1}{4} & \frac{1}{4} & \frac{1}{4} & \frac{1}{4} \\ \frac{1}{4} & \frac{1}{4} & -\frac{1}{4} & -\frac{1}{4} \\ \frac{1}{2} & -\frac{1}{2} & 0 & 0 \\ 0 & 0 & \frac{1}{2} & -\frac{1}{2} \end{bmatrix}.$$

In particular, if the revenue vector is $\mathbf{x} = \begin{bmatrix} x_1 \\ x_2 \\ x_3 \\ x_4 \end{bmatrix}$, then

$$\{\mathbf{x}\}_{\mathcal{B}} = \begin{bmatrix} \frac{1}{4}x_1 + \frac{1}{4}x_2 + \frac{1}{4}x_3 + \frac{1}{4}x_4 \\ \frac{1}{4}x_1 + \frac{1}{4}x_2 - \frac{1}{4}x_3 - \frac{1}{4}x_4 \\ \frac{1}{2}x_1 - \frac{1}{2}x_2 \\ \frac{1}{2}x_3 - \frac{1}{2}x_4 \end{bmatrix}.$$

Notice that the first component of $\{\mathbf{x}\}_{\mathcal{B}}$ is the average of the components of $\mathbf{x}$.

For our particular revenue vector $\mathbf{x} = \begin{bmatrix} 10.3 \\ 13.1 \\ 7.5 \\ 8.2 \end{bmatrix}$, we have

$$\{\mathbf{x}\}_{\mathcal{B}} = P_{\mathcal{B}}^{-1}\mathbf{x} = P_{\mathcal{B}}^{-1} \begin{bmatrix} 10.3 \\ 13.1 \\ 7.5 \\ 8.2 \end{bmatrix} = \begin{bmatrix} 9.775 \\ 1.925 \\ -1.400 \\ -0.350 \end{bmatrix}.$$

This means that our revenue vector is

$$\mathbf{x} = 9.775\mathbf{v}_1 + 1.925\mathbf{v}_2 - 1.400\mathbf{v}_3 - 0.350\mathbf{v}_4.$$

We will think about what these terms mean by adding them together one at a time.

The first term,

$$9.775\mathbf{v}_1 = \begin{bmatrix} 9.775 \\ 9.775 \\ 9.775 \\ 9.775 \end{bmatrix},$$

gives us the average revenue over the year.

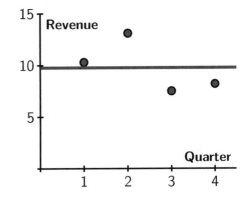

3.2. BASES AND COORDINATE SYSTEMS

The average revenue for the first two quarters is 11.7, which is 1.925 million dollars above the yearly average. Similarly, the average revenue for the last two quarters is 1.925 million dollars below the yearly average. This is recorded by the second term

$$1.925\mathbf{v}_2 = \begin{bmatrix} 1.925 \\ 1.925 \\ -1.925 \\ -1.925 \end{bmatrix}.$$

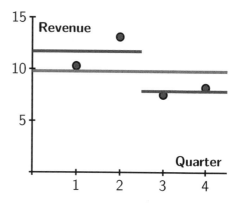

Finally, the first quarter's revenue is 1.400 million dollars below the average over the first two quarters and the second quarter's revenue is 1.400 million dollars above that average. This, and the corresponding data for the last two quarters, is captured by the last two terms:

$$-1.400\mathbf{v}_3 - 0.350\mathbf{v}_4 = \begin{bmatrix} -1.400 \\ 1.400 \\ -0.350 \\ 0.350 \end{bmatrix},$$

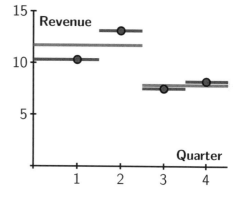

If we write $\{\mathbf{x}\}_{\mathcal{B}} = \begin{bmatrix} c_1 \\ c_2 \\ c_3 \\ c_4 \end{bmatrix}$, we see that the coefficient c_1 measures the average revenue over

the year, c_2 measures the deviation from the annual average in the first and second halves of the year, and c_3 measures how the revenue in the first and second quarter differs from the average in the first half of the year. In this way, the coefficients provide a view of the revenue over different time scales, from an annual summary to a finer view of quarterly behavior.

This basis is sometimes called a *Haar* wavelet basis, and the change of basis is known as a *Haar* wavelet transform. In the next section, we will see how this basis provides a useful way to store digital images.

Activity 3.2.4 Edge detection.. An important problem in the field of computer vision is to detect edges in a digital photograph, as is shown in Figure 3.2.12. Edge detection algorithms are useful when, say, we want a robot to locate an object in its field of view. Graphic designers also use these algorithms to create artistic effects.

Figure 3.2.12 A canyon wall in Capitol Reef National Park and the result of an edge detection algorithm.

We will consider a very simple version of an edge detection algorithm to give a sense of how this works. Rather than considering a two-dimensional photograph, we will think about a one-dimensional row of pixels in a photograph. The grayscale values of a pixel measure the brightness of a pixel; a grayscale value of 0 corresponds to black, and a value of 255 corresponds to white.

Suppose, for simplicity, that the grayscale values for a row of six pixels are represented by a vector $\mathbf{x}$ in $\mathbb{R}^6$:

$$\mathbf{x} = \begin{bmatrix} 25 \\ 34 \\ 30 \\ 45 \\ 190 \\ 200 \end{bmatrix}.$$

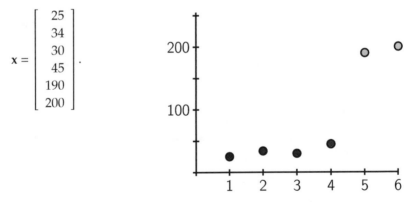

We can easily see that there is a jump in brightness between pixels 4 and 5, but how can we detect it computationally? We will introduce a new basis $\mathcal{B}$ for $\mathbb{R}^6$ with vectors:

$$\mathbf{v}_1 = \begin{bmatrix} 1 \\ 0 \\ 0 \\ 0 \\ 0 \\ 0 \end{bmatrix}, \mathbf{v}_2 = \begin{bmatrix} 1 \\ 1 \\ 0 \\ 0 \\ 0 \\ 0 \end{bmatrix}, \mathbf{v}_3 = \begin{bmatrix} 1 \\ 1 \\ 1 \\ 0 \\ 0 \\ 0 \end{bmatrix}, \mathbf{v}_4 = \begin{bmatrix} 1 \\ 1 \\ 1 \\ 1 \\ 0 \\ 0 \end{bmatrix}, \mathbf{v}_5 = \begin{bmatrix} 1 \\ 1 \\ 1 \\ 1 \\ 1 \\ 0 \end{bmatrix}, \mathbf{v}_6 = \begin{bmatrix} 1 \\ 1 \\ 1 \\ 1 \\ 1 \\ 1 \end{bmatrix}.$$

3.2. BASES AND COORDINATE SYSTEMS

a. Construct the matrix $P_\mathcal{B}$ that relates the standard coordinate system with the coordinates in the basis $\mathcal{B}$.

b. Determine the matrix $P_\mathcal{B}^{-1}$ that converts the representation of **x** in standard coordinates into the coordinate system defined by $\mathcal{B}$.

c. Suppose the vectors are expressed in general terms as

$$\mathbf{x} = \begin{bmatrix} x_1 \\ x_2 \\ x_3 \\ x_4 \\ x_5 \\ x_6 \end{bmatrix}, \quad \{\mathbf{x}\}_\mathcal{B} = \begin{bmatrix} c_1 \\ c_2 \\ c_3 \\ c_4 \\ c_5 \\ c_6 \end{bmatrix}.$$

Using the relationship $\{\mathbf{x}\}_\mathcal{B} = P_\mathcal{B}^{-1}\mathbf{x}$, determine an expression for the coefficient c_2 in terms of $x_1, x_2, \ldots, x_6$. What does c_2 measure in terms of the grayscale values of the pixels? What does c_4 measure in terms of the grayscale values of the pixels?

d. Now for the specific vector

$$\mathbf{x} = \begin{bmatrix} 25 \\ 34 \\ 30 \\ 45 \\ 190 \\ 200 \end{bmatrix},$$

determine the representation of **x** in the $\mathcal{B}$-coordinate system.

e. Explain how the coefficients in $\{\mathbf{x}\}_\mathcal{B}$ determine the location of the jump in brightness in the grayscale values represented by the vector **x**.

Readers who are familiar with calculus may recognize that this change of basis converts a vector **x** into $\{\mathbf{x}\}_\mathcal{B}$, the set of changes in **x**. This process is similar to differentiation in calculus. Similarly, the process of converting $\{\mathbf{x}\}_\mathcal{B}$ into the vector **x** adds together the changes in a process similar to integration. As a result, this change of basis represents a linear algebraic version of the Fundamental Theorem of Calculus.

3.2.4 Summary

We defined a basis to be a set of vectors $\mathcal{B} = \{\mathbf{v}_1, \mathbf{v}_2, \ldots, \mathbf{v}_n\}$ that is linearly independent and whose span is $\mathbb{R}^m$.

- A set of vectors forms a basis for $\mathbb{R}^m$ if and only if the matrix
$$A = \begin{bmatrix} \mathbf{v}_1 & \mathbf{v}_2 & \cdots & \mathbf{v}_n \end{bmatrix} \sim I.$$
This means there must be m vectors in a basis for $\mathbb{R}^m$.

- If $\mathbf{v}_1, \mathbf{v}_2, \ldots, \mathbf{v}_m$ forms a basis for $\mathbb{R}^m$, then any vector in $\mathbb{R}^m$ can be written as a linear combination of the vectors in exactly one way.

- We used the basis $\mathcal{B}$ to define a coordinate system in which $\{\mathbf{x}\}_\mathcal{B} = \begin{bmatrix} c_1 \\ c_2 \\ \vdots \\ c_n \end{bmatrix}$, the coordinates of $\mathbf{x}$ in the basis $\mathcal{B}$, are defined by
$$\mathbf{x} = c_1 \mathbf{v}_1 + c_2 \mathbf{v}_2 + \cdots + c_n \mathbf{v}_m.$$

- Forming the matrix $P_\mathcal{B}$ whose columns are the basis vectors, we can convert between coordinate systems:
$$x = P_\mathcal{B} \{\mathbf{x}\}_\mathcal{B},$$
$$P_\mathcal{B}^{-1} x = \{\mathbf{x}\}_\mathcal{B}.$$

3.2.5 Exercises

1. Shown in Figure 3.2.13 are two vectors $\mathbf{v}_1$ and $\mathbf{v}_2$ in the plane $\mathbb{R}^2$.

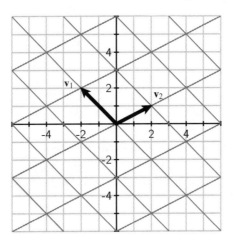

Figure 3.2.13 Vectors $\mathbf{v}_1$ and $\mathbf{v}_2$ in $\mathbb{R}^2$.

a. Explain why $\mathcal{B} = \{\mathbf{v}_1, \mathbf{v}_2\}$ is a basis for $\mathbb{R}^2$.

b. Using Figure 3.2.13, indicate the vectors $\mathbf{x}$ such that

1. $\{\mathbf{x}\}_\mathcal{B} = \begin{bmatrix} 2 \\ -1 \end{bmatrix}$

3.2. BASES AND COORDINATE SYSTEMS

2. $\{x\}_\mathcal{B} = \begin{bmatrix} -1 \\ -2 \end{bmatrix}$

3. $\{x\}_\mathcal{B} = \begin{bmatrix} 0 \\ 3 \end{bmatrix}$

c. Using Figure 3.2.13, find the representation $\{x\}_\mathcal{B}$ if

1. $x = \begin{bmatrix} -2 \\ -1 \end{bmatrix}$.

2. $x = \begin{bmatrix} 2 \\ 4 \end{bmatrix}$.

3. $x = \begin{bmatrix} 2 \\ -5 \end{bmatrix}$.

d. Find $\{x\}_\mathcal{B}$ if $x = \begin{bmatrix} 60 \\ 90 \end{bmatrix}$.

2. Consider vectors

$$v_1 = \begin{bmatrix} 1 \\ 2 \end{bmatrix}, v_2 = \begin{bmatrix} 1 \\ -3 \end{bmatrix}$$

$$w_1 = \begin{bmatrix} 2 \\ 3 \end{bmatrix}, w_2 = \begin{bmatrix} -1 \\ -2 \end{bmatrix}.$$

and let $\mathcal{B} = \{v_1, v_2\}$ and $\mathcal{C} = \{w_1, w_2\}$.

a. Explain why $\mathcal{B}$ and $\mathcal{C}$ are both bases of $\mathbb{R}^2$.

b. If $x = \begin{bmatrix} 5 \\ -3 \end{bmatrix}$, find $\{x\}_\mathcal{B}$ and $\{x\}_\mathcal{C}$.

c. If $\{x\}_\mathcal{B} = \begin{bmatrix} 2 \\ -4 \end{bmatrix}$, find x and $\{x\}_\mathcal{C}$.

d. If $\{x\}_\mathcal{C} = \begin{bmatrix} -3 \\ 2 \end{bmatrix}$, find x and $\{x\}_\mathcal{B}$.

e. Find a matrix Q such that $\{x\}_\mathcal{B} = Q\{x\}_\mathcal{C}$.

3. Consider the following vectors in $\mathbb{R}^4$:

$$v_1 = \begin{bmatrix} 1 \\ 1 \\ 1 \\ 1 \end{bmatrix}, v_2 = \begin{bmatrix} 0 \\ 1 \\ 1 \\ 1 \end{bmatrix}, v_3 = \begin{bmatrix} 0 \\ 0 \\ 1 \\ 1 \end{bmatrix}, v_4 = \begin{bmatrix} 0 \\ 0 \\ 0 \\ 1 \end{bmatrix}.$$

a. Explain why $\mathcal{B} = \{v_1, v_2, v_3, v_4\}$ forms a basis for $\mathbb{R}^4$.

b. Explain how to convert $\{x\}_\mathcal{B}$, the representation of a vector x in the coordinates defined by $\mathcal{B}$, into x, its representation in the standard coordinate system.

c. Explain how to convert the vector **x** into $\{x\}_{\mathcal{B}}$, its representation in the coordinate system defined by $\mathcal{B}$.

d. If $\mathbf{x} = \begin{bmatrix} 23 \\ 12 \\ 10 \\ 19 \end{bmatrix}$, find $\{x\}_{\mathcal{B}}$.

e. If $\{x\}_{\mathcal{B}} = \begin{bmatrix} 3 \\ 1 \\ -3 \\ -4 \end{bmatrix}$, find **x**.

4. Consider the following vectors in $\mathbb{R}^3$:

$$\mathbf{v}_1 = \begin{bmatrix} 1 \\ 3 \\ 2 \end{bmatrix}, \mathbf{v}_2 = \begin{bmatrix} 0 \\ 1 \\ 4 \end{bmatrix}, \mathbf{v}_3 = \begin{bmatrix} -2 \\ -5 \\ 0 \end{bmatrix}, \mathbf{v}_4 = \begin{bmatrix} -2 \\ -1 \\ -1 \end{bmatrix}, \mathbf{v}_5 = \begin{bmatrix} 1 \\ -2 \\ -1 \end{bmatrix}.$$

a. Do these vectors form a basis for $\mathbb{R}^3$? Explain your thinking.

b. Find a subset of these vectors that forms a basis of $\mathbb{R}^3$.

c. Suppose you have a set of vectors $\mathbf{v}_1, \mathbf{v}_2, \ldots, \mathbf{v}_6$ in $\mathbb{R}^4$ such that

$$\begin{bmatrix} \mathbf{v}_1 & \mathbf{v}_2 & \cdots & \mathbf{v}_6 \end{bmatrix} \sim \begin{bmatrix} 1 & 0 & -2 & 0 & 1 & 0 \\ 0 & 1 & 3 & 0 & -4 & 0 \\ 0 & 0 & 0 & 1 & 2 & 0 \\ 0 & 0 & 0 & 0 & 0 & 1 \end{bmatrix}.$$

Find a subset of the vectors that forms a basis for $\mathbb{R}^4$.

5. This exercise involves a simple Fourier transform, which will play an important role in the next section.

Suppose that we have the vectors

$$\mathbf{v}_1 = \begin{bmatrix} 1 \\ 1 \\ 1 \end{bmatrix}, \mathbf{v}_2 = \begin{bmatrix} \cos\left(\frac{\pi}{6}\right) \\ \cos\left(\frac{3\pi}{6}\right) \\ \cos\left(\frac{5\pi}{6}\right) \end{bmatrix}, \mathbf{v}_3 = \begin{bmatrix} \cos\left(\frac{2\pi}{6}\right) \\ \cos\left(\frac{6\pi}{6}\right) \\ \cos\left(\frac{10\pi}{6}\right) \end{bmatrix}.$$

a. Explain why $\mathcal{B} = \{\mathbf{v}_1, \mathbf{v}_2, \mathbf{v}_3\}$ is a basis for $\mathbb{R}^3$. Notice that you may enter $\cos\left(\frac{\pi}{6}\right)$ into Sage as cos(pi/6).

b. If $\mathbf{x} = \begin{bmatrix} 15 \\ 15 \\ 15 \end{bmatrix}$, find $\{x\}_{\mathcal{B}}$.

c. Find the matrices $P_\mathcal{B}$ and $P_\mathcal{B}^{-1}$. If $\mathbf{x} = \begin{bmatrix} x_1 \\ x_2 \\ x_3 \end{bmatrix}$ and $\{\mathbf{x}\}_\mathcal{B} = \begin{bmatrix} c_1 \\ c_2 \\ c_3 \end{bmatrix}$, explain why c_1 is the average of x_1, x_2, and x_3.

6. Determine whether the following statements are true or false and provide a justification for your response.

   a. If the columns of a matrix A form a basis for $\mathbb{R}^m$, then A is invertible.

   b. There must be 125 vectors in a basis for $\mathbb{R}^{125}$.

   c. If $\mathcal{B} = \{\mathbf{v}_1, \mathbf{v}_2, \ldots, \mathbf{v}_n\}$ is a basis of $\mathbb{R}^m$, then every vector in $\mathbb{R}^m$ can be expressed as a linear combination of basis vectors.

   d. The coordinates $\{\mathbf{x}\}_\mathcal{B}$ are the weights that form $\mathbf{x}$ as a linear combination of basis vectors.

   e. If the basis vectors form the columns of the matrix $P_\mathcal{B}$, then $\{\mathbf{x}\}_\mathcal{B} = P_\mathcal{B} \mathbf{x}$.

7. Provide a justification for your response to each of the following questions.

   a. Suppose you have m linearly independent vectors in $\mathbb{R}^m$. Can you guarantee that they form a basis of $\mathbb{R}^m$?

   b. If A is an invertible $m \times m$ matrix, do the columns necessarily form a basis of $\mathbb{R}^m$?

   c. Suppose we have an invertible $m \times m$ matrix A, and we perform a sequence of row operations on A to form a matrix B. Can you guarantee that the columns of B form a basis for $\mathbb{R}^m$?

   d. Suppose you have a set of 10 vectors in $\mathbb{R}^{10}$ and that every vector in $\mathbb{R}^{10}$ can be written as a linear combination of these vectors. Can you guarantee that this set of vectors is a basis for $\mathbb{R}^{10}$?

8. Crystallographers find it convenient to use coordinate systems that are adapted to the specific geometry of a crystal. As a two-dimensional example, consider a layer of graphite in which carbon atoms are arranged in regular hexagons to form the crystalline structure shown in Figure 3.2.14.

CHAPTER 3. INVERTIBILITY, BASES, AND COORDINATE SYSTEMS

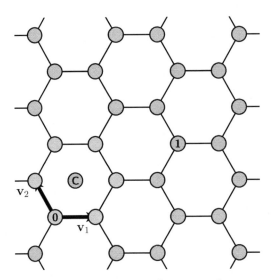

Figure 3.2.14 A layer of carbon atoms in a graphite crystal.

The origin of the coordinate system is at the carbon atom labeled by "0". It is convenient to choose the basis $\mathcal{B}$ defined by the vectors v_1 and v_2 and the coordinate system it defines.

a. Locate the points **x** for which

   i. $\{x\}_{\mathcal{B}} = \begin{bmatrix} 1 \\ 0 \end{bmatrix}$,

   ii. $\{x\}_{\mathcal{B}} = \begin{bmatrix} 0 \\ 1 \end{bmatrix}$,

   iii. $\{x\}_{\mathcal{B}} = \begin{bmatrix} 2 \\ 1 \end{bmatrix}$.

b. Find the coordinates $\{x\}_{\mathcal{B}}$ for all the carbon atoms in the hexagon whose lower left vertex is labeled "0".

c. What are the coordinates $\{x\}_{\mathcal{B}}$ of the center of that hexagon, which is labeled "C"?

d. How do the coordinates of the atoms in the hexagon whose lower left corner is labeled "1" compare to the coordinates in the hexagon whose lower left corner is labeled "0"?

e. Does the point **x** whose coordinates are $\{x\}_{\mathcal{B}} = \begin{bmatrix} 16 \\ 4 \end{bmatrix}$ correspond to a carbon atom or the center of a hexagon?

3.2. BASES AND COORDINATE SYSTEMS

9. Suppose that $A = \begin{bmatrix} 2 & 1 \\ 1 & 2 \end{bmatrix}$ and

$$\mathbf{v}_1 = \begin{bmatrix} 1 \\ 1 \end{bmatrix}, \quad \mathbf{v}_2 = \begin{bmatrix} 1 \\ -1 \end{bmatrix}.$$

   a. Explain why $\mathcal{B} = \{\mathbf{v}_1, \mathbf{v}_2\}$ is a basis for $\mathbb{R}^2$.

   b. Find $A\mathbf{v}_1$ and $A\mathbf{v}_2$.

   c. Use what you found in the previous part of this problem to find $\{A\mathbf{v}_1\}_{\mathcal{B}}$ and $\{A\mathbf{v}_2\}_{\mathcal{B}}$.

   d. If $\{\mathbf{x}\}_{\mathcal{B}} = \begin{bmatrix} 1 \\ -5 \end{bmatrix}$, find $\{A\mathbf{x}\}_{\mathcal{B}}$.

   e. Find a matrix D such that $\{A\mathbf{x}\}_{\mathcal{B}} = D\{\mathbf{x}\}_{\mathcal{B}}$.

   You should find that the matrix D is a very simple matrix, which means that this basis $\mathcal{B}$ is well suited to study the effect of multiplication by A. This observation is the central idea of the next chapter.

3.3 Image compression

Digital images, such as the photographs taken on your phone, are displayed as a rectangular array of pixels. For example, the photograph in Figure 3.3.1 is 1440 pixels wide and 1468 pixels high. If we were to zoom in on the photograph, we would be able to see individual pixels, such as those shown on the right.

Figure 3.3.1 An image stored as a 1440×1468 array of pixels along with a close-up of a smaller 8 × 8 array.

A lot of data is required to display this image. A quantity of digital data is frequently measured in bytes, where one byte is the amount of storage needed to record an integer between 0 and 255. As we will see shortly, each pixel requires three bytes to record that pixel's color. This means the amount of data required to display this image is $3 \times 1440 \times 1468 = 6,341,760$ bytes or about 6.3 megabytes.

Of course, we would like to store this image on a phone or computer and perhaps transmit it through our data plan to share it with others. If possible, we would like to find a way to represent this image using a smaller amount of data so that we don't run out of memory on our phone and quickly exhaust our data plan.

As we will see in this section, the JPEG compression algorithm provides a means for doing just that. This image, when stored in the JPEG format, requires only 467,359 bytes of data, which is about 7% of the 6.3 megabytes required to display the image. That is, when we display this image, we are reconstructing it from only 7% of the original data. This isn't too surprising since there is quite a bit of redundancy in the image; the left half of the image is almost uniformly blue. The JPEG algorithm detects this redundancy by representing the data using bases that are well-suited to the task.

Preview Activity 3.3.1. Since we will be using various bases and the coordinate systems they define, let's review how to translate between coordinate systems.

  a. Suppose that we have a basis $\mathcal{B} = \{\mathbf{v}_1, \mathbf{v}_2, \ldots, \mathbf{v}_m\}$ for $\mathbb{R}^m$. Explain what we

3.3. IMAGE COMPRESSION

mean by the representation $\{\mathbf{x}\}_{\mathcal{B}}$ of a vector $\mathbf{x}$ in the coordinate system defined by $\mathcal{B}$.

b. If we are given the representation $\{\mathbf{x}\}_{\mathcal{B}}$, how can we recover the vector $\mathbf{x}$?

c. If we are given the vector $\mathbf{x}$, how can we find $\{\mathbf{x}\}_{\mathcal{B}}$?

d. Suppose that
$$\mathcal{B} = \left\{ \begin{bmatrix} 1 \\ 3 \end{bmatrix}, \begin{bmatrix} 1 \\ 1 \end{bmatrix} \right\}$$
is a basis for $\mathbb{R}^2$. If $\{\mathbf{x}\}_{\mathcal{B}} = \begin{bmatrix} 1 \\ -2 \end{bmatrix}$, find the vector $\mathbf{x}$.

e. If $\mathbf{x} = \begin{bmatrix} 2 \\ -4 \end{bmatrix}$, find $\{\mathbf{x}\}_{\mathcal{B}}$.

3.3.1 Color models

A color is represented digitally by a vector in $\mathbb{R}^3$. There are different ways in which we can represent colors, however, depending on whether a computer or a human will be processing the color. We will describe two of these representations, called *color models*, and demonstrate how they are used in the JPEG compression algorithm.

Digital displays typically create colors by blending together various amounts of red, green, and blue. We can therefore describe a color by putting its constituent amounts of red, green, and blue into a vector $\begin{bmatrix} R \\ G \\ B \end{bmatrix}$. The quantities R, G, and B are each stored with one byte of information so they are integers between 0 and 255. This is called the *RGB* color model.

We define a basis $\mathcal{B} = \{\mathbf{v}_1, \mathbf{v}_2, \mathbf{v}_3\}$ where
$$\mathbf{v}_1 = \begin{bmatrix} 1 \\ 1 \\ 1 \end{bmatrix}, \quad \mathbf{v}_2 = \begin{bmatrix} 0 \\ -0.34413 \\ 1.77200 \end{bmatrix}, \quad \mathbf{v}_3 = \begin{bmatrix} 1.40200 \\ -0.71414 \\ 0 \end{bmatrix}$$
to define a new coordinate system with coordinates we denote Y, C_b, and C_r:
$$\left\{ \begin{bmatrix} R \\ G \\ B \end{bmatrix} \right\}_{\mathcal{B}} = \begin{bmatrix} Y \\ C_b \\ C_r \end{bmatrix}.$$

The coordinate Y is called *luminance* while C_b and C_r are called blue and red *chrominance*, respectively. In this coordinate system, luminance will vary from 0 to 255, while the chrominances vary between -127.5 and 127.5. This is known as the YC_bC_r color model. (To be completely accurate, we should add 127.5 to the chrominance values so that they lie between 0 and 255, but we won't worry about that here.)

Activity 3.3.2. This activity investigates these two color models, which we view as coordinate systems for describing colors.

a. First, we will explore the RGB color model.

There is an interactive diagram, available at the top of the page gvsu.edu/s/0Jc, that accompanies this activity.

Figure 3.3.2 The RGB color model.

1. What happens when $G = 0$, $B = 0$ (pushed all the way to the left), and R is allowed to vary?
2. What happens when $R = 0$, $G = 0$, and B is allowed to vary?
3. How can you create black in this color model?
4. How can you create white?

b. Next, we will explore the YC_bC_r color model.

There is an interactive diagram, available in the middle of the page gvsu.edu/s/0Jc, that accompanies this activity.

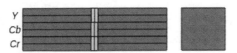

Figure 3.3.3 The YC_bC_r color model.

1. What happens when $C_b = 0$ and $C_r = 0$ (kept in the center) and Y is allowed to vary?
2. What happens when $Y = 0$ (pushed to the left), $C_r = 0$ (kept in the center), and C_b is allowed to increase between 0 and 127.5?
3. What happens when $Y = 0$, $C_b = 0$, and C_r is allowed to increase between 0 and 127.5?
4. How can you create black in this color model?
5. How can you create white?

c. Verify that $\mathcal{B}$ is a basis for $\mathbb{R}^3$.

d. Find the matrix $P_{\mathcal{B}}$ that converts from $\begin{bmatrix} Y \\ C_b \\ C_r \end{bmatrix}$ coordinates into $\begin{bmatrix} R \\ G \\ B \end{bmatrix}$ coordinates. Then find the matrix $P_{\mathcal{B}}^{-1}$ that converts from $\begin{bmatrix} R \\ G \\ B \end{bmatrix}$ coordinates back into

3.3. IMAGE COMPRESSION

$\begin{bmatrix} Y \\ C_b \\ C_r \end{bmatrix}$ coordinates.

e. Find the $\begin{bmatrix} Y \\ C_b \\ C_r \end{bmatrix}$ coordinates for the following colors and check, using the diagrams above, that the two representations agree.

1. Pure red is $\begin{bmatrix} R \\ G \\ B \end{bmatrix} = \begin{bmatrix} 255 \\ 0 \\ 0 \end{bmatrix}$.

2. Pure blue is $\begin{bmatrix} R \\ G \\ B \end{bmatrix} = \begin{bmatrix} 0 \\ 0 \\ 255 \end{bmatrix}$.

3. Pure white is $\begin{bmatrix} R \\ G \\ B \end{bmatrix} = \begin{bmatrix} 255 \\ 255 \\ 255 \end{bmatrix}$.

4. Pure black is $\begin{bmatrix} R \\ G \\ B \end{bmatrix} = \begin{bmatrix} 0 \\ 0 \\ 0 \end{bmatrix}$.

f. Find the $\begin{bmatrix} R \\ G \\ B \end{bmatrix}$ coordinates for the following colors and check, using the diagrams above, that the two representations agree.

1. $\begin{bmatrix} Y \\ C_b \\ C_r \end{bmatrix} = \begin{bmatrix} 128 \\ 0 \\ 0 \end{bmatrix}$.

2. $\begin{bmatrix} Y \\ C_b \\ C_r \end{bmatrix} = \begin{bmatrix} 128 \\ 60 \\ 0 \end{bmatrix}$.

3. $\begin{bmatrix} Y \\ C_b \\ C_r \end{bmatrix} = \begin{bmatrix} 128 \\ 0 \\ 60 \end{bmatrix}$.

g. Write an expression for

1. The luminance Y as it depends on R, G, and B.
2. The blue chrominance C_b as it depends on R, G, and B.
3. The red chrominance C_r as it depends on R, G, and B.

Explain how these quantities can be roughly interpreted by stating that

1. the luminance represents the brightness of the color.
2. the blue chrominance measures the amount of blue in the color.
3. the red chrominance measures the amount of red in the color.

These two color models provide us with two ways to represent colors, each of which is useful in a certain context. Digital displays, such as those in phones and computer monitors, create colors by combining various amounts of red, green, and blue. The RGB model is therefore most relevant in digital applications.

By contrast, the YC_bC_r color model was created based on research into human vision and aims to concentrate the most visually important data into a single coordinate, the luminance, to which our eyes are most sensitive. Of course, any basis of $\mathbb{R}^3$ must have three vectors so we need two more coordinates, blue and red chrominance, if we want to represent all colors.

To see this explicitly, shown in Figure 3.3.4 is the original image and the image as rendered with only the luminance. That is, on the right, the color of each pixel is represented by only one byte, which is the luminance. This image essentially looks like a grayscale version of the original image with all its visual detail. In fact, before digital television became the standard, television signals were broadcast using the YC_bC_r color model. When a signal was displayed on a black-and-white television, the luminance was displayed and the two chrominance values simply ignored.

Figure 3.3.4 The original image rendered with only the luminance values.

For comparison, shown in Figure 3.3.5 are the corresponding images created using only the blue chrominance and the red chrominance. Notice that the amount of visual detail is considerably less in these images.

3.3. IMAGE COMPRESSION

Figure 3.3.5 The original image rendered, on the left, with only blue chrominance and, on the right, with only red chrominance.

The aim of the JPEG compression algorithm is to represent an image using the smallest amount of data possible. By converting from the RGB color model to the YC_bC_r color model, we are concentrating the most visually important data into the luminance values. This is helpful because we can safely ignore some of the data in the chrominance values since that data is not as visually important.

3.3.2 The JPEG compression algorithm

The key to representing the image using a smaller amount of data is to detect redundancies in the data. To begin, we first break the image, which is composed of 1440×1468 pixels, into small 8×8 blocks of pixels. For example, we will consider the 8×8 block of pixels outlined in green in the original image, shown on the left of Figure 3.3.6. The image on the right zooms in on the block.

Figure 3.3.6 An 8 × 8 block of pixels outlined in green in the original image on the left. We see the same block on a smaller scale on the right.

Notice that this block, as seen in the original image, is very small. If we were to change some of the colors in this block slightly, our eyes would probably not notice.

Here we see a close-up of the block. The important point here is that the colors do not change too much over this block. In fact, we expect this to be true for most of the blocks. There will, of course, be some blocks that contain dramatic changes, such as where the sky and rock intersect, but they will be the exception.

Figure 3.3.7 The 8 × 8 block under consideration.

Following our earlier work, we will change the representation of colors from the RGB color model to the YC_bC_r model. This separates the colors into luminance and chrominance values that we will consider separately. In Figure 3.3.8, we see the luminance values of this block. Again, notice how these values do not vary significantly over the block.

3.3. IMAGE COMPRESSION

176	170	170	169	162	160	155	150
181	179	175	167	162	160	154	149
165	170	169	161	162	161	160	158
139	150	164	166	159	160	162	163
131	137	157	165	163	163	164	164
131	132	153	161	167	167	167	169
140	142	157	166	166	166	167	169
150	152	160	168	172	170	168	168

Figure 3.3.8 The luminance values in this block.

Our strategy in the compression algorithm is to perform a change of basis to take advantage of the fact that the luminance values do not change significantly over the block. Rather than recording the luminance of each of the pixels, this change of basis will allow us to record the average luminance along with some information about how the individual colors vary from the average.

Let's look at the first column of luminance values, which is a vector in $\mathbb{R}^8$:

$$\mathbf{x} = \begin{bmatrix} 176 \\ 181 \\ 165 \\ \vdots \\ 150 \end{bmatrix}.$$

We will perform a change of basis and describe this vector by the average of the luminance values and information about variations from the average.

The JPEG compression algorithm uses the *Discrete Fourier Transform*, which is defined using the basis $\mathcal{B}$ whose basis vectors are

$$\mathbf{v}_0 = \begin{bmatrix} \cos\left(\frac{(2\cdot 0+1)\cdot 0\pi}{16}\right) \\ \cos\left(\frac{(2\cdot 1+1)\cdot 0\pi}{16}\right) \\ \cos\left(\frac{(2\cdot 2+1)\cdot 0\pi}{16}\right) \\ \vdots \\ \cos\left(\frac{(2\cdot 7+1)\cdot 0\pi}{16}\right) \end{bmatrix}, \mathbf{v}_1 = \begin{bmatrix} \cos\left(\frac{(2\cdot 0+1)\cdot 1\pi}{16}\right) \\ \cos\left(\frac{(2\cdot 1+1)\cdot 1\pi}{16}\right) \\ \cos\left(\frac{(2\cdot 2+1)\cdot 1\pi}{16}\right) \\ \vdots \\ \cos\left(\frac{(2\cdot 7+1)\cdot 1\pi}{16}\right) \end{bmatrix},$$

$$\ldots, \mathbf{v}_6 = \begin{bmatrix} \cos\left(\frac{(2\cdot 0+1)\cdot 6\pi}{16}\right) \\ \cos\left(\frac{(2\cdot 1+1)\cdot 6\pi}{16}\right) \\ \cos\left(\frac{(2\cdot 2+1)\cdot 6\pi}{16}\right) \\ \vdots \\ \cos\left(\frac{(2\cdot 7+1)\cdot 6\pi}{16}\right) \end{bmatrix}, \mathbf{v}_7 = \begin{bmatrix} \cos\left(\frac{(2\cdot 0+1)\cdot 7\pi}{16}\right) \\ \cos\left(\frac{(2\cdot 1+1)\cdot 7\pi}{16}\right) \\ \cos\left(\frac{(2\cdot 2+1)\cdot 7\pi}{16}\right) \\ \vdots \\ \cos\left(\frac{(2\cdot 7+1)\cdot 7\pi}{16}\right) \end{bmatrix}.$$

On first glance, this probably looks intimidating, but we can make sense of it by looking at these vectors graphically. Shown in Figure 3.3.9 are four of these basis vectors. Notice that $\mathbf{v}_0$ is constantly 1, $\mathbf{v}_1$ varies relatively slowly, $\mathbf{v}_2$ varies a little more rapidly, and $\mathbf{v}_7$ varies quite rapidly. The main thing to notice is that: the basis vectors vary at different rates with the first vectors varying relatively slowly and the later vectors varying more rapidly.

3.3. IMAGE COMPRESSION

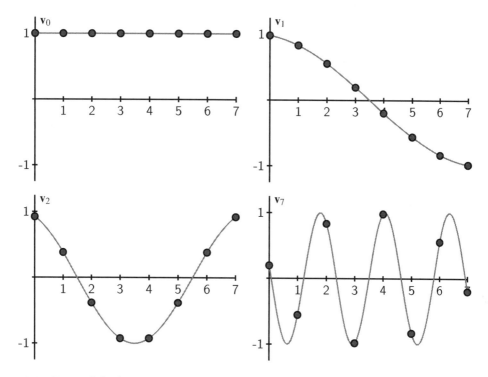

Figure 3.3.9 Four of the basis vectors v_0, v_1, v_2, and v_7.

These vectors form the basis $\mathcal{B}$ for $\mathbb{R}^8$. Remember that **x** is the vector of luminance values in the first column as seen on the right. We will write **x** in the new coordinates

$$\{x\}_\mathcal{B} = \begin{bmatrix} F_0 \\ F_1 \\ F_2 \\ \vdots \\ F_7 \end{bmatrix}.$$

The coordinates F_j are called the *Fourier coefficients* of the vector **x**.

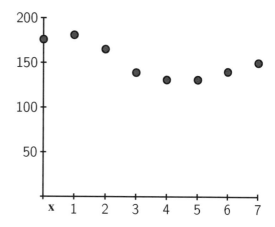

Activity 3.3.3. We will explore the influence that the Fourier coefficients have on the vector **x**.

a. To begin, we'll look at the Fourier coefficient F_0.

There is an interactive diagram that accompanies this part of the activity and that is available at the top of gvsu.edu/s/0Jd.

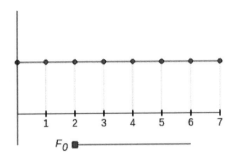

Figure 3.3.10 The effect of the Fourier coefficient F_0 on the vector $\mathbf{x} = F_0\mathbf{v}_0$.

Describe the effect that F_0 has on the vector $\mathbf{x}$. Would you describe the components in $\mathbf{x}$ as constant, slowly varying, or rapidly varying?

b. By comparison, let's see how the Fourier coefficient F_3 influences $\mathbf{x}$.

There is an interactive diagram that accompanies this part of the activity and that is available in the middle of gvsu.edu/s/0Jd.

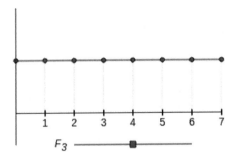

Figure 3.3.11 The effect of the Fourier coefficient F_3 on the vector $\mathbf{x} = F_3\mathbf{v}_3$.

Describe the effect that F_3 has on the vector $\mathbf{x}$. Would you describe the components in $\mathbf{x}$ as constant, slowly varying, or rapidly varying?

c. Let's now investigate how the Fourier coefficient F_7 influences the vector $\mathbf{x}$.

3.3. IMAGE COMPRESSION

There is an interactive diagram that accompanies this part of the activity and that is available at the bottom of gvsu.edu/s/0Jd.

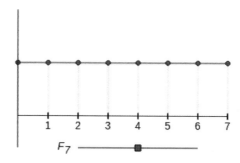

Figure 3.3.12 The effect of the Fourier coefficient F_0 on the vector $\mathbf{x} = F_7\mathbf{v}_7$.

Describe the effect that F_7 has on the vector $\mathbf{x}$. Would you describe the components in $\mathbf{x}$ as constant, slowly varying, or rapidly varying?

d. If the components of $\mathbf{x}$ vary relatively slowly, what would you expect to be true of the Fourier coefficients F_j?

e. The Sage cell below will construct the vector $P_\mathcal{B}$, which is denoted P, and its inverse $P_\mathcal{B}^{-1}$, which is denoted Pinv. Evaluate this Sage cell and notice that it prints the matrix $P_\mathcal{B}^{-1}$.

```
mat = [[cos((2*i+1)*j*pi/16) for j in range(8)] for i in
    range(8)]
P = matrix(mat).numerical_approx()
Pinv = P.inverse()
print (Pinv.numerical_approx(digits=3))
```

Now look at the form of $P_\mathcal{B}^{-1}$ and explain why F_0 is the average of the luminance values in the vector $\mathbf{x}$.

f. The Sage cell below defines the vector $\mathbf{x}$, which is the vector of luminance values in the first column, as seen in Figure 3.3.8. Use the cell below to find the vector $\mathbf{f}$ of Fourier coefficients $F_0, F_1, \ldots, F_7$. If you have evaluated the cell above, you will still be able to refer to P and Pinv in this cell.

```
x = vector([176,181,165,139,131,131,140,150])
#  find the vector of Fourier coefficients f below
f =
print (f.numerical_approx(digits=4))
```

Write the Fourier coefficients and discuss the relative sizes of the coefficients.

g. Let's see what happens when we simply ignore the coefficients F_6 and F_7. Form a new vector of Fourier coefficients by rounding the coefficients to the nearest integer and setting F_6 and F_7 to zero. This is an approximation to $\mathbf{f}$, the vector of Fourier coefficients. Use the approximation to **f** to form an approximation of the vector **x**.

```
# define fapprox below and then find xapprox
fapprox =
xapprox =
print ("x      =", x)
print ("xapprox=", xapprox.numerical_approx(digits=3))
```

How much does your approximation differ from the actual vector **x**?

h. When we ignore the Fourier coefficients corresponding to rapidly varying basis elements, we see that the vector **x** that we reconstruct is very close to the original one. In fact, the luminance values in the approximation differ by at most one or two from the actual luminance values. Our eyes are not sensitive enough to detect this difference.

So far, we have concentrated on only one column in our 8×8 block of luminance values. Let's now consider all of the columns. The following Sage cell defines a matrix called luminance, which is the 8×8 matrix of luminance values. Find the 8×8 matrix F whose columns are the Fourier coefficients of the columns of luminance values.

```
luminance = matrix(8,8, [176, 170, 170, 169, 162, 160, 155,
    150, 181,
179, 175, 167, 162, 160, 154, 149, 165, 170, 169, 161, 162,
    161, 160,
158, 139, 150, 164, 166, 159, 160, 162, 163, 131, 137, 157,
    165, 163,
163, 164, 164, 131, 132, 153, 161, 167, 167, 167, 169, 140,
    142, 157,
166, 166, 166, 167, 169, 150, 152, 160, 168, 172, 170, 168,
    168])
# define your matrix F below
F =
print (F.numerical_approx(digits=3))
```

i. Notice that the first row of this matrix consists of the Fourier coefficient F_0 for each of the columns. Just as we saw before, the entries in this row do not change significantly as we move across the row. In the Sage cell below, write these entries in the vector **y** and find the corresponding Fourier coefficients.

```
# define the vector y as the entries in the first row of F
y =
y_fourier =
print (y_fourier.numerical_approx(digits=3))
```

Up to this point, we have been working with the luminance values in one 8×8 block of our image. We formed the Fourier coefficients for each of the columns of this block. Once we

3.3. IMAGE COMPRESSION

notice that the Fourier coefficients across a row are relatively constant, it seems reasonable to find the Fourier coefficients of the *rows* of the matrix of Fourier coefficients. Doing so leads to the matrix

$$\begin{bmatrix} 160.6 & -4.0 & -4.8 & -1.7 & 0.0 & 0.9 & 0.8 & 0.3 \\ 2.7 & 14.7 & 3.8 & 1.1 & -1.6 & -0.3 & -0.3 & -0.4 \\ 3.8 & 7.0 & 2.1 & 2.9 & 0.8 & -0.2 & -0.3 & -0.3 \\ -2.4 & -3.9 & -1.9 & 0.1 & 1.2 & 1.2 & 0.7 & 0.1 \\ -0.6 & -1.4 & -1.5 & -0.9 & 0.2 & 0.6 & -0.2 & -0.5 \\ -0.7 & -1.6 & 0.0 & -1.1 & 0.0 & 0.3 & -0.1 & -0.2 \\ -0.0 & -1.4 & 0.4 & 0.9 & 0.1 & -0.5 & 0.0 & 0.5 \\ 0.0 & 0.2 & 0.3 & 0.3 & 0.0 & -0.0 & -0.2 & 0.0 \end{bmatrix}.$$

If we were to look inside a JPEG image file, we would see lots of matrices like this. For each 8×8 block, there would be three matrices of Fourier coefficients of the rows of Fourier coefficients, one matrix for each of the luminance, blue chrominance, and red chrominance values. However, we store these Fourier coefficients as integers inside the JPEG file so we need to round off the coefficients to the nearest integer, as shown here:

$$\begin{bmatrix} 161 & -4 & -5 & -2 & 0 & 1 & 1 & 0 \\ 3 & 15 & 4 & 1 & -2 & 0 & 0 & 0 \\ 4 & 7 & 2 & 3 & 1 & 0 & 0 & 0 \\ -2 & -4 & -2 & 0 & 1 & 1 & 1 & 0 \\ -1 & -1 & -1 & -1 & 0 & 1 & 0 & 0 \\ -1 & -2 & 0 & -1 & 0 & 0 & 0 & 0 \\ 0 & -1 & 0 & 1 & 0 & -1 & 0 & 1 \\ 0 & 0 & 0 & 0 & 0 & 0 & 0 & 0 \end{bmatrix}.$$

There are many zeroes in this matrix, and we can save space in a JPEG image file by only recording the *nonzero* Fourier coefficients.

In fact, when a JPEG file is created, there is a "quality" parameter that can be set, such as that shown in Figure 3.3.13. When the quality parameter is high, we will store many of the Fourier coefficients; when it is low, we will ignore more of them.

Figure 3.3.13 When creating a JPEG file, we choose a value of the "quality" parameter.

To see how this works, suppose the quality setting is relatively high. After rounding off the Fourier coefficients, we will set all of the coefficients whose absolute value is less than 2 to zero, which creates the matrix:

$$\begin{bmatrix} 161 & -4 & -5 & 0 & 0 & 0 & 0 & 0 \\ 3 & 15 & 4 & 0 & 0 & 0 & 0 & 0 \\ 4 & 7 & 2 & 3 & 0 & 0 & 0 & 0 \\ -2 & -4 & 0 & 0 & 0 & 0 & 0 & 0 \\ 0 & 0 & 0 & 0 & 0 & 0 & 0 & 0 \\ 0 & 0 & 0 & 0 & 0 & 0 & 0 & 0 \\ 0 & 0 & 0 & 0 & 0 & 0 & 0 & 0 \\ 0 & 0 & 0 & 0 & 0 & 0 & 0 & 0 \end{bmatrix}.$$

Notice that there are 12 nonzero Fourier coefficients, out of 64, that we need to record. Consequently, we only save $12/64 \approx 19\%$ of the data.

If instead, the quality setting is relatively low, we set all of the Fourier coefficients whose absolute value is less than 4 to zero, creating the matrix:

$$\begin{bmatrix} 161 & -4 & -5 & 0 & 0 & 0 & 0 & 0 \\ 0 & 15 & 0 & 0 & 0 & 0 & 0 & 0 \\ 0 & 7 & 0 & 0 & 0 & 0 & 0 & 0 \\ 0 & 0 & 0 & 0 & 0 & 0 & 0 & 0 \\ 0 & 0 & 0 & 0 & 0 & 0 & 0 & 0 \\ 0 & 0 & 0 & 0 & 0 & 0 & 0 & 0 \\ 0 & 0 & 0 & 0 & 0 & 0 & 0 & 0 \\ 0 & 0 & 0 & 0 & 0 & 0 & 0 & 0 \end{bmatrix}.$$

Notice that there are only 5 nonzero Fourier coefficients that we need to record now, meaning we save only $5/64 \approx 8\%$ of the data. This will result in a smaller JPEG file describing the image.

With a lower quality setting, we have thrown away more information about the Fourier coefficients so the image will not be reconstructed as accurately. To see this, we can reconstruct

3.3. IMAGE COMPRESSION

the luminance values from the Fourier coefficients by converting back into the standard coordinate system. Rather than showing the luminance values themselves, we will show the difference in the original luminance values and the reconstructed luminance values. When the quality setting was high and we stored 12 Fourier coefficients, we find this difference to be

$$\begin{bmatrix} -7 & -7 & -1 & 3 & -2 & -1 & 0 & -1 \\ 4 & 4 & 4 & -1 & -3 & 0 & -1 & -3 \\ 1 & 3 & 0 & -7 & -3 & 1 & 3 & 3 \\ -7 & -3 & 3 & 1 & -5 & -2 & 1 & 2 \\ 0 & -3 & 4 & 4 & -1 & -1 & -1 & -2 \\ 2 & -5 & 3 & 1 & 1 & -1 & -1 & 1 \\ 1 & -2 & 4 & 3 & -4 & -6 & -2 & 3 \\ 0 & -1 & 2 & 1 & -1 & -4 & -1 & 5 \end{bmatrix}.$$

When the quality setting is lower and we store only 5 Fourier coefficients, the difference is

$$\begin{bmatrix} 3 & -3 & -2 & 0 & 0 & 7 & 10 & 10 \\ 14 & 11 & 6 & -1 & -1 & 3 & 4 & 4 \\ 7 & 10 & 5 & -5 & -3 & -1 & 2 & 3 \\ -10 & -3 & 5 & 2 & -8 & -7 & -3 & -1 \\ -12 & -11 & 2 & 2 & -5 & -7 & -6 & -6 \\ -11 & -15 & -2 & -2 & -2 & -4 & -5 & -2 \\ -3 & -6 & 2 & 3 & -2 & -5 & -4 & -1 \\ 6 & 3 & 4 & 5 & 4 & 0 & -1 & 0 \end{bmatrix}.$$

This demonstrates the trade off. With a high quality setting, we require more storage to save more of the data, but the reconstructed image is closer to the original. With the lower quality setting, we require less storage, but the reconstructed image differs more from the original.

If we remember that the visual information stored by the blue and red chrominance values is not as important as that contained in the luminance values, we feel safer in discarding more of the Fourier coefficients for the chrominance values resulting in an even greater savings.

Shown in Figure 3.3.14 is the original image compared to a version stored with a very low quality setting. If you look carefully, you can individual 8×8 blocks.

Figure 3.3.14 The original image and the result of storing the image with a low quality setting.

This discussion of the JPEG compression algorithm is meant to explore the ideas that underlie its construction and demonstrate the importance of a choice of basis and its accompanying coordinate system. There are a few details, most notably about the rounding of the Fourier coefficients, that are not strictly accurate. The actual implementation is a little more complicated, but the presentation here conveys the spirit of the algorithm.

The JPEG compression algorithm allows us to store image files using only a fraction of the data. Similar ideas are used to efficiently store digital music and video files.

3.3.3 Summary

This section has explored how appropriate changes in bases help us reconstruct an image using only a fraction of its data. This is known as image compression.

- There are several ways of representing colors, all of which use vectors in $\mathbb{R}^3$. We explored the RGB color model, which is appropriate in digital applications, and the YC_bC_r model, in which the most important visual information is conveyed by the Y component, known as luminance.

- We also explored a change of basis called the Discrete Fourier Transform. In the coordinate system that results, the first coefficient measures the average of the components of a vector. Other coefficients measure variations in the components away from the average.

- We put both of these ideas to use in demonstrating the JPEG compression algorithm. An image is broken into 8×8 blocks, and the colors into luminance, blue chrominance, and red chrominance. Applying the Discrete Fourier Transform allows us to reconstruct a good approximation of the image using only a fraction of the original data.

3.3. IMAGE COMPRESSION

3.3.4 Exercises

1. Consider the vector $\mathbf{x} = \begin{bmatrix} 103 \\ 94 \\ 91 \\ 92 \\ 103 \\ 105 \\ 105 \\ 108 \end{bmatrix}$.

    a. In the Sage cell below is a copy of the change of basis matrices that define the Fourier transform. Find the Fourier coefficients of **x**.

    ```
    mat = [[cos((2*i+1)*j*pi/16) for j in range(8)] for i in
        range(8)]
    P = matrix(mat).numerical_approx()
    Pinv = P.inverse()
    print (Pinv.numerical_approx(digits=3))
    ```

    b. We will now form the vector **y**, which is an approximation of **x** by rounding all the Fourier coefficients of **x** to the nearest integer to obtain $\{\mathbf{y}\}_C$. Now find the vector **y** and compare this approximation to **x**. What is the error in this approximation?

    c. Repeat the last part of this problem, but set the rounded Fourier coefficients to zero if they have an absolute value less than five. Use it to create a second approximation of **x**. What is the error in this approximation?

    d. Compare the number of nonzero Fourier coefficients that you have in the two approximations and compare the accuracy of the approximations. Using a few sentences, discuss the comparisons that you find.

2. There are several steps to the JPEG compression algorithm. The following questions examine the motivation behind some of them.

    a. What is the overall goal of the JPEG compression algorithm?

    b. Why do we convert colors from the the RGB color model to the YC_bC_r model?

    c. Why do we decompose the image into a collection of 8×8 arrays of pixels?

    d. What role does the Discrete Fourier Transform play in the JPEG compression algorithm?

    e. Why is the information conveyed by the rapid-variation Fourier coefficients, generally speaking, less important than the slow-variation coefficients?

3. The Fourier transform that we used in this section is often called the Discrete Fourier Cosine Transform because it is defined using a basis C consisting of cosine functions. There is also a Fourier Sine Transform defined using a basis S consisting of sine func-

tions. For instance, in $\mathbb{R}^4$, the basis vectors of $\mathcal{S}$ are

$$\mathbf{v}_1 = \begin{bmatrix} \sin\left(\frac{1\cdot 1\pi}{8}\right) \\ \sin\left(\frac{3\cdot 1\pi}{8}\right) \\ \sin\left(\frac{5\cdot 1\pi}{8}\right) \\ \sin\left(\frac{7\cdot 1\pi}{8}\right) \end{bmatrix}, \mathbf{v}_2 = \begin{bmatrix} \sin\left(\frac{1\cdot 2\pi}{8}\right) \\ \sin\left(\frac{3\cdot 2\pi}{8}\right) \\ \sin\left(\frac{5\cdot 2\pi}{8}\right) \\ \sin\left(\frac{7\cdot 2\pi}{8}\right) \end{bmatrix},$$

$$\mathbf{v}_3 = \begin{bmatrix} \sin\left(\frac{1\cdot 3\pi}{8}\right) \\ \sin\left(\frac{3\cdot 3\pi}{8}\right) \\ \sin\left(\frac{5\cdot 3\pi}{8}\right) \\ \sin\left(\frac{7\cdot 3\pi}{8}\right) \end{bmatrix}, \mathbf{v}_4 = \begin{bmatrix} \sin\left(\frac{1\cdot 4\pi}{8}\right) \\ \sin\left(\frac{3\cdot 4\pi}{8}\right) \\ \sin\left(\frac{5\cdot 4\pi}{8}\right) \\ \sin\left(\frac{7\cdot 4\pi}{8}\right) \end{bmatrix}.$$

We can think of these vectors graphically, as shown in Figure 3.3.15.

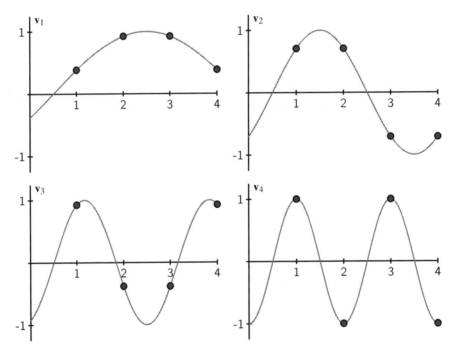

Figure 3.3.15 The vectors $\mathbf{v}_1, \mathbf{v}_2, \mathbf{v}_3, \mathbf{v}_4$ that form the basis $\mathcal{S}$.

a. The Sage cell below defines the matrix S whose columns are the vectors in the basis $\mathcal{S}$ as well as the matrix C whose columns form the basis $\mathcal{C}$ used in the Fourier Cosine Transform.

```
sinmat = [[sin((2*i+1)*j*pi/8) for j in range(1,5)] for i in
    range(4)]
cosmat = [[cos((2*i+1)*j*pi/8) for j in range(4)] for i in
    range(4)]
S = matrix(sinmat).numerical_approx()
C = matrix(cosmat).numerical_approx()
```

3.3. IMAGE COMPRESSION

In the 8×8 block of luminance values we considered in this section, the first column begins with the four entries 176, 181, 165, and 139, as seen in Figure 3.3.8. These form the vector $\mathbf{x} = \begin{bmatrix} 176 \\ 181 \\ 165 \\ 139 \end{bmatrix}$. Find both $\{\mathbf{x}\}_S$ and $\{\mathbf{x}\}_C$.

b. Write a sentence or two comparing the values for the Fourier Sine coefficients $\{\mathbf{x}\}_S$ and the Fourier Cosine coefficients $\{\mathbf{x}\}_C$.

c. Suppose now that $\mathbf{x} = \begin{bmatrix} 100 \\ 100 \\ 100 \\ 100 \end{bmatrix}$. Find the Fourier Sine coefficients $\{\mathbf{x}\}_S$ and the Fourier Cosine coefficients $\{\mathbf{x}\}_C$.

d. Write a few sentences explaining why we use the Fourier Cosine Transform in the JPEG compression algorithm rather than the Fourier Sine Transform.

4. In Example 3.2.9, we looked at a basis for $\mathbb{R}^4$ that we called the Haar wavelet basis. The basis vectors are

$$\mathbf{v}_1 = \begin{bmatrix} 1 \\ 1 \\ 1 \\ 1 \end{bmatrix}, \mathbf{v}_2 = \begin{bmatrix} 1 \\ 1 \\ -1 \\ -1 \end{bmatrix}, \mathbf{v}_3 = \begin{bmatrix} 1 \\ -1 \\ 0 \\ 0 \end{bmatrix}, \mathbf{v}_4 = \begin{bmatrix} 0 \\ 0 \\ 1 \\ -1 \end{bmatrix},$$

which may be understood graphically as in Figure 3.3.16. We will denote this basis by $\mathcal{W}$.

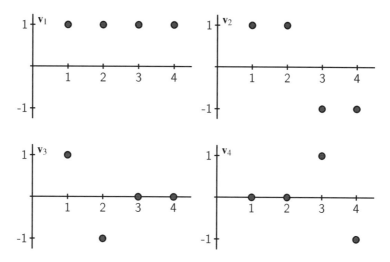

Figure 3.3.16 The Haar wavelet basis represented graphically.

The change of coordinates from a vector $\mathbf{x}$ in $\mathbb{R}^4$ to $\{\mathbf{x}\}_\mathcal{W}$ is called the Haar *wavelet transform* and we write

$$\{x\}_{\mathcal{W}} = \begin{bmatrix} H_1 \\ H_2 \\ H_3 \\ H_4 \end{bmatrix}.$$

The coefficients H_1, H_2, H_3, H_4 are called wavelet coefficients.

Let's work with the 4×4 block of luminance values in the upper left corner of our larger 8 × 8 block:

$$\begin{bmatrix} 176 & 170 & 170 & 169 \\ 181 & 179 & 175 & 167 \\ 165 & 170 & 169 & 161 \\ 139 & 150 & 164 & 166 \end{bmatrix}.$$

a. The following Sage cell defines the matrix W whose columns are the basis vectors in $\mathcal{W}$. If **x** is the first column of luminance values in the 4 × 4 block above, find the wavelet coefficients $\{x\}_{\mathcal{W}}$.

```
W = matrix(4,4,[1,1,1,0,1,1,-1,0,1,-1,0,1,1,-1,0,-1])
```

b. Notice that H_1 gives the average value of the components of **x** and H_2 describes how the averages of the first two and last two components differ from the overall average. The coefficients H_3 and H_4 describe small-scale variations between the first two components and last two components, respectively.

If we set the last wavelet coefficients $H_3 = 0$ and $H_4 = 0$, we obtain the wavelet coefficients $\{y\}_{\mathcal{W}}$ for a vector **y** that approximates **x**. Find the vector **y** and compare it to the original vector **x**.

c. What impact does the fact that $H_3 = 0$ and $H_4 = 0$ have on the form of the vector **y**? Explain how setting these coefficients to zero ignores the behavior of **x** on a small scale.

d. In the JPEG compression algorithm, we looked at the Fourier coefficients of all the columns of luminance values and then performed a Fourier transform on the rows. The Sage cell below will perform the same operation using the wavelet transform; that is, it will first find the wavelet coefficients of each of the columns and then perform the wavelet transform on the rows. You only need to evaluate the cell to find the wavelet coefficients obtained in this way.

```
luminance = matrix(4,4,[176, 170, 170, 169, 181, 179, 175,
    167, 165,
170, 169, 161, 139, 150,  164, 166])
Winv = W.inverse()
wavelet_transform =
    (Winv*(Winv*luminance).transpose()).transpose()
print (wavelet_transform.numerical_approx(digits=3))
```

3.3. IMAGE COMPRESSION

e. Now set all the wavelet coefficients equal to zero except those in the upper left 2×2 block and use them to define the matrix coeffs in the Sage cell below. This has the effect of ignoring all of the small-scale differences. Evaluating this cell will recover the approximate luminance values.

```
# define the matrix of coefficients below
coeffs = 
# this code will undo the wavelet transform
approx_luminance = W*((W*(coeffs.transpose())).transpose())
print (approx_luminance.numerical_approx(digits=3))
```

f. Explain how the wavelet transform and this approximation can be used to create a lower resolution version of the image.

This kind of wavelet transform is the basis of the JPEG 2000 compression algorithm, which is an alternative to the usual JPEG algorithm.

5. In this section, we looked at the RGB and YC_bC_r color models. In this exercise, we will look at the HSV color model where H is the hue, S is the saturation, and V is the value of the color. All three quantities vary between 0 and 255.

There is an interactive diagram, available at the bottom of the page gvsu.edu/s/0Jc, that accompanies this exercise.

Figure 3.3.17 The HSV color model.

a. If you leave S and V at some fixed values, what happens when you change the value of H?

b. Increase the value V to 255, which is on the far right. Describe what happens when you vary the saturation S using a fixed hue H and value V.

c. Describe what happens when H and S are fixed and V varies.

d. How can you create white in this color model?

e. How can you create black in this color model?

f. Find an approximate range of hues that correspond to blue.

g. Find an approximate range of hues that correspond to green.

The YC_bC_r color model concentrates the most important visual information in the luminance coordinate, which roughly measures the brightness of the color. The other two coordinates describe the hue of the color. By contrast, the HSV color model concentrates all the information about the hue in the H coordinate.

This is useful in computer vision applications. For instance, if we want a robot to detect a blue ball in its field of vision, we can specify a range of hue values to search for. If the lighting changes in the room, the saturation and value may change, but the hue will not. This increases the likelihood that the robot will still detect the blue ball across a wide range of lighting conditions.

3.4 Determinants

As invertibility plays a central role in this chapter, we need a criterion that tells us when a matrix is invertible. We already know that a square matrix is invertible if and only if it is row equivalent to the identity matrix. In this section, we will develop a second, numerical criterion that tells us when a square matrix is invertible.

To begin, let's consider a 2×2 matrix A whose columns are vectors $\mathbf{v}_1$ and $\mathbf{v}_2$. We have frequently drawn the vectors and studied the linear combinations they form using a figure such as Figure 3.4.1.

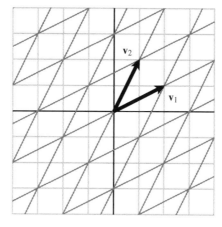

Figure 3.4.1 Linear combinations of two vectors $\mathbf{v}_1$ and $\mathbf{v}_2$ form a collection of congruent parallelograms.

Notice how the linear combinations form a set of congruent parallelograms in the plane. In this section, we will use the area of these parallelograms to define a numerical quantity called the determinant that tells us whether the matrix A is invertible.

To recall, the area of parallelogram is found by multiplying the length of one side by the perpendicular distance to its parallel side. Using the notation in the figure, the area of the parallelogram is bh.

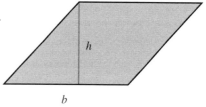

Preview Activity 3.4.1. We will explore the area formula in this preview activity.

    a. Find the area of the following parallelograms.

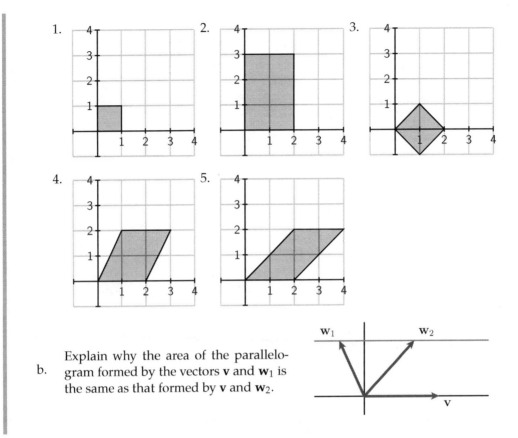

b. Explain why the area of the parallelogram formed by the vectors $\mathbf{v}$ and $\mathbf{w}_1$ is the same as that formed by $\mathbf{v}$ and $\mathbf{w}_2$.

3.4.1 Determinants of 2×2 matrices

We will begin by defining the determinant of a 2×2 matrix $A = \begin{bmatrix} \mathbf{v}_1 & \mathbf{v}_2 \end{bmatrix}$. First, however, we need to define the orientation of an ordered pair of vectors. As shown in Figure 3.4.2, an ordered pair of vectors $\mathbf{v}_1$ and $\mathbf{v}_2$ is called *positively* oriented if the angle, measured in the counterclockwise direction, from $\mathbf{v}_1$ to $\mathbf{v}_2$ is less than $180°$; we say the pair is *negatively* oriented if it is more than $180°$.

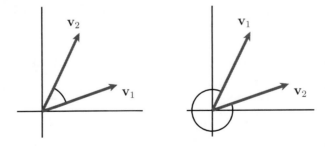

Figure 3.4.2 The vectors on the left are positively oriented while the ones on the right are negatively oriented.

3.4. DETERMINANTS

Definition 3.4.3 Suppose a 2×2 matrix A has columns $\mathbf{v}_1$ and $\mathbf{v}_2$. If the pair of vectors is positively oriented, then the *determinant* of A, denoted $\det(A)$, is the area of the parallelogram formed by $\mathbf{v}_1$ and $\mathbf{v}_2$. If the pair is negatively oriented, then $\det(A)$ is minus the area of the parallelogram.

Example 3.4.4 Consider the determinant of the identity matrix

$$I = \begin{bmatrix} 1 & 0 \\ 0 & 1 \end{bmatrix} = \begin{bmatrix} \mathbf{e}_1 & \mathbf{e}_2 \end{bmatrix}.$$

As seen on the left of Figure 3.4.5, the vectors $\mathbf{v}_1 = \mathbf{e}_1$ and $\mathbf{v}_2 = \mathbf{e}_2$ form a positively oriented pair. Since the parallelogram they form is a 1×1 square, we have $\det(I) = 1$.

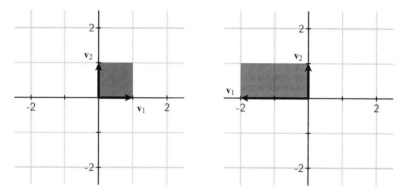

Figure 3.4.5 The determinant $\det(I) = 1$, as seen on the left. On the right, we see that $\det(A) = -2$ where A is the matrix whose columns are shown.

Now consider the matrix

$$A = \begin{bmatrix} -2 & 0 \\ 0 & 1 \end{bmatrix} = \begin{bmatrix} \mathbf{v}_1 & \mathbf{v}_2 \end{bmatrix}.$$

As seen on the right of Figure 3.4.5, the vectors $\mathbf{v}_1$ and $\mathbf{v}_2$ form a negatively oriented pair. The parallelogram they define is a 2×1 rectangle so we have $\det(A) = -2$.

Activity 3.4.2. In this activity, we will find the determinant of some simple 2×2 matrices and discover some important properties of determinants.

There is an interactive diagram at gvsu.edu/s/0J9 that accompanies this activity.

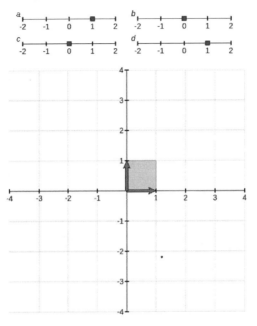

Figure 3.4.6 The geometric meaning of the determinant of a matrix.

a. Use the diagram to find the determinant of the matrix $\begin{bmatrix} -\frac{1}{2} & 0 \\ 0 & 2 \end{bmatrix}$. Along with Example 3.4.4, what does this lead you to believe is generally true about the determinant of a diagonal matrix?

b. Use the diagram to find the determinant of the matrix $\begin{bmatrix} 0 & 1 \\ 1 & 0 \end{bmatrix}$. What is the geometric effect of the matrix transformation defined by this matrix?

c. Use the diagram to find the determinant of the matrix $\begin{bmatrix} 2 & 1 \\ 0 & 1 \end{bmatrix}$. More generally, what do you notice about the determinant of any matrix of the form $\begin{bmatrix} 2 & k \\ 0 & 1 \end{bmatrix}$? What does this say about the determinant of an upper triangular matrix?

d. Use the diagram to find the determinant of any matrix of the form $\begin{bmatrix} 2 & 0 \\ k & 1 \end{bmatrix}$. What does this say about the determinant of a lower triangular matrix?

e. Use the diagram to find the determinant of the matrix $\begin{bmatrix} 1 & -1 \\ -2 & 2 \end{bmatrix}$. In general, what is the determinant of a matrix whose columns are linearly dependent?

3.4. DETERMINANTS

f. Consider the matrices

$$A = \begin{bmatrix} 2 & 1 \\ 2 & -1 \end{bmatrix}, \quad B = \begin{bmatrix} 1 & 0 \\ 0 & 2 \end{bmatrix}.$$

Use the diagram to find the determinants of A, B, and AB. What does this suggest is generally true about the relationship of $\det(AB)$ to $\det(A)$ and $\det(B)$?

Later in this section, we will learn an algebraic technique for computing determinants. In the meantime, we will simply note that we can define determinants for $n \times n$ matrices by measuring the volume of a box defined by the columns of the matrix, even if this box resides in $\mathbb{R}^n$ for some very large n.

For example, the columns of a 3×3 matrix A will form a parallelpiped, like the one shown here, and there is a means by which we can classify sets of such vectors as either positively or negatively oriented. Therefore, we can define the determinant $\det(A)$ in terms of the volume of the parallelpiped, but we will not worry about the details here.

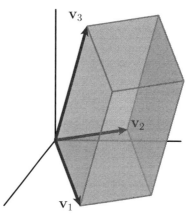

Though the previous activity deals with determinants of 2×2 matrices, it illustrates some important properties of determinants that are true more generally.

- If A is a triangular matrix, then $\det(A)$ equals the product of the entries on the diagonal. For example,

$$\det \begin{bmatrix} 2 & 2 \\ 0 & 3 \end{bmatrix} = 2 \cdot 3 = 6,$$

since the two parallelograms in Figure 3.4.7 have equal area.

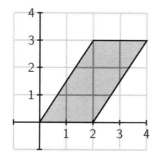
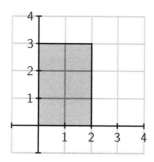

Figure 3.4.7 The determinant of a triangular matrix equals the product of its diagonal entries.

- We also saw that
$$\det\begin{bmatrix} 0 & 1 \\ 1 & 0 \end{bmatrix} = -1$$
because the columns form a negatively oriented pair. You may remember from Subsection 3.1.3 that a matrix such as this is obtained by interchanging two rows of the identity matrix.

- The determinant satisfies a multiplicative property, which says that
$$\det(AB) = \det(A)\det(B).$$

Rather than simply thinking of the determinant as the area of a parallelogram, we may also think of it as a factor by which areas are scaled under the matrix transformation defined by the matrix. Applying the matrix transformation defined by B will scale area by $\det(B)$. If we then compose B with the matrix transformation defined by A, area will scale a second time by the factor $\det(A)$. The net effect is that the matrix transformation defined by AB scales area by $\det(A)\det(B)$ so that $\det(AB) = \det(A)\det(B)$.

Proposition 3.4.8 *The determinant satisfies these properties:*

- *The determinant of a triangular matrix equals the product of its diagonal entries.*

- *If P is obtained by interchanging two rows of the identity matrix, then $\det(P) = -1$.*

- $\det(AB) = \det(A)\det(B).$

3.4.2 Determinants and invertibility

Perhaps the most important property of determinants also appeared in the previous activity. We saw that when the columns of the matrix A are linearly dependent, the parallelogram formed by those vectors folds down onto a line. For instance, if $A = \begin{bmatrix} 1 & 2 \\ -1 & -2 \end{bmatrix}$, then the resulting parallelogram, as shown in Figure 3.4.9, has zero area, which means that $\det(A) = 0$.

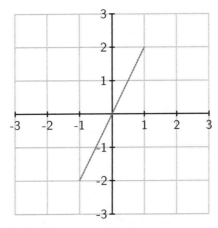

Figure 3.4.9 When the columns of A are linearly dependent, we find that $\det(A) = 0$.

3.4. DETERMINANTS

The condition that the columns of A are linearly dependent is precisely the same as the condition that A is not invertible. This leads us to believe that A is not invertible if and only if its determinant is zero. The following proposition expresses this thought.

Proposition 3.4.10 *The matrix A is invertible if and only if $\det(A) \neq 0$.*

To understand this proposition more fully, let's remember that the matrix A is invertible if and only if it is row equivalent to the identity matrix I. We will therefore consider how the determinant changes when we perform row operations on a matrix. Along the way, we will discover an effective means to compute the determinant.

In Subsection 3.1.3, we saw how to describe the three row operations, scaling, interchange, and replacement, using matrix multiplication. If we perform a row operation on the matrix A to obtain the matrix A', we would like to relate $\det(A)$ and $\det(A')$. To do so, remember that

- Scalings are performed by multiplying a matrix A by a diagonal matrix, such as

$$S = \begin{bmatrix} 1 & 0 & 0 \\ 0 & 3 & 0 \\ 0 & 0 & 1 \end{bmatrix},$$

which has the effect of multiplying the second row of A by 3 to obtain A'. Since S is diagonal, we know that its determinant is the product of its diagonal entries so that $\det(S) = 3$. This means that $A' = SA$ and therefore

$$\det(A') = \det(S)\det(A) = 3\det(A).$$

In general, if we scale a row of A by k, we have $\det(A') = k\det(A)$.

- Interchanges are performed by matrices such as

$$P = \begin{bmatrix} 0 & 1 & 0 \\ 1 & 0 & 0 \\ 0 & 0 & 1 \end{bmatrix},$$

which has the effect of interchanging the first and second rows of A. As we saw in Proposition 3.4.8, $\det(P) = -1$. Therefore, when $PA = A'$, we have

$$\det(A') = \det(P)\det(A) = -\det(A).$$

In other words, $\det(A') = -\det(A)$ when we perform an interchange.

- Row replacement operations are performed by matrices such as

$$R = \begin{bmatrix} 1 & 0 & 0 \\ 0 & 1 & 0 \\ -2 & 0 & 1 \end{bmatrix},$$

which multiplies the first row by -2 and adds the result to the third row. Since this is a lower triangular matrix, we know that the determinant is the product of the diagonal entries, which says that $\det(R) = 1$. This means that when $RA = A'$, we have $\det(A') = \det(R)\det(A) = \det(A)$. In other words, a row replacement does not change the determinant.

Proposition 3.4.11 The effect of row operations on the determinant.

- If A' is obtained from A by scaling a row by k, then $\det(A') = k \det(A)$.
- If A' is obtained from A by interchanging two rows, then $\det(A') = -\det(A)$.
- If A' is obtained from A by performing a row replacement operation, then $\det(A') = \det(A)$.

Activity 3.4.3. We will investigate the connection between the determinant of a matrix and its invertibility using Gaussian elimination.

a. Consider the two upper triangular matrices

$$U_1 = \begin{bmatrix} 1 & -1 & 2 \\ 0 & 2 & 4 \\ 0 & 0 & -2 \end{bmatrix}, \quad U_2 = \begin{bmatrix} 1 & -1 & 2 \\ 0 & 2 & 4 \\ 0 & 0 & 0 \end{bmatrix}.$$

Remembering Proposition 3.1.9, which of the matrices U_1 and U_2 are invertible? What are the determinants $\det(U_1)$ and $\det(U_2)$?

b. Explain why an upper triangular matrix is invertible if and only if its determinant is not zero.

c. Let's now consider the matrix

$$A = \begin{bmatrix} 1 & -1 & 2 \\ -2 & 2 & -6 \\ 3 & -1 & 10 \end{bmatrix}$$

and begin the Gaussian elimination process with a row replacement operation

$$A = \begin{bmatrix} 1 & -1 & 2 \\ -2 & 2 & -6 \\ 3 & -1 & 10 \end{bmatrix} \sim \begin{bmatrix} 1 & -1 & 2 \\ 0 & 0 & -2 \\ 3 & -1 & 10 \end{bmatrix} = A_1.$$

What is the relationship between $\det(A)$ and $\det(A_1)$?

d. Next we perform another row replacement operation:

$$A_1 = \begin{bmatrix} 1 & -1 & 2 \\ 0 & 0 & -2 \\ 3 & -1 & 10 \end{bmatrix} \sim \begin{bmatrix} 1 & -1 & 2 \\ 0 & 0 & -2 \\ 0 & 2 & 4 \end{bmatrix} = A_2.$$

What is the relationship between $\det(A)$ and $\det(A_2)$?

e. Finally, we perform an interchange:

$$A_2 = \begin{bmatrix} 1 & -1 & 2 \\ 0 & 0 & -2 \\ 0 & 2 & 4 \end{bmatrix} \sim \begin{bmatrix} 1 & -1 & 2 \\ 0 & 2 & 4 \\ 0 & 0 & -2 \end{bmatrix} = U$$

to arrive at an upper triangular matrix U. What is the relationship between $\det(A)$ and $\det(U)$?

3.4. DETERMINANTS

f. Since U is upper triangular, we can compute its determinant, which allows us to find $\det(A)$. What is $\det(A)$? Is A invertible?

g. Now consider the matrix

$$A = \begin{bmatrix} 1 & -1 & 3 \\ 0 & 2 & -2 \\ 2 & 1 & 3 \end{bmatrix}.$$

Perform a sequence of row operations to find an upper triangular matrix U that is row equivalent to A. Use this to determine $\det(A)$ and whether A is invertible.

h. Suppose we apply a sequence of row operations on a matrix A to obtain A'. Explain why $\det(A) \neq 0$ if and only if $\det(A') \neq 0$.

i. Explain why an $n \times n$ matrix A is invertible if and only if $\det(A) \neq 0$.

As seen in this activity, row operations can be used to compute the determinant of a matrix. More specifically, applying the forward substitution phase of Gaussian elimination to the matrix A leads us to an upper triangular matrix U so that $A \sim U$.

We know that U is invertible when all of its diagonal entries are nonzero. We also know that $\det(U) \neq 0$ under the same condition. This tells us U is invertible if and only if $\det(U) \neq 0$.

Now if $\det(A) \neq 0$, we also have $\det(U) \neq 0$ since applying a sequence of row operations to A only multiplies the determinant by a nonzero number. It then follows that U is invertible so $U \sim I$. Therefore, we also know that $A \sim I$ and so A must also be invertible.

This explains Proposition 3.4.10 and so we know that A is invertible if and only if $\det(A) \neq 0$.

Finally, notice that if A is invertible, we have $A^{-1}A = I$, which tells us that

$$\det(A^{-1}A) = \det(A^{-1})\det(A) = 1.$$

Therefore, $\det(A^{-1}) = 1/\det(A)$.

Proposition 3.4.12 *If A is an invertible matrix, then $\det(A^{-1}) = 1/\det(A)$.*

3.4.3 Cofactor expansions

We now have a technique for computing the determinant of a matrix using row operations. There is another way to compute determinants, using what are called *cofactor expansions*, that will be important for us in the next chapter. We will describe this method here.

To begin, the determinant of a 2×2 matrix is

$$\det \begin{bmatrix} a & b \\ c & d \end{bmatrix} = ad - bc.$$

With a little bit of work, it can be shown that this number is the same as the signed area of the parallelogram we introduced earlier.

Using a cofactor expansion to find the determinant of a more general $n \times n$ matrix is a little more work so we will demonstrate it with an example.

Example 3.4.13 We illustrate how to use a cofactor expansion to find the determinant of A where
$$A = \begin{bmatrix} 1 & -1 & 2 \\ -2 & 2 & -6 \\ 3 & -1 & 10 \end{bmatrix}.$$

To begin, we choose one row or column. It doesn't matter which we choose because the result will be the same in any case. Here, we choose the second row

$$\begin{bmatrix} 1 & -1 & 2 \\ -2 & 2 & -6 \\ 3 & -1 & 10 \end{bmatrix}.$$

The determinant will be found by creating a sum of terms, one for each entry in the row we have chosen. For each entry in the row, we form its term by multiplying

- $(-1)^{i+j}$ where i and j are the row and column numbers, respectively, of the entry,
- the entry itself, and
- the determinant of the entries left over when we have crossed out the row and column containing the entry.

Since we are computing the determinant of this matrix

$$\begin{bmatrix} 1 & -1 & 2 \\ -2 & 2 & -6 \\ 3 & -1 & 10 \end{bmatrix}$$

using the second row, the entry in the first column of this row is -2. Let's see how to form the term from this entry.

The term itself is -2, and the matrix that is left over when we cross out the second row and first column is
$$\begin{bmatrix} 1 & -1 & 2 \\ -2 & 2 & -6 \\ 3 & -1 & 10 \end{bmatrix}$$

whose determinant is
$$\det \begin{bmatrix} -1 & 2 \\ -1 & 10 \end{bmatrix} = -1(10) - 2(-1) = -8.$$

Since this entry is in the second row and first column, the term we construct is $(-1)^{2+1}(-2)(-8) = -16$.

3.4. DETERMINANTS

Putting this together, we find the determinant to be

$$\begin{bmatrix} 1 & -1 & 2 \\ -2 & 2 & -6 \\ 3 & -1 & 10 \end{bmatrix} = (-1)^{2+1}(-2)\det\begin{bmatrix} -1 & 2 \\ -1 & 10 \end{bmatrix}$$
$$+ (-1)^{2+2}(2)\det\begin{bmatrix} 1 & 2 \\ 3 & 10 \end{bmatrix}$$
$$+ (-1)^{2+3}(-6)\det\begin{bmatrix} -1 & -1 \\ 3 & -1 \end{bmatrix}.$$

$$= (-1)(-2)(-1(10) - 2(-1))$$
$$+ (1)(2)(1(10) - 2(3))$$
$$+ (-1)(-6)((-1)(-1) - (-1)3)$$

$$= -16 + 8 + 12$$
$$= 4$$

Notice that this agrees with the determinant that we found for this matrix using row operations in the last activity.

Activity 3.4.4. We will explore cofactor expansions through some examples.

a. Using a cofactor expansion, show that the determinant of the following matrix

$$\det\begin{bmatrix} 2 & 0 & -1 \\ 3 & 1 & 2 \\ -2 & 4 & -3 \end{bmatrix} = -36.$$

Remember that you can choose any row or column to create the expansion, but the choice of a particular row or column may simplify the computation.

b. Use a cofactor expansion to find the determinant of

$$\begin{bmatrix} -3 & 0 & 0 & 0 \\ 4 & 1 & 0 & 0 \\ -1 & 4 & -4 & 0 \\ 0 & 3 & 2 & 3 \end{bmatrix}.$$

Explain how the cofactor expansion technique shows that the determinant of a triangular matrix is equal to the product of its diagonal entries.

c. Use a cofactor expansion to determine whether the following vectors form a basis of $\mathbb{R}^3$:

$$\begin{bmatrix} 2 \\ -1 \\ -2 \end{bmatrix}, \begin{bmatrix} 1 \\ -1 \\ 2 \end{bmatrix}, \begin{bmatrix} 1 \\ 0 \\ -4 \end{bmatrix}.$$

d. Sage will compute the determinant of a matrix A with the command A.det(). Use Sage to find the determinant of the matrix

$$\begin{bmatrix} 2 & 1 & -2 & -3 \\ 3 & 0 & -1 & -2 \\ -3 & 4 & 1 & 2 \\ 1 & 3 & 3 & -1 \end{bmatrix}.$$

3.4.4 Summary

In this section, we associated a numerical quantity, the determinant, to a square matrix and showed how it tells us whether the matrix is invertible.

- The determinant of a matrix has a geometric interpretation. In particular, when $n = 2$, the determinant is the signed area of the parallelogram formed by the two columns of the matrix.

- The determinant satisfies many properties. For instance, $\det(AB) = \det(A)\det(B)$ and the determinant of a triangular matrix is equal to the product of its diagonal entries.

- These properties helped us compute the determinant of a matrix using row operations. This also led to the important observation that the determinant of a matrix is nonzero if and only if the matrix is invertible.

- Finally, we learned how to compute the determinant of a matrix using cofactor expansions, which will be a valuable tool for us in the next chapter.

We have seen three ways to compute the determinant: by interpreting the determinant as a signed area or volume; by applying appropriate row operations; and by using a cofactor expansion. It's worth spending a moment to think about the relative merits of these approaches.

The geometric definition of the determinant tells us that the determinant is measuring a natural geometric quantity, an insight that does not easily come through the other two approaches. The intuition we gain by thinking about the determinant geometrically makes it seem reasonable that the determinant should be zero for matrices that are not invertible: if the columns are linearly dependent, the vectors cannot create a positive volume.

Approaching the determinant through row operations provides an effective means of computing the determinant. In fact, this is what most computer programs do behind the scenes when they compute a determinant. This approach is also a useful theoretical tool for explaining why the determinant tells us whether a matrix is invertible.

The cofactor expansion method will be useful to us in the next chapter when we look at eigenvalues and eigenvectors. It is not, however, a practical way to compute a determinant. To see why, consider the fact that the determinant of a 2×2 matrix, written as $ad - bc$,

3.4. DETERMINANTS

requires us to compute two terms, ad and bc. To compute the determinant of a 3×3 matrix, we need to compute three 2×2 determinants, which involves $3 \cdot 2 = 6$ terms. For a 4×4 matrix, we need to compute four 3×3 determinants, which produces $4 \cdot 3 \cdot 2 = 24$ terms. Continuing in this way, we see that the cofactor expansion of a 10×10 matrix would involve $10 \cdot 9 \cdot 8 \ldots 3 \cdot 2 = 10! = 3628800$ terms.

By contrast, we have seen that the number of steps required to perform Gaussian elimination on an $n \times n$ matrix is proportional to n^3. When $n = 10$, we have $n^3 = 1000$, which points to the fact that finding the determinant using Gaussian elimination is considerably less work.

3.4.5 Exercises

1. Consider the matrices

$$A = \begin{bmatrix} 2 & 1 & 0 \\ -4 & -4 & 3 \\ 2 & 1 & -3 \end{bmatrix}, \quad B = \begin{bmatrix} -2 & 3 & 0 & 0 \\ 0 & 4 & 2 & 0 \\ 4 & -6 & -1 & 2 \\ 0 & 4 & 2 & -3 \end{bmatrix}.$$

   a. Find the determinants of A and B using row operations.

   b. Now find the determinants of A and B using cofactor expansions to verify your results

2. This exercise concerns rotations and reflections in $\mathbb{R}^2$.

   a. Suppose that A is the matrix that performs a counterclockwise rotation in $\mathbb{R}^2$. Draw a typical picture of the vectors that form the columns of A and use the geometric definition of the determinant to determine $\det(A)$.

   b. Suppose that B is the matrix that performs a reflection in a line passing through the origin. Draw a typical picture of the columns of B and use the geometric definition of the determinant to determine $\det(B)$.

   c. As we saw in Section 2.6, the matrices have the form

$$A = \begin{bmatrix} \cos\theta & -\sin\theta \\ \sin\theta & \cos\theta \end{bmatrix}, \quad B = \begin{bmatrix} \cos(2\theta) & \sin(2\theta) \\ \sin(2\theta) & -\cos(2\theta) \end{bmatrix}.$$

   Compute the determinants of A and B and verify that they agree with what you found in the earlier parts of this exercise.

3. In the next chapter, we will say that matrices A and B are *similar* if there is a matrix P such that $A = PBP^{-1}$.

   a. Suppose that A and B are matrices and that there is a matrix P such that $A = PBP^{-1}$. Explain why $\det(A) = \det(B)$.

b. Suppose that A is a 3×3 matrix and that there is a matrix P such that

$$A = P \begin{bmatrix} 2 & 0 & 0 \\ 0 & -5 & 0 \\ 0 & 0 & -3 \end{bmatrix} P^{-1}.$$

Find $\det(A)$.

4. Consider the matrix

$$A = \begin{bmatrix} -2 & 1 & k \\ 2 & 3 & 0 \\ 1 & 2 & 2 \end{bmatrix}$$

where k is a parameter.

  a. Find an expression for $\det(A)$ in terms of the parameter k.

  b. Use your expression for $\det(A)$ to determine the values of k for which the vectors

$$\begin{bmatrix} -2 \\ 2 \\ 1 \end{bmatrix}, \begin{bmatrix} 1 \\ 3 \\ 2 \end{bmatrix}, \begin{bmatrix} k \\ 0 \\ 2 \end{bmatrix}$$

are linearly independent.

5. Determine whether the following statements are true or false and explain your response.

  a. If we have a square matrix A and multiply the first row by 5 and add it to the third row to obtain A', then $\det(A') = 5\det(A)$.

  b. If we interchange two rows of a matrix, then the determinant is unchanged.

  c. If we scale a row of the matrix A by 17 to obtain A', then $\det(A') = 17\det(A)$.

  d. If A and A' are row equivalent and $\det(A') = 0$, then $\det(A) = 0$ also.

  e. If A is row equivalent to the identity matrix, then $\det(A) = \det(I) = 1$.

6. Suppose that A and B are 5×5 matrices such that $\det(A) = -2$ and $\det(B) = 5$. Find the following determinants:

  a. $\det(2A)$.

  b. $\det(A^3)$.

  c. $\det(AB)$.

  d. $\det(-A)$.

  e. $\det(AB^{-1})$.

7. Suppose that A and B are $n \times n$ matrices.

  a. If A and B are both invertible, use determinants to explain why AB is invertible.

  b. If AB is invertible, use determinants to explain why both A and B are invertible.

3.4. DETERMINANTS

8. Provide a justification for your responses to the following questions.

    a. If every entry in one row of a matrix is zero, what can you say about the determinant?

    b. If two rows of a square matrix are identical, what can you say about the determinant?

    c. If two columns of a square matrix are identical, what can you say about the determinant?

    d. If one column of a matrix is a linear combination of the others, what can you say about the determinant?

9. Consider the matrix
$$A = \begin{bmatrix} 0 & 1 & x \\ 2 & 2 & y \\ -1 & 0 & z \end{bmatrix}.$$

    a. Assuming that $\det(A) = 0$, rewrite the equation in terms of x, y, and z.

    b. Explain why $\mathbf{v}_1$ and $\mathbf{v}_2$, the first two columns of A, satisfy the equation you found in the previous part.

    c. Explain why the solution space of this equation is the plane spanned by $\mathbf{v}_1$ and $\mathbf{v}_2$.

10. In this section, we studied the effect of row operations on the matrix A. In this exercise, we will study the effect of analogous *column* operations.

    Suppose that A is the 3×3 matrix $A = \begin{bmatrix} \mathbf{v}_1 & \mathbf{v}_2 & \mathbf{v}_3 \end{bmatrix}$. Also consider elementary matrices
$$R = \begin{bmatrix} 1 & 0 & 0 \\ 0 & 1 & 0 \\ -3 & 0 & 1 \end{bmatrix}, \quad S = \begin{bmatrix} 1 & 0 & 0 \\ 0 & 3 & 0 \\ 0 & 0 & 1 \end{bmatrix}, \quad P = \begin{bmatrix} 0 & 0 & 1 \\ 0 & 1 & 0 \\ 1 & 0 & 0 \end{bmatrix}.$$

    a. Explain why the matrix AR is obtained from A by replacing the first column $\mathbf{v}_1$ by $\mathbf{v}_1 - 3\mathbf{v}_3$. We call this a column replacement operation. Explain why column replacement operations do not change the determinant.

    b. Explain why the matrix AS is obtained from A by multiplying the second column by 3. Explain the effect that scaling a column has on the determinant of a matrix.

    c. Explain why the matrix AP is obtained from A by interchanging the first and third columns. What is the effect of this operation on the determinant?

    d. Use column operations to compute the determinant of
$$A = \begin{bmatrix} 0 & -3 & 1 \\ 1 & 1 & 4 \\ 1 & 1 & 0 \end{bmatrix}.$$

11. Consider the matrices

$$A = \begin{bmatrix} 0 & 1 & 0 & 0 \\ 0 & 0 & 1 & 0 \\ 0 & 0 & 0 & 1 \\ 1 & 0 & 0 & 0 \end{bmatrix}, \quad B = \begin{bmatrix} 0 & 1 & 0 & 0 \\ 1 & 0 & 0 & 0 \\ 0 & 0 & 0 & 1 \\ 0 & 0 & 1 & 0 \end{bmatrix}, \quad C = \begin{bmatrix} 0 & 0 & 0 & a \\ 0 & 0 & b & 0 \\ 0 & c & 0 & 0 \\ d & 0 & 0 & 0 \end{bmatrix}.$$

Use row operations to find the determinants of these matrices.

12. Consider the matrices

$$A = \begin{bmatrix} 0 & 1 \\ 1 & 0 \end{bmatrix}, \quad B = \begin{bmatrix} 0 & 1 & 0 \\ 1 & 0 & 1 \\ 0 & 1 & 0 \end{bmatrix},$$

$$C = \begin{bmatrix} 0 & 1 & 0 & 0 \\ 1 & 0 & 1 & 0 \\ 0 & 1 & 0 & 1 \\ 0 & 0 & 1 & 0 \end{bmatrix}, \quad D = \begin{bmatrix} 0 & 1 & 0 & 0 & 0 \\ 1 & 0 & 1 & 0 & 0 \\ 0 & 1 & 0 & 1 & 0 \\ 0 & 0 & 1 & 0 & 1 \\ 0 & 0 & 0 & 1 & 0 \end{bmatrix}.$$

a. Use row (and/or column) operations to find the determinants of these matrices.

b. Write the 6×6 and 7×7 matrices that follow in this pattern and state their determinants based on what you have seen.

13. The following matrix is called a *Vandermond* matrix:

$$V = \begin{bmatrix} 1 & a & a^2 \\ 1 & b & b^2 \\ 1 & c & c^2 \end{bmatrix}.$$

a. Use row operations to explain why $\det(V) = (b-a)(c-a)(c-b)$.

b. Explain why V is invertible if and only if a, b, and c are all distinct real numbers.

c. There is a natural way to generalize this to a 4×4 matrix with parameters $a, b, c,$ and d. Write this matrix and state its determinant based on your previous work.

This matrix appeared in Exercise 1.4.4.9 when we were found a polynomial that passed through a given set of points.

3.5 Subspaces

In this chapter, we have been looking at bases for $\mathbb{R}^p$, sets of vectors that are linearly independent and span $\mathbb{R}^p$. Frequently, however, we focus on only a subset of $\mathbb{R}^p$. In particular, if we are given an $m \times n$ matrix A, we have been interested in both the span of the columns of A and the solution space to the homogeneous equation $A\mathbf{x} = \mathbf{0}$. In this section, we will expand the concept of basis to describe sets like these.

> **Preview Activity 3.5.1.** Let's consider the following matrix A and its reduced row echelon form.
> $$A = \begin{bmatrix} 2 & -1 & 2 & 3 \\ 1 & 0 & 0 & 2 \\ -2 & 2 & -4 & -2 \end{bmatrix} \sim \begin{bmatrix} 1 & 0 & 0 & 2 \\ 0 & 1 & -2 & 1 \\ 0 & 0 & 0 & 0 \end{bmatrix}.$$
>
> a. Are the columns of A linearly independent? Is the span of the columns $\mathbb{R}^3$?
>
> b. Give a parametric description of the solution space to the homogeneous equation $A\mathbf{x} = \mathbf{0}$.
>
> c. Explain how this parametric description produces two vectors $\mathbf{w}_1$ and $\mathbf{w}_2$ whose span is the solution space to the equation $A\mathbf{x} = \mathbf{0}$.
>
> d. What can you say about the linear independence of the set of vectors $\mathbf{w}_1$ and $\mathbf{w}_2$?
>
> e. Let's denote the columns of A as $\mathbf{v}_1$, $\mathbf{v}_2$, $\mathbf{v}_3$, and $\mathbf{v}_4$. Explain why $\mathbf{v}_3$ and $\mathbf{v}_4$ can be written as linear combinations of $\mathbf{v}_1$ and $\mathbf{v}_2$.
>
> f. Explain why $\mathbf{v}_1$ and $\mathbf{v}_2$ are linearly independent and
> $$\text{Span}\{\mathbf{v}_1, \mathbf{v}_2\} = \text{Span}\{\mathbf{v}_1, \mathbf{v}_2, \mathbf{v}_3, \mathbf{v}_4\}.$$

3.5.1 Subspaces

Our goal is to develop a common framework for describing subsets like the span of the columns of a matrix and the solution space to a homogeneous equation. That leads us to the following definition.

Definition 3.5.1 A **subspace** of $\mathbb{R}^p$ is a subset of $\mathbb{R}^p$ that is the span of a set of vectors.

Since we have explored the concept of span in some detail, this definition just gives us a new word to describe something familiar. Let's look at some examples.

Example 3.5.2 Subspaces of $\mathbb{R}^3$. In Activity 2.3.3 and the following discussion, we looked at subspaces in $\mathbb{R}^3$ without explicitly using that language. Let's recall some of those examples.

- Suppose we have a single nonzero vector $\mathbf{v}$. The span of $\mathbf{v}$ is a subspace, which we'll write as $S = \text{Span}\{\mathbf{v}\}$. As we have seen, the span of a single vector consists of all scalar multiples of that vector, and these form a line passing through the origin.

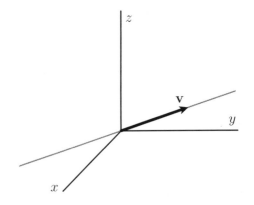

- If instead we have two linearly independent vectors $\mathbf{v}_1$ and $\mathbf{v}_2$, the subspace $S = \text{Span}\{\mathbf{v}_1, \mathbf{v}_2\}$ is a plane passing through the origin.

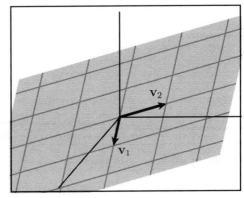

- Consider the three vectors $\mathbf{e}_1$, $\mathbf{e}_2$, and $\mathbf{e}_3$. Since we know that every 3-dimensional vector can be written as a linear combination, we have $S = \text{Span}\{\mathbf{e}_1, \mathbf{e}_2, \mathbf{e}_3\} = \mathbb{R}^3$.

- One more subspace worth mentioning is $S = \text{Span}\{\mathbf{0}\}$. Since any linear combination of the zero vector is itself the zero vector, this subspace consists of a single vector, $\mathbf{0}$.

In fact, any subspace of $\mathbb{R}^3$ is one of these types: the origin, a line, a plane, or all of $\mathbb{R}^3$.

Activity 3.5.2. We will look at some sets of vectors and the subspaces they form.

a. If $\mathbf{v}_1, \mathbf{v}_2, \ldots, \mathbf{v}_n$ is a set of vectors in $\mathbb{R}^m$, explain why $\mathbf{0}$ can be expressed as a linear combination of these vectors. Use this fact to explain why the zero vector $\mathbf{0}$ belongs to any subspace in $\mathbb{R}^m$.

b. Explain why the line on the left of Figure 3.5.3 is not a subspace of $\mathbb{R}^2$ and why the line on the right is.

3.5. SUBSPACES

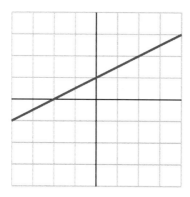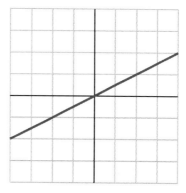

Figure 3.5.3 Two lines in $\mathbb{R}^2$, one of which is a subspace and one of which is not.

c. Consider the vectors

$$\mathbf{v}_1 = \begin{bmatrix} 1 \\ 0 \\ 1 \end{bmatrix}, \quad \mathbf{v}_2 = \begin{bmatrix} 0 \\ 1 \\ 1 \end{bmatrix}, \quad \mathbf{v}_3 = \begin{bmatrix} 1 \\ 1 \\ 0 \end{bmatrix},$$

and describe the subspace $S = \text{Span}\{\mathbf{v}_1, \mathbf{v}_2, \mathbf{v}_3\}$ of $\mathbb{R}^3$.

d. Consider the vectors

$$\mathbf{w}_1 = \begin{bmatrix} 2 \\ 1 \\ 0 \end{bmatrix}, \quad \mathbf{w}_2 = \begin{bmatrix} -1 \\ 1 \\ -1 \end{bmatrix}, \quad \mathbf{w}_3 = \begin{bmatrix} 0 \\ 3 \\ -2 \end{bmatrix}$$

1. Write $\mathbf{w}_3$ as a linear combination of $\mathbf{w}_1$ and $\mathbf{w}_2$.
2. Explain why $\text{Span}\{\mathbf{w}_1, \mathbf{w}_2, \mathbf{w}_3\} = \text{Span}\{\mathbf{w}_1, \mathbf{w}_2\}$.
3. Describe the subspace $S = \text{Span}\{\mathbf{w}_1, \mathbf{w}_2, \mathbf{w}_3\}$ of $\mathbb{R}^3$.

e. Suppose that $\mathbf{v}_1, \mathbf{v}_2, \mathbf{v}_3$, and $\mathbf{v}_4$ are four vectors in $\mathbb{R}^3$ and that

$$\begin{bmatrix} \mathbf{v}_1 & \mathbf{v}_2 & \mathbf{v}_3 & \mathbf{v}_4 \end{bmatrix} \sim \begin{bmatrix} 1 & 2 & 0 & -2 \\ 0 & 0 & 1 & 1 \\ 0 & 0 & 0 & 0 \end{bmatrix}.$$

Give a description of the subspace $S = \text{Span}\{\mathbf{v}_1, \mathbf{v}_2, \mathbf{v}_3, \mathbf{v}_4\}$ of $\mathbb{R}^3$.

As the activity shows, it is possible to represent some subspaces as the span of more than one set of vectors. We are particularly interested in representing a subspace as the span of a linearly independent set of vectors.

Definition 3.5.4 A *basis* for a subspace S of $\mathbb{R}^p$ is a set of vectors in S that are linearly independent and whose span is S. We say that the *dimension* of the subspace S, denoted $\dim S$, is the number of vectors in any basis.

Example 3.5.5 A subspace of $\mathbb{R}^4$. Suppose we have the 4-dimensional vectors $\mathbf{v}_1$, $\mathbf{v}_2$, and $\mathbf{v}_3$ that define the subspace $S = \text{Span}\{\mathbf{v}_1, \mathbf{v}_2, \mathbf{v}_3\}$ of $\mathbb{R}^4$. Suppose also that

$$\begin{bmatrix} \mathbf{v}_1 & \mathbf{v}_2 & \mathbf{v}_3 \end{bmatrix} \sim \begin{bmatrix} 1 & -1 & 0 \\ 0 & 0 & 1 \\ 0 & 0 & 0 \\ 0 & 0 & 0 \end{bmatrix}.$$

From the reduced row echelon form of the matrix, we see that $\mathbf{v}_2 = -\mathbf{v}$. Therefore, any linear combination of $\mathbf{v}_1$, $\mathbf{v}_2$, and $\mathbf{v}_3$ can be rewritten

$$c_1 \mathbf{v}_1 + c_2 \mathbf{v}_2 + c_3 \mathbf{v}_3 = (c_1 - c_2) \mathbf{v}_1 + c_2 \mathbf{v}_3$$

as a linear combination of $\mathbf{v}_1$ and $\mathbf{v}_3$. This tells us that

$$S = \text{Span}\{\mathbf{v}_1, \mathbf{v}_2, \mathbf{v}_3\} = \text{Span}\{\mathbf{v}_1, \mathbf{v}_3\}.$$

Furthermore, the reduced row echelon form of the matrix shows that $\mathbf{v}_1$ and $\mathbf{v}_3$ are linearly independent. Therefore, $\{\mathbf{v}_1, \mathbf{v}_3\}$ is a basis for S, which means that S is a two-dimensional subspace of $\mathbb{R}^4$.

Subspaces of $\mathbb{R}^3$ are either

- 0-dimensional, consisting of the single vector $\mathbf{0}$,
- a 1-dimensional line,
- a 2-dimensional plane, or
- the 3-dimensional subspace $\mathbb{R}^3$.

There is no 4-dimensional subspace of $\mathbb{R}^3$ because there is no linearly independent set of four vectors in $\mathbb{R}^3$.

There are two important subspaces associated to any matrix, each of which springs from one of our two fundamental questions, as we will now see.

3.5.2 The column space of A

The first subspace associated to a matrix that we'll consider is its column space.

Definition 3.5.6 If A is an $m \times n$ matrix, we call the span of its columns the *column space* of A and denote it as $\text{Col}(A)$.

Notice that the columns of A are vectors in $\mathbb{R}^m$, which means that any linear combination of the columns is also in $\mathbb{R}^m$. Since the column space is described as the span of a set of vectors, we see that $\text{Col}(A)$ is a subspace of $\mathbb{R}^m$.

3.5. SUBSPACES

Activity 3.5.3. We will explore some column spaces in this activity.

a. Consider the matrix
$$A = \begin{bmatrix} \mathbf{v}_1 & \mathbf{v}_2 & \mathbf{v}_3 \end{bmatrix} = \begin{bmatrix} 1 & 3 & -1 \\ -2 & 0 & -4 \\ 1 & 2 & 0 \end{bmatrix}.$$

Since Col(A) is the span of the columns, we have
$$\text{Col}(A) = \text{Span}\{\mathbf{v}_1, \mathbf{v}_2, \mathbf{v}_3\}.$$

Explain why $\mathbf{v}_3$ can be written as a linear combination of $\mathbf{v}_1$ and $\mathbf{v}_2$ and why Col(A) = Span$\{\mathbf{v}_1, \mathbf{v}_2\}$.

b. Explain why the vectors $\mathbf{v}_1$ and $\mathbf{v}_2$ form a basis for Col(A) and why Col(A) is a 2-dimensional subspace of $\mathbb{R}^3$ and therefore a plane.

c. Now consider the matrix B and its reduced row echelon form:
$$B = \begin{bmatrix} -2 & -4 & 0 & 6 \\ 1 & 2 & 0 & -3 \end{bmatrix} \sim \begin{bmatrix} 1 & 2 & 0 & -3 \\ 0 & 0 & 0 & 0 \end{bmatrix}.$$

Explain why Col(B) is a 1-dimensional subspace of $\mathbb{R}^2$ and is therefore a line.

d. For a general matrix A, what is the relationship between the dimension dim Col(A) and the number of pivot positions in A?

e. How does the location of the pivot positions indicate a basis for Col(A)?

f. If A is an invertible 9×9 matrix, what can you say about the column space Col(A)?

g. Suppose that A is an 8×10 matrix and that Col(A) = $\mathbb{R}^8$. If $\mathbf{b}$ is an 8-dimensional vector, what can you say about the equation $A\mathbf{x} = \mathbf{b}$?

Example 3.5.7 Consider the matrix A and its reduced row echelon form:
$$A = \begin{bmatrix} 2 & 0 & -4 & -6 & 0 \\ -4 & -1 & 7 & 11 & 2 \\ 0 & -1 & -1 & -1 & 2 \end{bmatrix} \sim \begin{bmatrix} 1 & 0 & -2 & -3 & 0 \\ 0 & 1 & 1 & 1 & -2 \\ 0 & 0 & 0 & 0 & 0 \end{bmatrix},$$

and denote the columns of A as $\mathbf{v}_1, \mathbf{v}_2, \ldots, \mathbf{v}_5$.

It is certainly true that Col(A) = Span$\{\mathbf{v}_1, \mathbf{v}_2, \ldots, \mathbf{v}_5\}$ by the definition of the column space. However, the reduced row echelon form of the matrix shows us that the vectors are not linearly independent so $\mathbf{v}_1, \mathbf{v}_2, \ldots, \mathbf{v}_5$ do not form a basis for Col(A).

From the reduced row echelon form, however, we can see that
$$\mathbf{v}_3 = -2\mathbf{v}_1 + \mathbf{v}_2$$
$$\mathbf{v}_4 = -3\mathbf{v}_1 + \mathbf{v}_2.$$
$$\mathbf{v}_5 = -2\mathbf{v}_2$$

This means that any linear combination of $v_1, v_2, \ldots, v_5$ can be written as a linear combination of just v_1 and v_2. Therefore, we see that $Col(A) = Span\{v_1, v_2\}$.

Moreover, the reduced row echelon form shows that v_1 and v_2 are linearly independent, which implies that they form a basis for $Col(A)$. This means that $Col(A)$ is a 2-dimensional subspace of $\mathbb{R}^3$, which is a plane in $\mathbb{R}^3$, having basis

$$\begin{bmatrix} 2 \\ -4 \\ 0 \end{bmatrix}, \begin{bmatrix} 0 \\ -1 \\ 1 \end{bmatrix}.$$

In general, a column without a pivot position can be written as a linear combination of the columns that have pivot positions. This means that a basis for $Col(A)$ will always be given by the columns of A having pivot positions. This leads us to the following definition and proposition.

Definition 3.5.8 The *rank* of a matrix A is the number of pivot positions in A and is denoted by $rank(A)$.

Proposition 3.5.9 *If A is an $m \times n$ matrix, then $Col(A)$ is a subspace of $\mathbb{R}^m$ whose dimension equals $rank(A)$. The columns of A that contain pivot positions form a basis for $Col(A)$.*

For example, the rank of the matrix A in Example 3.5.7 is two because there are two pivot positions. A basis for $Col(A)$ is given by the first two columns of A since those columns have pivot positions.

As a note of caution, we determine the pivot positions by looking at the reduced row echelon form of A. However, we form a basis of $Col(A)$ from the columns of A rather than the columns of the reduced row echelon matrix.

3.5.3 The null space of A

The second subspace associated to a matrix is its null space.

Definition 3.5.10 If A is an $m \times n$ matrix, we call the subset of vectors x in $\mathbb{R}^n$ satisfying $Ax = 0$ the *null space* of A and denote it by $Nul(A)$.

Remember that a subspace is a subset that can be represented as the span of a set of vectors. The column space of A, which is simply the span of the columns of A, fits this definition. It may not be immediately clear how the null space of A, which is the solution space of the equation $Ax = 0$, does, but we will see that $Nul(A)$ is a subspace of $\mathbb{R}^n$.

> **Activity 3.5.4.** We will explore some null spaces in this activity and see why $Nul(A)$ satisfies the definition of a subspace.
>
> a. Consider the matrix
> $$A = \begin{bmatrix} 1 & 3 & -1 & 2 \\ -2 & 0 & -4 & 2 \\ 1 & 2 & 0 & 1 \end{bmatrix}$$
> and give a parametric description of the solution space to the equation $Ax = 0$.

3.5. SUBSPACES

In other words, give a parametric description of Nul(A).

b. This parametric description shows that the vectors satisfying the equation $A\mathbf{x} = \mathbf{0}$ can be written as a linear combination of a set of vectors. In other words, this description shows why Nul(A) is the span of a set of vectors and is therefore a subspace. Identify a set of vectors whose span is Nul(A).

c. Use this set of vectors to find a basis for Nul(A) and state the dimension of Nul(A).

d. The null space Nul(A) is a subspace of $\mathbb{R}^p$ for which value of p?

e. Now consider the matrix B whose reduced row echelon form is given by

$$B \sim \begin{bmatrix} 1 & 2 & 0 & -3 \\ 0 & 0 & 0 & 0 \end{bmatrix}.$$

Give a parametric description of Nul(B).

f. The parametric description gives a set of vectors that span Nul(B). Explain why this set of vectors is linearly independent and hence forms a basis. What is the dimension of Nul(B)?

g. For a general matrix A, how does the number of pivot positions indicate the dimension of Nul(A)?

h. Suppose that the columns of a matrix A are linearly independent. What can you say about Nul(A)?

Example 3.5.11 Consider the matrix A along with its reduced row echelon form:

$$A = \begin{bmatrix} 2 & 0 & -4 & -6 & 0 \\ -4 & -1 & 7 & 11 & 2 \\ 0 & -1 & -1 & -1 & 2 \end{bmatrix} \sim \begin{bmatrix} 1 & 0 & -2 & -3 & 0 \\ 0 & 1 & 1 & 1 & -2 \\ 0 & 0 & 0 & 0 & 0 \end{bmatrix}.$$

To find a parametric description of the solution space to $A\mathbf{x} = \mathbf{0}$, imagine that we augment both A and its reduced row echelon form by a column of zeroes, which leads to the equations

$$\begin{aligned} x_1 \quad - 2x_3 - 3x_4 \quad &= 0 \\ x_2 + x_3 + x_4 - 2x_5 &= 0. \end{aligned}$$

Notice that x_3, x_4, and x_5 are free variables so we rewrite these equations as

$$\begin{aligned} x_1 &= 2x_3 + 3x_4 \\ x_2 &= -x_3 - x_4 + 2x_5. \end{aligned}$$

In vector form, we have

$$\mathbf{x} = \begin{bmatrix} x_1 \\ x_2 \\ x_3 \\ x_4 \\ x_5 \end{bmatrix} = \begin{bmatrix} 2x_3 + 3x_4 \\ -x_3 - x_4 + 2x_5 \\ x_3 \\ x_4 \\ x_5 \end{bmatrix}$$

$$= x_3 \begin{bmatrix} 2 \\ -1 \\ 1 \\ 0 \\ 0 \end{bmatrix} + x_4 \begin{bmatrix} 3 \\ -1 \\ 0 \\ 1 \\ 0 \end{bmatrix} + x_5 \begin{bmatrix} 0 \\ 2 \\ 0 \\ 0 \\ 1 \end{bmatrix}.$$

This expression says that any vector $\mathbf{x}$ satisfying $A\mathbf{x} = \mathbf{0}$ is a linear combination of the vectors

$$\mathbf{v}_1 = \begin{bmatrix} 2 \\ -1 \\ 1 \\ 0 \\ 0 \end{bmatrix}, \quad \mathbf{v}_2 = \begin{bmatrix} 3 \\ -1 \\ 0 \\ 1 \\ 0 \end{bmatrix}, \quad \mathbf{v}_3 = \begin{bmatrix} 0 \\ 2 \\ 0 \\ 0 \\ 1 \end{bmatrix}.$$

It is straightforward to check that these vectors are linearly independent, which means that $\mathbf{v}_1$, $\mathbf{v}_2$, and $\mathbf{v}_3$ form a basis for Nul(A), a 3-dimensional subspace of $\mathbb{R}^5$.

As illustrated in this example, the dimension of Nul(A) is equal to the number of free variables in the equation $A\mathbf{x} = \mathbf{0}$, which equals the number of columns of A without pivot positions or the number of columns of A minus the number of pivot positions.

Proposition 3.5.12 *If A is an $m \times n$ matrix, then Nul(A) is a subspace of $\mathbb{R}^n$ whose dimension is*

$$\dim \text{Nul}(A) = n - \text{rank}(A).$$

Combining Proposition 3.5.9 and Proposition 3.5.12 shows that

Proposition 3.5.13 *If A is an $m \times n$ matrix, then*

$$\dim \text{Col}(A) + \dim \text{Nul}(A) = n.$$

3.5.4 Summary

Once again, we find ourselves revisiting our two fundamental questions concerning the existence and uniqueness of solutions to linear systems. The column space Col(A) contains all the vectors $\mathbf{b}$ for which the equation $A\mathbf{x} = \mathbf{b}$ is consistent. The null space Nul(A) is the solution space to the equation $A\mathbf{x} = \mathbf{0}$, which reflects on the uniqueness of solutions to this and other equations.

- A subspace S of $\mathbb{R}^p$ is a subset of $\mathbb{R}^p$ that can be represented as the span of a set of

3.5. SUBSPACES

vectors. A basis of S is a linearly independent set of vectors whose span is S.

- If A is an $m \times n$ matrix, the column space Col(A) is the span of the columns of A and forms a subspace of $\mathbb{R}^m$.

- A basis for Col(A) is found from the columns of A that have pivot positions. The dimension is therefore dim Col(A) = rank(A).

- The null space Nul(A) is the solution space to the homogeneous equation $A\mathbf{x} = \mathbf{0}$ and is a subspace of $\mathbb{R}^n$.

- A basis for Nul(A) is found through a parametric description of the solution space of $A\mathbf{x} = \mathbf{0}$, and we have that dim Nul(A) = n − rank(A).

3.5.5 Exercises

1. Suppose that A and its reduced row echelon form are

$$A = \begin{bmatrix} 0 & 2 & 0 & -4 & 0 & 6 \\ 0 & -4 & -1 & 7 & 0 & -16 \\ 0 & 6 & 0 & -12 & 3 & 15 \\ 0 & 4 & -1 & -9 & 0 & 8 \end{bmatrix} \sim \begin{bmatrix} 0 & 1 & 0 & -2 & 0 & 3 \\ 0 & 0 & 1 & 1 & 0 & 4 \\ 0 & 0 & 0 & 0 & 1 & -1 \\ 0 & 0 & 0 & 0 & 0 & 0 \end{bmatrix}.$$

   a. The null space Nul(A) is a subspace of $\mathbb{R}^p$ for what p? The column space Col(A) is a subspace of $\mathbb{R}^p$ for what p?

   b. What are the dimensions dim Nul(A) and dim Col(A)?

   c. Find a basis for the column space Col(A).

   d. Find a basis for the null space Nul(A).

2. Suppose that

$$A = \begin{bmatrix} 2 & 0 & -2 & -4 \\ -2 & -1 & 1 & 2 \\ 0 & -1 & -1 & -2 \end{bmatrix}.$$

   a. Is the vector $\begin{bmatrix} 0 \\ -1 \\ -1 \end{bmatrix}$ in Col(A)?

   b. Is the vector $\begin{bmatrix} 2 \\ 1 \\ 0 \\ 2 \end{bmatrix}$ in Col(A)?

   c. Is the vector $\begin{bmatrix} 2 \\ -2 \\ 0 \end{bmatrix}$ in Nul(A)?

d. Is the vector $\begin{bmatrix} 1 \\ -1 \\ 3 \\ -1 \end{bmatrix}$ in Nul(A)?

e. Is the vector $\begin{bmatrix} 1 \\ 0 \\ 1 \\ -1 \end{bmatrix}$ in Nul(A)?

3. Determine whether the following statements are true or false and provide a justification for your response. Unless otherwise stated, assume that A is an $m \times n$ matrix.

   a. If A is a 127×341 matrix, then Nul(A) is a subspace of $\mathbb{R}^{127}$.

   b. If dim Nul(A) = 0, then the columns of A are linearly independent.

   c. If Col(A) = $\mathbb{R}^m$, then A is invertible.

   d. If A has a pivot position in every column, then Nul(A) = $\mathbb{R}^n$.

   e. If Col(A) = $\mathbb{R}^m$ and Nul(A) = $\{\mathbf{0}\}$, then A is invertible.

4. Explain why the following statements are true.

   a. If B is invertible, then Nul(BA) = Nul(A).

   b. If B is invertible, then Col(AB) = Col(A).

   c. If $A \sim A'$, then Nul(A) = Nul(A').

5. For each of the following conditions, construct a 3×3 matrix having the given properties.

   a. dim Nul(A) = 0.

   b. dim Nul(A) = 1.

   c. dim Nul(A) = 2.

   d. dim Nul(A) = 3.

6. Suppose that A is a 3×4 matrix.

   a. Is it possible that dim Nul(A) = 0?

   b. If dim Nul(A) = 1, what can you say about Col(A)?

   c. If dim Nul(A) = 2, what can you say about Col(A)?

   d. If dim Nul(A) = 3, what can you say about Col(A)?

   e. If dim Nul(A) = 4, what can you say about Col(A)?

3.5. SUBSPACES

7. Suppose we have the vectors

$$\mathbf{v}_1 = \begin{bmatrix} 2 \\ 3 \\ -1 \end{bmatrix}, \mathbf{v}_2 = \begin{bmatrix} -1 \\ 2 \\ 4 \end{bmatrix}, \mathbf{w}_1 = \begin{bmatrix} 3 \\ -1 \\ 1 \\ 0 \end{bmatrix}, \mathbf{w}_2 = \begin{bmatrix} -2 \\ 4 \\ 0 \\ 1 \end{bmatrix}$$

   and that A is a matrix such that $\text{Col}(A) = \text{Span}\{\mathbf{v}_1, \mathbf{v}_2\}$ and $\text{Nul}(A) = \text{Span}\{\mathbf{w}_1, \mathbf{w}_2\}$.

   a. What are the dimensions of A?

   b. Find such a matrix A.

8. Suppose that A is an 8×8 matrix and that $\det A = 14$.

   a. What can you conclude about $\text{Nul}(A)$?

   b. What can you conclude about $\text{Col}(A)$?

9. Suppose that A is a matrix and there is an invertible matrix P such that

$$A = P \begin{bmatrix} 2 & 0 & 0 \\ 0 & -3 & 0 \\ 0 & 0 & 1 \end{bmatrix} P^{-1}.$$

   a. What can you conclude about $\text{Nul}(A)$?

   b. What can you conclude about $\text{Col}(A)$?

10. In this section, we saw that the solution space to the homogeneous equation $A\mathbf{x} = \mathbf{0}$ is a subspace of $\mathbb{R}^p$ for some p. In this exercise, we will investigate whether the solution space to another equation $A\mathbf{x} = \mathbf{b}$ can form a subspace.

    Let's consider the matrix

$$A = \begin{bmatrix} 2 & -4 \\ -1 & 2 \end{bmatrix}.$$

   a. Find a parametric description of the solution space to the homogeneous equation $A\mathbf{x} = \mathbf{0}$.

b. Graph the solution space to the homogeneous equation to the right.

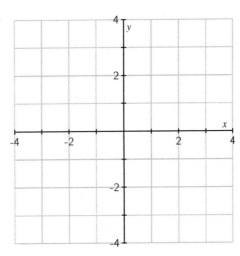

c. Find a parametric description of the solution space to the equation $A\mathbf{x} = \begin{bmatrix} 4 \\ -2 \end{bmatrix}$ and graph it above.

d. Is the solution space to the equation $A\mathbf{x} = \begin{bmatrix} 4 \\ -2 \end{bmatrix}$ a subspace of $\mathbb{R}^2$?

e. Find a parametric description of the solution space to the equation $A\mathbf{x} = \begin{bmatrix} -8 \\ 4 \end{bmatrix}$ and graph it above.

f. What can you say about all the solution spaces to equations of the form $A\mathbf{x} = \mathbf{b}$ when $\mathbf{b}$ is a vector in $\mathrm{Col}(A)$?

g. Suppose that the solution space to the equation $A\mathbf{x} = \mathbf{b}$ forms a subspace. Explain why it must be true that $\mathbf{b} = \mathbf{0}$.

CHAPTER 4

Eigenvalues and eigenvectors

Our primary concern so far has been to develop an understanding of solutions to linear systems $A\mathbf{x} = \mathbf{b}$. In this way, our two fundamental questions about the existence and uniqueness of solutions led us to the concepts of span and linear independence.

We saw that some linear systems are easier to understand than others. For instance, given the two matrices

$$A = \begin{bmatrix} 3 & 0 \\ 0 & -1 \end{bmatrix}, \quad B = \begin{bmatrix} 1 & 2 \\ 2 & 1 \end{bmatrix},$$

we would much prefer working with the diagonal matrix A. Solutions to linear systems $A\mathbf{x} = \mathbf{b}$ are easily determined, and the geometry of the matrix transformation defined by A is easily described.

We saw in the last chapter, however, that some problems become simpler when we look at them in a new basis. Is it possible that questions about the non-diagonal matrix B become simpler when viewed in a different basis? We will see that the answer is "yes," and see how the theory of eigenvalues and eigenvectors, which will be developed in this chapter, provides the key. We will see how this theory provides an appropriate change of basis so that questions about the non-diagonal matrix B are equivalent to questions about the diagonal matrix A. In fact, we will see that these two matrices are, in some sense, equivalent to one another.

4.1 An introduction to eigenvalues and eigenvectors

This section introduces the concept of eigenvalues and eigenvectors and offers an example that motivates our interest in them. The point here is to develop an intuitive understanding of eigenvalues and eigenvectors and explain how they can be used to simplify some problems that we have previously encountered. In the rest of this chapter, we will develop this concept into a richer theory and illustrate its use with more meaningful examples.

Preview Activity 4.1.1. Before we introduce the definition of eigenvectors and eigenvalues, it will be helpful to remember some ideas we have seen previously.

a. Suppose that **v** is the vector shown in the figure. Sketch the vector 2**v** and the vector −**v**.

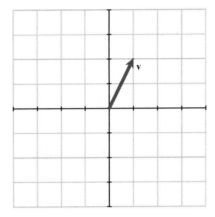

b. State the geometric effect that scalar multiplication has on the vector **v**. Then sketch all the vectors of the form $\lambda\mathbf{v}$ where λ is a scalar.

c. State the geometric effect of the matrix transformation defined by

$$\begin{bmatrix} 3 & 0 \\ 0 & -1 \end{bmatrix}.$$

d. Suppose that A is a 2×2 matrix and that $\mathbf{v}_1$ and $\mathbf{v}_2$ are vectors such that

$$A\mathbf{v}_1 = 3\mathbf{v}_1, \qquad A\mathbf{v}_2 = -\mathbf{v}_2.$$

Use the linearity of matrix multiplication to express the following vectors in terms of $\mathbf{v}_1$ and $\mathbf{v}_2$.

1. $A(4\mathbf{v}_1)$.
2. $A(\mathbf{v}_1 + \mathbf{v}_2)$.
3. $A(4\mathbf{v}_1 - 3\mathbf{v}_2)$.
4. $A^2\mathbf{v}_1$.
5. $A^2(4\mathbf{v}_1 - 3\mathbf{v}_2)$.
6. $A^4\mathbf{v}_1$.

4.1.1 A few examples

We will now introduce the definition of eigenvalues and eigenvectors and then look at a few simple examples.

Definition 4.1.1 Given a square $n \times n$ matrix A, we say that a nonzero vector **v** is an *eigenvector* of A if there is a scalar λ such that
$$A\mathbf{v} = \lambda\mathbf{v}.$$

4.1. AN INTRODUCTION TO EIGENVALUES AND EIGENVECTORS

The scalar λ is called the *eigenvalue* associated to the eigenvector **v**.

At first glance, there is a lot going on in this definition so let's look at an example.

Example 4.1.2 Consider the matrix $A = \begin{bmatrix} 7 & 6 \\ 6 & -2 \end{bmatrix}$ and the vector $\mathbf{v} = \begin{bmatrix} 2 \\ 1 \end{bmatrix}$. We find that

$$A\mathbf{v} = \begin{bmatrix} 7 & 6 \\ 6 & -2 \end{bmatrix} \begin{bmatrix} 2 \\ 1 \end{bmatrix} = \begin{bmatrix} 20 \\ 10 \end{bmatrix} = 10 \begin{bmatrix} 2 \\ 1 \end{bmatrix} = 10\mathbf{v}.$$

In other words, $A\mathbf{v} = 10\mathbf{v}$, which says that **v** is an eigenvector of the matrix A with associated eigenvalue $\lambda = 10$.

Similarly, if $\mathbf{w} = \begin{bmatrix} -1 \\ 2 \end{bmatrix}$, we find that

$$A\mathbf{w} = \begin{bmatrix} 7 & 6 \\ 6 & -2 \end{bmatrix} \begin{bmatrix} -1 \\ 2 \end{bmatrix} = \begin{bmatrix} 5 \\ -10 \end{bmatrix} = -5 \begin{bmatrix} -1 \\ 2 \end{bmatrix} = -5\mathbf{w}.$$

Here again, we have $A\mathbf{w} = -5\mathbf{w}$ showing that **w** is an eigenvector of A with associated eigenvalue $\lambda = -5$.

> **Activity 4.1.2.** This definition has an important geometric interpretation that we will investigate here.
>
> a. Suppose that **v** is a nonzero vector and that λ is a scalar. What is the geometric relationship between **v** and $\lambda\mathbf{v}$?
>
> b. Let's now consider the eigenvector condition: $A\mathbf{v} = \lambda\mathbf{v}$. Here we have two vectors, **v** and $A\mathbf{v}$. If $A\mathbf{v} = \lambda\mathbf{v}$, what is the geometric relationship between **v** and $A\mathbf{v}$?

c. There is an interactive diagram, available at gvsu.edu/s/0Ja, that accompanies this activity.

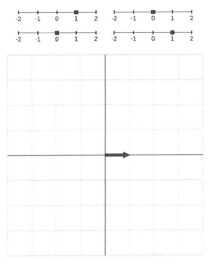

Figure 4.1.3 A geometric interpretation of the eigenvalue-eigenvector condition $A\mathbf{v} = \lambda\mathbf{v}$.

Choose the matrix $A = \begin{bmatrix} 1 & 2 \\ 2 & 1 \end{bmatrix}$. Move the vector $\mathbf{v}$ so that the eigenvector condition holds. What is the eigenvector $\mathbf{v}$ and what is the associated eigenvalue?

d. By algebraically computing $A\mathbf{v}$, verify that the eigenvector condition holds for the vector $\mathbf{v}$ that you found.

e. If you multiply the eigenvector $\mathbf{v}$ that you found by 2, do you still have an eigenvector? If so, what is the associated eigenvalue?

f. Are you able to find another eigenvector $\mathbf{v}$ that is not a scalar multiple of the first one that you found? If so, what is the eigenvector and what is the associated eigenvalue?

g. Now consider the matrix $A = \begin{bmatrix} 2 & 1 \\ 0 & 2 \end{bmatrix}$. Use the diagram to describe any eigenvectors and associated eigenvalues.

h. Finally, consider the matrix $A = \begin{bmatrix} 0 & -1 \\ 1 & 0 \end{bmatrix}$. Use the diagram to describe any eigenvectors and associated eigenvalues. What geometric transformation does this matrix perform on vectors? How does this explain the presence of any eigenvectors?

Let's consider the ideas we saw in the activity in some more depth. To be an eigenvector of A, the vector $\mathbf{v}$ must satisfy $A\mathbf{v} = \lambda\mathbf{v}$ for some scalar λ. This means that $\mathbf{v}$ and $A\mathbf{v}$ are scalar multiples of each other so they must lie on the same line.

Consider now the matrix $A = \begin{bmatrix} 1 & 2 \\ 2 & 1 \end{bmatrix}$. On the left of Figure 4.1.4, we see that $\mathbf{v} = \begin{bmatrix} 1 \\ 0 \end{bmatrix}$ is not an eigenvector of A since the vectors $\mathbf{v}$ and $A\mathbf{v}$ do not lie on the same line. On the right, however, we see that $\mathbf{v} = \begin{bmatrix} 1 \\ 1 \end{bmatrix}$ is an eigenvector. In fact, $A\mathbf{v}$ is obtained from $\mathbf{v}$ by stretching $\mathbf{v}$ by a factor of 3. Therefore, $\mathbf{v}$ is an eigenvector of A with eigenvalue $\lambda = 3$.

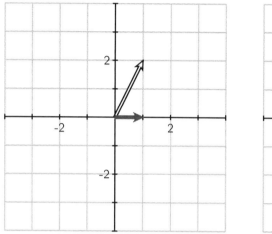
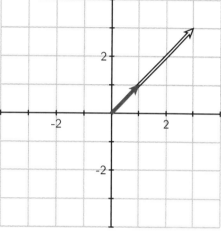

Figure 4.1.4 On the left, the vector $\mathbf{v}$ is not an eigenvector. On the right, the vector $\mathbf{v}$ is an eigenvector with eigenvalue $\lambda = 3$.

It is not difficult to see that any multiple of $\begin{bmatrix} 1 \\ 1 \end{bmatrix}$ is also an eigenvector of A with eigenvalue $\lambda = 3$. Indeed, we will see later that all the eigenvectors associated to a given eigenvalue form a subspace of $\mathbb{R}^n$.

In Figure 4.1.5, we see that $\mathbf{v} = \begin{bmatrix} -1 \\ 1 \end{bmatrix}$ is also an eigenvector with eigenvalue $\lambda = -1$.

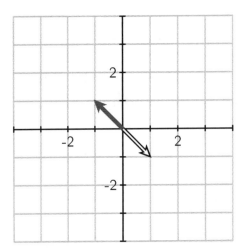

Figure 4.1.5 Here we see another eigenvector **v** with eigenvalue $\lambda = -1$.

The interactive diagram we used in the activity is meant to convey the fact that the eigenvectors of a matrix A are special vectors. Most of the time, the vectors **v** and $A\mathbf{v}$ appear visually unrelated. For certain vectors, however, **v** and $A\mathbf{v}$ line up with one another. Something important is going on when that happens so we call attention to these vectors by calling them eigenvectors. For these vectors, the operation of multiplying by A reduces to the much simpler operation of scalar multiplying by λ. The reason eigenvectors are important is because it is extremely convenient to be able to replace matrix multiplication by scalar multiplication.

4.1.2 The usefulness of eigenvalues and eigenvectors

In the next section, we will introduce an algebraic technique for finding the eigenvalues and eigenvectors of a matrix. Before doing that, however, we would like to discuss why eigenvalues and eigenvectors are so useful.

Let's continue looking at the example $A = \begin{bmatrix} 1 & 2 \\ 2 & 1 \end{bmatrix}$. We have seen that $\mathbf{v}_1 = \begin{bmatrix} 1 \\ 1 \end{bmatrix}$ is an eigenvector with eigenvalue $\lambda = 3$ and $\mathbf{v}_2 = \begin{bmatrix} -1 \\ 1 \end{bmatrix}$ is an eigenvector with eigenvalue $\lambda = -1$. This means that $A\mathbf{v}_1 = 3\mathbf{v}_1$ and $A\mathbf{v}_2 = -\mathbf{v}_2$. By the linearity of matrix multiplication, we can determine what happens when we multiply a linear combination of $\mathbf{v}_1$ and $\mathbf{v}_2$ by A:

$$A(c_1\mathbf{v}_1 + c_2\mathbf{v}_2) = 3c_1\mathbf{v}_1 - c_2\mathbf{v}_2.$$

For instance, if we consider the vector $\mathbf{x} = \mathbf{v}_1 - 2\mathbf{v}_2$, we find that

$$A\mathbf{x} = A(\mathbf{v}_1 - 2\mathbf{v}_2)$$
$$A\mathbf{x} = 3\mathbf{v}_1 + 2\mathbf{v}_2$$

as seen in the figure.

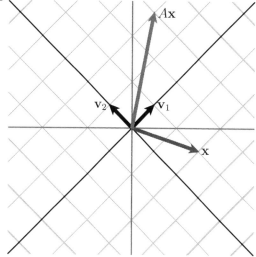

In other words, multiplying by A has the effect of stretching a vector $\mathbf{x}$ in the $\mathbf{v}_1$ direction by a factor of 3 and flipping $\mathbf{x}$ in $\mathbf{v}_2$ direction.

We can draw an analogy with the more familiar example of the diagonal matrix $D = \begin{bmatrix} 3 & 0 \\ 0 & -1 \end{bmatrix}$. As we have seen, the matrix transformation defined by D combines a horizontal stretching by a factor of 3 with a reflection across the horizontal axis, as is illustrated in Figure 4.1.6.

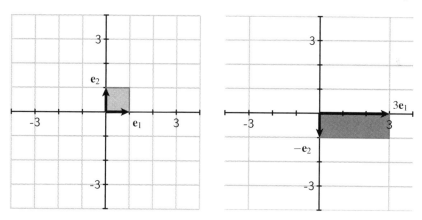

Figure 4.1.6 The diagonal matrix D stretches vectors horizontally by a factor of 3 and flips vectors vertically.

The matrix $A = \begin{bmatrix} 1 & 2 \\ 2 & 1 \end{bmatrix}$ has a similar effect when viewed in the basis defined by the eigenvectors $\mathbf{v}_1$ and $\mathbf{v}_2$, as seen in Figure 4.1.7.

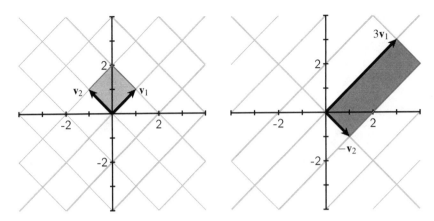

Figure 4.1.7 The matrix A has the same geometric effect as the diagonal matrix D when expressed in the coordinate system defined by the basis of eigenvectors.

In a sense that will be made precise later, having a set of eigenvectors of A that forms a basis of $\mathbb{R}^2$ enables us to think of A as being equivalent to a diagonal matrix D. Of course, as the other examples in the previous activity show, it may not always be possible to form a basis from the eigenvectors of a matrix. For example, the only eigenvectors of the matrix $\begin{bmatrix} 2 & 1 \\ 0 & 2 \end{bmatrix}$, which represents a shear, have the form $\begin{bmatrix} x \\ 0 \end{bmatrix}$. In this example, we are not able to create a basis for $\mathbb{R}^2$ consisting of eigenvectors of the matrix. This is also true for the matrix $\begin{bmatrix} 0 & -1 \\ 1 & 0 \end{bmatrix}$, which represents a 90° rotation.

> **Activity 4.1.3.** Let's consider an example that illustrates how we can put these ideas to use.
>
> Suppose that we work for a car rental company that has two locations, P and Q. When a customer rents a car at one location, they have the option to return it to either location at the end of the day. After doing some market research, we determine:
>
> - 80% of the cars rented at location P are returned to P and 20% are returned to Q.
> - 40% of the cars rented at location Q are returned to Q and 60% are returned to P.
>
> a. Suppose that there are 1000 cars at location P and no cars at location Q on Monday morning. How many cars are there are locations P and Q at the end of the day on Monday?
>
> b. How many are at locations P and Q at end of the day on Tuesday?
>
> c. If we let P_k and Q_k be the number of cars at locations P and Q, respectively, at

the end of day k, we then have

$$P_{k+1} = 0.8P_k + 0.6Q_k$$
$$Q_{k+1} = 0.2P_k + 0.4Q_k.$$

We can write the vector $\mathbf{x}_k = \begin{bmatrix} P_k \\ Q_k \end{bmatrix}$ to reflect the number of cars at the two locations at the end of day k, which says that

$$\mathbf{x}_{k+1} = \begin{bmatrix} 0.8 & 0.6 \\ 0.2 & 0.4 \end{bmatrix} \mathbf{x}_k$$

or $\mathbf{x}_{k+1} = A\mathbf{x}_k$ where $A = \begin{bmatrix} 0.8 & 0.6 \\ 0.2 & 0.4 \end{bmatrix}$.

Suppose that

$$\mathbf{v}_1 = \begin{bmatrix} 3 \\ 1 \end{bmatrix}, \quad \mathbf{v}_2 = \begin{bmatrix} -1 \\ 1 \end{bmatrix}.$$

Compute $A\mathbf{v}_1$ and $A\mathbf{v}_2$ to demonstrate that $\mathbf{v}_1$ and $\mathbf{v}_2$ are eigenvectors of A. What are the associated eigenvalues λ_1 and λ_2?

d. We said that 1000 cars are initially at location P and none at location Q. This means that the initial vector describing the number of cars is $\mathbf{x}_0 = \begin{bmatrix} 1000 \\ 0 \end{bmatrix}$. Write $\mathbf{x}_0$ as a linear combination of $\mathbf{v}_1$ and $\mathbf{v}_2$.

e. Remember that $\mathbf{v}_1$ and $\mathbf{v}_2$ are eigenvectors of A. Use the linearity of matrix multiplication to write the vector $\mathbf{x}_1 = A\mathbf{x}_0$, describing the number of cars at the two locations at the end of the first day, as a linear combination of $\mathbf{v}_1$ and $\mathbf{v}_2$.

f. Write the vector $\mathbf{x}_2 = A\mathbf{x}_1$ as a linear combination of $\mathbf{v}_1$ and $\mathbf{v}_2$. Then write the next few vectors as linear combinations of $\mathbf{v}_1$ and $\mathbf{v}_2$:

  1. $\mathbf{x}_3 = A\mathbf{x}_2$.
  2. $\mathbf{x}_4 = A\mathbf{x}_3$.
  3. $\mathbf{x}_5 = A\mathbf{x}_4$.
  4. $\mathbf{x}_6 = A\mathbf{x}_5$.

g. What will happen to the number of cars at the two locations after a very long time? Explain how writing $\mathbf{x}_0$ as a linear combination of eigenvectors helps you determine the long-term behavior.

This activity is important and motivates much of our work with eigenvalues and eigenvectors so it's worth reviewing to make sure we have a clear understanding of the concepts.

First, we compute

$$A\mathbf{v}_1 = \begin{bmatrix} 0.8 & 0.6 \\ 0.2 & 0.4 \end{bmatrix} \begin{bmatrix} 3 \\ 1 \end{bmatrix} = \begin{bmatrix} 3 \\ 1 \end{bmatrix} = 1\mathbf{v}_1$$

$$A\mathbf{v}_2 = \begin{bmatrix} 0.8 & 0.6 \\ 0.2 & 0.4 \end{bmatrix} \begin{bmatrix} -1 \\ 1 \end{bmatrix} = \begin{bmatrix} -0.2 \\ 0.2 \end{bmatrix} = 0.2\mathbf{v}_2.$$

This shows that $\mathbf{v}_1$ is an eigenvector of A with eigenvalue $\lambda_1 = 1$ and $\mathbf{v}_2$ is an eigenvector of A with eigenvalue $\lambda_2 = 0.2$.

By the linearity of matrix matrix multiplication, we have

$$A(c_1\mathbf{v}_1 + c_2\mathbf{v}_2) = c_1\mathbf{v}_1 + 0.2c_2\mathbf{v}_2.$$

Therefore, we will write the vector describing the initial distribution of cars $\mathbf{x}_0 = \begin{bmatrix} 1000 \\ 0 \end{bmatrix}$ as a linear combination of $\mathbf{v}_1$ and $\mathbf{v}_2$; that is, $\mathbf{x}_0 = c_1\mathbf{v}_2 + c_2\mathbf{v}_2$. To do, we form the augmented matrix and row reduce:

$$\begin{bmatrix} \mathbf{v}_1 & \mathbf{v}_2 & | & \mathbf{x}_0 \end{bmatrix} = \begin{bmatrix} 3 & -1 & | & 1000 \\ 1 & 1 & | & 0 \end{bmatrix} \sim \begin{bmatrix} 1 & 0 & | & 250 \\ 0 & 1 & | & -250 \end{bmatrix}.$$

Therefore, $\mathbf{x}_0 = 250\mathbf{v}_1 - 250\mathbf{v}_2$.

To determine the distribution of cars on subsequent days, we will repeatedly multiply by A. We find that

$$\mathbf{x}_1 = A\mathbf{x}_0 = A(250\mathbf{v}_1 - 250\mathbf{v}_2) = 250\mathbf{v}_1 - (0.2)250\mathbf{v}_2$$
$$\mathbf{x}_2 = A\mathbf{x}_1 = A(250\mathbf{v}_1 - (0.2)250\mathbf{v}_2) = 250\mathbf{v}_1 - (0.2)^2 250\mathbf{v}_2$$
$$\mathbf{x}_3 = A\mathbf{x}_2 = A(250\mathbf{v}_1 - (0.2)^2 250\mathbf{v}_2) = 250\mathbf{v}_1 - (0.2)^3 250\mathbf{v}_2.$$
$$\mathbf{x}_4 = A\mathbf{x}_3 = A(250\mathbf{v}_1 - (0.2)^3 250\mathbf{v}_2) = 250\mathbf{v}_1 - (0.2)^4 250\mathbf{v}_2$$
$$\mathbf{x}_5 = A\mathbf{x}_4 = A(250\mathbf{v}_1 - (0.2)^4 250\mathbf{v}_2) = 250\mathbf{v}_1 - (0.2)^5 250\mathbf{v}_2$$

In particular, this shows us that

$$\mathbf{x}_5 = 250\mathbf{v}_1 - (0.2)^5 250\mathbf{v}_2 = \begin{bmatrix} 250 \cdot 3 - (0.2)^5 250 \cdot (-1) \\ 250 \cdot 1 - (0.2)^5 250 \cdot 1 \end{bmatrix} = \begin{bmatrix} 750.09 \\ 249.92 \end{bmatrix}.$$

Taking notice of the pattern, we may write

$$\mathbf{x}_k = 250\mathbf{v}_1 - (0.2)^k 250\mathbf{v}_2.$$

Multiplying a number by 0.2 is the same as taking 20% of that number. As each day goes by, the second term is multiplied by 0.2 so the coefficient of $\mathbf{v}_2$ in the expression for $\mathbf{x}_k$ will eventually become extremely small. We therefore see that the distribution of cars will stabilize at $\mathbf{x} = 250\mathbf{v}_1 = \begin{bmatrix} 750 \\ 250 \end{bmatrix}$.

Notice how our understanding of the eigenvectors of the matrix allows us to replace matrix multiplication with the simpler operation of scalar multiplication. As a result, we can look far into the future without having to repeatedly perform matrix multiplication.

4.1. AN INTRODUCTION TO EIGENVALUES AND EIGENVECTORS

Furthermore, notice how this example relies on the fact that we can express the initial vector $\mathbf{x}_0$ as a linear combination of eigenvectors. For this reason, we would like, when given an $n \times n$ matrix, to be able to create a basis of $\mathbb{R}^n$ that consists of its eigenvectors. We will frequently return to this question in later sections.

Question 4.1.8 If A is an $n \times n$ matrix, can we form a basis of $\mathbb{R}^n$ consisting of eigenvectors of A?

4.1.3 Summary

We defined an eigenvector of a square matrix A to be a nonzero vector $\mathbf{v}$ such that $A\mathbf{v} = \lambda\mathbf{v}$ for some scalar λ, which is called the eigenvalue associated to $\mathbf{v}$.

- If $\mathbf{v}$ is an eigenvector, then matrix multiplication by A reduces to the simpler operation of scalar multiplication by λ.

- Scalar multiples of an eigenvector are also eigenvectors. In fact, we will see that the eigenvectors associated to an eigenvalue λ form a subspace.

- If we can form a basis for $\mathbb{R}^n$ consisting of eigenvectors of A, then A is, in some sense, equivalent to a diagonal matrix.

- Rewriting a vector $\mathbf{x}$ as a linear combination of eigenvectors of A simplifies the process of repeatedly multiplying $\mathbf{x}$ by A.

4.1.4 Exercises

1. Consider the matrix and vectors

$$A = \begin{bmatrix} 8 & -10 \\ 5 & -7 \end{bmatrix}, \quad \mathbf{v}_1 = \begin{bmatrix} 2 \\ 1 \end{bmatrix}, \quad \mathbf{v}_2 = \begin{bmatrix} 1 \\ 1 \end{bmatrix}.$$

   a. Show that $\mathbf{v}_1$ and $\mathbf{v}_2$ are eigenvectors of A and find their associated eigenvalues.

   b. Express the vector $\mathbf{x} = \begin{bmatrix} -4 \\ -1 \end{bmatrix}$ as a linear combination of $\mathbf{v}_1$ and $\mathbf{v}_2$.

   c. Use this expression to compute $A\mathbf{x}$, $A^2\mathbf{x}$, and $A^{-1}\mathbf{x}$ as a linear combination of eigenvectors.

2. Consider the matrix and vectors

$$A = \begin{bmatrix} -5 & -2 & 2 \\ 24 & 14 & -10 \\ 21 & 14 & -10 \end{bmatrix}, \quad \mathbf{v}_1 = \begin{bmatrix} 1 \\ -2 \\ -1 \end{bmatrix}, \mathbf{v}_2 = \begin{bmatrix} 2 \\ -3 \\ 0 \end{bmatrix}, \mathbf{v}_3 = \begin{bmatrix} 0 \\ -1 \\ -1 \end{bmatrix}.$$

   a. Show that the vectors $\mathbf{v}_1$, $\mathbf{v}_2$, and $\mathbf{v}_3$ are eigenvectors of A and find their associated eigenvalues.

b. Express the vector $\mathbf{x} = \begin{bmatrix} 0 \\ -3 \\ -4 \end{bmatrix}$ as a linear combination of the eigenvectors.

c. Use this expression to compute $A\mathbf{x}$, $A^2\mathbf{x}$, and $A^{-1}\mathbf{x}$ as a linear combination of eigenvectors.

3. Suppose that A is an $n \times n$ matrix.

   a. Explain why $\lambda = 0$ is an eigenvalue of A if and only if there is a nonzero solution to the homogeneous equation $A\mathbf{x} = 0$.

   b. Explain why A is not invertible if and only if $\lambda = 0$ is an eigenvalue.

   c. If $\mathbf{v}$ is an eigenvector of A having associated eigenvalue λ, explain why $\mathbf{v}$ is also an eigenvector of A^2 with associated eigenvalue λ^2.

   d. If A is invertible and $\mathbf{v}$ is eigenvector of A having associated eigenvalue λ, explain why $\mathbf{v}$ is also an eigenvector of A^{-1} with associated eigenvalue λ^{-1}.

   e. The matrix $A = \begin{bmatrix} 1 & 2 \\ 2 & 1 \end{bmatrix}$ has eigenvectors $\mathbf{v}_1 = \begin{bmatrix} 1 \\ 1 \end{bmatrix}$ and $\mathbf{v}_2 = \begin{bmatrix} -1 \\ 1 \end{bmatrix}$ and associated eigenvalues $\lambda_1 = 3$ and $\lambda = -1$. What are some eigenvectors and associated eigenvalues for A^5?

4. Suppose that A is a matrix with eigenvectors $\mathbf{v}_1$ and $\mathbf{v}_2$ and eigenvalues $\lambda_1 = -1$ and $\lambda_2 = 2$ as shown in Figure 4.1.9.

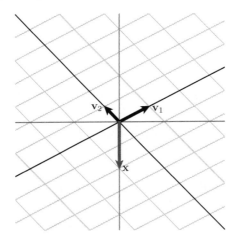

Figure 4.1.9 The vectors $\mathbf{v}_1$ and $\mathbf{v}_2$ are eigenvectors of A.

Sketch the vectors $A\mathbf{x}$, $A^2\mathbf{x}$, and $A^{-1}\mathbf{x}$.

5. For the following matrices, find the eigenvectors and associated eigenvalues by thinking geometrically about the corresponding matrix transformation.

   a. $\begin{bmatrix} 3 & 0 \\ 0 & 3 \end{bmatrix}$.

4.1. AN INTRODUCTION TO EIGENVALUES AND EIGENVECTORS

b. $\begin{bmatrix} -2 & 0 \\ 0 & 4 \end{bmatrix}$.

c. What are the eigenvectors and associated eigenvalues of the identity matrix?

d. What are the eigenvectors and associated eigenvalues of a diagonal matrix with distinct diagonal entries?

6. Suppose that A is a 2×2 matrix having eigenvectors

$$\mathbf{v}_1 = \begin{bmatrix} 2 \\ 1 \end{bmatrix}, \quad \mathbf{v}_2 = \begin{bmatrix} -1 \\ 2 \end{bmatrix}$$

and associated eigenvalues $\lambda_1 = 2$ and $\lambda_2 = -3$.

   a. If $\mathbf{x} = \begin{bmatrix} 5 \\ 0 \end{bmatrix}$, find the vector $A^4\mathbf{x}$.

   b. Find the vectors $A\begin{bmatrix} 1 \\ 0 \end{bmatrix}$ and $A\begin{bmatrix} 0 \\ 1 \end{bmatrix}$.

   c. What is the matrix A?

7. Determine whether the following statements are true or false and provide a justification for your response.

   a. The eigenvalues of a diagonal matrix are equal to the entries on the diagonal.

   b. If $A\mathbf{v} = \lambda \mathbf{v}$, then $A^2\mathbf{v} = \lambda \mathbf{v}$ as well.

   c. Every vector is an eigenvector of the identity matrix.

   d. If $\lambda = 0$ is an eigenvalue of A, then A is invertible.

   e. For every $n \times n$ matrix A, it is possible to find a basis of $\mathbb{R}^n$ consisting of eigenvectors of A.

8. Suppose that A is an $n \times n$ matrix.

   a. Assuming that $\mathbf{v}$ is an eigenvector of A whose associated eigenvector is nonzero, explain why $\mathbf{v}$ is in Col(A).

   b. Assuming that $\mathbf{v}$ is an eigenvector of A whose associated eigenvector is zero, explain why $\mathbf{v}$ is in Nul(A).

   c. Consider the two special matrices below and find their eigenvectors and associated eigenvalues.

$$A = \begin{bmatrix} 1 & 1 & 1 \\ 1 & 1 & 1 \\ 1 & 1 & 1 \end{bmatrix}, \quad B = \begin{bmatrix} 1 & 1 & 1 \\ 2 & 2 & 2 \\ 3 & 3 & 3 \end{bmatrix}.$$

9. For each of the following matrix transformations, describe the eigenvalues and eigenvectors of the corresponding matrix A.

   a. A reflection in $\mathbb{R}^2$ in the line $y = x$.

b. A 180° rotation in $\mathbb{R}^2$.

c. A 180° rotation in $\mathbb{R}^3$ about the y-axis.

d. A 90° rotation in $\mathbb{R}^3$ about the x-axis.

10. Suppose we have two species, P and Q, where species P preys on Q. Their populations, in millions, in year k are denoted by P_k and Q_k and satisfy

$$P_{k+1} = 0.8P_k + 0.2Q_k$$
$$Q_{k+1} = -0.3P_k + 1.5Q_k.$$

We will keep track of the populations in year k using the vector $\mathbf{x}_k = \begin{bmatrix} P_k \\ Q_k \end{bmatrix}$ so that

$$\mathbf{x}_{k+1} = A\mathbf{x}_k = \begin{bmatrix} 0.8 & 0.2 \\ -0.3 & 1.5 \end{bmatrix} \mathbf{x}_k.$$

a. Show that $\mathbf{v}_1 = \begin{bmatrix} 1 \\ 3 \end{bmatrix}$ and $\mathbf{v}_2 = \begin{bmatrix} 2 \\ 1 \end{bmatrix}$ are eigenvectors of A and find their associated eigenvalues.

b. Suppose that the initial populations are described by the vector $\mathbf{x}_0 = \begin{bmatrix} 38 \\ 44 \end{bmatrix}$. Express $\mathbf{x}_0$ as a linear combination of $\mathbf{v}_1$ and $\mathbf{v}_2$.

c. Find the populations after one year, two years, and three years by writing the vectors $\mathbf{x}_1$, $\mathbf{x}_2$, and $\mathbf{x}_3$ as linear combinations of $\mathbf{v}_1$ and $\mathbf{v}_2$.

d. What is the general form for $\mathbf{x}_k$?

e. After a very long time, what is the ratio of P_k to Q_k?

4.2 Finding eigenvalues and eigenvectors

The last section introduced eigenvalues and eigenvectors, presented the underlying geometric intuition behind their definition, and demonstrated their use in understanding the long-term behavior of certain systems. We will now develop a more algebraic understanding of eigenvalues and eigenvectors. In particular, we will find an algebraic method for determining the eigenvalues and eigenvectors of a square matrix.

> **Preview Activity 4.2.1.** Let's begin by reviewing some important ideas that we have seen previously.
>
> a. Suppose that A is a square matrix and that the nonzero vector $\mathbf{x}$ is a solution to the homogeneous equation $A\mathbf{x} = \mathbf{0}$. What can we conclude about the invertibility of A?
>
> b. How does the determinant $\det(A)$ tell us if there is a nonzero solution to the homogeneous equation $A\mathbf{x} = \mathbf{0}$?
>
> c. Suppose that
> $$A = \begin{bmatrix} 3 & -1 & 1 \\ 0 & 2 & 4 \\ 1 & 1 & 3 \end{bmatrix}.$$
> Find the determinant $\det(A)$. What does this tell us about the solution space to the homogeneous equation $A\mathbf{x} = \mathbf{0}$?
>
> d. Find a basis for $\text{Nul}(A)$.
>
> e. What is the relationship between the rank of a matrix and the dimension of its null space?

4.2.1 The characteristic polynomial

We will first see that the eigenvalues of a square matrix appear as the roots of a particular polynomial. To begin, notice that we originally defined an eigenvector as a nonzero vector $\mathbf{v}$ that satisfies the equation $A\mathbf{v} = \lambda\mathbf{v}$. We will rewrite this as

$$A\mathbf{v} = \lambda\mathbf{v}$$
$$A\mathbf{v} - \lambda\mathbf{v} = \mathbf{0}$$
$$A\mathbf{v} - \lambda I\mathbf{v} = \mathbf{0}$$
$$(A - \lambda I)\mathbf{v} = \mathbf{0}.$$

In other words, an eigenvector $\mathbf{v}$ is a solution of the homogeneous equation $(A - \lambda I)\mathbf{v} = \mathbf{0}$. This puts us in the familiar territory explored in the next activity.

Activity 4.2.2. The eigenvalues of a square matrix are defined by the condition that there be a nonzero solution to the homogeneous equation $(A - \lambda I)\mathbf{v} = \mathbf{0}$.

a. If there is a nonzero solution to the homogeneous equation $(A - \lambda I)\mathbf{v} = \mathbf{0}$, what can we conclude about the invertibility of the matrix $A - \lambda I$?

b. If there is a nonzero solution to the homogeneous equation $(A - \lambda I)\mathbf{v} = \mathbf{0}$, what can we conclude about the determinant $\det(A - \lambda I)$?

c. Let's consider the matrix
$$A = \begin{bmatrix} 1 & 2 \\ 2 & 1 \end{bmatrix}$$
from which we construct
$$A - \lambda I = \begin{bmatrix} 1 & 2 \\ 2 & 1 \end{bmatrix} - \lambda \begin{bmatrix} 1 & 0 \\ 0 & 1 \end{bmatrix} = \begin{bmatrix} 1-\lambda & 2 \\ 2 & 1-\lambda \end{bmatrix}.$$
Find the determinant $\det(A - \lambda I)$. What kind of equation do you obtain when we set this determinant to zero to obtain $\det(A - \lambda I) = 0$?

d. Use the determinant you found in the previous part to find the eigenvalues λ by solving the equation $\det(A - \lambda I) = 0$. We considered this matrix in Activity 4.1.2 so we should find the same eigenvalues for A that we found by reasoning geometrically there.

e. Consider the matrix $A = \begin{bmatrix} 2 & 1 \\ 0 & 2 \end{bmatrix}$ and find its eigenvalues by solving the equation $\det(A - \lambda I) = 0$.

f. Consider the matrix $A = \begin{bmatrix} 0 & -1 \\ 1 & 0 \end{bmatrix}$ and find its eigenvalues by solving the equation $\det(A - \lambda I) = 0$.

g. Find the eigenvalues of the triangular matrix $\begin{bmatrix} 3 & -1 & 4 \\ 0 & -2 & 3 \\ 0 & 0 & 1 \end{bmatrix}$. What is generally true about the eigenvalues of a triangular matrix?

This activity demonstrates a technique that enables us to find the eigenvalues of a square matrix A. Since an eigenvalue λ is a scalar for which the equation $(A - \lambda I)\mathbf{v} = \mathbf{0}$ has a nonzero solution, it must be the case that $A - \lambda I$ is not invertible. Therefore, its determinant is zero. This gives us the equation

$$\det(A - \lambda I) = 0$$

whose solutions are the eigenvalues of A. This equation is called the *characteristic equation* of A.

Example 4.2.1 If we write the characteristic equation for the matrix $A = \begin{bmatrix} -4 & 4 \\ -12 & 10 \end{bmatrix}$, we see

4.2. FINDING EIGENVALUES AND EIGENVECTORS

that
$$\det(A - \lambda I) = 0$$

$$\det \begin{bmatrix} -4-\lambda & 4 \\ -12 & 10-\lambda \end{bmatrix} = 0$$

$$(-4-\lambda)(10-\lambda) + 48 = 0$$
$$\lambda^2 - 6\lambda + 8 = 0$$
$$(\lambda - 4)(\lambda - 2) = 0.$$

This shows us that the eigenvalues are $\lambda = 4$ and $\lambda = 2$.

In general, the expression $\det(A - \lambda I)$ is a polynomial in λ, which is called the *characteristic polynomial* of A. If A is an $n \times n$ matrix, the degree of the characteristic polynomial is n. For instance, if A is a 2×2 matrix, then $\det(A - \lambda I)$ is a quadratic polynomial; if A is a 3×3 matrix, then $\det(A - \lambda I)$ is a cubic polynomial.

The matrix in Example 4.2.1 has a characteristic polynomial with two real and distinct roots. This will not always be the case, as demonstrated in the next two examples.

Example 4.2.2 Consider the matrix $A = \begin{bmatrix} 5 & -1 \\ 4 & 1 \end{bmatrix}$, whose characteristic equation is

$$\lambda^2 - 6\lambda + 9 = (\lambda - 3)^2 = 0.$$

In this case, the characteristic polynomial has one real root, which means that this matrix has a single real eigenvalue, $\lambda = 3$.

Example 4.2.3 To find the eigenvalues of a triangular matrix, we remember that the determinant of a triangular matrix is the product of the entries on the diagonal. For instance, the following triangular matrix has the characteristic equation

$$\det\left(\begin{bmatrix} 4 & 2 & 3 \\ 0 & -2 & -1 \\ 0 & 0 & 3 \end{bmatrix} - \lambda I \right) = \det \begin{bmatrix} 4-\lambda & 2 & 3 \\ 0 & -2-\lambda & -1 \\ 0 & 0 & 3-\lambda \end{bmatrix}$$

$$= (4-\lambda)(-2-\lambda)(3-\lambda) = 0,$$

showing that the eigenvalues are the diagonal entries $\lambda = 4, -2, 3$.

4.2.2 Finding eigenvectors

Now that we can find the eigenvalues of a square matrix A by solving the characteristic equation $\det(A - \lambda I) = 0$, we will turn to the question of finding the eigenvectors associated to an eigenvalue λ. The key, as before, is to note that an eigenvector is a nonzero solution to the homogeneous equation $(A - \lambda I)\mathbf{v} = \mathbf{0}$. In other words, the eigenvectors associated to an eigenvalue λ form the null space $\text{Nul}(A - \lambda I)$.

This shows that the eigenvectors associated to an eigenvalue form a subspace of $\mathbb{R}^n$. We will denote the subspace of eigenvectors of a matrix A associated to the eigenvalue λ by E_λ and note that
$$E_\lambda = \text{Nul}(A - \lambda I).$$
We say that E_λ is the *eigenspace* of A associated to the eigenvalue λ.

Activity 4.2.3. In this activity, we will find the eigenvectors of a matrix as the null space of the matrix $A - \lambda I$.

a. Let's begin with the matrix $A = \begin{bmatrix} 1 & 2 \\ 2 & 1 \end{bmatrix}$. We have seen that $\lambda = 3$ is an eigenvalue. Form the matrix $A - 3I$ and find a basis for the eigenspace $E_3 = \text{Nul}(A-3I)$. What is the dimension of this eigenspace? For each of the basis vectors $\mathbf{v}$, verify that $A\mathbf{v} = 3\mathbf{v}$.

b. We also saw that $\lambda = -1$ is an eigenvalue. Form the matrix $A - (-1)I$ and find a basis for the eigenspace E_{-1}. What is the dimension of this eigenspace? For each of the basis vectors $\mathbf{v}$, verify that $A\mathbf{v} = -\mathbf{v}$.

c. Is it possible to form a basis of $\mathbb{R}^2$ consisting of eigenvectors of A?

d. Now consider the matrix $A = \begin{bmatrix} 3 & 0 \\ 0 & 3 \end{bmatrix}$. Write the characteristic equation for A and use it to find the eigenvalues of A. For each eigenvalue, find a basis for its eigenspace E_λ. Is it possible to form a basis of $\mathbb{R}^2$ consisting of eigenvectors of A?

e. Next, consider the matrix $A = \begin{bmatrix} 2 & 1 \\ 0 & 2 \end{bmatrix}$. Write the characteristic equation for A and use it to find the eigenvalues of A. For each eigenvalue, find a basis for its eigenspace E_λ. Is it possible to form a basis of $\mathbb{R}^2$ consisting of eigenvectors of A?

f. Finally, find the eigenvalues and eigenvectors of the diagonal matrix $A = \begin{bmatrix} 4 & 0 \\ 0 & -1 \end{bmatrix}$. Explain your result by considering the geometric effect of the matrix transformation defined by A.

Once we find an eigenvalue of a matrix A, describing the associated eigenspace E_λ amounts to the familiar task of describing the null space $\text{Nul}(A - \lambda I)$.

Example 4.2.4 Revisiting the matrix $A = \begin{bmatrix} -4 & 4 \\ -12 & 10 \end{bmatrix}$ from Example 4.2.1, we recall that we found eigenvalues $\lambda = 4$ and $\lambda = 2$.

Considering the eigenvalue $\lambda = 4$, we have
$$A - 4I = \begin{bmatrix} -8 & 4 \\ -12 & 6 \end{bmatrix} \sim \begin{bmatrix} 1 & -1/2 \\ 0 & 0 \end{bmatrix}.$$

4.2. FINDING EIGENVALUES AND EIGENVECTORS

Since the eigenvectors $\mathbf{v} = \begin{bmatrix} v_1 \\ v_2 \end{bmatrix}$ are the solutions of the equation $(A - 4I)\mathbf{v} = \mathbf{0}$, we see that they are determined by the single equation $v_1 - \frac{1}{2}v_2 = 0$ or $v_1 = \frac{1}{2}v_2$. Therefore the eigenvectors in E_4 have the form

$$\mathbf{v} = \begin{bmatrix} v_1 \\ v_2 \end{bmatrix} = \begin{bmatrix} \frac{1}{2}v_2 \\ v_2 \end{bmatrix} = v_2 \begin{bmatrix} 1/2 \\ 1 \end{bmatrix}.$$

In other words, E_4 is a one-dimensional subspace of $\mathbb{R}^2$ with basis vector $\begin{bmatrix} 1/2 \\ 1 \end{bmatrix}$ or basis vector $\begin{bmatrix} 1 \\ 2 \end{bmatrix}$. In the same way, we find that a basis for the eigenspace E_2 is $\begin{bmatrix} 2 \\ 3 \end{bmatrix}$.

We note that, for this matrix, it is possible to construct a basis of $\mathbb{R}^2$ consisting of eigenvectors, namely,

$$\mathcal{B} = \left\{ \begin{bmatrix} 1 \\ 2 \end{bmatrix}, \begin{bmatrix} 2 \\ 3 \end{bmatrix} \right\}.$$

Example 4.2.5 Consider the matrix $A = \begin{bmatrix} 1 & 1 \\ -1 & 3 \end{bmatrix}$ whose characteristic equation is

$$\det(A - \lambda I) = \lambda^2 - 4\lambda + 4 = (\lambda - 2)^2 = 0.$$

There is a single eigenvalue $\lambda = 2$, and we find that

$$A - 2\lambda = \begin{bmatrix} -1 & 1 \\ -1 & 1 \end{bmatrix} \sim \begin{bmatrix} 1 & -1 \\ 0 & 0 \end{bmatrix}.$$

Therefore, the eigenspace $E_2 = \text{Nul}(A - 2I)$ is one-dimensional with a basis vector $\begin{bmatrix} 1 \\ 1 \end{bmatrix}$.

Example 4.2.6 If $A = \begin{bmatrix} -1 & 0 \\ 0 & -1 \end{bmatrix}$, then

$$\det(A - \lambda I) = (\lambda + 1)^2 = 0,$$

which implies that there is a single eigenvalue $\lambda = -1$. We find that

$$A - (-1)I = \begin{bmatrix} 0 & 0 \\ 0 & 0 \end{bmatrix},$$

which says that every two-dimensional vector $\mathbf{v}$ satisfies $(A - (-1)I)\mathbf{v} = \mathbf{0}$. Therefore, every vector is an eigenvector and so $E_{-1} = \mathbb{R}^2$. This eigenspace is two-dimensional.

We can see this in another way. The matrix transformation defined by A rotates vectors by $180°$, which says that $A\mathbf{x} = -\mathbf{x}$ for every vector $\mathbf{x}$. In other words, every two-dimensional vector is an eigenvector with associated eigenvalue $\lambda = -1$.

These last two examples illustrate two types of behavior when there is a single eigenvalue.

In one case, we are able to construct a basis of $\mathbb{R}^2$ using eigenvectors; in the other, we are not. We will explore this behavior more in the next subsection.

> **A check on our work.**
>
> When finding eigenvalues and their associated eigenvectors in this way, we first find eigenvalues λ by solving the characteristic equation. If λ is a solution to the characteristic equation, then $A - \lambda I$ is not invertible and, consequently, $A - \lambda I$ must contain a row without a pivot position.
>
> This serves as a check on our work. If we row reduce $A - \lambda I$ and find the identity matrix, then we have made an error either in solving the characteristic equation or in finding $\text{Nul}(A - \lambda I)$.

4.2.3 The characteristic polynomial and the dimension of eigenspaces

Given a square $n \times n$ matrix A, we saw in the previous section the value of being able to express any vector in $\mathbb{R}^n$ as a linear combination of eigenvectors of A. For this reason, Question 4.1.8 asks when we can construct a basis of $\mathbb{R}^n$ consisting of eigenvectors. We will explore this question more fully now.

As we saw above, the eigenvalues of A are the solutions of the characteristic equation $\det(A - \lambda I) = 0$. The examples we have considered demonstrate some different types of behavior. For instance, we have seen the characteristic equations

- $(4 - \lambda)(-2 - \lambda)(3 - \lambda) = 0$, which has real and distinct roots,
- $(2 - \lambda)^2 = 0$, which has repeated roots, and
- $\lambda^2 + 1 = (i - \lambda)(-i - \lambda) = 0$, which has complex roots.

If A is an $n \times n$ matrix, then the characteristic polynomial is a degree n polynomial, and this means that it has n roots. Therefore, the characteristic equation can be written as

$$\det(A - \lambda I) = (\lambda_1 - \lambda)(\lambda_2 - \lambda)\ldots(\lambda_n - \lambda) = 0$$

giving eigenvalues $\lambda_1, \lambda_2, \ldots, \lambda_n$. As we have seen, some of the eigenvalues may be complex. Moreover, some of the eigenvalues may appear in this list more than once. However, we can always write the characteristic equation in the form

$$(\lambda_1 - \lambda)^{m_1}(\lambda_2 - \lambda)^{m_2}\ldots(\lambda_p - \lambda)^{m_p} = 0.$$

The number of times that $\lambda_j - \lambda$ appears as a factor in the characteristic polynomial, is called the *multiplicity* of the eigenvalue λ_j.

Example 4.2.7 We have seen that the matrix $A = \begin{bmatrix} 1 & 1 \\ -1 & 3 \end{bmatrix}$ has the characteristic equation $(2 - \lambda)^2 = 0$. This matrix has a single eigenvalue $\lambda = 2$, which has multiplicity 2.

4.2. FINDING EIGENVALUES AND EIGENVECTORS

Example 4.2.8 If a matrix has the characteristic equation

$$(4 - \lambda)^2(-5 - \lambda)(1 - \lambda)^7(3 - \lambda)^2 = 0,$$

then that matrix has four eigenvalues: $\lambda = 4$ having multiplicity 2; $\lambda = -5$ having multiplicity 1; $\lambda = 1$ having multiplicity 7; and $\lambda = 3$ having multiplicity 2. The degree of the characteristic polynomial is the sum of the multiplicities $2 + 1 + 7 + 2 = 12$ so this matrix must be a 12×12 matrix.

The multiplicities of the eigenvalues are important because they influence the dimension of the eigenspaces. We know that the dimension of an eigenspace must be at least one; the following proposition also tells us the dimension of an eigenspace can be no larger than the multiplicity of its associated eigenvalue.

Proposition 4.2.9 *If λ is a real eigenvalue of the matrix A with multiplicity m, then*

$$1 \leq \dim E_\lambda \leq m.$$

Example 4.2.10 The diagonal matrix $\begin{bmatrix} -1 & 0 \\ 0 & -1 \end{bmatrix}$ has the characteristic equation $(-1-\lambda)^2 = 0$. There is a single eigenvalue $\lambda = -1$ having multiplicity $m = 2$, and we saw earlier that $\dim E_{-1} = 2 \leq m = 2$.

Example 4.2.11 The matrix $\begin{bmatrix} 1 & 1 \\ -1 & 3 \end{bmatrix}$ has the characteristic equation $(2 - \lambda)^2 = 0$. This tells us that there is a single eigenvalue $\lambda = 2$ having multiplicity $m = 2$. In contrast with the previous example, we have $\dim E_2 = 1 \leq m = 2$.

Example 4.2.12 We saw earlier that the matrix $\begin{bmatrix} 4 & 2 & 3 \\ 0 & -2 & -1 \\ 0 & 0 & 3 \end{bmatrix}$ has the characteristic equation

$$(4 - \lambda)(-2 - \lambda)(3 - \lambda) = 0.$$

There are three eigenvalues $\lambda = 3, -2, 1$ each having multiplicity 1. By the proposition, we are guaranteed that the dimension of each eigenspace is 1; that is,

$$\dim E_3 = \dim E_{-2} = \dim E_1 = 1.$$

It turns out that this is enough to guarantee that there is a basis of $\mathbb{R}^3$ consisting of eigenvectors.

Example 4.2.13 If a 12×12 matrix has the characteristic equation

$$(4 - \lambda)^2(-5 - \lambda)(1 - \lambda)^7(3 - \lambda)^2 = 0,$$

we know there are four eigenvalues $\lambda = 4, -5, 1, 3$. Without more information, all we can say about the dimensions of the eigenspaces is

$$1 \leq \dim E_4 \leq 2$$
$$1 \leq \dim E_{-5} \leq 1$$
$$1 \leq \dim E_1 \leq 7$$
$$1 \leq \dim E_3 \leq 2.$$

We can guarantee that dim $E_{-5} = 1$, but we cannot be more specific about the dimensions of the other eigenspaces.

Fortunately, if we have an $n \times n$ matrix, it frequently happens that the characteristic equation has the form
$$(\lambda_1 - \lambda)(\lambda_2 - \lambda)\ldots(\lambda_n - \lambda) = 0$$
where there are n distinct real eigenvalues, each of which has multiplicity 1. In this case, the dimension of each of the eigenspaces dim $E_{\lambda_j} = 1$. With a little work, it can be seen that choosing a basis vector $\mathbf{v}_j$ for each of the eigenspaces produces a basis for $\mathbb{R}^n$. We therefore have the following proposition.

Proposition 4.2.14 *If A is an $n \times n$ matrix having n distinct real eigenvalues, then there is a basis of $\mathbb{R}^n$ consisting of eigenvectors of A.*

This proposition provides one answer to our Question 4.1.8. The next activity explores this question further.

Activity 4.2.4.

a. Identify the eigenvalues, and their multiplicities, of an $n \times n$ matrix whose characteristic polynomial is $(2 - \lambda)^3(-3 - \lambda)^{10}(5 - \lambda)$. What can you conclude about the dimensions of the eigenspaces? What is the shape of the matrix? Do you have enough information to guarantee that there is a basis of $\mathbb{R}^n$ consisting of eigenvectors?

b. Find the eigenvalues of $\begin{bmatrix} 0 & -1 \\ 4 & -4 \end{bmatrix}$ and state their multiplicities. Can you find a basis of $\mathbb{R}^2$ consisting of eigenvectors of this matrix?

c. Consider the matrix $A = \begin{bmatrix} -1 & 0 & 2 \\ -2 & -2 & -4 \\ 0 & 0 & -2 \end{bmatrix}$ whose characteristic equation is
$$(-2 - \lambda)^2(-1 - \lambda) = 0.$$

1. Identify the eigenvalues and their multiplicities.
2. For each eigenvalue λ, find a basis of the eigenspace E_λ and state its dimension.
3. Is there a basis of $\mathbb{R}^3$ consisting of eigenvectors of A?

d. Now consider the matrix $A = \begin{bmatrix} -5 & -2 & -6 \\ -2 & -2 & -4 \\ 2 & 1 & 2 \end{bmatrix}$ whose characteristic equation is also
$$(-2 - \lambda)^2(-1 - \lambda) = 0.$$

1. Identify the eigenvalues and their multiplicities.
2. For each eigenvalue λ, find a basis of the eigenspace E_λ and state its dimension.

4.2. FINDING EIGENVALUES AND EIGENVECTORS

3. Is there a basis of $\mathbb{R}^3$ consisting of eigenvectors of A?

e. Consider the matrix $A = \begin{bmatrix} -5 & -2 & -6 \\ 4 & 1 & 8 \\ 2 & 1 & 2 \end{bmatrix}$ whose characteristic equation is

$$(-2 - \lambda)(1 - \lambda)(-1 - \lambda) = 0.$$

1. Identify the eigenvalues and their multiplicities.
2. For each eigenvalue λ, find a basis of the eigenspace E_λ and state its dimension.
3. Is there a basis of $\mathbb{R}^3$ consisting of eigenvectors of A?

4.2.4 Using Sage to find eigenvalues and eigenvectors

We can use Sage to find the characteristic polynomial, eigenvalues, and eigenvectors of a matrix. As we will see, however, some care is required when dealing with matrices whose entries include decimals.

Activity 4.2.5. We will use Sage to find the eigenvalues and eigenvectors of a matrix. Let's begin with the matrix $A = \begin{bmatrix} -3 & 1 \\ 0 & -3 \end{bmatrix}$.

a. We can find the characteristic polynomial of A by writing `A.charpoly('lambda')`. Notice that we have to give Sage a variable in which to write the polynomial; here, we use `lambda` though x works just as well.

```
A = matrix(2,2,[-3,1,0,-3])
A.charpoly('lambda')
```

The factored form of the characteristic polynomial may be more useful since it will tell us the eigenvalues and their multiplicities. The factored characteristic polynomial is found with `A.fcp('lambda')`.

```
A = matrix(2,2,[-3,1,0,-3])
A.fcp('lambda')
```

b. If we only want the eigenvalues, we can use `A.eigenvalues()`.

```
A = matrix(2,2,[-3,1,0,-3])
A.eigenvalues()
```

Notice that the multiplicity of an eigenvalue is the number of times it is repeated in the list of eigenvalues.

c. Finally, we can find eigenvectors by A.eigenvectors_right(). (We are looking for *right* eigenvalues since the vector **v** appears to the right of A in the definition $A\mathbf{v} = \lambda\mathbf{v}$.)

```
A = matrix(2,2,[-3,1,0,-3])
A.eigenvectors_right()
```

At first glance, the result of this command can be a little confusing to interpret. What we see is a list with one entry for each eigenvalue. For each eigenvalue, there is a triple consisting of (i) the eigenvalue λ, (ii) a basis for E_λ, and (iii) the multiplicity of λ.

d. When working with decimal entries, which are called *floating point numbers* in computer science, we must remember that computers perform only approximate arithmetic. This is a problem when we wish to find the eigenvectors of such a matrix. To illustrate, consider the matrix $A = \begin{bmatrix} 0.4 & 0.3 \\ 0.6 & 0.7 \end{bmatrix}$.

1. Without using Sage, find the eigenvalues of this matrix.
2. What do you find for the reduced row echelon form of $A - I$?
3. Let's now use Sage to determine the reduced row echelon form of $A - I$:

```
A = matrix(2,2,[0.4,0.3,0.6,0.7])
(A-identity_matrix(2)).rref()
```

What result does Sage report for the reduced row echelon form? Why is this result not correct?

4. Because the arithmetic Sage performs with floating point entries is only approximate, we are not able to find the eigenspace E_1. In this next chapter, we will learn how to address this issue. In the meantime, we can get around this problem by writing the entries in the matrix as rational numbers:

```
A = matrix(2,2,[4/10,3/10,6/10,7/10])
A.eigenvectors_right()
```

4.2.5 Summary

In this section, we developed a technique for finding the eigenvalues and eigenvectors of an $n \times n$ matrix A.

- The expression $\det(A - \lambda I)$ is a degree n polynomial, known as the characteristic polynomial of A. The eigenvalues of A are the roots of the characteristic polynomial found by solving the characteristic equation $\det(A - \lambda I) = 0$.

- The set of eigenvectors associated to the eigenvalue λ forms a subspace of $\mathbb{R}^n$, the eigenspace $E_\lambda = \text{Nul}(A - \lambda I)$.

4.2. FINDING EIGENVALUES AND EIGENVECTORS

- If the factor $(\lambda_j - \lambda)$ appears m_j times in the characteristic polynomial, we say that the eigenvalue λ_j has multiplicity m_j and note that

$$1 \leq \dim E_{\lambda_j} \leq m_j.$$

- If each of the eigenvalues is real and has multiplicity 1, then we can form a basis of $\mathbb{R}^n$ consisting of eigenvectors of A.

- We can use Sage to find the eigenvalues and eigenvalues of matrices. However, we need to be careful working with floating point numbers since floating point arithmetic is only an approximation.

4.2.6 Exercises

1. For each of the following matrices, find its characteristic polynomial, its eigenvalues, and the multiplicity of each eigenvalue.

   a. $A = \begin{bmatrix} 4 & -1 \\ 4 & 0 \end{bmatrix}.$

   b. $A = \begin{bmatrix} 3 & -1 & 3 \\ 0 & 4 & 0 \\ 0 & 0 & -6 \end{bmatrix}.$

   c. $A = \begin{bmatrix} -2 & 0 \\ 0 & -2 \end{bmatrix}.$

   d. $A = \begin{bmatrix} -1 & 2 \\ 2 & 2 \end{bmatrix}.$

2. Given an $n \times n$ matrix A, an important question, Question 4.1.8, asks whether we can find a basis of $\mathbb{R}^n$ consisting of eigenvectors of A. For each of the matrices in the previous exercise, find a basis of $\mathbb{R}^n$ consisting of eigenvectors or state why such a basis does not exist.

3. Determine whether the following statements are true or false and provide a justification for your response.

   a. The eigenvalues of a matrix A are the entries on the diagonal of A.

   b. If λ is an eigenvalue of multiplicity 1, then E_λ is one-dimensional.

   c. If a matrix A is invertible, then $\lambda = 0$ cannot be an eigenvalue.

   d. If A is a 13×13 matrix, the characteristic polynomial has degree less than 13.

   e. The eigenspace E_λ of A is the same as the null space $\text{Nul}(A - \lambda I)$.

4. Provide a justification for your response to the following questions.

   a. Suppose that A is a 3×3 matrix having eigenvalues $\lambda = -3, 3, -5$. What are the eigenvalues of $2A$?

   b. Suppose that D is a diagonal 3×3 matrix. Why can you guarantee that there is a basis of $\mathbb{R}^3$ consisting of eigenvectors of D?

c. If A is a 3×3 matrix whose eigenvalues are $\lambda = -1, 3, 5$, can you guarantee that there is a basis of $\mathbb{R}^3$ consisting of eigenvectors of A?

d. Suppose that the characteristic polynomial of a matrix A is
$$\det(A - \lambda I) = -\lambda^3 + 4\lambda.$$
What are the eigenvalues of A? Is A invertible? Is there a basis of $\mathbb{R}^n$ consisting of eigenvectors of A?

e. If the characteristic polynomial of A is
$$\det(A - \lambda I) = (4 - \lambda)(-2 - \lambda)(1 - \lambda),$$
what is the characteristic polynomial of A^2? what is the characteristic polynomial of A^{-1}?

5. For each of the following matrices, use Sage to determine its eigenvalues, their multiplicities, and a basis for each eigenspace. For which matrices is it possible to construct a basis for $\mathbb{R}^3$ consisting of eigenvectors?

a. $A = \begin{bmatrix} -4 & 12 & -6 \\ 4 & -5 & 4 \\ 11 & -20 & 13 \end{bmatrix}$

b. $A = \begin{bmatrix} 1 & -3 & 1 \\ -4 & 8 & -5 \\ -8 & 17 & -10 \end{bmatrix}$

c. $A = \begin{bmatrix} 3 & -8 & 4 \\ -2 & 3 & -2 \\ -6 & 12 & -7 \end{bmatrix}$

6. There is a relationship between the determinant of a matrix and the product of its eigenvalues.

a. We have seen that the eigenvalues of the matrix $A = \begin{bmatrix} 1 & 2 \\ 2 & 1 \end{bmatrix}$ are $\lambda = 3, -1$. What is $\det A$? What is the product of the eigenvalues of A?

b. Consider the triangular matrix $A = \begin{bmatrix} 2 & 0 & 0 \\ -1 & -3 & 0 \\ 3 & 1 & -2 \end{bmatrix}$. What are the eigenvalues of A? What is $\det A$? What is the product of the eigenvalues of A?

c. Based on these examples, what do you think is the relationship between the determinant of a matrix and the product of its eigenvalues?

4.2. FINDING EIGENVALUES AND EIGENVECTORS

d. Suppose the characteristic polynomial is written as

$$\det(A - \lambda I) = (\lambda_1 - \lambda)(\lambda_2 - \lambda)\ldots(\lambda_n - \lambda).$$

By substituting $\lambda = 0$ into this equation, explain why the determinant of a matrix equals the product of its eigenvalues.

7. Consider the matrix $A = \begin{bmatrix} 0.5 & 0.6 \\ -0.3 & 1.4 \end{bmatrix}$.

   a. Find the eigenvalues of A and a basis for their associated eigenspaces.

   b. Suppose that $\mathbf{x}_0 = \begin{bmatrix} 11 \\ 6 \end{bmatrix}$. Express $\mathbf{x}_0$ as a linear combination of eigenvectors of A.

   c. Define the vectors
   $$\mathbf{x}_1 = A\mathbf{x}_0$$
   $$\mathbf{x}_2 = A\mathbf{x}_1 = A^2\mathbf{x}_0$$
   $$\mathbf{x}_3 = A\mathbf{x}_2 = A^3\mathbf{x}_0$$
   $$\vdots = \vdots$$

   Write $\mathbf{x}_1, \mathbf{x}_2$, and $\mathbf{x}_3$ as a linear combination of eigenvectors of A.

   d. What happens to $\mathbf{x}_k$ as k grows larger and larger?

8. Consider the matrix $A = \begin{bmatrix} 0.4 & 0.3 \\ 0.6 & 0.7 \end{bmatrix}$.

   a. Find the eigenvalues of A and a basis for their associated eigenspaces.

   b. Suppose that $\mathbf{x}_0 = \begin{bmatrix} 0 \\ 1 \end{bmatrix}$. Express $\mathbf{x}_0$ as a linear combination of eigenvectors of A.

   c. Define the vectors
   $$\mathbf{x}_1 = A\mathbf{x}_0$$
   $$\mathbf{x}_2 = A\mathbf{x}_1 = A^2\mathbf{x}_0$$
   $$\mathbf{x}_3 = A\mathbf{x}_2 = A^3\mathbf{x}_0$$
   $$\vdots = \vdots$$

   Write $\mathbf{x}_1, \mathbf{x}_2$, and $\mathbf{x}_3$ as a linear combination of eigenvectors of A.

   d. What happens to $\mathbf{x}_k$ as k grows larger and larger?

4.3 Diagonalization, similarity, and powers of a matrix

The first example we considered in this chapter was the matrix $A = \begin{bmatrix} 1 & 2 \\ 2 & 1 \end{bmatrix}$, which has eigenvectors $\mathbf{v}_1 = \begin{bmatrix} 1 \\ 1 \end{bmatrix}$ and $\mathbf{v}_2 = \begin{bmatrix} -1 \\ 1 \end{bmatrix}$ and associated eigenvalues $\lambda_1 = 3$ and $\lambda_2 = -1$. In Subsection 4.1.2, we described how A is, in some sense, equivalent to the diagonal matrix $D = \begin{bmatrix} 3 & 0 \\ 0 & -1 \end{bmatrix}$.

This equivalence is summarized by Figure 4.3.1. The diagonal matrix D has the geometric effect of stretching vectors horizontally by a factor of 3 and flipping vectors vertically. The matrix A has the geometric effect of stretching vectors by a factor of 3 in the $\mathbf{v}_1$ direction and flipping them in the $\mathbf{v}_2$ direction. That is, the geometric effect of A is the same as that of D when viewed in a basis of eigenvectors of A.

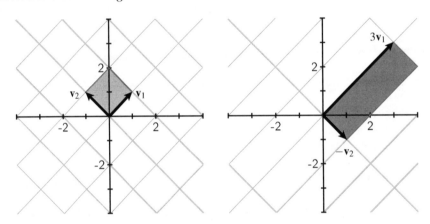

Figure 4.3.1 The matrix A has the same geometric effect as the diagonal matrix D when viewed in the basis of eigenvectors.

Our goal in this section is to express this geometric observation in algebraic terms. In doing so, we will make precise the sense in which A and D are equivalent.

> **Preview Activity 4.3.1.** In this preview activity, we will review some familiar properties about matrix multiplication that appear in this section.
>
> a. Remember that matrix-vector multiplication constructs linear combinations of the columns of the matrix. For instance, if $A = \begin{bmatrix} \mathbf{a}_1 & \mathbf{a}_2 \end{bmatrix}$, express the product $A \begin{bmatrix} 2 \\ -3 \end{bmatrix}$ in terms of $\mathbf{a}_1$ and $\mathbf{a}_2$.
>
> b. What is the product $A \begin{bmatrix} 4 \\ 0 \end{bmatrix}$ in terms of $\mathbf{a}_1$ and $\mathbf{a}_2$?
>
> c. Next, remember how matrix-matrix multiplication is defined. Suppose that we

have matrices A and B and that $B = \begin{bmatrix} \mathbf{b}_1 & \mathbf{b}_2 \end{bmatrix}$. How can we express the matrix product AB in terms of the columns of B?

d. Suppose that A is a matrix having eigenvectors $\mathbf{v}_1$ and $\mathbf{v}_2$ with associated eigenvalues $\lambda_1 = 4$ and $\lambda_2 = -1$. Express the product $A(2\mathbf{v}_1 + 3\mathbf{v}_2)$ in terms of $\mathbf{v}_1$ and $\mathbf{v}_2$.

e. Suppose that A is the matrix from the previous part and that $P = \begin{bmatrix} \mathbf{v}_1 & \mathbf{v}_2 \end{bmatrix}$. What is the matrix product

$$AP = A \begin{bmatrix} \mathbf{v}_1 & \mathbf{v}_2 \end{bmatrix}?$$

4.3.1 Diagonalization of matrices

When working with an $n \times n$ matrix A, Subsection 4.1.2 demonstrated the value of having a basis of $\mathbb{R}^n$ consisting of eigenvectors of A. In fact, Proposition 4.2.9 tells us that if the eigenvalues of A are real and distinct, then there is a such a basis. As we'll see later, there are other conditions on A that guarantee a basis of eigenvectors. For now, suffice it to say that we can find a basis of eigenvectors for many matrices. With this assumption, we will see how the matrix A is equivalent to a diagonal matrix D.

Activity 4.3.2. Suppose that A is a 2×2 matrix having eigenvectors $\mathbf{v}_1$ and $\mathbf{v}_2$ with associated eigenvalues $\lambda_1 = 3$ and $\lambda_2 = -6$. Because the eigenvalues are real and distinct, we know by Proposition 4.2.9 that these eigenvectors form a basis of $\mathbb{R}^2$.

a. What are the products $A\mathbf{v}_1$ and $A\mathbf{v}_2$ in terms of $\mathbf{v}_1$ and $\mathbf{v}_2$?

b. If we form the matrix $P = \begin{bmatrix} \mathbf{v}_1 & \mathbf{v}_2 \end{bmatrix}$, what is the product AP in terms of $\mathbf{v}_1$ and $\mathbf{v}_2$?

c. Use the eigenvalues to form the diagonal matrix $D = \begin{bmatrix} 3 & 0 \\ 0 & -6 \end{bmatrix}$ and determine the product PD in terms of $\mathbf{v}_1$ and $\mathbf{v}_2$.

d. The results from the previous two parts of this activity demonstrate that $AP = PD$. Using the fact that the eigenvectors $\mathbf{v}_1$ and $\mathbf{v}_2$ form a basis of $\mathbb{R}^2$, explain why P is invertible and that we must have $A = PDP^{-1}$.

e. Suppose that $A = \begin{bmatrix} -3 & 6 \\ 3 & 0 \end{bmatrix}$. Verify that $\mathbf{v}_1 = \begin{bmatrix} 1 \\ 1 \end{bmatrix}$ and $\mathbf{v}_2 = \begin{bmatrix} 2 \\ -1 \end{bmatrix}$ are eigenvectors of A with eigenvalues $\lambda_1 = 3$ and $\lambda_2 = -6$.

f. Use the Sage cell below to define the matrices P and D and then verify that $A = PDP^{-1}$.

```
# enter the matrices P and D below
P = 
D = 
P*D*P.inverse()
```

More generally, suppose that we have an $n \times n$ matrix A and that there is a basis of $\mathbb{R}^n$ consisting of eigenvectors $\mathbf{v}_1, \mathbf{v}_2, \ldots, \mathbf{v}_n$ of A with associated eigenvalues $\lambda_1, \lambda_2, \ldots, \lambda_n$. If we use the eigenvectors to form the matrix

$$P = \begin{bmatrix} \mathbf{v}_1 & \mathbf{v}_2 & \cdots & \mathbf{v}_n \end{bmatrix}$$

and the eigenvalues to form the diagonal matrix

$$D = \begin{bmatrix} \lambda_1 & 0 & \cdots & 0 \\ 0 & \lambda_2 & \cdots & 0 \\ \vdots & \vdots & \ddots & 0 \\ 0 & 0 & \cdots & \lambda_n \end{bmatrix}$$

and apply the same reasoning demonstrated in the activity, we find that $AP = PD$ and hence

$$A = PDP^{-1}.$$

We have now seen the following proposition.

Proposition 4.3.2 *If A is an $n \times n$ matrix and there is a basis $\{\mathbf{v}_1, \mathbf{v}_2, \ldots, \mathbf{v}_n\}$ of $\mathbb{R}^n$ consisting of eigenvectors of A having associated eigenvalues $\lambda_1, \lambda_2, \ldots, \lambda_n$, then we can write $A = PDP^{-1}$ where D is the diagonal matrix whose diagonal entries are the eigenvalues of A*

$$D = \begin{bmatrix} \lambda_1 & 0 & \cdots & 0 \\ 0 & \lambda_2 & \cdots & 0 \\ \vdots & \vdots & \ddots & 0 \\ 0 & 0 & \cdots & \lambda_n \end{bmatrix}$$

and the matrix $P = \begin{bmatrix} \mathbf{v}_1 & \mathbf{v}_2 & \cdots & \mathbf{v}_n \end{bmatrix}$.

Example 4.3.3 We have seen that $A = \begin{bmatrix} 1 & 2 \\ 2 & 1 \end{bmatrix}$ has eigenvectors $\mathbf{v}_1 = \begin{bmatrix} 1 \\ 1 \end{bmatrix}$ and $\mathbf{v}_2 = \begin{bmatrix} -1 \\ 1 \end{bmatrix}$ with associated eigenvalues $\lambda_1 = 3$ and $\lambda_2 = -1$. Forming the matrices

$$P = \begin{bmatrix} \mathbf{v}_1 & \mathbf{v}_2 \end{bmatrix} = \begin{bmatrix} 1 & -1 \\ 1 & 1 \end{bmatrix}, \quad D = \begin{bmatrix} 3 & 0 \\ 0 & -1 \end{bmatrix},$$

we see that $A = PDP^{-1}$.

This is the sense in which we mean that A is equivalent to a diagonal matrix D. The expression $A = PDP^{-1}$ says that A, expressed in the basis defined by the columns of P, has the same geometric effect as D, expressed in the standard basis $\mathbf{e}_1, \mathbf{e}_2, \ldots, \mathbf{e}_n$.

4.3. DIAGONALIZATION, SIMILARITY, AND POWERS OF A MATRIX

Definition 4.3.4 We say that the matrix A is *diagonalizable* if there is a diagonal matrix D and invertible matrix P such that
$$A = PDP^{-1}.$$

Example 4.3.5 We will try to find a diagonalization of $A = \begin{bmatrix} -5 & 6 \\ -3 & 4 \end{bmatrix}$ whose characteristic equation is
$$\det(A - \lambda I) = (-5 - \lambda)(4 - \lambda) + 18 = (-2 - \lambda)(1 - \lambda) = 0.$$

This shows that the eigenvalues of A are $\lambda_1 = -2$ and $\lambda_2 = 1$.

By constructing $\text{Nul}(A - (-2)I)$, we find a basis for E_{-2} consisting of the vector $\mathbf{v}_1 = \begin{bmatrix} 2 \\ 1 \end{bmatrix}$. Similarly, a basis for E_1 consists of the vector $\mathbf{v}_2 = \begin{bmatrix} 1 \\ 1 \end{bmatrix}$. This shows that we can construct a basis $\{\mathbf{v}_1, \mathbf{v}_2\}$ of $\mathbb{R}^2$ consisting of eigenvectors of A.

We now form the matrices
$$D = \begin{bmatrix} -2 & 0 \\ 0 & 1 \end{bmatrix}, \quad P = \begin{bmatrix} \mathbf{v}_1 & \mathbf{v}_2 \end{bmatrix} = \begin{bmatrix} 2 & 1 \\ 1 & 1 \end{bmatrix}$$

and verify that
$$PDP^{-1} = \begin{bmatrix} 2 & 1 \\ 1 & 1 \end{bmatrix} \begin{bmatrix} -2 & 0 \\ 0 & 1 \end{bmatrix} \begin{bmatrix} 1 & -1 \\ -1 & 2 \end{bmatrix} = \begin{bmatrix} -5 & 6 \\ -3 & 4 \end{bmatrix} = A.$$

There are, in fact, many ways to diagonalize A. For instance, we could change the order of the eigenvalues and eigenvectors and write
$$D = \begin{bmatrix} 1 & 0 \\ 0 & -2 \end{bmatrix}, \quad P = \begin{bmatrix} \mathbf{v}_2 & \mathbf{v}_1 \end{bmatrix} = \begin{bmatrix} 1 & 2 \\ 1 & 1 \end{bmatrix}.$$

If we choose a different basis for the eigenspaces, we will also find a different matrix P that diagonalizes A. The point is that there are many ways in which A can be written in the form $A = PDP^{-1}$.

Example 4.3.6 We will try to find a diagonalization of $A = \begin{bmatrix} 0 & 4 \\ -1 & 4 \end{bmatrix}$.

Once again, we find the eigenvalues by solving the characteristic equation:
$$\det(A - \lambda I) = -\lambda(4 - \lambda) + 4 = (2 - \lambda)^2 = 0.$$

In this case, there is a single eigenvalue $\lambda = 2$.

We find a basis for the eigenspace E_2 by describing $\text{Nul}(A - 2I)$:
$$A - 2I = \begin{bmatrix} -2 & 4 \\ -1 & 2 \end{bmatrix} \sim \begin{bmatrix} 1 & -2 \\ 0 & 0 \end{bmatrix}.$$

This shows that the eigenspace E_2 is one-dimensional with $\mathbf{v}_1 = \begin{bmatrix} 2 \\ 1 \end{bmatrix}$ forming a basis.

In this case, there is not a basis of $\mathbb{R}^2$ consisting of eigenvectors of A, which tells us that A is not diagonalizable.

In fact, if we only know that $A = PDP^{-1}$, we can say that the columns of P are eigenvectors of A and that the diagonal entries of D are the associated eigenvalues.

Proposition 4.3.7 *An $n \times n$ matrix A is diagonalizable if and only if there is a basis of $\mathbb{R}^n$ consisting of eigenvectors of A.*

Example 4.3.8 Suppose we know that $A = PDP^{-1}$ where

$$D = \begin{bmatrix} 2 & 0 \\ 0 & -2 \end{bmatrix}, \quad P = \begin{bmatrix} \mathbf{v}_2 & \mathbf{v}_1 \end{bmatrix} = \begin{bmatrix} 1 & 1 \\ 1 & 2 \end{bmatrix}.$$

The columns of P form eigenvectors of A so that $\mathbf{v}_1 = \begin{bmatrix} 1 \\ 1 \end{bmatrix}$ is an eigenvector of A with eigenvalue $\lambda_1 = 2$ and $\mathbf{v}_2 = \begin{bmatrix} 1 \\ 2 \end{bmatrix}$ is an eigenvector with eigenvalue $\lambda_2 = -2$.

We can verify this by computing

$$A = PDP^{-1} = \begin{bmatrix} 6 & -4 \\ 8 & -6 \end{bmatrix}$$

and checking that $A\mathbf{v}_1 = \begin{bmatrix} 1 \\ 1 \end{bmatrix} = 2\mathbf{v}_1$ and $A\mathbf{v}_2 = \begin{bmatrix} 1 \\ 2 \end{bmatrix} = -2\mathbf{v}_2$.

Activity 4.3.3.

a. Find a diagonalization of A, if one exists, when

$$A = \begin{bmatrix} 3 & -2 \\ 6 & -5 \end{bmatrix}.$$

b. Can the diagonal matrix

$$A = \begin{bmatrix} 2 & 0 \\ 0 & -5 \end{bmatrix}$$

be diagonalized? If so, explain how to find the matrices P and D.

c. Find a diagonalization of A, if one exists, when

$$A = \begin{bmatrix} -2 & 0 & 0 \\ 1 & -3 & 0 \\ 2 & 0 & -3 \end{bmatrix}.$$

4.3. DIAGONALIZATION, SIMILARITY, AND POWERS OF A MATRIX

d. Find a diagonalization of A, if one exists, when

$$A = \begin{bmatrix} -2 & 0 & 0 \\ 1 & -3 & 0 \\ 2 & 1 & -3 \end{bmatrix}.$$

e. Suppose that $A = PDP^{-1}$ where

$$D = \begin{bmatrix} 3 & 0 \\ 0 & -1 \end{bmatrix}, \quad P = \begin{bmatrix} \mathbf{v}_2 & \mathbf{v}_1 \end{bmatrix} = \begin{bmatrix} 2 & 2 \\ 1 & -1 \end{bmatrix}.$$

1. Explain why A is invertible.
2. Find a diagonalization of A^{-1}.
3. Find a diagonalization of A^3.

4.3.2 Powers of a diagonalizable matrix

In several earlier examples, we have been interested in computing powers of a given matrix. For instance, in Activity 4.1.3, we had the matrix $A = \begin{bmatrix} 0.8 & 0.6 \\ 0.2 & 0.4 \end{bmatrix}$ and an initial vector $\mathbf{x}_0 = \begin{bmatrix} 1000 \\ 0 \end{bmatrix}$, and we wanted to compute

$$\mathbf{x}_1 = A\mathbf{x}_0$$
$$\mathbf{x}_2 = A\mathbf{x}_1 = A^2\mathbf{x}_0$$
$$\mathbf{x}_3 = A\mathbf{x}_2 = A^3\mathbf{x}_0.$$

In particular, we wanted to find $\mathbf{x}_k = A^k\mathbf{x}_0$ and determine what happens as k becomes very large. If a matrix A is diagonalizable, writing $A = PDP^{-1}$ can help us understand powers of A more easily.

Activity 4.3.4.

a. Let's begin with the diagonal matrix

$$D = \begin{bmatrix} 2 & 0 \\ 0 & -1 \end{bmatrix}.$$

Find the powers D^2, D^3, and D^4. What is D^k for a general value of k?

b. Suppose that A is a matrix with eigenvector $\mathbf{v}$ and associated eigenvalue λ; that is, $A\mathbf{v} = \lambda\mathbf{v}$. By considering $A^2\mathbf{v}$, explain why $\mathbf{v}$ is also an eigenvector of A with eigenvalue λ^2.

c. Suppose that $A = PDP^{-1}$ where
$$D = \begin{bmatrix} 2 & 0 \\ 0 & -1 \end{bmatrix}.$$

Remembering that the columns of P are eigenvectors of A, explain why A^2 is diagonalizable and find a diagonalization in terms of P and D.

d. Give another explanation of the diagonalizability of A^2 by writing
$$A^2 = (PDP^{-1})(PDP^{-1}) = PD(P^{-1}P)DP^{-1}.$$

e. In the same way, find a diagonalization of A^3, A^4, and A^k.

f. Suppose that A is a diagonalizable 2×2 matrix with eigenvalues $\lambda_1 = 0.5$ and $\lambda_2 = 0.1$. What happens to A^k as k becomes very large?

If A is diagonalizable, the activity demonstrates that any power of A is as well.

Proposition 4.3.9 *If $A = PDP^{-1}$, then $A^k = PD^kP^{-1}$. When A is invertible, we also have $A^{-1} = PD^{-1}P^{-1}$.*

Example 4.3.10 Let's revisit Activity 4.1.3 where we had the matrix $A = \begin{bmatrix} 0.8 & 0.6 \\ 0.2 & 0.4 \end{bmatrix}$ and the initial vector $\mathbf{x}_0 = \begin{bmatrix} 1000 \\ 0 \end{bmatrix}$. We were interested in understanding the sequence of vectors $\mathbf{x}_{k+1} = A\mathbf{x}_k$, which means that $\mathbf{x}_k = A^k\mathbf{x}_0$.

We can verify that $\mathbf{v}_1 = \begin{bmatrix} 3 \\ 1 \end{bmatrix}$ and $\mathbf{v}_2 = \begin{bmatrix} -1 \\ 1 \end{bmatrix}$ are eigenvectors of A having associated eigenvalues $\lambda_1 = 1$ and $\lambda_2 = 0.2$. This means that $A = PDP^{-1}$ where
$$P = \begin{bmatrix} 3 & -1 \\ 1 & 1 \end{bmatrix}, \quad D = \begin{bmatrix} 1 & 0 \\ 0 & 0.2 \end{bmatrix}.$$

Therefore, the powers of A have the form $A^k = PD^kP^{-1}$.

Notice that $D^k = \begin{bmatrix} 1^k & 0 \\ 0 & 0.2^k \end{bmatrix} = \begin{bmatrix} 1 & 0 \\ 0 & 0.2^k \end{bmatrix}$. As k increases, 0.2^k becomes closer and closer to zero. This means that for very large powers k, we have
$$D^k \approx \begin{bmatrix} 1 & 0 \\ 0 & 0 \end{bmatrix}$$
and therefore
$$A^k = PD^kP^{-1} \approx P \begin{bmatrix} 1 & 0 \\ 0 & 0 \end{bmatrix} P^{-1} = \begin{bmatrix} \frac{3}{4} & \frac{3}{4} \\ \frac{1}{4} & \frac{1}{4} \end{bmatrix}.$$

Beginning with the vector $\mathbf{x}_0 = \begin{bmatrix} 1000 \\ 0 \end{bmatrix}$, we find that $\mathbf{x}_k = A^k\mathbf{x}_0 \approx \begin{bmatrix} 750 \\ 250 \end{bmatrix}$ when k is very large.

4.3. DIAGONALIZATION, SIMILARITY, AND POWERS OF A MATRIX

4.3.3 Similarity and complex eigenvalues

We have been interested in diagonalizing a matrix A because doing so relates a matrix A to a simpler diagonal matrix D. In particular, the effect of multiplying a vector by $A = PDP^{-1}$, viewed in the basis defined by the columns of P, is the same as the effect of multiplying by D in the standard basis.

While many matrices are diagonalizable, there are some that are not. For example, if a matrix has complex eigenvalues, it is not possible to find a basis of $\mathbb{R}^n$ consisting of eigenvectors, which means that the matrix is not diagonalizable. In this case, however, we can still relate the matrix to a simpler form that explains the geometric effect this matrix has on vectors.

Definition 4.3.11 We say that A is *similar* to B if there is an invertible matrix P such that $A = PBP^{-1}$.

Notice that a matrix is diagonalizable if and only if it is similar to a diagonal matrix. In case a matrix A has complex eigenvalues, we will find a simpler matrix C that is similar to A and note that $A = PCP^{-1}$ has the same effect, when viewed in the basis defined by the columns of P, as C, when viewed in the standard basis.

To begin, suppose that A is a 2×2 matrix having a complex eigenvalue $\lambda = a + bi$. It turns out that A is similar to $C = \begin{bmatrix} a & -b \\ b & a \end{bmatrix}$.

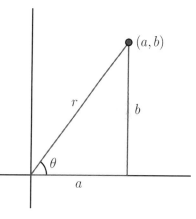

The next activity shows that C has a simple geometric effect on $\mathbb{R}^2$. First, however, we will use polar coordinates to rewrite C. As shown in the figure, the point (a, b) defines r, the distance from the origin, and θ, the angle formed with the positive horizontal axis. We then have

$$a = r\cos\theta$$
$$b = r\sin\theta.$$

Notice that the Pythagorean theorem says that $r = \sqrt{a^2 + b^2}$.

Activity 4.3.5. We begin by rewriting C in terms of r and θ and noting that

$$C = \begin{bmatrix} a & -b \\ b & a \end{bmatrix} = \begin{bmatrix} r\cos\theta & -r\sin\theta \\ r\sin\theta & r\cos\theta \end{bmatrix} = \begin{bmatrix} r & 0 \\ 0 & r \end{bmatrix}\begin{bmatrix} \cos\theta & -\sin\theta \\ \sin\theta & \cos\theta \end{bmatrix}.$$

a. Explain why C has the geometric effect of rotating vectors by θ and scaling them by a factor of r.

b. Let's now consider the matrix

$$A = \begin{bmatrix} -2 & 2 \\ -5 & 4 \end{bmatrix}$$

whose eigenvalues are $\lambda_1 = 1 + i$ and $\lambda_2 = 1 - i$. We will choose to focus on one of the eigenvalues $\lambda_1 = a + bi = 1 + i$.

Form the matrix C using these values of a and b. Then rewrite the point (a, b) in polar coordinates by identifying the values of r and θ. Explain the geometric effect of multiplying vectors by C.

c. Suppose that $P = \begin{bmatrix} 1 & 1 \\ 2 & 1 \end{bmatrix}$. Verify that $A = PCP^{-1}$.

```
C =
P =
P*C*P.inverse()
```

d. Explain why $A^k = PC^kP^{-1}$.

e. We formed the matrix C by choosing the eigenvalue $\lambda_1 = 1 + i$. Suppose we had instead chosen $\lambda_2 = 1 - i$. Form the matrix C' and use polar coordinates to describe the geometric effect of C.

f. Using the matrix $P' = \begin{bmatrix} 1 & -1 \\ 2 & -1 \end{bmatrix}$, show that $A = P'C'P'^{-1}$.

If the 2×2 matrix A has a complex eigenvalue $\lambda = a + bi$, it turns out that A is always similar to the matrix $C = \begin{bmatrix} a & -b \\ b & a \end{bmatrix}$, whose geometric effect on vectors can be described in terms of a rotation and a scaling. There is, in fact, a method for finding the matrix P so that $A = PCP^{-1}$ that we'll see in Exercise 4.3.5.8. For now, we note that A has the same geometric effect as C, when viewed in the basis provided by the columns of P. We will put this fact to use in the next section to understand certain dynamical systems.

Proposition 4.3.12 *If A is a 2×2 matrix with a complex eigenvalue $\lambda = a + bi$, then A is similar to $C = \begin{bmatrix} a & -b \\ b & a \end{bmatrix}$; that is, there is a matrix P such that $A = PCP^{-1}$.*

4.3.4 Summary

Our goal in this section has been to use the eigenvalues and eigenvectors of a matrix A to relate A to a simpler matrix.

- We said that A is diagonalizable if we can write $A = PDP^{-1}$ where D is a diagonal matrix. The columns of P consist of eigenvectors of A and the diagonal entries of D are the associated eigenvalues.

- An $n \times n$ matrix A is diagonalizable if and only if there is a basis of $\mathbb{R}^n$ consisting of eigenvectors of A.

4.3. DIAGONALIZATION, SIMILARITY, AND POWERS OF A MATRIX

- We said that A and B are similar if there is an invertible matrix P such that $A = PBP^{-1}$. In this case, $A^k = PB^kP^{-1}$.

- If A is a 2×2 matrix with complex eigenvalue $\lambda = a + bi$, then A is similar to $C = \begin{bmatrix} a & -b \\ b & a \end{bmatrix}$. Writing the point (a, b) in polar coordinates r and θ, we see that C rotates vectors through an angle θ and scales them by a factor of $r = \sqrt{a^2 + b^2}$.

4.3.5 Exercises

1. Determine whether the following matrices are diagonalizable. If so, find matrices D and P such that $A = PDP^{-1}$.

    a. $A = \begin{bmatrix} -2 & -2 \\ -2 & 1 \end{bmatrix}$.

    b. $A = \begin{bmatrix} -1 & 1 \\ -1 & -3 \end{bmatrix}$.

    c. $A = \begin{bmatrix} 3 & -4 \\ 2 & -1 \end{bmatrix}$.

    d. $A = \begin{bmatrix} 1 & 0 & 0 \\ 2 & -2 & 0 \\ 0 & 1 & 4 \end{bmatrix}$.

    e. $A = \begin{bmatrix} 1 & 2 & 2 \\ 2 & 1 & 2 \\ 2 & 2 & 1 \end{bmatrix}$.

2. Determine whether the following matrices have complex eigenvalues. If so, find the matrix C such that $A = PCP^{-1}$.

    a. $A = \begin{bmatrix} -2 & -2 \\ -2 & 1 \end{bmatrix}$.

    b. $A = \begin{bmatrix} -1 & 1 \\ -1 & -3 \end{bmatrix}$.

    c. $A = \begin{bmatrix} 3 & -4 \\ 2 & -1 \end{bmatrix}$.

3. Determine whether the following statements are true or false and provide a justification for your response.

    a. If A is invertible, then A is diagonalizable.

    b. If A and B are similar and A is invertible, then B is also invertible.

    c. If A is a diagonalizable $n \times n$ matrix, then there is a basis of $\mathbb{R}^n$ consisting of eigenvectors of A.

d. If A is diagonalizable, then A^{10} is also diagonalizable.

e. If A is diagonalizable, then A is invertible.

4. Provide a justification for your response to the following questions.

   a. If A is a 3×3 matrix having eigenvalues $\lambda = 2, 3, -4$, can you guarantee that A is diagonalizable?

   b. If A is a 2×2 matrix with a complex eigenvalue, can you guarantee that A is diagonalizable?

   c. If A is similar to the matrix $B = \begin{bmatrix} -5 & 0 & 0 \\ 0 & -5 & 0 \\ 0 & 0 & 3 \end{bmatrix}$, is A diagonalizable?

   d. What can you say about a matrix that is similar to the identity matrix?

   e. If A is a diagonalizable 2×2 matrix with a single eigenvalue $\lambda = 4$, what is A?

5. Describe geometric effect that the following matrices have on $\mathbb{R}^2$:

   a. $A = \begin{bmatrix} 2 & 0 \\ 0 & 2 \end{bmatrix}$

   b. $A = \begin{bmatrix} 4 & 2 \\ 0 & 4 \end{bmatrix}$

   c. $A = \begin{bmatrix} 3 & -6 \\ 6 & 3 \end{bmatrix}$

   d. $A = \begin{bmatrix} 4 & 0 \\ 0 & -2 \end{bmatrix}$

   e. $A = \begin{bmatrix} 1 & 3 \\ 3 & 1 \end{bmatrix}$

6. We say that A is similar to B if there is a matrix P such that $A = PBP^{-1}$.

   a. If A is similar to B, explain why B is similar to A.

   b. If A is similar to B and B is similar to C, explain why A is similar to C.

   c. If A is similar to B and B is diagonalizable, explain why A is diagonalizable.

   d. If A and B are similar, explain why A and B have the same characteristic polynomial; that is, explain why $\det(A - \lambda I) = \det(B - \lambda I)$.

   e. If A and B are similar, explain why A and B have the same eigenvalues.

7. Suppose that $A = PDP^{-1}$ where
$$D = \begin{bmatrix} 1 & 0 \\ 0 & 0 \end{bmatrix}, \quad P = \begin{bmatrix} 1 & -2 \\ 2 & 1 \end{bmatrix}.$$

   a. Explain the geometric effect that D has on vectors in $\mathbb{R}^2$.

4.3. DIAGONALIZATION, SIMILARITY, AND POWERS OF A MATRIX

b. Explain the geometric effect that A has on vectors in $\mathbb{R}^2$.

c. What can you say about A^2 and other powers of A?

d. Is A invertible?

8. When A is a 2×2 matrix with a complex eigenvalue $\lambda = a + bi$, we have said that there is a matrix P such that $A = PCP^{-1}$ where $C = \begin{bmatrix} a & -b \\ b & a \end{bmatrix}$. In this exercise, we will learn how to find the matrix P. As an example, we will consider the matrix $A = \begin{bmatrix} 2 & 2 \\ -1 & 4 \end{bmatrix}$.

   a. Show that the eigenvalues of A are complex.

   b. Choose one of the complex eigenvalues $\lambda = a + bi$ and construct the usual matrix C.

   c. Using the same eigenvalue, we will find an eigenvector $\mathbf{v}$ where the entries of $\mathbf{v}$ are complex numbers. As always, we will describe $\text{Nul}(A - \lambda I)$ by constructing the matrix $A - \lambda I$ and finding its reduced row echelon form. In doing so, we will necessarily need to use complex arithmetic.

   d. We have now found a complex eigenvector $\mathbf{v}$. Write $\mathbf{v} = \mathbf{v}_1 - i\mathbf{v}_2$ to identify vectors $\mathbf{v}_1$ and $\mathbf{v}_2$ having real entries.

   e. Construct the matrix $P = \begin{bmatrix} \mathbf{v}_1 & \mathbf{v}_2 \end{bmatrix}$ and verify that $A = PCP^{-1}$.

9. For each of the following matrices, sketch the vector $\mathbf{x} = \begin{bmatrix} 1 \\ 0 \end{bmatrix}$ and powers $A^k \mathbf{x}$ for $k = 1, 2, 3, 4$.

   a. $A = \begin{bmatrix} 0 & -1.4 \\ 1.4 & 0 \end{bmatrix}$.

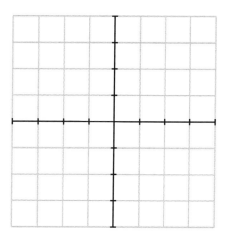

b. $A = \begin{bmatrix} 0 & -0.8 \\ 0.8 & 0 \end{bmatrix}$.

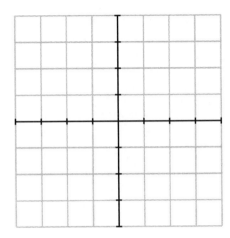

c. $A = \begin{bmatrix} 0 & -1 \\ 1 & 0 \end{bmatrix}$.

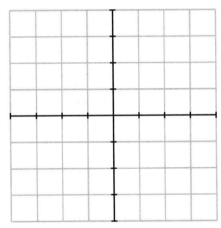

d. Consider a matrix of the form $C = \begin{bmatrix} a & -b \\ b & a \end{bmatrix}$ with $r = \sqrt{a^2 + b^2}$. What happens when k becomes very large when

  1. $r < 1$.
  2. $r = 1$.
  3. $r > 1$.

10. For each of the following matrices and vectors, sketch the vector **x** along with $A^k \mathbf{x}$ for

4.3. DIAGONALIZATION, SIMILARITY, AND POWERS OF A MATRIX

$k = 1, 2, 3, 4.$

a.
$$A = \begin{bmatrix} 1.4 & 0 \\ 0 & 0.7 \end{bmatrix}$$

$$\mathbf{x} = \begin{bmatrix} 1 \\ 2 \end{bmatrix}.$$

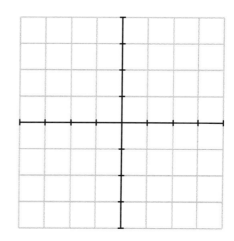

b.
$$A = \begin{bmatrix} 0.6 & 0 \\ 0 & 0.9 \end{bmatrix}$$

$$\mathbf{x} = \begin{bmatrix} 4 \\ 3 \end{bmatrix}.$$

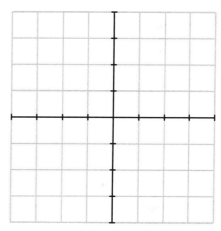

c.
$$A = \begin{bmatrix} 1.2 & 0 \\ 0 & 1.4 \end{bmatrix}$$

$$\mathbf{x} = \begin{bmatrix} 2 \\ 1 \end{bmatrix}.$$

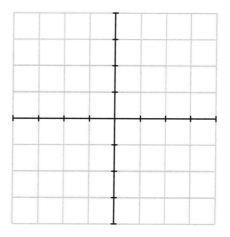

d.
$$A = \begin{bmatrix} 0.95 & 0.25 \\ 0.25 & 0.95 \end{bmatrix}$$

$$\mathbf{x} = \begin{bmatrix} 3 \\ 0 \end{bmatrix}.$$

Find the eigenvalues and eigenvectors of A to create your sketch.

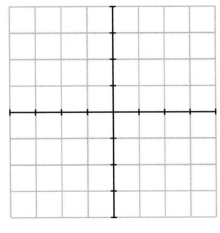

e. If A is a 2×2 matrix with eigenvalues $\lambda_1 = 0.7$ and $\lambda_2 = 0.5$ and $\mathbf{x}$ is any vector, what happens to $A^k\mathbf{x}$ when k becomes very large?

4.4 Dynamical systems

The last section demonstrated ways in which we may relate a matrix, and the effect that multiplication has on vectors, to a simpler form. For instance, if there is a basis of $\mathbb{R}^n$ consisting of eigenvectors of A, we saw that A is similar to a diagonal matrix D. As a result, the effect of multiplying vectors by A, when expressed using the basis of eigenvectors, is the same as multiplying by D.

In this section, we will put these ideas to use as we explore discrete dynamical systems, first encountered in Subsection 2.5.3. Recall that we used a state vector $\mathbf{x}$ to characterize the state of some system at a particular time, such as the distribution of delivery trucks between two locations. A matrix A described the transition of the state vector with $A\mathbf{x}$ characterizing the state of the system at a later time. Since we would like to understand how the state vector evolves over time, we are interested in studying the sequence of vectors $A^k\mathbf{x}$.

Our goal in this section is to describe the types of behaviors that dynamical systems exhibit and to develop a means of detecting these behaviors.

> **Preview Activity 4.4.1.** Suppose that we have a diagonalizable matrix $A = PDP^{-1}$ where
> $$P = \begin{bmatrix} 1 & -1 \\ 1 & 2 \end{bmatrix}, \quad D = \begin{bmatrix} 2 & 0 \\ 0 & -3 \end{bmatrix}.$$
>
> a. Find the eigenvalues of A and find a basis for the associated eigenspaces.
>
> b. Form a basis of $\mathbb{R}^2$ consisting of eigenvectors of A and write the vector $\mathbf{x} = \begin{bmatrix} 1 \\ 4 \end{bmatrix}$ as a linear combination of basis vectors.
>
> c. Write $A\mathbf{x}$ as a linear combination of basis vectors.
>
> d. For some power k, write $A^k\mathbf{x}$ as a linear combination of basis vectors.
>
> e. Find the vector $A^5\mathbf{x}$.

4.4.1 A first example

We will begin with a dynamical system that illustrates how the ideas we've been developing can help us understand the populations of two interacting species. There are several possible ways in which two species may interact. For example, wolves on Isle Royale in northern Michigan prey on moose so this interaction is often called a predator-prey relationship. Other interactions between species, such as bees and flowering plants, are mutually beneficial for both species.

> **Activity 4.4.2.** Suppose we have two species R and S that interact with each another and that we record the change in their populations from year to year. When we begin

our study, the populations, measured in thousands, are R_0 and S_0; after k years, the populations are R_k and S_k.

If we know the populations in one year, suppose that the populations in the following year are determined by the expressions

$$R_{k+1} = 0.9R_k + 0.8S_k$$
$$S_{k+1} = 0.2R_k + 0.9S_k.$$

This is an example of a mutually beneficial relationship between two species. If species S is not present, then $R_{k+1} = 0.9R_k$, which means that the population of species R decreases every year. However, species R benefits from the presence of species S, which helps R to grow by 80% of the population of species S. In the same way, S benefits from the presence of R.

We will record the populations in a vector $\mathbf{x}_k = \begin{bmatrix} R_k \\ S_k \end{bmatrix}$ and note that $\mathbf{x}_{k+1} = A\mathbf{x}_k$ where $A = \begin{bmatrix} 0.9 & 0.8 \\ 0.2 & 0.9 \end{bmatrix}$.

a. Verify that
$$\mathbf{v}_1 = \begin{bmatrix} 2 \\ 1 \end{bmatrix}, \quad \mathbf{v}_2 = \begin{bmatrix} -2 \\ 1 \end{bmatrix}$$
are eigenvectors of A and find their respective eigenvalues.

b. Suppose that initially $\mathbf{x}_0 = \begin{bmatrix} 2 \\ 3 \end{bmatrix}$. Write $\mathbf{x}_0$ as a linear combination of the eigenvectors $\mathbf{v}_1$ and $\mathbf{v}_2$.

c. Write the vectors $\mathbf{x}_1, \mathbf{x}_2,$ and $\mathbf{x}_3$ as linear combinations of the eigenvectors $\mathbf{v}_1$ and $\mathbf{v}_2$.

d. What happens to $\mathbf{x}_k$ after a very long time?

e. When k becomes very large, what happens to the ratio of the populations R_k/S_k?

f. After a very long time, by approximately what factor does the population of R grow every year? By approximately what factor does the population of S grow every year?

g. If we begin instead with $\mathbf{x}_0 = \begin{bmatrix} 4 \\ 4 \end{bmatrix}$, what eventually happens to the ratio R_k/S_k as k becomes very large?

This activity demonstrates the type of systems we will be considering. In particular, we will have vectors $\mathbf{x}_k$ that describe the state of the system at time k and a matrix A that describes how the state evolves from one time to the next: $\mathbf{x}_{k+1} = A\mathbf{x}_k$. The eigenvalues and eigenvectors of A provide the key that helps us understand how the vectors $\mathbf{x}_k$ evolve and that enables us to make long-range predictions.

4.4. DYNAMICAL SYSTEMS

Let's look at the specific example in the previous activity more carefully. We see that

$$\mathbf{x}_{k+1} = A\mathbf{x}_k = \begin{bmatrix} 0.9 & 0.8 \\ 0.2 & 0.9 \end{bmatrix} \mathbf{x}_k$$

and that the matrix A has eigenvectors $\mathbf{v}_1 = \begin{bmatrix} 2 \\ 1 \end{bmatrix}$ and $\mathbf{v}_2 = \begin{bmatrix} -2 \\ 1 \end{bmatrix}$ with associated eigenvalues $\lambda_1 = 1.3$ and $\lambda_2 = 0.5$.

With initial populations $\mathbf{x}_0 = \begin{bmatrix} 2 \\ 3 \end{bmatrix}$, we have

$$\mathbf{x}_0 = 2\mathbf{v}_1 + 1\mathbf{v}_2$$
$$\mathbf{x}_1 = 1.3 \cdot 2\mathbf{v}_1 + 0.5 \cdot 1\mathbf{v}_2$$
$$\mathbf{x}_2 = 1.3^2 \cdot 2\mathbf{v}_1 + 0.5^2 \cdot 1\mathbf{v}_2$$
$$\mathbf{x}_k = 1.3^k \cdot 2\mathbf{v}_1 + 0.5^k \cdot 1\mathbf{v}_2.$$

Let's shift our perspective slightly. The eigenvectors $\mathbf{v}_1$ and $\mathbf{v}_2$ form a basis $\mathcal{B}$ of $\mathbb{R}^2$, which says that A is diagonalizable; that is, $A = PDP^{-1}$ where

$$P = \begin{bmatrix} \mathbf{v}_1 & \mathbf{v}_2 \end{bmatrix} = \begin{bmatrix} 2 & -2 \\ 1 & 1 \end{bmatrix}, \qquad D = \begin{bmatrix} 1.3 & 0 \\ 0 & 0.5 \end{bmatrix}.$$

The coordinate system defined by the basis $\mathcal{B}$ can be used to express the state vectors. For instance, we can write the initial state vector $\mathbf{x}_0 = \begin{bmatrix} 2 \\ 3 \end{bmatrix} = 2\mathbf{v}_1 + \mathbf{v}_2$, which means that $\{\mathbf{x}_0\}_\mathcal{B} = \begin{bmatrix} 2 \\ 1 \end{bmatrix}$. Moreover, $\mathbf{x}_1 = A\mathbf{x}_0 = (1.3) \cdot 2\mathbf{v}_1 + (0.5) \cdot 1\mathbf{v}_2$ so that

$$\{\mathbf{x}_1\}_\mathcal{B} = \begin{bmatrix} 1.3 \cdot 2 \\ 0.5 \cdot 1 \end{bmatrix} = D \begin{bmatrix} 2 \\ 1 \end{bmatrix} = D\{\mathbf{x}_0\}_\mathcal{B}.$$

In the same way,

$$\{\mathbf{x}_1\}_\mathcal{B} = D\{\mathbf{x}_0\}_\mathcal{B} = \begin{bmatrix} 1.3 \cdot 2 \\ 0.5 \cdot 1 \end{bmatrix}$$

$$\{\mathbf{x}_2\}_\mathcal{B} = D\{\mathbf{x}_1\}_\mathcal{B} = \begin{bmatrix} 1.3^2 \cdot 2 \\ 0.5^2 \cdot 1 \end{bmatrix}$$

$$\{\mathbf{x}_3\}_\mathcal{B} = D\{\mathbf{x}_2\}_\mathcal{B} = \begin{bmatrix} 1.3^3 \cdot 2 \\ 0.5^3 \cdot 1 \end{bmatrix}$$

$$\{\mathbf{x}_k\}_\mathcal{B} = \begin{bmatrix} 1.3^k \cdot 2 \\ 0.5^k \cdot 1 \end{bmatrix}.$$

More generally, we have
$$\{A\mathbf{x}\}_\mathcal{B} = D\{\mathbf{x}\}_\mathcal{B},$$
which is a restatement of the fact that A is similar to D.

Thinking about this geometrically, we begin with the vector $\{\mathbf{x}_0\}_\mathcal{B} = \begin{bmatrix} 2 \\ 1 \end{bmatrix}$. Subsequent vectors $\{\mathbf{x}_k\}_\mathcal{B}$ are obtained by scaling horizontally by a factor of 1.3 and scaling vertically by a factor 0.5. Notice how the points move along a curve away from the origin becoming ever closer to the horizontal axis. After a very long time, $\{\mathbf{x}_k\}_\mathcal{B} \approx \begin{bmatrix} 1.3^k \cdot 2 \\ 0 \end{bmatrix}$.

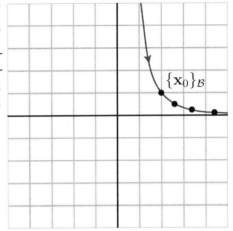

To recover the behavior of the sequence $\mathbf{x}_0, \mathbf{x}_1, \mathbf{x}_2, \ldots$, we change coordinate systems using the basis defined by $\mathbf{v}_1$ and $\mathbf{v}_2$. Here, the points move along a curve away from the origin becoming ever closer to the line defined by $\mathbf{v}_1$.

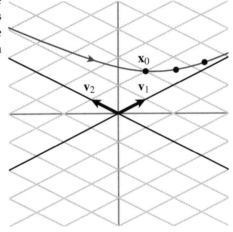

Eventually, the vectors become practically indistinguishable from a scalar multiple of $\mathbf{v}_1 = \begin{bmatrix} 2 \\ 1 \end{bmatrix}$ since $\mathbf{x}_k \approx 1.3^k \cdot 2\mathbf{v}_1$. This means that

$$\mathbf{x}_k = \begin{bmatrix} R_k \\ S_k \end{bmatrix} \approx 1.3^k \cdot 2\mathbf{v}_1 = \begin{bmatrix} 1.3^k \cdot 4 \\ 1.3^k \cdot 2 \end{bmatrix}.$$

This shows that

$$R_k/S_k \approx (1.3^k \cdot 4)/(1.3^k \cdot 2) = 2$$

so that we expect the population of species R to eventually be about twice that of species S.

In addition, $\mathbf{x}_{k+1} \approx 1.3\mathbf{x}_k$ so that $R_{k+1} \approx 1.3 R_k$ and $S_{k+1} \approx 1.3 S_k$, which tells us that both populations are multiplied by 1.3 every year meaning the annual growth rate for both populations is about 30%.

In the same way, we can consider other possible initial populations $\mathbf{x}_0$ as shown in Figure 4.4.1. Regardless of $\mathbf{x}_0$, the population vectors, in the coordinates defined by $\mathcal{B}$, are scaled horizontally by a factor of 1.3 and vertically by a factor of 0.5. The sequence of points

4.4. DYNAMICAL SYSTEMS

$\{\mathbf{x}_k\}_\mathcal{B}$, called *trajectories*, move along the curves, as shown on the left. In the standard coordinate system, we see that the trajectories converge to the eigenspace $E_{1.3}$.

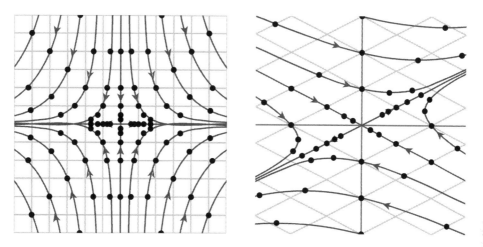

Figure 4.4.1 The trajectories of the dynamical system formed by the matrix A in the coordinate system defined by $\mathcal{B}$, on the left, and in the standard coordinate system, on the right.

We conclude that, regardless of the initial populations, the ratio of the populations R_k/S_k will approach 2 to 1 and that the growth rate for both populations approaches 30%. This example demonstrates the power of using eigenvalues and eigenvectors to rewrite the problem in terms of a new coordinate system. By doing so, we are able to predict the long-term behavior of the populations independently of the initial populations.

Diagrams like those shown in Figure 4.4.1 are called *phase portraits*. On the left of Figure 4.4.1 is the phase portrait of the diagonal matrix $D = \begin{bmatrix} 1.3 & 0 \\ 0 & 0.5 \end{bmatrix}$ while the right of that figure shows the phase portrait of $A = \begin{bmatrix} 0.9 & 0.8 \\ 0.2 & 0.9 \end{bmatrix}$. The phase portrait of D is relatively easy to understand because it is determined only by the two eigenvalues. Once we have the phase portrait of D, however, the phase portrait of A has a similar appearance with the eigenvectors $\mathbf{v}_j$ replacing the standard basis vectors $\mathbf{e}_j$.

4.4.2 Classifying dynamical systems

In the previous example, we were able to make predictions about the behavior of trajectories $\mathbf{x}_k = A^k \mathbf{x}_0$ by considering the eigenvalues and eigenvectors of the matrix A. The next activity looks at a collection of matrices that demonstrate the types of behavior a 2×2 dynamical system can exhibit.

Activity 4.4.3. We will now look at several more examples of dynamical systems. If $P = \begin{bmatrix} 1 & -1 \\ 1 & 1 \end{bmatrix}$, we note that the columns of P form a basis $\mathcal{B}$ of $\mathbb{R}^2$. Given below are several matrices A written in the form $A = PEP^{-1}$ for some matrix E. For each matrix, state the eigenvalues of A and sketch a phase portrait for the matrix E on the left and a phase portrait for A on the right. Describe the behavior of $A^k \mathbf{x}_0$ as k becomes very large for a typical initial vector $\mathbf{x}_0$.

a. $A = PEP^{-1}$ where $E = \begin{bmatrix} 1.3 & 0 \\ 0 & 1.5 \end{bmatrix}$.

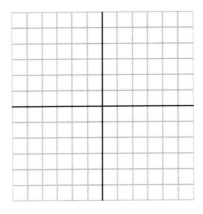
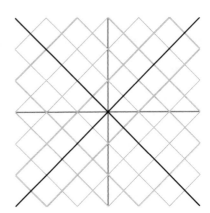

b. $A = PEP^{-1}$ where $E = \begin{bmatrix} 0 & -1 \\ 1 & 0 \end{bmatrix}$.

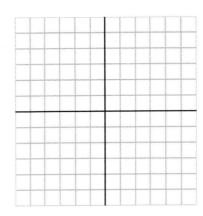
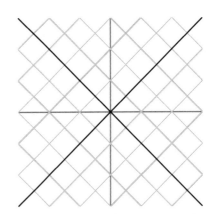

c. $A = PEP^{-1}$ where $E = \begin{bmatrix} 0.7 & 0 \\ 0 & 1.5 \end{bmatrix}$.

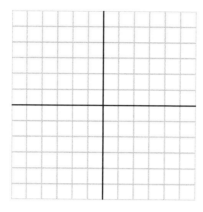 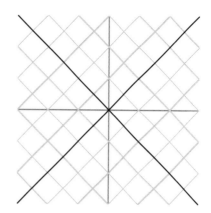

d. $A = PEP^{-1}$ where $E = \begin{bmatrix} 0.3 & 0 \\ 0 & 0.7 \end{bmatrix}$.

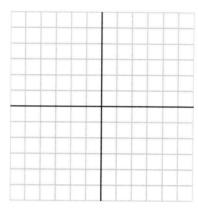 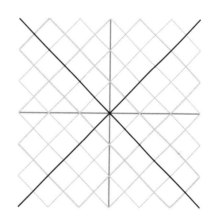

e. $A = PEP^{-1}$ where $E = \begin{bmatrix} 1 & -0.9 \\ 0.9 & 1 \end{bmatrix}$.

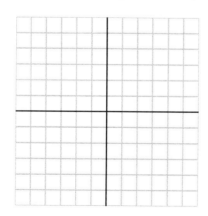 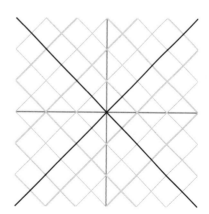

f. $A = PEP^{-1}$ where $E = \begin{bmatrix} 0.6 & -0.2 \\ 0.2 & 0.6 \end{bmatrix}$.

278 CHAPTER 4. EIGENVALUES AND EIGENVECTORS

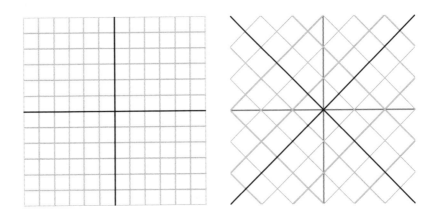

This activity demonstrates six possible types of dynamical systems, which are determined by the eigenvalues of A.

- Suppose that A has two real eigenvalues λ_1 and λ_2 and that both $|\lambda_1|, |\lambda_2| > 1$. In this case, any nonzero vector $\mathbf{x}_0$ forms a trajectory that moves away from the origin so we say that the origin is a *repellor*. This is illustrated in Figure 4.4.2.

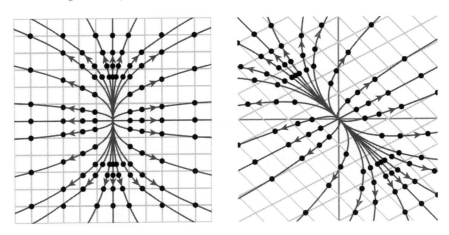

Figure 4.4.2 The origin is a repellor when $|\lambda_1|, |\lambda_2| > 1$.

- Suppose that A has two real eigenvalues λ_1 and λ_2 and that $|\lambda_1| > 1 > |\lambda_2|$. In this case, most nonzero vectors $\mathbf{x}_0$ form trajectories that converge to the eigenspace E_{λ_1}. In this case, we say that the origin is a *saddle* as illustrated in Figure 4.4.3.

4.4. DYNAMICAL SYSTEMS

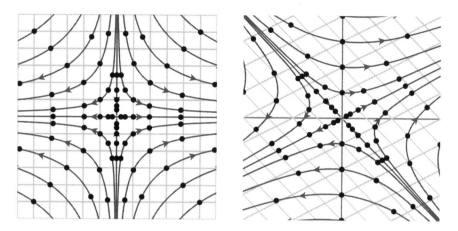

Figure 4.4.3 The origin is a saddle when $|\lambda_1| > 1 > |\lambda_2|$.

- Suppose that A has two real eigenvalues λ_1 and λ_2 and that both $|\lambda_1|, |\lambda_2| < 1$. In this case, any nonzero vector $\mathbf{x}_0$ forms a trajectory that moves into the origin so we say that the origin is an *attractor*. This is illustrated in Figure 4.4.4.

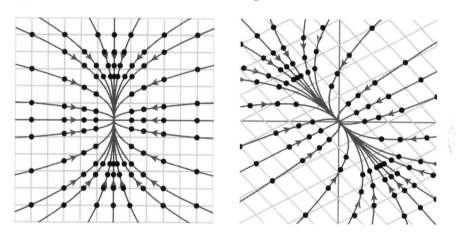

Figure 4.4.4 The origin is an attractor when $|\lambda_1|, |\lambda_2| < 1$.

- Suppose that A has a complex eigenvalue $\lambda = a + bi$ where $|\lambda| > 1$. In this case, a nonzero vector $\mathbf{x}_0$ forms a trajectory that spirals away from the origin. We say that the origin is a *spiral repellor*, as illustrated in Figure 4.4.5.

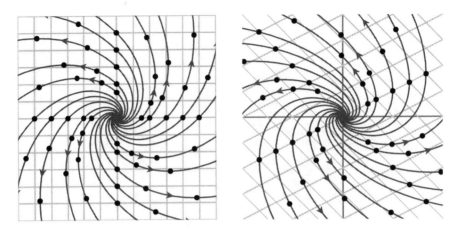

Figure 4.4.5 The origin is a spiral repellor when A has an eigenvalue $\lambda = a + bi$ with $a^2 + b^2 > 1$.

- Suppose that A has a complex eigenvalue $\lambda = a + bi$ where $|\lambda| = 1$. In this case, a nonzero vector $\mathbf{x}_0$ forms a trajectory that moves on a closed curve around the origin. We say that the origin is a *center*, as illustrated in Figure 4.4.6.

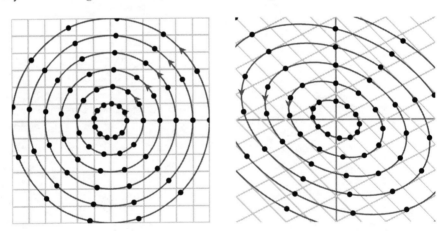

Figure 4.4.6 The origin is a center when A has an eigenvalue $\lambda = a + bi$ with $a^2 + b^2 = 1$.

- Suppose that A has a complex eigenvalue $\lambda = a + bi$ where $|\lambda| < 1$. In this case, a nonzero vector $\mathbf{x}_0$ forms a trajectory that spirals into the origin. We say that the origin is a *spiral attractor*, as illustrated in Figure 4.4.7.

4.4. DYNAMICAL SYSTEMS

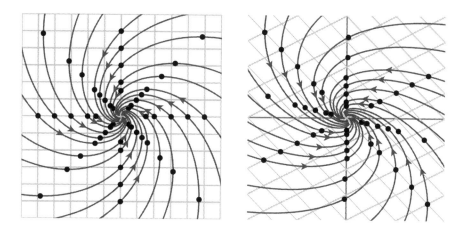

Figure 4.4.7 The origin is a spiral attractor when A has an eigenvalue $\lambda = a + bi$ with $a^2 + b^2 < 1$.

This list includes many types of expected behavior, but there are other possibilities if, for instance, one of the eigenvalues is 0. The next section explores the situation when one of the eigenvalues is 1.

Activity 4.4.4. In this activity, we will consider several ways in which two species might interact with one another. Throughout, we will consider two species R and S whose populations in year k form a vector $\mathbf{x}_k = \begin{bmatrix} R_k \\ S_k \end{bmatrix}$ and which evolve according to the rule
$$\mathbf{x}_{k+1} = A\mathbf{x}_k.$$

a. Suppose that $A = \begin{bmatrix} 0.7 & 0 \\ 0 & 1.6 \end{bmatrix}$.

Explain why the species do not interact with one another. Which of the six types of dynamical systems do we have? What happens to both species after a long time?

b. Suppose now that $A = \begin{bmatrix} 0.7 & 0.3 \\ 0 & 1.6 \end{bmatrix}$.

Explain why S is a beneficial species for R. Which of the six types of dynamical systems do we have? What happens to both species after a long time?

c. If $A = \begin{bmatrix} 0.7 & 0.5 \\ -0.4 & 1.6 \end{bmatrix}$, explain why this describes a predator-prey system. Which of the species is the predator and which is the prey? Which of the six types of dynamical systems do we have? What happens to both species after a long time?

d. Suppose that $A = \begin{bmatrix} 0.5 & 0.2 \\ -0.4 & 1.1 \end{bmatrix}$. Compare this predator-prey system to the one in the previous part. Which of the six types of dynamical systems do we have? What happens to both species after a long time?

4.4.3 A 3×3 system

Up to this point, we have focused on 2×2 systems. In fact, the general case is quite similar. As an example, consider a 3×3 system $\mathbf{x}_{k+1} = A\mathbf{x}_k$ where the matrix A has eigenvalues $\lambda_1 = 0.6$, $\lambda_2 = 0.8$, and $\lambda_3 = 1.1$. Since the eigenvalues are real and distinct, there is a basis $\mathcal{B}$ consisting of eigenvectors of A so we can look at the trajectories $\{\mathbf{x}_k\}_\mathcal{B}$ in the coordinate system defined by $\mathcal{B}$. The phase portraits in Figure 4.4.8 show how some representative trajectories will evolve. We see that all the trajectories will converge into the eigenspace $E_{1.1}$.

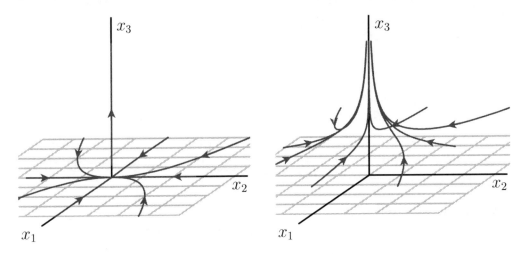

Figure 4.4.8 In a 3×3 system with $\lambda_1 = 0.6$, $\lambda_2 = 0.8$, and $\lambda_3 = 1.1$, the trajectories $\{\mathbf{x}_k\}_\mathcal{B}$ move along the curves shown above.

In the same way, suppose we have a 3×3 system with complex eigenvalues $\lambda = 0.8 \pm 0.5i$ and $\lambda_3 = 1.1$. Since the complex eigenvalues satisfy $|\lambda| < 1$, there is a two-dimensional subspace in which the trajectories spiral in toward the origin. The phase portraits in Figure 4.4.9 show some of the trajectories. Once again, we see that all the trajectories converge into the eigenspace $E_{1.1}$.

4.4. DYNAMICAL SYSTEMS

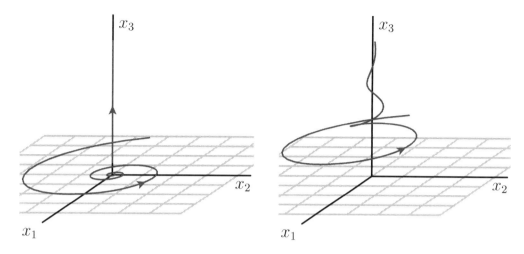

Figure 4.4.9 In a 3×3 system with complex eigenvalues $\lambda = a \pm bi$ with $|\lambda| < 1$ and $\lambda_3 = 1.1$, the trajectories $\{\mathbf{x}_k\}_{\mathcal{B}}$ move along the curves shown above.

Activity 4.4.5. The following type of analysis has been used to study the population of a bison herd. We will divide the population of female bison into three groups: juveniles who are less than one year old; yearlings between one and two years old; and adults who are older than two years.

Each year,

- 80% of the juveniles survive to become yearlings.
- 90% of the yearlings survive to become adults.
- 80% of the adults survive.
- 40% of the adults give birth to a juvenile.

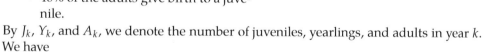

By J_k, Y_k, and A_k, we denote the number of juveniles, yearlings, and adults in year k. We have

$$J_{k+1} = 0.4 A_k.$$

a. Find similar expressions for Y_{k+1} and A_{k+1} in terms of J_k, Y_k, and A_k.

b. As is usual, we write the matrix $\mathbf{x}_k = \begin{bmatrix} J_k \\ Y_k \\ A_k \end{bmatrix}$. Write the matrix A such that $\mathbf{x}_{k+1} = A\mathbf{x}_k$ and find its eigenvalues.

c. We can write $A = PEP^{-1}$ where the matrices E and P are approximately:

$$E = \begin{bmatrix} 1.058 & 0 & 0 \\ 0 & -0.128 & -0.506 \\ 0 & 0.506 & -0.128 \end{bmatrix},$$

$$P = \begin{bmatrix} 1 & 1 & 0 \\ 0.756 & -0.378 & 1.486 \\ 2.644 & -0.322 & -1.264 \end{bmatrix}.$$

Make a prediction about the long-term behavior of $\mathbf{x}_k$. For instance, at what rate does it grow? For every 100 adults, how many juveniles, and yearlings are there?

d. Suppose that the birth rate decreases so that only 30% of adults give birth to a juvenile. How does this affect the long-term growth rate of the herd?

e. Suppose that the birth rate decreases further so that only 20% of adults give birth to a juvenile. How does this affect the long-term growth rate of the herd?

f. Find the smallest birth rate that supports a stable population.

4.4.4 Summary

We have been exploring discrete dynamical systems in which an initial state vector $\mathbf{x}_0$ evolves over time according to the rule $\mathbf{x}_{k+1} = A\mathbf{x}_k$. The eigenvalues and eigenvectors of A help us understand the behavior of the state vectors. In the 2×2 case, we saw that

- $|\lambda_1|, |\lambda_2| < 1$ produces an attractor so that trajectories are pulled in toward the origin.
- $|\lambda_1| > 1$ and $|\lambda_2| < 1$ produces a saddle in which most trajectories are pushed away from the origin and in the direction of E_{λ_1}.
- $|\lambda_1|, |\lambda_2| > 1$ produces a repellor in which trajectories are pushed away from the origin.

The same kind of reasoning allows us to analyze $n \times n$ systems as well.

4.4.5 Exercises

1. For each of the 2×2 matrices below, find the eigenvalues and, when appropriate, the eigenvectors to classify the dynamical system $\mathbf{x}_{k+1} = A\mathbf{x}_k$. Use this information to

4.4. DYNAMICAL SYSTEMS

sketch the phase portraits.

a. $A = \begin{bmatrix} 3 & 1 \\ 1 & 3 \end{bmatrix}$.

b. $A = \begin{bmatrix} 3 & -2 \\ 4 & -1 \end{bmatrix}$.

c. $A = \begin{bmatrix} 1.9 & 1.4 \\ -0.7 & -0.2 \end{bmatrix}$.

d. $A = \begin{bmatrix} 1.1 & -0.2 \\ 0.4 & 0.5 \end{bmatrix}$.

2. We will consider matrices that have the form $A = PDP^{-1}$ where

$$D = \begin{bmatrix} p & 0 \\ 0 & \frac{1}{2} \end{bmatrix}, P = \begin{bmatrix} 2 & -2 \\ 1 & 1 \end{bmatrix}$$

where p is a parameter that we will vary. Sketch phase portraits for D and A below when

a. $p = \frac{1}{2}$.

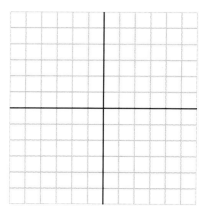 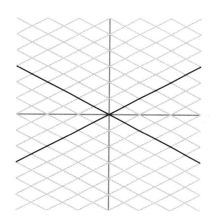

b. $p = 1$.

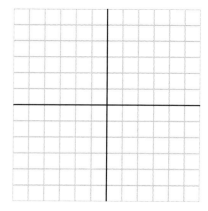 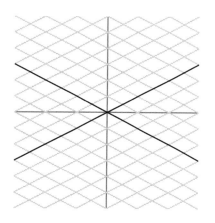

c. $p = 2$.

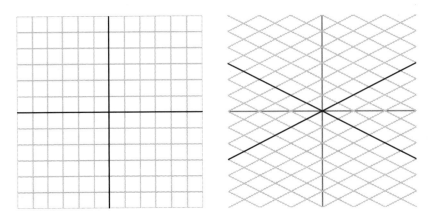

d. For the different values of p, determine which types of dynamical system results. For what range of p values do we have an attractor? For what range of p values do we have a saddle? For what value does the transition between the two types occur?

3. Suppose that the populations of two species interact according to the relationships

$$R_{k+1} = \frac{1}{2}R_k + \frac{1}{2}S_k$$
$$S_{k+1} = -pR_k + 2S_k$$

where p is a parameter. As we saw in the text, this dynamical system represents a typical predator-prey relationship, and the parameter p represents the rate at which species R preys on S. We will denote the matrix $A = \begin{bmatrix} \frac{1}{2} & \frac{1}{2} \\ -p & 2 \end{bmatrix}$.

a. If $p = 0$, determine the eigenvectors and eigenvalues of the system and classify it as one of the six types. Sketch the phase portraits for the diagonal matrix D to which A is similar as well as the phase portrait for A.

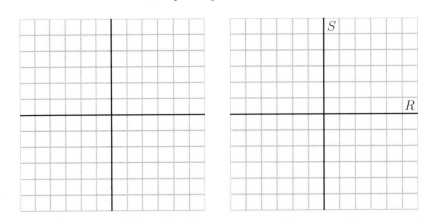

b. If $p = 1$, determine the eigenvectors and eigenvalues of the system. Sketch the phase portraits for the diagonal matrix D to which A is similar as well as the phase portrait for A.

4.4. DYNAMICAL SYSTEMS

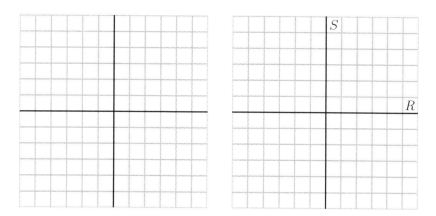

c. For what values of p is the origin a saddle? What can you say about the populations when this happens?

d. Describe the evolution of the dynamical system as p begins at 0 and increases to $p = 1$.

4. Consider the matrices

$$A = \begin{bmatrix} 3 & 2 \\ -5 & -3 \end{bmatrix}, \quad B = \begin{bmatrix} 5 & 7 \\ -3 & -4 \end{bmatrix}.$$

a. Find the eigenvalues of A. To which of the six types does the system $x_{k+1} = Ax_k$ belong?

b. Using the eigenvalues of A, we can write $A = PEP^{-1}$ for some matrices E and P. What is the matrix E and what geometric effect does multiplication by E have on vectors in the plane?

c. If we remember that $A^k = PE^kP^{-1}$, determine the smallest positive value of k for which $A^k = I$.

d. Find the eigenvalues of B.

e. Find a matrix E such that $B = PEP^{-1}$ for some matrix P. What geometric effect does multiplication by E have on vectors in the plane?

f. Determine the smallest positive value of k for which $B^k = I$.

5. Suppose we have the female population of a species is divided into juveniles, yearlings, and adults and that each year

  - 90% of the juveniles live to be yearlings.
  - 80% of the yearlings live to be adults.
  - 60% of the adults survive to the next year.
  - 50% of the adults give birth to a juvenile.

  a. Set up a system of the form $x_{k+1} = Ax_k$ that describes this situation.

b. Find the eigenvalues of the matrix A.

c. What prediction can you make about these populations after a very long time?

d. If the birth rate goes up to 80%, what prediction can you make about these populations after a very long time? For every 100 adults, how many juveniles, and yearlings are there?

6. Determine whether the following statements are true or false and provide a justification for your response. In each case, we are considering a dynamical system of the form $\mathbf{x}_{k+1} = A\mathbf{x}_k$.

   a. If the 2×2 matrix A has a complex eigenvalue, we cannot make a prediction about the behavior of the trajectories.

   b. If A has eigenvalues whose absolute value is smaller than 1, then all the trajectories are pulled in toward the origin.

   c. If the origin is a repellor, then it is an attractor for the system $\mathbf{x}_{k+1} = A^{-1}\mathbf{x}_k$.

   d. If a 4×4 matrix has complex eigenvalues $\lambda_1, \lambda_2, \lambda_3$, and λ_4, all of which satisfy $|\lambda_j| > 1$, then all the trajectories are pushed away from the origin.

   e. If the origin is a saddle, then all the trajectories are pushed away from the origin.

7. The Fibonacci numbers form the sequence of numbers that begins $0, 1, 1, 2, 3, 5, 8, 13, \ldots$. If we let F_n denote the nth Fibonacci number, then

$$F_0 = 0, F_1 = 1, F_2 = 1, F_3 = 2, F_4 = 3, \ldots.$$

In general, a Fibonacci number is the sum of the previous two Fibonacci numbers; that is, $F_{n+2} = F_n + F_{n+1}$ so that we have

$$F_{n+2} = F_n + F_{n+1}$$
$$F_{n+1} = F_{n+1}.$$

   a. If we write $\mathbf{x}_n = \begin{bmatrix} F_{n+1} \\ F_n \end{bmatrix}$, find the matrix A such that $\mathbf{x}_{n+1} = A\mathbf{x}_n$.

   b. Show that A has eigenvalues

$$\lambda_1 = \frac{1 + \sqrt{5}}{2} \approx 1.61803\ldots$$

$$\lambda_2 = \frac{1 - \sqrt{5}}{2} \approx -0.61803\ldots$$

   with associated eigenvectors $\mathbf{v}_1 = \begin{bmatrix} \lambda_1 \\ 1 \end{bmatrix}$ and $\mathbf{v}_2 = \begin{bmatrix} \lambda_2 \\ 1 \end{bmatrix}$.

   c. Classify this dynamical system as one of the six types that we have seen in this section. What happens to $\mathbf{x}_n$ as n becomes very large?

4.4. DYNAMICAL SYSTEMS

d. Write the initial vector $\mathbf{x}_0 = \begin{bmatrix} 1 \\ 0 \end{bmatrix}$ as a linear combination of eigenvectors $\mathbf{v}_1$ and $\mathbf{v}_2$.

e. Write the vector $\mathbf{x}_n$ as a linear combinations of $\mathbf{v}_1$ and $\mathbf{v}_2$.

f. Explain why the nth Fibonacci number

$$F_n = \frac{1}{\sqrt{5}}\left[\left(\frac{1+\sqrt{5}}{2}\right)^n - \left(\frac{1-\sqrt{5}}{2}\right)^n\right].$$

g. Use this relationship to compute F_{20}.

h. Explain why $F_{n+1}/F_n \approx \lambda_1$ when n is very large.

The number $\lambda_1 = \frac{1+\sqrt{5}}{2} = \phi$ is called the *golden ratio* and is one of mathematics' special numbers.

8. This exercise is a continuation of the previous one.

The Lucas numbers L_n are defined by the same relationship as the Fibonacci numbers: $L_{n+2} = L_{n+1} + L_n$. However, we begin with $L_0 = 2$ and $L_1 = 1$, which leads to the sequence $2, 1, 3, 4, 7, 11, \ldots$.

a. As before, form the vector $\mathbf{x}_n = \begin{bmatrix} L_{n+1} \\ L_n \end{bmatrix}$ so that $\mathbf{x}_{n+1} = A\mathbf{x}_n$. Express $\mathbf{x}_0$ as a linear combination of $\mathbf{v}_1$ and $\mathbf{v}_2$, eigenvectors of A.

b. Explain why

$$L_n = \left(\frac{1+\sqrt{5}}{2}\right)^n + \left(\frac{1-\sqrt{5}}{2}\right)^n.$$

c. Explain why L_n is the closest integer to ϕ^n when n is large, where $\phi = \lambda_1$ is the golden ratio.

d. Use this observation to find L_{20}.

9. Gil Strang defines the *Gibonacci numbers* G_n as follows. We begin with $G_0 = 0$ and $G_1 = 1$. A subsequent Gibonacci number is the average of the two previous; that is, $G_{n+2} = \frac{1}{2}(G_n + G_{n+1})$. We then have

$$G_{n+2} = \frac{1}{2}G_n + \frac{1}{2}G_{n+1}$$
$$G_{n+1} = G_{n+1}.$$

a. If $\mathbf{x}_n = \begin{bmatrix} G_{n+1} \\ G_n \end{bmatrix}$, find the matrix A such that $\mathbf{x}_{n+1} = A\mathbf{x}_n$.

b. Find the eigenvalues and associated eigenvectors of A.

c. Explain why this dynamical system does not neatly fit into one of the six types that we saw in this section.

d. Write x_0 as a linear combination of eigenvectors of A.

e. Write x_n as a linear combination of eigenvectors of A.

f. What happens to G_n as n becomes very large?

10. Consider a small rodent that lives for three years. Once again, we can separate a population of females into juveniles, yearlings, and adults. Suppose that, each year,

    - Half of the juveniles live to be yearlings.
    - One quarter of the yearlings live to be adults.
    - Adult females produce eight female offspring.
    - None of the adults survive to the next year.

    a. Writing the populations of juveniles, yearlings, and adults in year k using the vector $x_k = \begin{bmatrix} J_k \\ Y_k \\ A_k \end{bmatrix}$, find the matrix A such that $x_{k+1} = Ax_k$.

    b. Show that $A^3 = I$.

    c. What are the eigenvalues of A^3? What does this say about the eigenvalues of A?

    d. Verify your observation by finding the eigenvalues of A.

    e. What can you say about the trajectories of this dynamical system?

    f. What does this mean about the population of rodents?

    g. Find a population vector x_0 that is unchanged from year to year.

4.5 Markov chains and Google's PageRank algorithm

In the last section, we used our understanding of eigenvalues and eigenvectors to describe the long-term behavior of some discrete dynamical systems. The state of the system, which could record, say, the populations of a few interacting species, at one time is described by a vector $\mathbf{x}_k$. The state vector then evolves according to a linear rule $\mathbf{x}_{k+1} = A\mathbf{x}_k$.

This section continues this exploration by looking at *Markov chains*, which form a specific type of discrete dynamical system. For instance, we could be interested in a rental car company that rents cars from several locations. From one day to the next, the number of cars at different locations can change, but the total number of cars stays the same. Once again, an understanding of eigenvalues and eigenvectors will help us make predictions about the long-term behavior of the system.

Preview Activity 4.5.1. Suppose that our rental car company rents from two locations P and Q. We find that 80% of the cars rented from location P are returned to P while the other 20% are returned to Q. For cars rented from location Q, 60% are returned to Q and 40% to P.

We will use P_k and Q_k to denote the number of cars at the two locations on day k. The following day, the number of cars at P equals 80% of P_k and 40% of Q_k. This shows that
$$P_{k+1} = 0.8P_k + 0.4Q_k$$
$$Q_{k+1} = 0.2P_k + 0.6Q_k.$$

a. If we use the vector $\mathbf{x}_k = \begin{bmatrix} P_k \\ Q_k \end{bmatrix}$ to represent the distribution of cars on day k, find a matrix A such that $\mathbf{x}_{k+1} = A\mathbf{x}_k$.

b. Find the eigenvalues and associated eigenvectors of A.

c. Suppose that there are initially 1500 cars, all of which are at location P. Write the vector $\mathbf{x}_0$ as a linear combination of eigenvectors of A.

d. Write the vectors $\mathbf{x}_k$ as a linear combination of eigenvectors of A.

e. What happens to the distribution of cars after a long time?

4.5.1 A first example

In the preview activity, the distribution of rental cars was described by the discrete dynamical system
$$\mathbf{x}_{k+1} = A\mathbf{x}_k = \begin{bmatrix} 0.8 & 0.4 \\ 0.2 & 0.6 \end{bmatrix} \mathbf{x}_k.$$

This matrix has some special properties. First, each entry represents the probability that a car rented at one location is returned to another. For instance, there is an 80% chance that a car rented at P is returned to P, which explains the entry of 0.8 in the upper left corner. Therefore, the entries of the matrix are between 0 and 1.

Second, a car rented at one location must be returned to one of the locations. For example, since 80% of the cars rented at P are returned to P, it follows that the other 20% of cars rented at P are returned to Q. This implies that the entries in each column must add to 1. This will occur frequently in our discussion so we introduce the following definitions.

Definition 4.5.1 A vector whose entries are nonnegative and add to 1 is called a *probability vector*. A square matrix whose columns are probability vectors is called a *stochastic* matrix.

Activity 4.5.2. Suppose you live in a country with three political parties P, Q, and R. We use P_k, Q_k, and R_k to denote the percentage of voters voting for that party in election k.

Voters will change parties from one election to the next as shown in the figure. We see that 60% of voters stay with the same party. However, 40% of those who vote for party P will vote for party Q in the next election.

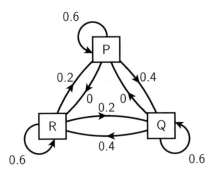

a. Write expressions for P_{k+1}, Q_{k+1}, and R_{k+1} in terms of P_k, Q_k, and R_k.

b. If we write $\mathbf{x}_k = \begin{bmatrix} P_k \\ Q_k \\ R_k \end{bmatrix}$, find the matrix A such that $\mathbf{x}_{k+1} = A\mathbf{x}_k$.

c. Explain why A is a stochastic matrix.

d. Suppose that initially 40% of citizens vote for party P, 30% vote for party Q, and 30% vote for party R. Form the vector $\mathbf{x}_0$ and explain why $\mathbf{x}_0$ is a probability vector.

e. Find $\mathbf{x}_1$, the percentages who vote for the three parties in the next election. Verify that $\mathbf{x}_1$ is also a probability vector and explain why $\mathbf{x}_k$ will be a probability vector for every k.

f. Find the eigenvalues of the matrix A and explain why the eigenspace E_1 is a one-dimensional subspace of $\mathbb{R}^3$. Then verify that $\mathbf{v} = \begin{bmatrix} 1 \\ 2 \\ 2 \end{bmatrix}$ is a basis vector for E_1.

g. As every vector in E_1 is a scalar multiple of $\mathbf{v}$, find a probability vector in E_1 and explain why it is the only probability vector in E_1.

h. Describe what happens to $\mathbf{x}_k$ after a very long time.

The previous activity illustrates some important points that we wish to emphasize.

First, to determine P_{k+1}, we note that in election $k+1$, party P retains 60% of its voters from the previous election and adds 20% of those who voted for party R. In this way, we see that

$$P_{k+1} = 0.6P_k \qquad\qquad + 0.2R_k$$
$$Q_{k+1} = 0.4P_k + 0.6Q_k + 0.2R_k$$
$$R_{k+1} = \qquad\qquad 0.4Q_k + 0.6R_k$$

We therefore define the matrix

$$A = \begin{bmatrix} 0.6 & 0 & 0.2 \\ 0.4 & 0.6 & 0.2 \\ 0 & 0.4 & 0.6 \end{bmatrix}$$

and note that $\mathbf{x}_{k+1} = A\mathbf{x}_k$.

If we consider the first column of A, we see that the entries represent the percentages of party P's voters in the last election who vote for each of the three parties in the next election. Since everyone who voted for party P previously votes for one of the three parties in the next election, the sum of these percentages must be 1. This is true for each of the columns of A, which explains why A is a stochastic matrix.

We begin with the vector $\mathbf{x}_0 = \begin{bmatrix} 0.4 \\ 0.3 \\ 0.3 \end{bmatrix}$, the entries of which represent the percentage of voters voting for each of the three parties. Since every voter votes for one of the three parties, the sum of these entries must be 1, which means that $\mathbf{x}_0$ is a probability vector. We then find that

$$\mathbf{x}_1 = \begin{bmatrix} 0.300 \\ 0.400 \\ 0.300 \end{bmatrix}, \quad \mathbf{x}_2 = \begin{bmatrix} 0.240 \\ 0.420 \\ 0.340 \end{bmatrix}, \quad \mathbf{x}_3 = \begin{bmatrix} 0.212 \\ 0.416 \\ 0.372 \end{bmatrix}, \quad \ldots,$$

$$\mathbf{x}_5 = \begin{bmatrix} 0.199 \\ 0.404 \\ 0.397 \end{bmatrix}, \quad \ldots, \quad \mathbf{x}_{10} = \begin{bmatrix} 0.200 \\ 0.400 \\ 0.400 \end{bmatrix}, \quad \ldots$$

Notice that the vectors $\mathbf{x}_k$ are also probability vectors and that the sequence $\mathbf{x}_k$ seems to be converging to $\begin{bmatrix} 0.2 \\ 0.4 \\ 0.4 \end{bmatrix}$. It is this behavior that we would like to understand more fully by investigating the eigenvalues and eigenvectors of A.

We find that the eigenvalues of A are

$$\lambda_1 = 1, \qquad \lambda_2 = 0.4 + 0.2i, \qquad \lambda_3 = 0.4 - 0.2i.$$

Notice that if **v** is an eigenvector of A with associated eigenvalue $\lambda_1 = 1$, then $A\mathbf{v} = 1\mathbf{v} = \mathbf{v}$. That is, **v** is unchanged when we multiply it by A.

Otherwise, we have $A = PEP^{-1}$ where

$$E = \begin{bmatrix} 1 & 0 & 0 \\ 0 & 0.4 & -0.2 \\ 0 & 0.2 & 0.4 \end{bmatrix}$$

Notice that $|\lambda_2| = |\lambda_3| < 1$ so the trajectories $\mathbf{x}_k$ spiral into the eigenspace E_1 as indicated in the figure.

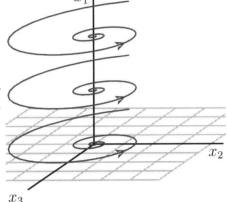

This tells us that the sequence $\mathbf{x}_k$ converges to a vector in E_1. In the usual way, we see that $\mathbf{v} = \begin{bmatrix} 1 \\ 2 \\ 2 \end{bmatrix}$ is a basis vector for E_1 because $A\mathbf{v} = \mathbf{v}$ so we expect that $\mathbf{x}_k$ will converge to a scalar multiple of **v**. Indeed, since the vectors $\mathbf{x}_k$ are probability vectors, we expect them to converge to a probability vector in E_1.

We can find the probability vector in E_1 by finding the appropriate scalar multiple of **v**. Notice that $c\mathbf{v} = \begin{bmatrix} c \\ 2c \\ 2c \end{bmatrix}$ is a probability vector when $c + 2c + 2c = 5c = 1$, which implies that $c = 1/5$. Therefore, $\mathbf{q} = \begin{bmatrix} 0.2 \\ 0.4 \\ 0.4 \end{bmatrix}$ is the unique probability vector in E_1. Since the sequence $\mathbf{x}_k$ converges to a probability vector in E_1, we see that $\mathbf{x}_k$ converges to **q**, which agrees with the computations we showed above.

The role of the eigenvalues is important in this example. Since $\lambda_1 = 1$, we can find a probability vector **q** that is unchanged by multiplication by A. Also, the other eigenvalues satisfy $|\lambda_j| < 1$, which means that all the trajectories get pulled in to the eigenspace E_1. Since $\mathbf{x}_k$ is a sequence of probability vectors, these vectors converge to the probability vector **q** as they are pulled into E_1.

4.5.2 Markov chains

If we have a stochastic matrix A and a probability vector $\mathbf{x}_0$, we can form the sequence $\mathbf{x}_k$ where $\mathbf{x}_{k+1} = A\mathbf{x}_k$. We call this sequence of vectors a *Markov chain*. Exercise 4.5.5.6 explains why we can guarantee that the vectors $\mathbf{x}_k$ are probability vectors.

In the example that studied voting patterns, we constructed a Markov chain that described how the percentages of voters choosing different parties changed from one election to the

4.5. MARKOV CHAINS AND GOOGLE'S PAGERANK ALGORITHM

next. We saw that the Markov chain converges to $\mathbf{q} = \begin{bmatrix} 0.2 \\ 0.4 \\ 0.4 \end{bmatrix}$, a probability vector in the eigenspace E_1. In other words, $\mathbf{q}$ is a probability vector that is unchanged under multiplication by A; that is, $A\mathbf{q} = \mathbf{q}$. This implies that, after a long time, 20% of voters choose party P, 40% choose Q, and 40% choose R.

Definition 4.5.2 If A is a stochastic matrix, we say that a probability vector $\mathbf{q}$ is a *steady-state* or *stationary* vector if $A\mathbf{q} = \mathbf{q}$.

An important question that arises from our previous example is

Question 4.5.3 If A is a stochastic matrix and $\mathbf{x}_k$ a Markov chain, does $\mathbf{x}_k$ converge to a steady-state vector?

Activity 4.5.3. Consider the matrices

$$A = \begin{bmatrix} 0 & 1 \\ 1 & 0 \end{bmatrix}, \quad B = \begin{bmatrix} 0.4 & 0.3 \\ 0.6 & 0.7 \end{bmatrix}.$$

a. Verify that both A and B are stochastic matrices.

b. Find the eigenvalues of A and then find a steady-state vector for A.

c. We will form the Markov chain beginning with the vector $\mathbf{x}_0 = \begin{bmatrix} 1 \\ 0 \end{bmatrix}$ and defining $\mathbf{x}_{k+1} = A\mathbf{x}_k$. The Sage cell below constructs the first N terms of the Markov chain with the command markov_chain(A, x0, N). Define the matrix A and vector x0 and evaluate the cell to find the first 10 terms of the Markov chain.

```
def markov_chain(A, x0, N):
    for i in range(N):
        x0 = A*x0
        print (x0)
## define the matrix A and x0
A =
x0 =
markov_chain(A, x0, 10)
```

What do you notice about the Markov chain? Does it converge to the steady-state vector for A?

d. Now find the eigenvalues of B along with a steady-state vector for B.

e. As before, find the first 10 terms in the Markov chain beginning with $\mathbf{x}_0 = \begin{bmatrix} 1 \\ 0 \end{bmatrix}$ and $\mathbf{x}_{k+1} = B\mathbf{x}_k$. What do you notice about the Markov chain? Does it converge to the steady-state vector for B?

f. What condition on the eigenvalues of a stochastic matrix will guarantee that a Markov chain will converge to a steady-state vector?

As this activity implies, the eigenvalues of a stochastic matrix tell us whether a Markov chain will converge to a steady-state vector. Here are a few important facts about the eigenvalues of a stochastic matrix.

- As is demonstrated in Exercise 4.5.5.8, $\lambda = 1$ is an eigenvalue of any stochastic matrix. We usually order the eigenvalues so it is the first eigenvalue meaning that $\lambda_1 = 1$.
- All other eigenvalues satisfy the property that $|\lambda_j| \leq 1$.
- Any stochastic matrix has at least one steady-state vector $\mathbf{q}$.

As illustrated in the activity, a Markov chain could fail to converge to a steady-state vector if $|\lambda_2| = 1$. This happens for the matrix $A = \begin{bmatrix} 0 & 1 \\ 1 & 0 \end{bmatrix}$, whose eigenvalues are $\lambda_1 = 1$ and $\lambda_2 = -1$.

However, if all but the first eigenvalue satisfy $|\lambda_j| < 1$, then there is a unique steady-state vector $\mathbf{q}$ and any Markov chain will converge to $\mathbf{q}$. This was the case for the matrix $B = \begin{bmatrix} 0.4 & 0.3 \\ 0.6 & 0.7 \end{bmatrix}$, whose eigenvalues are $\lambda_1 = 1$ and $\lambda_2 = 0.1$. In this case, any Markov chain will converge to the unique steady-state vector $\mathbf{q} = \begin{bmatrix} \frac{1}{3} \\ \frac{2}{3} \end{bmatrix}$.

In this way, we see that the eigenvalues of a stochastic matrix tell us whether a Markov chain will converge to a steady-state vector. However, it is somewhat inconvenient to compute the eigenvalues to answer this question. Is there some way to conclude that every Markov chain will converge to a steady-state vector without actually computing the eigenvalues? It turns out that there is a simple condition on the matrix A that guarantees this.

Definition 4.5.4 We say that a matrix A is *positive* if either A or some power A^k has all positive entries.

Example 4.5.5 The matrix $A = \begin{bmatrix} 0 & 1 \\ 1 & 0 \end{bmatrix}$ is not positive. We can see this because some of the entries of A are zero and therefore not positive. In addition, we see that $A^2 = I$, $A^3 = A$ and so forth. Therefore, every power of A also has some zero entries, which means that A is not positive.

The matrix $B = \begin{bmatrix} 0.4 & 0.3 \\ 0.6 & 0.7 \end{bmatrix}$ is positive because every entry of B is positive.

Also, the matrix $C = \begin{bmatrix} 0 & 0.5 \\ 1 & 0.5 \end{bmatrix}$ clearly has a zero entry. However, $C^2 = \begin{bmatrix} 0.5 & 0.25 \\ 0.5 & 0.75 \end{bmatrix}$, which has all positive entries. Therefore, we see that C is a positive matrix.

Positive matrices are important because of the following theorem.

Theorem 4.5.6 Perron-Frobenius. *If A is a positive stochastic matrix, then the eigenvalues satisfy $\lambda_1 = 1$ and $|\lambda_j| < 1$ for $j > 1$. This means that A has a unique positive, steady-state vector $\mathbf{q}$ and that every Markov chain defined by A will converge to $\mathbf{q}$.*

4.5. MARKOV CHAINS AND GOOGLE'S PAGERANK ALGORITHM

Activity 4.5.4. We will explore the meaning of the Perron-Frobenius theorem in this activity.

a. Consider the matrix $C = \begin{bmatrix} 0 & 0.5 \\ 1 & 0.5 \end{bmatrix}$. This is a positive matrix, as we saw in the previous example. Find the eigenvectors of C and verify there is a unique steady-state vector.

b. Using the Sage cell below, construct the Markov chain with initial vector $x_0 = \begin{bmatrix} 1 \\ 0 \end{bmatrix}$ and describe what happens to x_k as k becomes large.

```
def markov_chain(A, x0, N):
    for i in range(N):
        x0 = A*x0
        print (x0)
## define the matrix C and x0
C =
x0 =
markov_chain(C, x0, 10)
```

c. Construct another Markov chain with initial vector $x_0 = \begin{bmatrix} 0.2 \\ 0.8 \end{bmatrix}$ and describe what happens to x_k as k becomes large.

d. Consider the matrix $D = \begin{bmatrix} 0 & 0.5 & 0 \\ 1 & 0.5 & 0 \\ 0 & 0 & 1 \end{bmatrix}$ and compute several powers of D below.

Determine whether D is a positive matrix.

e. Find the eigenvalues of D and then find the steady-state vectors. Is there a unique steady-state vector?

f. What happens to the Markov chain defined by D with initial vector $x_0 = \begin{bmatrix} 1 \\ 0 \\ 0 \end{bmatrix}$?

What happens to the Markov chain with initial vector $x_0 = \begin{bmatrix} 0 \\ 0 \\ 1 \end{bmatrix}$.

g. Explain how the matrices C and D, which we have considered in this activity, relate to the Perron-Frobenius theorem.

4.5.3 Google's PageRank algorithm

Markov chains and the Perron-Frobenius theorem are the central ingredients in Google's PageRank algorithm, developed by Google to assess the quality of web pages.

Suppose we enter "linear algebra" into Google's search engine. Google responds by telling us there are 138 million web pages containing those terms. On the first page, however, there are links to ten web pages that Google judges to have the highest quality and to be the ones we are most likely to be interested in. How does Google assess the quality of web pages?

At the time this is being written, Google is tracking 35 trillion web pages. Clearly, this is too many for humans to evaluate. Plus, human evaluators may inject their own biases into their evaluations, perhaps even unintentionally. Google's idea is to use the structure of the Internet to assess the quality of web pages without any human intervention. For instance, if a web page has quality content, other web pages will link to it. This means that the number of links to a page reflect the quality of that page. In addition, we would expect a page to have even higher quality content if those links are coming from pages that are themselves assessed to have high quality. Simply said, if many quality pages link to a page, that page must itself be of high quality. This is the essence of the PageRank algorithm, which we introduce in the next activity.

Activity 4.5.5.

We will consider a simple model of the Internet that has three pages and links between them as shown here. For instance, page 1 links to both pages 2 and 3, but page 2 only links to page 1.

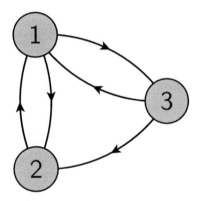

Figure 4.5.7 Our first Internet.

We will measure the quality of the jth page with a number x_j, which is called the PageRank of page j. The PageRank is determined by the following rule: each page divides its PageRank into equal pieces, one for each outgoing link, and gives one piece to each of the pages it links to. A page's PageRank is the sum of all the PageRank it receives from pages linking to it.

For instance, page 3 has two outgoing links. It therefore divides its PageRank x_3 in half and gives half to page 1. Page 2 has only one outgoing link so it gives all of its

PageRank x_2 to page 1. We therefore have

$$x_1 = x_2 + \frac{1}{2}x_3.$$

a. Find similar expressions for x_2 and x_3.

b. We now form the PageRank vector $\mathbf{x} = \begin{bmatrix} x_1 \\ x_2 \\ x_3 \end{bmatrix}$. Find a matrix G such that the expressions for x_1, x_2, and x_3 can be written in the form $G\mathbf{x} = \mathbf{x}$. The matrix G is called the "Google matrix".

c. Explain why G is a stochastic matrix.

d. Since $\mathbf{x}$ is defined by the equation $G\mathbf{x} = \mathbf{x}$, any vector in the eigenspace E_1 satisfies this equation. So that we might work with a specific vector, we will define the PageRank vector to be the steady-state vector of the stochastic matrix G. Find this steady state vector.

e. The PageRank vector $\mathbf{x}$ is composed of the PageRanks for each of the three pages. Which page of the three is assessed to have the highest quality? By referring to the structure of this small model of the Internet, explain why this is a good choice.

f. If we begin with the initial vector $\mathbf{x}_0 = \begin{bmatrix} 1 \\ 0 \\ 0 \end{bmatrix}$ and form the Markov chain $\mathbf{x}_{k+1} = G\mathbf{x}_k$, what does the Perron-Frobenius theorem tell us about the long-term behavior of the Markov chain?

g. Verify that this Markov chain converges to the steady-state PageRank vector.

```
def markov_chain(A, x0, N):
    for i in range(N):
        x0 = A*x0
        print (x0.numerical_approx(digits=3))
## define the matrix G and x0
G =
x0 =
markov_chain(G, x0, 20)
```

This activity shows us two ways to find the PageRank vector. In the first, we determine a steady-state vector directly by finding a description of the eigenspace E_1 and then finding the appropriate scalar multiple of a basis vector that gives us the steady-state vector. To find a description of the eigenspace E_1, however, we need to find the null space $\text{Nul}(G - I)$. Remember that the real Internet has 35 trillion pages so finding $\text{Nul}(G - I)$ requires us to row reduce a matrix with 35 trillion rows and columns. As we saw in Subsection 1.3.3, that is not computationally feasible.

As suggested by the activity, the second way to find the PageRank vector is to use a Markov chain that converges to the PageRank vector. Since multiplying a vector by a matrix is significantly less work than row reducing the matrix, this approach is computationally feasible, and it is, in fact, how Google computes the PageRank vector.

Activity 4.5.6. Consider the Internet with eight web pages, shown in Figure 4.5.8.

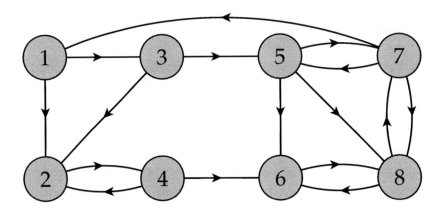

Figure 4.5.8 A simple model of the Internet with eight web pages.

a. Construct the Google matrix G for this Internet. Then use a Markov chain to find the steady-state PageRank vector **x**.

```
def markov_chain(A, x0, N):
    for i in range(N):
        x0 = A*x0
        print (x0.numerical_approx(digits=3))
## define the matrix G and x0
G = 
x0 = 
markov_chain(G, x0, 20)
```

b. What does this vector tell us about the relative quality of the pages in this Internet? Which page has the highest quality and which the lowest?

c. Now consider the Internet with five pages, shown in Figure 4.5.9.

4.5. MARKOV CHAINS AND GOOGLE'S PAGERANK ALGORITHM

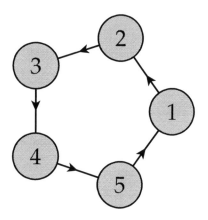

Figure 4.5.9 A model of the Internet with five web pages.

What happens when you begin the Markov chain with the vector $\mathbf{x}_0 = \begin{bmatrix} 1 \\ 0 \\ 0 \\ 0 \\ 0 \end{bmatrix}$?

Explain why this behavior is consistent with the Perron-Frobenius theorem.

d. What do you think the PageRank vector for this Internet should be? Is any one page of a higher quality than another?

e. Now consider the Internet with eight web pages, shown in Figure 4.5.10.

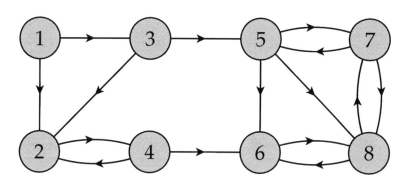

Figure 4.5.10 Another model of the Internet with eight web pages.

Notice that this version of the Internet is identical to the first one that we saw in this activity, except that a single link from page 7 to page 1 has been removed. We can therefore find its Google matrix G by slightly modifying the earlier matrix.

What is the long-term behavior of a Markov chain defined by G and why is this behavior not desirable? How is this behavior consistent with the Perron-Frobenius theorem?

The Perron-Frobenius theorem Theorem 4.5.6 tells us that a Markov chain $\mathbf{x}_{k+1} = G\mathbf{x}_k$ converges to a unique steady-state vector when the matrix G is positive. This means that G or some power of G should have only positive entries. Clearly, this is not the case for the matrix formed from the Internet in Figure 4.5.9.

We can understand the problem with the Internet shown in Figure 4.5.10 by adding a box around some of the pages as shown in Figure 4.5.11. Here we see that the pages outside of the box give up all of their PageRank to the pages inside the box. This is not desirable because the PageRanks of the pages outside of the box are found to be zero. Once again, the Google matrix G is not a positive matrix.

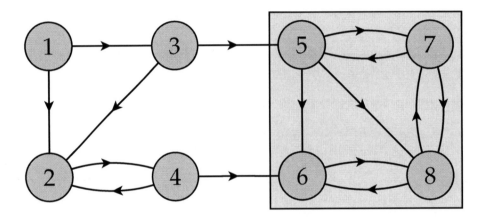

Figure 4.5.11 The pages outside the box give up all of their PageRank to the pages inside the box.

Google solves this problem by slightly modifying the Google matrix G to obtain a positive matrix G'. To understand this, think of the entries in the Google matrix as giving the probability that an Internet user follows a link from one page of another. To create a positive matrix, we will allow that user to randomly jump to any other page on the Internet with a small probability.

To make sense of this, suppose that there are N pages on our internet. The matrix

$$H_n = \begin{bmatrix} \frac{1}{n} & \frac{1}{n} & \cdots & \frac{1}{n} \\ \frac{1}{n} & \frac{1}{n} & \cdots & \frac{1}{n} \\ \vdots & \vdots & \ddots & \vdots \\ \frac{1}{n} & \frac{1}{n} & \cdots & \frac{1}{n} \end{bmatrix}$$

is a positive stochastic matrix describing a process where we can move from any page to another with equal probability. To form the modified Google matrix G', we choose a parameter

4.5. MARKOV CHAINS AND GOOGLE'S PAGERANK ALGORITHM

α that is used to mix G and H_n together; that is, G' is the positive stochastic matrix

$$G' = \alpha G + (1 - \alpha)H_n.$$

In practice, it is thought that Google uses a value of $\alpha = 0.85$ (Google doesn't publish this number as it is a trade secret) so that we have

$$G' = 0.85G + 0.15H_n.$$

Intuitively, this means that an Internet user will randomly follow a link from one page to another 85% of the time and will randomly jump to any other page on the Internet 15% of the time. Since the matrix G' is positive, the Perron-Frobenius theorem tells us that any Markov chain will converge to a unique steady-state vector that we call the PageRank vector.

Activity 4.5.7. The following Sage cell will generate the Markov chain for the modified Google matrix G if you simply enter the original Google matrix G in the appropriate line.

```
def modified_markov_chain(A, x0, N):
    r = A.nrows()
    A = 0.85*A + 0.15*matrix(r,r,[1.0/r]*(r*r))
    for i in range(N):
        x0 = A*x0
        print (x0.numerical_approx(digits=3))
## Define original Google matrix G and initial vector x0.
## The function above finds the modified Google matrix
## and resulting Markov chain
G =
x0 =
modified_markov_chain(G, x0, 20)
```

a. Consider the original Internet with three pages shown in Figure 4.5.7 and find the PageRank vector **x** using the modified Google matrix in the Sage cell above. How does this modified PageRank vector compare to the vector we found using the original Google matrix G?

b. Find the modified PageRank vector for the Internet shown in Figure 4.5.9. Explain why this vector seems to be the correct one.

c. Find the modified PageRank vector for the Internet shown in Figure 4.5.10. Explain why this modified PageRank vector fixes the problem that appeared with the original PageRank vector.

The ability to access almost anything we want to know through the Internet is something we take for granted in today's society. Without Google's PageRank algorithm, however, the Internet would be a chaotic place indeed; imagine trying to find a useful web page among the 30 trillion available pages without it. (There are, of course, other search algorithms, but Google's is the most widely used.) The fundamental role that Markov chains and the Perron-Frobenius theorem play in Google's algorithm demonstrates the vast power that mathematics has to shape our society.

4.5.4 Summary

This section explored stochastic matrices and Markov chains.

- A probability vector is one whose entries are nonnegative and whose columns add to 1. A stochastic matrix is a square matrix whose columns are probability vectors.

- A Markov chain is formed from a stochastic matrix A and an initial probability vector x_0 using the rule $x_{k+1} = Ax_k$. We may think of the sequence x_k as describing the evolution of some conserved quantity, such as the number of rental cars or voters, among a number of possible states over time.

- A steady-state vector q for a stochastic matrix A is a probability vector that satisfies $Aq = q$.

- The Perron-Frobenius theorem tells us that, if A is a positive stochastic matrix, then every Markov chain defined by A converges to a unique, positive steady-state vector.

- Google's PageRank algorithm uses Markov chains and the Perron-Frobenius theorem to assess the relative quality of web pages on the Internet.

4.5.5 Exercises

1. Consider the following 2×2 stochastic matrices.

   For each, make a copy of the diagram and label each edge to indicate the probability of that transition. Then find all the steady-state vectors and describe what happens to a Markov chain defined by that matrix.

   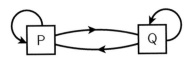

   a. $\begin{bmatrix} 1 & 1 \\ 0 & 0 \end{bmatrix}$.

   b. $\begin{bmatrix} 0.8 & 1 \\ 0.2 & 0 \end{bmatrix}$.

   c. $\begin{bmatrix} 1 & 0 \\ 0 & 1 \end{bmatrix}$.

   d. $\begin{bmatrix} 0.7 & 0.6 \\ 0.3 & 0.4 \end{bmatrix}$.

2. Every year, people move between urban (U), suburban (S), and rural (R) populations with the probabilities given in Figure 4.5.12.

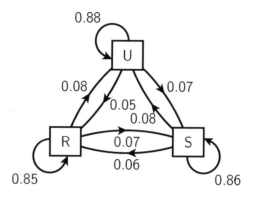

Figure 4.5.12 The flow between urban, suburban, and rural populations.

a. Construct the stochastic matrix A describing the movement of people.

b. Explain what the Perron-Frobenius theorem tells us about the existence of a steady-state vector **q** and the behavior of a Markov chain.

c. Use the Sage cell below to find the some terms of a Markov chain.

```
def markov_chain(A, x0, N):
    for i in range(N):
        x0 = A*x0
        print (x0.numerical_approx(digits=3))
## define the matrix G and x0
A =
x0 =
markov_chain(A, x0, 20)
```

d. Describe the long-term distribution of people among urban, suburban, and rural populations.

3. Determine whether the following statements are true or false and provide a justification of your response.

   a. Every stochastic matrix has a steady-state vector.

   b. If A is a stochastic matrix, then any Markov chain defined by A converges to a steady-state vector.

   c. If A is a stochastic matrix, then $\lambda = 1$ is an eigenvalue and all the other eigenvalues satisfy $|\lambda| < 1$.

   d. A positive stochastic matrix has a unique steady-state vector.

   e. If A is an invertible stochastic matrix, then so is A^{-1}.

4. Consider the stochastic matrix

$$A = \begin{bmatrix} 1 & 0.2 & 0.2 \\ 0 & 0.6 & 0.2 \\ 0 & 0.2 & 0.6 \end{bmatrix}.$$

   a. Find the eigenvalues of A.

   b. Do the conditions of the Perron-Frobenius theorem apply to this matrix?

   c. Find the steady-state vectors of A.

   d. What can we guarantee about the long-term behavior of a Markov chain defined by the matrix A?

5. Explain your responses to the following.

   a. Why does Google use a Markov chain to compute the PageRank vector?

   b. Describe two problems that can happen when Google constructs a Markov chain using the Google matrix G.

   c. Describe how these problems are consistent with the Perron-Frobenius theorem.

   d. Describe why the Perron-Frobenius theorem suggests creating a Markov chain using the modified Google matrix $G' = \alpha G + (1 - \alpha)H_n$.

In the next few exercises, we will consider the $1 \times n$ matrix $S = \begin{bmatrix} 1 & 1 & \cdots & 1 \end{bmatrix}$.

6. Suppose that A is a stochastic matrix and that $\mathbf{x}$ is a probability vector. We would like to explain why the product $A\mathbf{x}$ is a probability vector.

   a. Explain why $\mathbf{x} = \begin{bmatrix} 0.4 \\ 0.5 \\ 0.1 \end{bmatrix}$ is a probability vector and then find the product $S\mathbf{x}$.

   b. More generally, if $\mathbf{x}$ is any probability vector, what is the product $S\mathbf{x}$?

   c. If A is a stochastic matrix, explain why $SA = S$.

   d. Explain why $A\mathbf{x}$ is a probability vector by considering the product $SA\mathbf{x}$.

7. Using the results of the previous exercise, we would like to explain why A^2 is a stochastic matrix if A is stochastic.

   a. Suppose that A and B are stochastic matrices. Explain why the product AB is a stochastic matrix by considering the product SAB.

   b. Explain why A^2 is a stochastic matrix.

   c. How do the steady-state vectors of A^2 compare to the steady-state vectors of A?

8. This exercise explains why $\lambda = 1$ is an eigenvalue of a stochastic matrix A. To conclude that $\lambda = 1$ is an eigenvalue, we need to know that $A - I$ is not invertible.

   a. What is the product $S(A - I)$?

4.5. MARKOV CHAINS AND GOOGLE'S PAGERANK ALGORITHM

b. What is the product $S\mathbf{e}_1$?

c. Consider the equation $(A - I)\mathbf{x} = \mathbf{e}_1$. Explain why this equation cannot be consistent by multiplying by S to obtain $S(A - I)\mathbf{x} = S\mathbf{e}_1$.

d. What can you say about the span of the columns of $A - I$?

e. Explain why we can conclude that $A - I$ is not invertible and that $\lambda = 1$ is an eigenvalue of A.

9. We saw a couple of model Internets in which a Markov chain defined by the Google matrix G did not converge to an appropriate PageRank vector. For this reason, Google defines the matrix

$$H_n = \begin{bmatrix} \frac{1}{n} & \frac{1}{n} & \cdots & \frac{1}{n} \\ \frac{1}{n} & \frac{1}{n} & \cdots & \frac{1}{n} \\ \vdots & \vdots & \ddots & \vdots \\ \frac{1}{n} & \frac{1}{n} & \cdots & \frac{1}{n} \end{bmatrix},$$

where n is the number of web pages, and constructs a Markov chain from the modified Google matrix

$$G' = \alpha G + (1 - \alpha)H_n.$$

Since G' is positive, the Markov chain is guaranteed to converge to a unique steady-state vector.

We said that Google chooses $\alpha = 0.85$ so we might wonder why this is a good choice. We will explore the role of α in this exercise. Let's consider the model Internet described in Figure 4.5.9 and construct the Google matrix G. In the Sage cell below, you can enter the matrix G and choose a value for α.

```
def modified_markov_chain(A, x0, N):
    r = A.nrows()
    A = alpha*A + (1-alpha)*matrix(r,r,[1.0/r]*(r*r))
    for i in range(N):
        x0 = A*x0
        print (x0.numerical_approx(digits=3))
## Define the matrix original Google matrix G and choose alpha.
## The function above finds the modified Google matrix
## and resulting Markov chain
alpha = 0
G =
x0 = vector([1,0,0,0,0])
modified_markov_chain(G, x0, 20)
```

a. Let's begin with $\alpha = 0$. With this choice, what is the matrix $G' = \alpha G + (1 - \alpha)H_n$? Construct a Markov chain using the Sage cell above. How many steps are required for the Markov chain to converge to the accuracy with which the vectors $\mathbf{x}_k$ are displayed?

b. Now choose $\alpha = 0.25$. How many steps are required for the Markov chain to converge to the accuracy at which the vectors $\mathbf{x}_k$ are displayed?

c. Repeat this experiment with $\alpha = 0.5$ and $\alpha = 0.75$.

d. What happens if $\alpha = 1$?

This experiment gives some insight into the choice of α. The smaller α is, the faster the Markov chain converges. This is important; since the matrix G' that Google works with is so large, we would like to minimize the number of terms in the Markov chain that we need to compute. On the other hand, as we lower α, the matrix $G' = \alpha G + (1-\alpha)H_n$ begins to resemble H_n more and G less. The value $\alpha = 0.85$ is chosen so that the matrix G' sufficiently resembles G while having the Markov chain converge in a reasonable amount of steps.

10. This exercise will analyze the board game *Chutes and Ladders*, or at least a simplified version of it.

    The board for this game consists of 100 squares arranged in a 10×10 grid and numbered 1 to 100. There are pairs of squares joined by a ladder and pairs joined by a chute. All players begin in square 1 and take turns rolling a die. On their turn, a player will move ahead the number of squares indicated on the die. If they arrive at a square at the bottom of a ladder, they move to the square at the top of the ladder. If they arrive at a square at the top of a chute, they move down to the square at the bottom of the chute. The winner is the first player to reach square 100.

    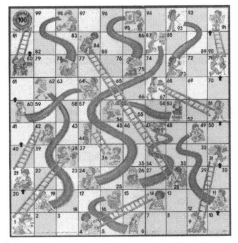

    a. We begin by playing a simpler version of this game with only eight squares laid out in a row as shown in Figure 4.5.13 and containing neither chutes nor ladders. Rather than a six-sided die, we will toss a coin and move ahead one or two squares depending on the result of the coin toss. If we are on square 7, we move ahead to square 8 regardless of the coin flip, and if we are on square 8, we will stay there forever.

    Figure 4.5.13 A simple version of Chutes and Ladders with neither chutes nor ladders.

    Construct the 8×8 matrix A that records the probability that a player moves from one square to another on one move. For instance, if a player is on square 2, there is a 50% chance they move to square 3 and a 50% chance they move to square 4 on the next move.

    Since we begin the game on square 1, the initial vector $\mathbf{x}_0 = \mathbf{e}_1$. Generate a few terms of the Markov chain $\mathbf{x}_{k+1} = A\mathbf{x}_k$.

4.5. MARKOV CHAINS AND GOOGLE'S PAGERANK ALGORITHM

```
def markov_chain(A, x0, N):
    for i in range(N):
        x0 = A*x0
        print (x0.numerical_approx(digits=3))
## define the matrix A and x0
A =
x0 =
markov_chain(A, x0, 10)
```

What is the probability that we arrive at square 8 by the fourth move? By the sixth move? By the seventh move?

b. We will now modify the game by adding one chute and one ladder as shown in Figure 4.5.14.

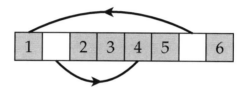

Figure 4.5.14 A version of Chutes and Ladders with one chute and one ladder.

Even though there are eight squares, we only need to consider six of them. For instance, if we arrive at the first white square, we move up to square 4. Similarly, if we arrive at the second white square, we move down to square 1.

Once again, construct the 6 × 6 stochastic matrix that records the probability that we move from one square to another on a given turn and generate some terms in the Markov chain that begins with $x_0 = e_1$.

```
def markov_chain(A, x0, N):
    for i in range(N):
        x0 = A*x0
        print (x0.numerical_approx(digits=3))
## define the matrix A and x0
A =
x0 =
markov_chain(A, x0, 10)
```

1. What is the smallest number of moves we can make and arrive at square 6? What is the probability that we arrive at square 6 using this number of moves?

2. What is the probability that we arrive at square 6 after five moves?

3. What is the probability that we are still on square 1 after five moves? After seven moves? After nine moves?

4. After how many moves do we have a 90% chance of having arrived at square 6?

5. Find the steady-state vector and discuss what this vector implies about the game.

One can analyze the full version of Chutes and Ladders having 100 squares in the same way. Without any chutes or ladders, one finds that the average number of moves required to reach square 100 is 29.0. Once we add the chutes and ladders back in, the average number of moves required to reach square 100 is 27.1. This shows that the average number of moves does not change significantly when we add the chutes and ladders. There is, however, much more variation in the possibilities because it is possible to reach square 100 much more quickly and much more slowly.

CHAPTER 5

Linear algebra and computing

Our principal tool for finding solutions to linear systems has been Gaussian elimination, which we first met back in Section 1.2. When presented with a linear system, we frequently find the reduced row echelon form of the system's augmented matrix to read off the solution.

While this is a convenient approach for learning linear algebra, people rarely use the reduced row echelon form of a matrix. In fact, many linear algebra software packages do not include functions for finding the reduced row echelon form. In this chapter, we will describe why this is the case and then explore some alternatives. The intent of this chapter is to demonstrate how linear algebraic computations are handled in practice. More specifically, we will improve our techniques for solving linear systems and for finding eigenvectors through Gaussian elimination.

5.1 Gaussian elimination revisited

In this section, we revisit Gaussian elimination and explore some problems with implementing it in the straightforward way that we described back in Section 1.2. In particular, we will see how the fact that computers only approximate arithmetic operations can lead us to find solutions that are far from the actual solutions. Second, we will explore how much work is required to implement Gaussian elimination and devise a more efficient means of implementing it when we want to solve equations $A\mathbf{x} = \mathbf{b}$ for several different vectors $\mathbf{b}$.

> **Preview Activity 5.1.1.** To begin, let's recall how we implemented Gaussian elimination by considering the matrix
>
> $$A = \begin{bmatrix} 1 & 2 & -1 & 2 \\ 1 & 0 & -2 & 1 \\ 3 & 2 & 1 & 0 \end{bmatrix}$$
>
> a. What is the first row operation we perform? If the resulting matrix is A_1, find a matrix E_1 such that $E_1 A = A_1$.
>
> b. What is the matrix inverse E_1^{-1}? You can find this using your favorite technique for finding a matrix inverse. However, it may be easier to think about the effect that the row operation has and how it can be undone.

c. Perform the next two steps in the Gaussian elimination algorithm to obtain A_3. Represent these steps using multiplication by matrices E_2 and E_3 so that

$$E_3 E_2 E_1 A = A_3.$$

d. Suppose we need to scale the second row by -2. What is the 3×3 matrix that perfoms this row operation by left multiplication?

e. Suppose that we need to interchange the first and second rows. What is the 3×3 matrix that performs this row operation by left multiplication?

5.1.1 Partial pivoting

The first issue that we address is the fact that computers do not perform arithemtic operations exactly. For instance, Python will evaluate `0.1 + 0.2` and report `0.30000000000000004` even though we know that the true value is 0.3. There are a couple of reasons for this.

First, computers perform arithmetic using base 2 numbers, which means that numbers we enter in decimal form, such as 0.1, must be converted to base 2. Even though 0.1 has a simple decimal form, its representation in base 2 is the repeating decimal

$$0.00011001100110011001100110011001100110011\ldots,$$

To accurately represent this number inside a computer would require infinitely many digits. Since a computer can only hold a finite number of digits, we are necessarily using an approximation just by representing this number in a computer.

In addition, arithmetic operations, such as addition, are prone to error. To keep things simple, suppose we have a computer that represents numbers using only three decimal digits. For instance, the number 1.023 would be represented as `1.02` while 0.023421 would be `0.0234`. If we add these numbers, we have $1.023 + 0.023421 = 1.046421$; the computer reports this sum as `1.02 + 0.0234 = 1.04`, whose last digit is not correctly rounded. Generally speaking, we will see this problem, which is called *round off error*, whenever we add numbers of signficantly different magnitudes.

Remember that Gaussian elimination, when applied to an $n \times n$ matrix, requires approximately $\frac{2}{3}n^3$ operations. If we have a 1000×1000 matrix, performing Gaussian elimination requires roughly a billion operations, and the errors introduced in each operation could accumulate. How can we have confidence in the final result? We can never completely avoid these errors, but we can take steps to mitigate them. The next activity will introduce one such technique.

Activity 5.1.2. Suppose we have a hypothetical computer that represents numbers using only three decimal digits. We will consider the linear system

$$0.0001x + y = 1$$
$$x + y = 2.$$

5.1. GAUSSIAN ELIMINATION REVISITED

a. Show that this system has the unique solution

$$x = \frac{10000}{9999} = 1.00010001\ldots,$$
$$y = \frac{9998}{9999} = 0.99989998\ldots.$$

b. If we represent this solution inside our computer that only holds 3 decimal digits, what do we find for the solution? This is the best that we can hope to find using our computer.

c. Let's imagine that we use our computer to find the solution using Gaussian elimination; that is, after every arithmetic operation, we keep only three decimal digits. Our first step is to multiply the first equation by 10000 and subtract it from the second equation. If we represent numbers using only three decimal digits, what does this give for the value of y?

d. By substituting our value for y into the first equation, what do we find for x?

e. Compare the solution we find on our computer with the actual solution and assess the quality of the approximation.

f. Let's now modify the linear system by simplying interchanging the equations:

$$x + y = 2$$
$$0.0001x + y = 1.$$

Of course, this doesn't change the actual solution. Let's imagine we use our computer to find the solution using Gaussian elimination. Perform the first step where we multiply the first equation by 0.0001 and subtract from the second equation. What does this give for y if we represent numbers using only three decimal digits?

g. Substitute the value you found for y into the first equation and solve for x. Then compare the approximate solution found with our hypothetical computer to the exact solution.

h. Which approach produces the most accurate approximation?

This activity demonstrates how the practical aspects of computing differ from the theoretical. We know that the order in which we write the equations has no effect on the solution space; row interchange is one of our three allowed row operations in the Gaussian elimination algorithm. However, when we are only able to perform arithmetic operations approximately, applying row interchanges can dramatically improve the accuracy of our approximations.

If we could compute the solution exactly, we find

$$x = 1.00010001\ldots, \qquad y = 0.99989998\ldots.$$

Since our hypothetical computer represents numbers using only three decimal digits, our

computer finds
$$x \approx 1.00, \qquad y \approx 1.00.$$
This is the best we can hope to do with our computer since it is impossible to represent the solution exactly.

When the equations are written in their original order and we multiply the first equation by 10000 and subtract from the second, we find
$$(1 - 10000)y = 2 - 10000$$
$$-9999y = -9998$$
$$-10000y \approx -10000$$
$$y \approx 1.00.$$

In fact, we find the same value for y when we interchange the equations. Here we multiply the first equation by 0.0001 and subtract from the second equation. We then find
$$(1 - 0.0001)y = 2 - 0.0001$$
$$-0.9999y = -0.9998$$
$$-y \approx -1.00$$
$$y \approx 1.00.$$

The difference occurs when we substitute $y \approx 1$ into the first equation. When the equations are written in their original order, we have
$$0.0001x + 1.00 \approx 1.00$$
$$0.0001x \approx 0.00$$
$$x \approx 0.00.$$

When the equations are written in their original order, we find the solution $x \approx 0.00$, $y \approx 1.00$.

When we write the equation in the opposite order, however, substituting $y \approx 1$ into the first equation gives
$$x + 1.00 \approx 2.00$$
$$x \approx 1.00.$$

In this case, we find the approximate solution $x \approx 1.00$, $y \approx 1.00$, which is the most accurate solution that our hypothetical computer can find. Simply interchanging the order of the equation produces a much more accurate solution.

5.1. GAUSSIAN ELIMINATION REVISITED

We can understand why this works graphically. Each equation represents a line in the plane, and the solution is the intersection point. Notice that the slopes of these lines differ considerably.

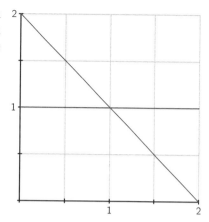

When the equations are written in their original order, we substitute $y \approx 1$ into the equation $0.00001x + y = 1$, which is a nearly horizontal line. Along this line, a small change in y leads to a large change in x. The slight difference in our approximation $y \approx 1$ from the exact value $y = 0.9998999\ldots$ leads to a large difference in the approximation $x \approx 0$ from the exact value $x = 1.00010001\ldots$.

If we exchange the order in which the equations are written, we substitute our approximation $y \approx 1$ into the equation $x + y = 2$. Notice that the slope of the associated line is -1. On this line, a small change in y leads to a relatively small change in x as well. Therefore, the difference in our approximation $y \approx 1$ from the exact value leads to only a small difference in the approximation $x \approx 1$ from the exact value.

This example motivates the technique that computers usually use to perform Gaussian elimination. We only need to perform a row interchange when a zero occurs in a pivot position, such as

$$\begin{bmatrix} 1 & -1 & 2 & 2 \\ 0 & 0 & -3 & 1 \\ 0 & 2 & 2 & -3 \end{bmatrix}.$$

However, we will perform a row interchange to put the entry having the largest possible absolute value into the pivot position. For instance, when performing Gaussian elimination on the following matrix, we begin by interchanging the first and third rows so that the upper left entry has the largest possible absolute value.

$$\begin{bmatrix} 2 & 1 & 2 & 3 \\ 1 & -3 & -2 & 1 \\ -3 & 2 & 3 & -2 \end{bmatrix} \sim \begin{bmatrix} -3 & 2 & 3 & -2 \\ 1 & -3 & -2 & 1 \\ 2 & 1 & 2 & 3 \end{bmatrix}.$$

This technique is called *partial pivoting*, and it means that, in practice, we will perform many more row interchange operations than we typically do when computing exactly by hand.

5.1.2 *LU* factorizations

In Subsection 1.3.3, we saw that the number of arithmetic operations needed to perform Gaussian elimination on an $n \times n$ matrix is about $\frac{2}{3}n^3$. This means that a 1000×1000 matrix, requires about two thirds of a billion operations.

Suppose that we have two equations, $A\mathbf{x} = \mathbf{b}_1$ and $A\mathbf{x} = \mathbf{b}_2$, that we would like to solve. Usually, we would form augmented matrices $\begin{bmatrix} A & | & \mathbf{b}_1 \end{bmatrix}$ and $\begin{bmatrix} A & | & \mathbf{b}_2 \end{bmatrix}$ and apply Gaussian elimination. Of course, the steps we perform in these two computations are nearly identical. Is there a way to store some of the computation we perform in reducing $\begin{bmatrix} A & | & \mathbf{b}_1 \end{bmatrix}$ and reuse it in solving subsequent equations? The next activity will point us in the right direction.

Activity 5.1.3. We will consider the matrix

$$A = \begin{bmatrix} 1 & 2 & 1 \\ -2 & -3 & -2 \\ 3 & 7 & 4 \end{bmatrix}$$

and begin performing Gaussian elimination without using partial pivoting.

a. Perform two row replacement operations to find the row equivalent matrix

$$A' = \begin{bmatrix} 1 & 2 & 1 \\ 0 & 1 & 0 \\ 0 & 1 & 1 \end{bmatrix}.$$

Find elementary matrices E_1 and E_2 that perform these two operations so that $E_2 E_1 A = A'$.

b. Perform a third row replacement to find the upper triangular matrix

$$U = \begin{bmatrix} 1 & 2 & 1 \\ 0 & 1 & 0 \\ 0 & 0 & 1 \end{bmatrix}.$$

Find the elementary matrix E_3 such that $E_3 E_2 E_1 A = U$.

c. We can write $A = E_1^{-1} E_2^{-1} E_3^{-1} U$. Find the inverse matrices E_1^{-1}, E_2^{-1}, and E_3^{-1} and the product $L = E_1^{-1} E_2^{-1} E_3^{-1}$. Then verify that $A = LU$.

d. Suppose that we want to solve the equation $A\mathbf{x} = \mathbf{b} = \begin{bmatrix} 4 \\ -7 \\ 12 \end{bmatrix}$. We will write

$$A\mathbf{x} = LU\mathbf{x} = L(U\mathbf{x}) = \mathbf{b}$$

and introduce an unknown vector $\mathbf{c}$ such that $U\mathbf{x} = \mathbf{c}$. Find $\mathbf{c}$ by noting that $L\mathbf{c} = \mathbf{b}$ and solving this equation.

e. Now that we have found $\mathbf{c}$, find $\mathbf{x}$ by solving $U\mathbf{x} = \mathbf{c}$.

f. Using the factorization $A = LU$ and this two-step process, solve the equation

$$A\mathbf{x} = \begin{bmatrix} 2 \\ -2 \\ 7 \end{bmatrix}.$$

5.1. GAUSSIAN ELIMINATION REVISITED

This activity introduces a method for factoring a matrix A as a product of two triangular matrices, $A = LU$, where L is lower triangular and U is upper triangular. The key to finding this factorization is to represent the row operations that we apply in the Gaussian elimination algorithm through multiplication by elementary matrices.

Example 5.1.1 Suppose we have the equation

$$\begin{bmatrix} 2 & -3 & 1 \\ -4 & 5 & 0 \\ 2 & -2 & 2 \end{bmatrix} \mathbf{x} = \begin{bmatrix} 8 \\ -13 \\ 8 \end{bmatrix},$$

which we write in the form $A\mathbf{x} = \mathbf{b}$. We begin by applying the Gaussian elimination algorithm to find an LU factorization of A.

The first step is to multiply the first row of A by 2 and add it to the second row. The elementary matrix

$$E_1 = \begin{bmatrix} 1 & 0 & 0 \\ 2 & 1 & 0 \\ 0 & 0 & 0 \end{bmatrix}$$

performs this operation so that $E_1 A = \begin{bmatrix} 2 & -3 & 1 \\ 0 & -1 & 2 \\ 2 & -2 & 2 \end{bmatrix}$.

We next apply matrices

$$E_2 = \begin{bmatrix} 1 & 0 & 0 \\ 0 & 1 & 0 \\ -1 & 0 & 1 \end{bmatrix}, \quad E_3 = \begin{bmatrix} 1 & 0 & 0 \\ 0 & 1 & 0 \\ 0 & 1 & 1 \end{bmatrix}$$

to obtain the upper triangular matrix $U = E_3 E_2 E_1 A = \begin{bmatrix} 2 & -3 & 1 \\ 0 & -1 & 2 \\ 0 & 0 & 3 \end{bmatrix}$.

We can write $U = (E_3 E_2 E_1) A$, which tells us that

$$A = (E_3 E_2 E_1)^{-1} U = \begin{bmatrix} 1 & 0 & 0 \\ -2 & 1 & 0 \\ 1 & -1 & 1 \end{bmatrix} U = LU.$$

That is, we have

$$A = LU = \begin{bmatrix} 1 & 0 & 0 \\ -2 & 1 & 0 \\ 1 & -1 & 1 \end{bmatrix} \begin{bmatrix} 2 & -3 & 1 \\ 0 & -1 & 2 \\ 0 & 0 & 3 \end{bmatrix}.$$

Notice that the matrix L is lower triangular, a result of the fact that the elementary matrices E_1, E_2, and E_3 are lower triangular.

Now that we have factored $A = LU$ into two triangular matrices, we can solve the equation $A\mathbf{x} = \mathbf{b}$ by solving two triangular systems. We write

$$A\mathbf{x} = L(U\mathbf{x}) = \mathbf{b}$$

and define the unknown vector $c = Ux$, which is determined by the equation $Lc = b$. Because L is lower triangular, we find the solution using forward substitution, $c = \begin{bmatrix} 8 \\ 3 \\ 3 \end{bmatrix}$. Finally, we find x, the solution to our original system $Ax = b$, by applying back substitution to solve $Ux = c$. This gives $x = \begin{bmatrix} 2 \\ -1 \\ 1 \end{bmatrix}$.

If we want to solve $Ax = b$ for a different right-hand side b, we can simply repeat this two-step process.

An LU factorization allow us to trade in one equation $Ax = b$ for two simpler equations

$$Lc = b$$
$$Ux = c.$$

For instance, the equation $Lc = b$ in our example has the form

$$\begin{bmatrix} 1 & 0 & 0 \\ -2 & 1 & 0 \\ 1 & -1 & 1 \end{bmatrix} c = \begin{bmatrix} 8 \\ -13 \\ 8 \end{bmatrix}.$$

Because L is a lower-triangular matrix, we can read off the first component of c directly from the equations: $c_1 = 8$. We then have $-2c_1 + c_2 = -13$, which gives $c_2 = 3$, and $c_1 - c_2 + c_3 = 8$, which gives $c_3 = 3$. Solving a triangular system is simplified because we only need to perform a sequence of substitutions.

In fact, solving an equation with an $n \times n$ triangular matrix requires approximately $\frac{1}{2}n^2$ operations. Once we have the factorization $A = LU$, we solve the equation $Ax = b$ by solving two equations involving triangular matrices, which requires about n^2 operations. For example, if A is a 1000×1000 matrix, we solve the equation $Ax = b$ using about one million steps. The compares with roughly a billion operations needed to perform Gaussian elimination, which represents a significant savings. Of course, we have to first find the LU factorization of A and this requires roughly the same amount of work as performing Gaussian elimination. However, once we have the LU factorization, we can use it to solve $Ax = b$ for different right hand sides b.

Our discussion so far has ignored one issue, however. Remember that we sometimes have to perform row interchange operations in addition to row replacement. A typical row interchange is represented by multiplication by a matrix such as

$$P = \begin{bmatrix} 0 & 0 & 1 \\ 0 & 1 & 0 \\ 1 & 0 & 0 \end{bmatrix},$$

which has the effect of interchanging the first and third rows. Notice that this matrix is not triangular so performing a row interchange will disrupt the structure of the LU factorization we seek. Without giving the details, we simply note that linear algebra software packages provide a matrix P that describes how the rows are permuted in the Gaussian elimination

5.1. GAUSSIAN ELIMINATION REVISITED

process. In particular, we will write $PA = LU$, where P is a permutation matrix, L is lower triangular, and U is upper triangular.

Therefore, to solve the equation $A\mathbf{x} = \mathbf{b}$, we first multiply both sides by P to obtain

$$PA\mathbf{x} = LU\mathbf{x} = P\mathbf{b}.$$

That is, we multiply $\mathbf{b}$ by P and then find $\mathbf{x}$ using the factorization: $L\mathbf{c} = P\mathbf{b}$ and $U\mathbf{x} = \mathbf{c}$.

Activity 5.1.4. Sage will create LU factorizations; once we have a matrix A, we write P, L, U = A.LU() to obtain the matrices P, L, and U such that $PA = LU$.

a. In Example 5.1.1, we found the LU factorization

$$A = \begin{bmatrix} 2 & -3 & 1 \\ -4 & 5 & 0 \\ 2 & -2 & 2 \end{bmatrix} = \begin{bmatrix} 1 & 0 & 0 \\ -2 & 1 & 0 \\ 1 & -1 & 1 \end{bmatrix} \begin{bmatrix} 2 & -3 & 1 \\ 0 & -1 & 2 \\ 0 & 0 & 3 \end{bmatrix} = LU.$$

Using Sage, define the matrix A, and then ask Sage for the LU factorization. What are the matrices P, L, and U?

Notice that Sage finds a different LU factorization than we found in the previous activity because Sage uses partial pivoting, as described in the previous section, when it performs Gaussian elimination.

b. Define the vector $\mathbf{b} = \begin{bmatrix} 8 \\ -13 \\ 8 \end{bmatrix}$ in Sage and compute $P\mathbf{b}$.

c. Use the matrices L and U to solve $L\mathbf{c} = P\mathbf{b}$ and $U\mathbf{x} = \mathbf{c}$. You should find the same solution $\mathbf{x}$ that you found in the previous activity.

d. Use the factorization to solve the equation $A\mathbf{x} = \begin{bmatrix} 9 \\ -16 \\ 10 \end{bmatrix}$.

e. How does the factorization show us that A is invertible and that, therefore, every equation $A\mathbf{x} = \mathbf{b}$ has a unique solution?

f. Suppose that we have the matrix

$$B = \begin{bmatrix} 3 & -1 & 2 \\ 2 & -1 & 1 \\ 2 & 1 & 3 \end{bmatrix}.$$

Use Sage to find the LU factorization. Explain how the factorization shows that B is not invertible.

g. Consider the matrix
$$C = \begin{bmatrix} -2 & 1 & 2 & -1 \\ 1 & -1 & 0 & 2 \\ 3 & 2 & -1 & 0 \end{bmatrix}$$
and find its LU factorization. Explain why C and U have the same null space and use this observation to find a basis for $\text{Nul}(A)$.

5.1.3 Summary

We returned to Gaussian elimination, which we have used as a primary tool for finding solutions to linear systems, and explored its practicality, both in terms of numerical accuracy and computational effort.

- We saw that the accuracy of computations implemented on a computer could be improved using *partial pivoting*, a technique that performs row interchanges so that the entry in a pivot position has the largest possible magnitude.

- Beginning with a matrix A, we used the Gaussian elimination algorithm to write $PA = LU$, where P is a permutation matrix, L is lower triangular, and U is upper triangular.

- Finding this factorization involves roughly as much work as performing Gaussian elimination. However, once we have the factorization, we are able to quickly solve equations of the form $A\mathbf{x} = \mathbf{b}$ by first solving $L\mathbf{c} = P\mathbf{b}$ and then $U\mathbf{x} = \mathbf{c}$.

5.1.4 Exercises

1. In this section, we saw that errors made in computer arithmetic can produce approximate solutions that are far from the exact solutions. Here is another example in which this can happen. Consider the matrix
$$A = \begin{bmatrix} 1 & 1 \\ 1 & 1.0001 \end{bmatrix}.$$

   a. Find the exact solution to the equation $A\mathbf{x} = \begin{bmatrix} 2 \\ 2 \end{bmatrix}$.

   b. Suppose that this linear system arises in the midst of a larger computation except that, due to some error in the computation of the right hand side of the equation, our computer thinks we want to solve $A\mathbf{x} = \begin{bmatrix} 2 \\ 2.0001 \end{bmatrix}$. Find the solution to this equation and compare it to the solution of the equation in the previous part of this exericse.

   Notice how a small change in the right hand side of the equation leads to a large change in the solution. In this case, we say that the matrix A is *ill-conditioned* because the solutions are extremely sensitive to small changes in the right hand side of the equation.

5.1. GAUSSIAN ELIMINATION REVISITED

Though we will not do so here, it is possible to create a measure of the matrix that tells us when a matrix is ill-conditioned. Regrettably, there is not much we can do to remedy this problem.

2. In this section, we found the LU factorization of the matrix

$$A = \begin{bmatrix} 1 & 2 & 1 \\ -2 & -3 & -2 \\ 3 & 7 & 4 \end{bmatrix}$$

in one of the activities, without using partial pivoting. Apply a sequence of row operations, now using partial pivoting, to find an upper triangular matrix U that is row equivalent to A.

3. In the following exercises, use the given LU factorizations to solve the equations $A\mathbf{x} = \mathbf{b}$.

   a. Solve the equation

   $$A\mathbf{x} = \begin{bmatrix} 1 & 0 \\ -2 & 1 \end{bmatrix} \begin{bmatrix} 3 & 1 \\ 0 & -2 \end{bmatrix} \mathbf{x} = \begin{bmatrix} -3 \\ 0 \end{bmatrix}.$$

   b. Solve the equation

   $$A\mathbf{x} = \begin{bmatrix} 1 & 0 & 0 \\ -2 & 1 & 0 \\ -1 & 2 & 1 \end{bmatrix} \begin{bmatrix} 2 & 1 & 0 \\ 0 & -1 & 3 \\ 0 & 0 & 1 \end{bmatrix} \mathbf{x} = \begin{bmatrix} 5 \\ -5 \\ 7 \end{bmatrix}.$$

4. Use Sage to solve the following equation by finding an LU factorization:

$$\begin{bmatrix} 3 & 4 & -1 \\ 2 & 4 & 1 \\ -3 & 1 & 4 \end{bmatrix} \mathbf{x} = \begin{bmatrix} -3 \\ -3 \\ -4 \end{bmatrix}.$$

5. Here is another problem with approximate computer arithmetic that we will encounter in the next section. Consider the matrix

$$A = \begin{bmatrix} 0.2 & 0.2 & 0.4 \\ 0.2 & 0.3 & 0.1 \\ 0.6 & 0.5 & 0.5 \end{bmatrix}.$$

   a. Notice that this is a positive stochastic matrix. What do we know about the eigenvalues of this matrix?

   b. Use Sage to define the matrix A using decimals such as 0.2 and the 3×3 identity matrix I. Ask Sage to compute $B = A - I$ and find the reduced row echelon form of B.

   c. Why is the computation that Sage performed incorrect?

d. Explain why using a computer to find the eigenvectors of a matrix A by finding a basis for $\text{Nul}(A - \lambda I)$ is problematic.

6. In practice, one rarely finds the inverse of a matrix A. It requires considerable effort to compute, and we can solve any equation of the form $A\mathbf{x} = \mathbf{b}$ using an LU factorization, which means that the inverse isn't necessary. In any case, the best way to compute an inverse is using an LU factorization, as this exericse demonstrates.

   a. Suppose that $PA = LU$. Explain why $A^{-1} = U^{-1}L^{-1}P$.

   Since L and U are triangular, finding their inverses is relatively efficient. That makes this an effective means of finding A^{-1}.

   b. Consider the matrix
   $$A = \begin{bmatrix} 3 & 4 & -1 \\ 2 & 4 & 1 \\ -3 & 1 & 4 \end{bmatrix}.$$
   Find the LU factorization of A and use it to find A^{-1}.

7. Consider the matrix
   $$A = \begin{bmatrix} a & a & a & a \\ a & b & b & b \\ a & b & c & c \\ a & b & c & d \end{bmatrix}.$$

   a. Find the LU factorization of A.

   b. What conditions on a, b, c, and d guarantee that A is invertible?

8. In the LU factorization of a matrix, the diagonal entries of L are all 1 while the diagonal entries of U are not necessarily 1. This exercise will explore that observation by considering the matrix
   $$A = \begin{bmatrix} 3 & 1 & 1 \\ -6 & -4 & -1 \\ 0 & -4 & 1 \end{bmatrix}.$$

   a. Perform Gaussian elimination without partial pivoting to find U, an upper triangular matrix that is row equivalent to A.

   b. The diagonal entries of U are called *pivots*. Explain why $\det A$ equals the product of the pivots.

   c. What is $\det A$ for our matrix A?

   d. More generally, if we have $PA = LU$, explain why $\det A$ equals plus or minus the product of the pivots.

9. Please provide a justification to your responses to these questions.

   a. In this section, our hypothetical computer could only store numbers using 3 decimal places. Most computers can store numbers using 15 or more decimal places. Why do we still need to be concerned about the accuracy of our computations when solving systems of linear equations?

5.1. GAUSSIAN ELIMINATION REVISITED

b. Finding the LU factorization of a matrix A is roughly the same amount of work as finding its reduced row echelon form. Why is the LU factorization useful then?

c. How can we detect whether a matrix is invertible from its LU factorization?

10. Consider the matrix
$$A = \begin{bmatrix} -1 & 1 & 0 & 0 \\ 1 & -2 & 1 & 0 \\ 0 & 1 & -2 & 1 \\ 0 & 0 & -1 & 1 \end{bmatrix}.$$

a. Find the LU factorization of A.

b. Use the factorization to find a basis for $\text{Nul}(A)$.

c. We have seen that $\text{Nul}(A) = \text{Nul}(U)$. Is it true that $\text{Col}(A) = \text{Col}(L)$?

5.2 Finding eigenvectors numerically

We have typically found eigenvalues of a square matrix A as the roots of the characteristic polynomial $\det(A - \lambda I) = 0$ and the associated eigenvectors as the null space $\text{Nul}(A - \lambda I)$. Unfortunately, this approach is not practical when we are working with large matrices. First, finding the charactertic polynomial of a large matrix requires considerable computation, as does finding the roots of that polynomial. Second, finding the null space of a singular matrix is plagued by numerical problems, as we will see in the preview activity.

For this reason, we will explore a technique called the *power method* that finds numerical approximations to the eigenvalues and eigenvectors of a square matrix.

> **Preview Activity 5.2.1.** Let's recall some earlier observations about eigenvalues and eigenvectors.
>
> a. How are the eigenvalues and associated eigenvectors of A related to those of A^{-1}?
>
> b. How are the eigenvalues and associated eigenvectors of A related to those of $A - 3I$?
>
> c. If λ is an eigenvalue of A, what can we say about the pivot positions of $A - \lambda I$?
>
> d. Suppose that $A = \begin{bmatrix} 0.8 & 0.4 \\ 0.2 & 0.6 \end{bmatrix}$. Explain how we know that 1 is an eigenvalue of A and then explain why the following Sage computation is incorrect.
>
> ```
> A = matrix(2,2,[0.8, 0.4, 0.2, 0.6])
> I = matrix(2,2,[1, 0, 0, 1])
> (A-I).rref()
> ```
>
> e. Suppose that $\mathbf{x}_0 = \begin{bmatrix} 1 \\ 0 \end{bmatrix}$, and we define a sequence $\mathbf{x}_{k+1} = A\mathbf{x}_k$; in other words, $\mathbf{x}_k = A^k \mathbf{x}_0$. What happens to $\mathbf{x}_k$ as k grows increasingly large?
>
> f. Explain how the eigenvalues of A are responsible for the behavior noted in the previous question.

5.2.1 The power method

Our goal is to find a technique that produces numerical approximations to the eigenvalues and associated eigenvectors of a matrix A. We begin by searching for the eigenvalue having the largest absolute value, which is called the *dominant* eigenvalue. The next two examples demonstrate this technique.

Example 5.2.1 Let's begin with the positive stochastic matrix $A = \begin{bmatrix} 0.7 & 0.6 \\ 0.3 & 0.4 \end{bmatrix}$. We spent

5.2. FINDING EIGENVECTORS NUMERICALLY 325

quite a bit of time studying this type of matrix in Section 4.5; in particular, we saw that any Markov chain will converge to the unique steady-state vector. Let's rephrase this statement in terms of the eigenvectors of A.

This matrix has eigenvalues $\lambda_1 = 1$ and $\lambda_2 = 0.1$ so the dominant eigenvalue is $\lambda_1 = 1$. The associated eigenvectors are $\mathbf{v}_1 = \begin{bmatrix} 2 \\ 1 \end{bmatrix}$ and $\mathbf{v}_2 = \begin{bmatrix} -1 \\ 1 \end{bmatrix}$. Suppose we begin with the vector

$$\mathbf{x}_0 = \begin{bmatrix} 1 \\ 0 \end{bmatrix} = \frac{1}{3}\mathbf{v}_1 - \frac{1}{3}\mathbf{v}_2$$

and find

$$\mathbf{x}_1 = A\mathbf{x}_0 = \frac{1}{3}\mathbf{v}_1 - \frac{1}{3}(0.1)\mathbf{v}_2$$

$$\mathbf{x}_2 = A^2\mathbf{x}_0 = \frac{1}{3}\mathbf{v}_1 - \frac{1}{3}(0.1)^2\mathbf{v}_2$$

$$\mathbf{x}_3 = A^3\mathbf{x}_0 = \frac{1}{3}\mathbf{v}_1 - \frac{1}{3}(0.1)^3\mathbf{v}_2$$

$$\vdots$$

$$\mathbf{x}_k = A^k\mathbf{x}_0 = \frac{1}{3}\mathbf{v}_1 - \frac{1}{3}(0.1)^k\mathbf{v}_2$$

and so forth. Notice that the powers 0.1^k become increasingly small as k grows so that $\mathbf{x}_k \approx \frac{1}{3}\mathbf{v}_1$ when k is large. Therefore, the vectors $\mathbf{x}_k$ become increasingly close to a vector in the eigenspace E_1, the eigenspace associated to the dominant eigenvalue. If we did not know the eigenvector $\mathbf{v}_1$, we could use a Markov chain in this way to find a basis vector for E_1, which, as seen in Section 4.5, is essentially how the Google PageRank algorithm works.

Example 5.2.2 Let's now look at the matrix $A = \begin{bmatrix} 2 & 1 \\ 1 & 2 \end{bmatrix}$, which has eigenvalues $\lambda_1 = 3$ and $\lambda_2 = 1$. The dominant eigenvalue is $\lambda_1 = 3$, and the associated eigenvectors are $\mathbf{v}_1 = \begin{bmatrix} 1 \\ 1 \end{bmatrix}$ and $\mathbf{v}_2 = \begin{bmatrix} -1 \\ 1 \end{bmatrix}$. Once again, begin with the vector $\mathbf{x}_0 = \begin{bmatrix} 1 \\ 0 \end{bmatrix} = \frac{1}{2}\mathbf{v}_1 - \frac{1}{2}\mathbf{v}_2$ so that

$$\mathbf{x}_1 = A\mathbf{x}_0 = 3\frac{1}{2}\mathbf{v}_1 - \frac{1}{2}\mathbf{v}_2$$

$$\mathbf{x}_2 = A^2\mathbf{x}_0 = 3^2\frac{1}{3}\mathbf{v}_1 - \frac{1}{2}\mathbf{v}_2$$

$$\mathbf{x}_3 = A^3\mathbf{x}_0 = 3^3\frac{1}{3}\mathbf{v}_1 - \frac{1}{2}\mathbf{v}_2$$

$$\vdots$$

$$\mathbf{x}_k = A^k\mathbf{x}_0 = 3^k\frac{1}{3}\mathbf{v}_1 - \frac{1}{2}\mathbf{v}_2.$$

As the figure shows, the vectors $\mathbf{x}_k$ are stretched by a factor of 3 in the $\mathbf{v}_1$ direction and not at all in the $\mathbf{v}_2$ direction. Consequently, the vectors $\mathbf{x}_k$ become increasingly long, but their direction becomes closer to the direction of the eigenvector $\mathbf{v}_1 = \begin{bmatrix} 1 \\ 1 \end{bmatrix}$ associated to the dominant eigenvalue.

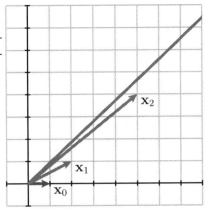

To find an eigenvector associated to the dominant eigenvalue, we will prevent the length of the vectors $\mathbf{x}_k$ from growing arbitrarily large by multiplying by an appropriate scaling constant. Here is one way to do this. Given the vector $\mathbf{x}_k$, we identify its component having the largest absolute value and call it m_k. We then define $\overline{\mathbf{x}}_k = \frac{1}{m_k}\mathbf{x}_k$, which means that the component of $\overline{\mathbf{x}}_k$ having the largest absolute value is 1.

For example, beginning with $\mathbf{x}_0 = \begin{bmatrix} 1 \\ 0 \end{bmatrix}$, we find $\mathbf{x}_1 = A\mathbf{x}_0 = \begin{bmatrix} 2 \\ 1 \end{bmatrix}$. The component of $\mathbf{x}_1$ having the largest absolute value is $m_1 = 2$ so we multiply by $\frac{1}{m_1} = \frac{1}{2}$ to obtain $\overline{\mathbf{x}}_1 = \begin{bmatrix} 1 \\ \frac{1}{2} \end{bmatrix}$. Then $\mathbf{x}_2 = A\overline{\mathbf{x}}_1 = \begin{bmatrix} \frac{5}{2} \\ 2 \end{bmatrix}$. Now the component having the largest absolute value is $m_2 = \frac{5}{2}$ so we multiply by $\frac{2}{5}$ to obtain $\overline{\mathbf{x}}_2 = \begin{bmatrix} 1 \\ \frac{4}{5} \end{bmatrix}$.

The resulting sequence of vectors $\overline{\mathbf{x}}_k$ is shown in the figure. Notice how the vectors $\overline{\mathbf{x}}_k$ now approach the eigenvector $\mathbf{v}_1$, which gives us a way to find the eigenvector $\mathbf{v} = \begin{bmatrix} 1 \\ 1 \end{bmatrix}$. This is the *power method* for finding an eigenvector associated to the dominant eigenvalue of a matrix.

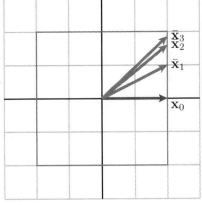

Activity 5.2.2. Let's begin by considering the matrix $A = \begin{bmatrix} 0.5 & 0.2 \\ 0.4 & 0.7 \end{bmatrix}$ and the initial

5.2. FINDING EIGENVECTORS NUMERICALLY

vector $x_0 = \begin{bmatrix} 1 \\ 0 \end{bmatrix}$.

a. Compute the vector $x_1 = Ax_0$.

b. Find m_1, the component of x_1 that has the largest absolute value. Then form $\bar{x}_1 = \frac{1}{m_1}x_1$. Notice that the component having the largest absolute value of $\bar{x}_1$ is 1.

c. Find the vector $x_2 = A\bar{x}_1$. Identify the component m_2 of x_2 having the largest absolute value. Then form $\bar{x}_2 = \frac{1}{m_2}\bar{x}_1$ to obtain a vector in which the component with the largest absolute value is 1.

d. The Sage cell below defines a function that implements the power method. Define the matrix A and initial vector x_0 below. The command power(A, x0, N) will print out the multiplier m and the vectors $\bar{x}_k$ for N steps of the power method.

```
def power(A, x, N):
    for i in range(N):
        x = A*x
        m = max([comp for comp in x],
                key=abs).numerical_approx(digits=14)
        x = 1/float(m)*x
        print (m, x)

### Define the matrix A and initial vector x0 below
A = 
x0 = 
power(A, x0, 20)
```

How does this computation identify an eigenvector of the matrix A?

e. What is the corresponding eigenvalue of this eigenvector?

f. How do the values of the multipliers m_k tell us the eigenvalue associated to the eigenvector we have found?

g. Consider now the matrix $A = \begin{bmatrix} -5.1 & 5.7 \\ -3.8 & 4.4 \end{bmatrix}$. Use the power method to find the dominant eigenvalue of A and an associated eigenvector.

Notice that the power method gives us not only an eigenvector v but also its associated eigenvalue. As in the activity, consider the matrix $A = \begin{bmatrix} -5.1 & 5.7 \\ -3.8 & 4.4 \end{bmatrix}$, which has eigenvector $v = \begin{bmatrix} 3 \\ 2 \end{bmatrix}$. The first component has the largest absolute value so we multiply by $\frac{1}{3}$ to obtain $\bar{\mathbf{v}} = \begin{bmatrix} 1 \\ \frac{2}{3} \end{bmatrix}$. When we multiply by A, we have $A\bar{\mathbf{v}} = \begin{bmatrix} -1.30 \\ -0.86 \end{bmatrix}$. Notice that the first component still has the largest absolute value so that the multiplier $m = -1.3$ is the eigenvalue λ corresponding to the eigenvector. This demonstrates the fact that the multipliers m_k approach the eigenvalue λ having the largest absolute value.

Notice that the power method requires us to choose an initial vector $\mathbf{x}_0$. For most choices, this method will find the eigenvalue having the largest absolute value. However, an unfortunate choice of $\mathbf{x}_0$ may not. For instance, if we had chosen $\mathbf{x}_0 = \mathbf{v}_2$ in our example above, the vectors in the sequence $\mathbf{x}_k = A^k \mathbf{x}_0 = \lambda_2^k \mathbf{v}_2$ will not detect the eigenvector $\mathbf{v}_1$. However, it usually happens that our initial guess $\mathbf{x}_0$ has some contribution from $\mathbf{v}_1$ that enables us to find it.

The power method, as presented here, will fail for certain unlucky matrices. This is examined in Exercise 5.2.4.5 along with a means to improve the power method to work for all matrices.

5.2.2 Finding other eigenvalues

The power method gives a technique for finding the dominant eigenvalue of a matrix. We can modify the method to find the other eigenvalues as well.

> **Activity 5.2.3.** The key to finding the eigenvalue of A having the smallest absolute value is to note that the eigenvectors of A are the same as those of A^{-1}.
>
> a. If $\mathbf{v}$ is an eigenvector of A with associated eigenvector λ, explain why $\mathbf{v}$ is an eigenvector of A^{-1} with associated eigenvalue λ^{-1}.
>
> b. Explain why the eigenvalue of A having the smallest absolute value is the reciprocal of the dominant eigenvalue of A^{-1}.
>
> c. Explain how to use the power method applied to A^{-1} to find the eigenvalue of A having the smallest absolute value.
>
> d. If we apply the power method to A^{-1}, we begin with an intial vector $\mathbf{x}_0$ and generate the sequence $\mathbf{x}_{k+1} = A^{-1} \mathbf{x}_k$. It is not computationally efficient to compute A^{-1}, however, so instead we solve the equation $A\mathbf{x}_{k+1} = \mathbf{x}_k$. Explain why an LU factorization of A is useful for implementing the power method applied to A^{-1}.
>
> e. The following Sage cell defines a command called inverse_power that applies the power method to A^{-1}. That is, inverse_power(A, x0, N) prints the vectors $\mathbf{x}_k$, where $\mathbf{x}_{k+1} = A^{-1}\mathbf{x}_k$, and multipliers $\frac{1}{m_k}$, which approximate the eigenvalue of A. Use it to find the eigenvalue of $A = \begin{bmatrix} -5.1 & 5.7 \\ -3.8 & 4.4 \end{bmatrix}$ having the smallest absolute value.

5.2. FINDING EIGENVECTORS NUMERICALLY

```
def inverse_power(A, x, N):
    for i in range(N):
        x = A \ x
        m = max([comp for comp in x],
            key=abs).numerical_approx(digits=14)
        x = 1/float(m)*x
        print (1/float(m), x)
### define the matrix A and vector x0
A =
x0 =
inverse_power(A, x0, 20)
```

f. The inverse power method only works if A is invertible. If A is not invertible, what is its eigenvalue having the smallest absolute value?

g. Use the power method and the inverse power method to find the eigenvalues and associated eigenvectors of the matrix $A = \begin{bmatrix} -0.23 & -2.33 \\ -1.16 & 1.08 \end{bmatrix}$.

With the power method and the inverse power method, we can now find the eigenvalues of a matrix A having the largest and smallest absolute values. With one more modification, we can find all the eigenvalues of A.

Activity 5.2.4. Remember that the absolute value of a number tells us how far that number is from 0 on the real number line. We may therefore think of the inverse power method as telling us the eigenvalue closest to 0.

a. If $\mathbf{v}$ is an eigenvector of A with associated eigenvalue λ, explain why $\mathbf{v}$ is an eigenvector of $A - sI$ where s is some scalar.

b. What is the eigenvalue of $A - sI$ associated to the eigenvector $\mathbf{v}$?

c. Explain why the eigenvalue of A closest to s is the eigenvalue of $A - sI$ closest to 0.

d. Explain why applying the inverse power method to $A - sI$ gives the eigenvalue of A closest to s.

e. Consider the matrix $A = \begin{bmatrix} 3.6 & 1.6 & 4.0 & 7.6 \\ 1.6 & 2.2 & 4.4 & 4.1 \\ 3.9 & 4.3 & 9.0 & 0.6 \\ 7.6 & 4.1 & 0.6 & 5.0 \end{bmatrix}$. If we use the power method and inverse power method, we find two eigenvalues, $\lambda_1 = 16.35$ and $\lambda_2 = 0.75$. Viewing these eigenvalues on a number line, we know that the other eigenvalues lie in the range between $-\lambda_1$ and λ_1, as shaded in Figure 5.2.3.

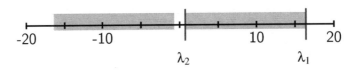

Figure 5.2.3 The range of eigenvalues of A.

The Sage cell below has a function find_closest_eigenvalue(A, s, x, N) that implements N steps of the inverse power method using the matrix $A - sI$ and an initial vector x. This function prints approximations to the eigenvalue of A closest to s and its associated eigenvector. By trying different values of s in the shaded regions of the number line shown in Figure 5.2.3, find the other two eigenvalues of A.

```
def find_closest_eigenvalue(A, s, x, N):
    B = A-s*identity_matrix(A.nrows())
    for i in range(N):
        x = B \ x
        m = max([comp for comp in x],
            key=abs).numerical_approx(digits=14)
        x = 1/float(m)*x
        print (1/float(m)+s, x)
### define the matrix A and vector x0
A =
x0 =
find_closest_eigenvalue(A, 2, x0, 20)
```

f. Write a list of the four eigenvalues of A in increasing order.

There are some restrictions on the matrices to which this technique applies as we have assumed that the eigenvalues of A are real and distinct. If A has repeated or complex eigenvalues, this technique will need to be modified, as explored in some of the exercises.

5.2.3 Summary

We have explored the power method as a tool for numerically approximating the eigenvalues and eigenvectors of a matrix.

- After choosing an initial vector x_0, we define the sequence $x_{k+1} = Ax_k$. As k grows larger, the direction of the vectors x_k closely approximates the direction of the eigenspace corresponding to the eigenvalue λ_1 having the largest absolute value.

- We normalize the vectors x_k by multiplying by $\frac{1}{m_k}$, where m_k is the component having the largest absolute value. In this way, the vectors $\bar{x}_k$ approach an eigenvector associated to λ_1, and the multipliers m_k approach the eigenvalue λ_1.

- To find the eigenvalue having the smallest absolute value, we apply the power method using the matrix A^{-1}.

- To find the eigenvalue closest to some number s, we apply the power method using the matrix $(A - sI)^{-1}$.

5.2.4 Exercises

This Sage cell has the commands power, inverse_power, and find_closest_eigenvalue that we have developed in this section. After evaluating this cell, these commands will be available in any other cell on this page.

```
def power(A, x, N):
    for i in range(N):
        x = A*x
        m = max([comp for comp in x],
            key=abs).numerical_approx(digits=14)
        x = 1/float(m)*x
        print (m, x)
def find_closest_eigenvalue(A, s, x, N):
    B = A-s*identity_matrix(A.nrows())
    for i in range(N):
        x = B \ x
        m = max([comp for comp in x],
            key=abs).numerical_approx(digits=14)
        x = 1/float(m)*x
        print (1/float(m)+s, x)
def inverse_power(A, x, N):
    find_closest_eigenvalue(A, 0, x, N)
```

1. Suppose that A is a matrix having eigenvalues $-3, -0.2, 1$, and 4.

    a. What are the eigenvalues of A^{-1}?

    b. What are the eigenvalues of $A + 7I$?

2. Use the commands power, inverse_power, and find_closest_eigenvalue to approximate the eigenvalues and associated eigenvectors of the following matrices.

    a. $A = \begin{bmatrix} -2 & -2 \\ -8 & -2 \end{bmatrix}$.

    b. $A = \begin{bmatrix} 0.6 & 0.7 \\ 0.5 & 0.2 \end{bmatrix}$.

    c. $A = \begin{bmatrix} 1.9 & -16.0 & -13.0 & 27.0 \\ -2.4 & 20.3 & 4.6 & -17.7 \\ -0.51 & -11.7 & -1.4 & 13.1 \\ -2.1 & 15.3 & 6.9 & -20.5 \end{bmatrix}$.

3. Use the techniques we have seen in this section to find the eigenvalues of the matrix

$$A = \begin{bmatrix} -14.6 & 9.0 & -14.1 & 5.8 & 13.0 \\ 27.8 & -4.2 & 16.0 & 0.9 & -21.3 \\ -5.5 & 3.4 & 3.4 & 3.3 & 1.1 \\ -25.4 & 11.3 & -15.4 & 4.7 & 20.3 \\ -33.7 & 14.8 & -22.5 & 9.7 & 26.6 \end{bmatrix}.$$

```
A = matrix(5,5, [-14.6,   9.0,  -14.1,   5.8,   13.0,
                 27.8,   -4.2,   16.0,   0.9,  -21.3,
                 -5.5,    3.4,    3.4,   3.3,    1.1,
                -25.4,   11.3,  -15.4,   4.7,   20.3,
                -33.7,   14.8,  -22.5,   9.7,   26.6])
```

4. Consider the matrix $A = \begin{bmatrix} 0 & -1 \\ -4 & 0 \end{bmatrix}$.

   a. Describe what happens if we apply the power method and the inverse power method using the initial vector $\mathbf{x}_0 = \begin{bmatrix} 1 \\ 0 \end{bmatrix}$.

   b. Find the eigenvalues of this matrix and explain this observed behavior.

   c. How can we apply the techniques of this section to find the eigenvalues of A?

5. We have seen that the matrix $A = \begin{bmatrix} 1 & 2 \\ 2 & 1 \end{bmatrix}$ has eigenvalues $\lambda_1 = 3$ and $\lambda_2 = -1$ and associated eigenvectors $\mathbf{v}_1 = \begin{bmatrix} 1 \\ 1 \end{bmatrix}$ and $\mathbf{v}_2 = \begin{bmatrix} -1 \\ 1 \end{bmatrix}$.

   a. Describe what happens when we apply the inverse power method using the initial vector $\mathbf{x}_0 = \begin{bmatrix} 1 \\ 0 \end{bmatrix}$.

   b. Explain why this is happening and provide a contrast with how the power method usually works.

   c. How can we modify the power method to give the dominant eigenvalue in this case?

6. Suppose that A is a 2×2 matrix with eigenvalues 4 and −3 and that B is a 2×2 matrix with eigenvalues 4 and 1. If we apply the power method to find the dominant eigenvalue of these matrices to the same degree of accuracy, which matrix will require more steps in the algorithm? Explain your response.

7. Suppose that we apply the power method to the matrix A with an initial vector $\mathbf{x}_0$ and find the eigenvalue $\lambda = 3$ and eigenvector $\mathbf{v}$. Suppose that we then apply the power method again with a different initial vector and find the same eigenvalue $\lambda = 3$ but a different eigenvector $\mathbf{w}$. What can we conclude about the matrix A in this case?

5.2. FINDING EIGENVECTORS NUMERICALLY

8. The power method we have developed only works if the matrix has real eigenvalues. Suppose that A is a 2×2 matrix that has a complex eigenvalue $\lambda = 2 + 3i$. What would happen if we apply the power method to A?

9. Consider the matrix $A = \begin{bmatrix} 1 & 1 \\ 0 & 1 \end{bmatrix}$.

    a. Find the eigenvalues and associated eigenvectors of A.

    b. Make a prediction about what happens if we apply the power method and the inverse power method to find eigenvalues of A.

    c. Verify your prediction using Sage.

CHAPTER 6

Orthogonality and Least Squares

We introduced vectors as a means to develop visual intuition about our basic questions concerning linear systems. For example, vectors allow us to reinterpret questions about the existence of solutions to linear systems as questions about the span of a set of vectors. Questions about the uniqueness of solutions led to the concept of linear independence.

In this chapter, we will begin to think of vectors as geometric objects that have lengths and that form angles. In some cases, this will simplify our search for solutions to a linear system. Perhaps more importantly, we will be able to measure the distance between vectors. This means that if a system $A\mathbf{x} = \mathbf{b}$ is inconsistent, we can look for $\widehat{\mathbf{x}}$, the vector for which $A\widehat{\mathbf{x}}$ is as close to $\mathbf{b}$ as possible. This leads to the method of *least squares*, which underpins regression, a key tool in data science.

6.1 The dot product

In this section, we introduce a simple algebraic operation, known as the *dot product*, that helps us measure the length of vectors and the angle formed by a pair of vectors. For two-dimensional vectors $\mathbf{v}$ and $\mathbf{w}$, their dot product $\mathbf{v} \cdot \mathbf{w}$ is the scalar defined to be

$$\mathbf{v} \cdot \mathbf{w} = \begin{bmatrix} v_1 \\ v_2 \end{bmatrix} \cdot \begin{bmatrix} w_1 \\ w_2 \end{bmatrix} = v_1 w_1 + v_2 w_2.$$

For instance,

$$\begin{bmatrix} 2 \\ -3 \end{bmatrix} \cdot \begin{bmatrix} 4 \\ 1 \end{bmatrix} = 2 \cdot 4 + (-3) \cdot 1 = 5.$$

Preview Activity 6.1.1.

a. Compute the dot product
$$\begin{bmatrix} 3 \\ 4 \end{bmatrix} \cdot \begin{bmatrix} 2 \\ -2 \end{bmatrix}.$$

b. Sketch the vector $\mathbf{v} = \begin{bmatrix} 3 \\ 4 \end{bmatrix}$ below. Then use the Pythagorean theorem to find the length of $\mathbf{v}$.

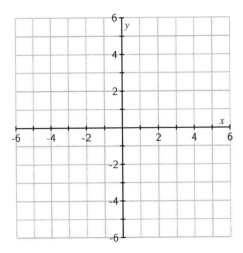

Figure 6.1.1 Sketch the vector **v** and find its length.

c. Compute the dot product $\mathbf{v} \cdot \mathbf{v}$. How is the dot product related to the length of **v**?

d. Remember that the matrix $\begin{bmatrix} 0 & -1 \\ 1 & 0 \end{bmatrix}$ represents the matrix transformation that rotates vectors counterclockwise by 90°. Beginning with the vector $\mathbf{v} = \begin{bmatrix} 3 \\ 4 \end{bmatrix}$, find **w**, the result of rotating **v** by 90°, and sketch it above.

e. What is the dot product $\mathbf{v} \cdot \mathbf{w}$?

f. Suppose that $\mathbf{v} = \begin{bmatrix} a \\ b \end{bmatrix}$. Find the vector **w** that results from rotating **v** by 90° and find the dot product $\mathbf{v} \cdot \mathbf{w}$.

g. Suppose that **v** and **w** are two perpendicular vectors. What do you think their dot product $\mathbf{v} \cdot \mathbf{w}$ is?

6.1.1 The geometry of the dot product

The dot product is defined, more generally, for any two m-dimensional vectors:

$$\mathbf{v} \cdot \mathbf{w} = \begin{bmatrix} v_1 \\ v_2 \\ \vdots \\ v_m \end{bmatrix} \cdot \begin{bmatrix} w_1 \\ w_2 \\ \vdots \\ w_m \end{bmatrix} = v_1 w_1 + v_2 w_2 + \ldots + v_m w_m.$$

The important thing to remember is that the dot product will produce a scalar. In other words, the two vectors are combined in such a way as to create a number, and, as we'll see, this number conveys useful geometric information.

6.1. THE DOT PRODUCT

Example 6.1.2 We compute the dot product between two four-dimensional vectors as

$$\begin{bmatrix} 2 \\ 0 \\ -3 \\ 1 \end{bmatrix} \cdot \begin{bmatrix} -1 \\ 3 \\ 1 \\ 2 \end{bmatrix} = 2(-1) + 0(3) + (-3)(1) + 1(2) = -3.$$

> **Properties of dot products.**
>
> As with ordinary multiplication, the dot product enjoys some familiar algebraic properties, such as commutativity and distributivity. More specifically, it doesn't matter in which order we compute the dot product of two vectors:
>
> $$\mathbf{v} \cdot \mathbf{w} = \mathbf{w} \cdot \mathbf{v}.$$
>
> If s is a scalar, we have
>
> $$(s\mathbf{v}) \cdot \mathbf{w} = s(\mathbf{v} \cdot \mathbf{w}).$$
>
> We may also distribute the dot product across linear combinations:
>
> $$(c_1 \mathbf{v}_1 + c_2 \mathbf{v}_2) \cdot \mathbf{w} = c_1 \mathbf{v}_1 \cdot \mathbf{w} + c_2 \mathbf{v}_2 \cdot \mathbf{w}.$$

Example 6.1.3 Suppose that $\mathbf{v}_1 \cdot \mathbf{w} = 4$ and $\mathbf{v}_2 \cdot \mathbf{w} = -7$. Then

$$(2\mathbf{v}_1) \cdot \mathbf{w} = 2(\mathbf{v}_1 \cdot \mathbf{w}) = 2(4) = 8$$
$$(-3\mathbf{v}_1 + 2\mathbf{v}_2) \cdot \mathbf{w} = -3(\mathbf{v}_1 \cdot \mathbf{w}) + 2(\mathbf{v}_2 \cdot \mathbf{w}) = -3(4) + 2(-7) = -26.$$

The most important property of the dot product, and the real reason for our interest in it, is that it gives us geometric information about vectors and their relationship to one another. Let's first think about the length of a vector by looking at the vector $\mathbf{v} = \begin{bmatrix} 3 \\ 2 \end{bmatrix}$ as shown in Figure 6.1.4

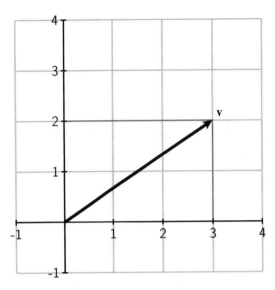

Figure 6.1.4 The vector $\mathbf{v} = \begin{bmatrix} 3 \\ 2 \end{bmatrix}$.

We may find the length of this vector using the Pythagorean theorem since the vector forms the hypotenuse of a right triangle having a horizontal leg of length 3 and a vertical leg of length 2. The length of $\mathbf{v}$, which we denote as $|\mathbf{v}|$, is therefore $|\mathbf{v}| = \sqrt{3^2 + 2^2} = \sqrt{13}$. Now notice that the dot product of $\mathbf{v}$ with itself is

$$\mathbf{v} \cdot \mathbf{v} = 3(3) + 2(2) = 13 = |\mathbf{v}|^2.$$

This is true in general; that is, we have

$$\mathbf{v} \cdot \mathbf{v} = |\mathbf{v}|^2.$$

More than that, the dot product of two vectors records information about the angle between them. Consider Figure 6.1.5.

6.1. THE DOT PRODUCT

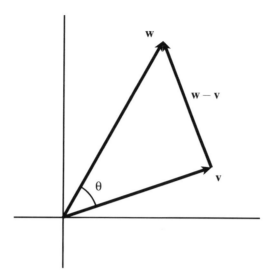

Figure 6.1.5 The dot product $\mathbf{v} \cdot \mathbf{w}$ measures the angle θ.

To see this, we will apply the Law of Cosines, which says that

$$|\mathbf{w} - \mathbf{v}|^2 = |\mathbf{v}|^2 + |\mathbf{w}|^2 - 2|\mathbf{v}||\mathbf{w}|\cos\theta$$
$$(\mathbf{w} - \mathbf{v}) \cdot (\mathbf{w} - \mathbf{v}) = \mathbf{v} \cdot \mathbf{v} + \mathbf{w} \cdot \mathbf{w} - 2|\mathbf{v}||\mathbf{w}|\cos\theta$$
$$\mathbf{w} \cdot \mathbf{w} + \mathbf{v} \cdot \mathbf{v} - 2\mathbf{v} \cdot \mathbf{w} = \mathbf{v} \cdot \mathbf{v} + \mathbf{w} \cdot \mathbf{w} - 2|\mathbf{v}||\mathbf{w}|\cos\theta$$
$$-2\mathbf{v} \cdot \mathbf{w} = -2|\mathbf{v}||\mathbf{w}|\cos\theta$$
$$\mathbf{v} \cdot \mathbf{w} = |\mathbf{v}||\mathbf{w}|\cos\theta$$

The upshot of this reasoning is that

$$\mathbf{v} \cdot \mathbf{w} = |\mathbf{v}||\mathbf{w}|\cos\theta.$$

To summarize:

> **Geometric properties of the dot product.**
>
> The dot product gives us the following geometric information:
>
> $$\mathbf{v} \cdot \mathbf{v} = |\mathbf{v}|^2$$
> $$\mathbf{v} \cdot \mathbf{w} = |\mathbf{v}||\mathbf{w}|\cos\theta$$
>
> where θ is the angle between $\mathbf{v}$ and $\mathbf{w}$.

Activity 6.1.2.

a. Sketch the vectors $\mathbf{v} = \begin{bmatrix} 3 \\ 2 \end{bmatrix}$ and $\mathbf{w} = \begin{bmatrix} -1 \\ 3 \end{bmatrix}$ using Figure 6.1.6.

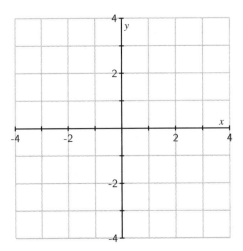

Figure 6.1.6 Sketch the vectors **v** and **w** here.

b. Find the lengths $|\mathbf{v}|$ and $|\mathbf{w}|$ using the dot product.

c. Find the dot product $\mathbf{v} \cdot \mathbf{w}$ and use it to find the angle between **v** and **w**.

d. Consider the vector $\mathbf{x} = \begin{bmatrix} -2 \\ 3 \end{bmatrix}$. Include it in your sketch in Figure 6.1.6 and find the angle between **v** and **x**.

e. If two vectors are perpendicular, what can you say about their dot product? Explain your thinking.

f. For what value of k is the vector $\begin{bmatrix} 6 \\ k \end{bmatrix}$ perpendicular to **w**?

g. Sage can be used to find lengths of vectors and their dot products. For instance, if v and w are vectors, then v.norm() gives the length of v and v * w gives $\mathbf{v} \cdot \mathbf{w}$.

Suppose that
$$\mathbf{v} = \begin{bmatrix} 2 \\ 0 \\ 3 \\ -2 \end{bmatrix}, \quad \mathbf{w} = \begin{bmatrix} 1 \\ -3 \\ 4 \\ 1 \end{bmatrix}.$$

Use the Sage cell below to find $|\mathbf{v}|$, $|\mathbf{w}|$, $\mathbf{v} \cdot \mathbf{w}$, and the angle between **v** and **w**. You may use arccos to find the angle's measure expressed in radians.

As we move forward, it will be important for us to recognize when vectors are perpendicular to one another. For instance, when vectors **v** and **w** are perpendicular, the angle between them $\theta = 90°$ and we have

$$\mathbf{v} \cdot \mathbf{w} = |\mathbf{v}|\,|\mathbf{w}| \cos \theta = |\mathbf{v}|\,|\mathbf{w}| \cos 90° = 0.$$

Therefore, the dot product between perpendicular vectors must be zero. This leads to the following definition.

6.1. THE DOT PRODUCT

Definition 6.1.7 We say that vectors **v** and **w** are orthogonal if $\mathbf{v} \cdot \mathbf{w} = 0$.

In practical terms, two perpendicular vectors are orthogonal. However, the concept of orthogonality is somewhat more general because it allows one or both of the vectors to be the zero vector **0**.

We've now seen that the dot product gives us geometric information about vectors. It also provides a way to compare vectors. For example, consider the vectors **u**, **v**, and **w**, shown in Figure 6.1.8. The vectors **v** and **w** seem somewhat similar as the directions they define are nearly the same. By comparison, **u** appears rather dissimilar to both **v** and **w**. We will measure the similarity of vectors by finding the angle between them; the smaller the angle, the more similar the vectors.

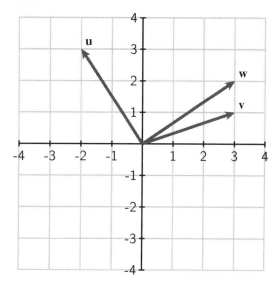

Figure 6.1.8 Which of the vectors are most similar?

Activity 6.1.3. This activity explores two further uses of the dot product beginning with the similarity of vectors.

a. Our first task is to assess the similarity between various Wikipedia articles by forming vectors from each of five articles. In particular, one may download the text from a Wikipedia article, remove common words, such as "the" and "and", count the number of times the remaining words appear in the article, and represent these counts in a vector.

For example, evaluate the following cell that loads some special commands along with the vectors constructed from the Wikipedia articles on Veteran's Day, Memorial Day, Labor Day, the Golden Globe Awards, and the Super Bowl. For each of the five articles, you will see a list of the number of times 10 words appear in these articles. For instance, the word "act" appears 3 times in the Veteran's Day article and 0 times in the Labor Day article.

```
url='https://raw.githubusercontent.com/davidaustinm/'
url+='ula_modules/master/dot_similarity.py'
sage.repl.load.load(url, globals())
events.head(int(10))
```

For each of the five articles, we obtain 604-dimensional vectors, which are named veterans, memorial, labor, golden, and super.

1. Suppose that two articles have no words in common. What is the value of the dot product between their corresponding vectors? What does this say about the angle between these vectors?

2. Suppose there are two articles on the same subject, yet one article is twice as long. What approximate relationship would you expect to hold between the two vectors? What does this say about the angle between them?

3. Use the Sage cell below to find the angle between the vector veterans and the other four vectors. To express the angle in degrees, use the degrees(x) command, which gives the number of degrees in x radians.

4. Compare the four angles you have found and discuss what they mean about the similarity between the Veteran's Day article and the other four. How do your findings reflect the nature of these five events?

b. Vectors are often used to represent how a quantity changes over time. For instance, the vector $\mathbf{s} = \begin{bmatrix} 78.3 \\ 81.2 \\ 82.1 \\ 79.0 \end{bmatrix}$ might represent the value of a company's stock on four consecutive days. When interpreted in this way, we call the vector a *time series*. Evaluate the Sage cell below to see a representation of two time series $\mathbf{s}_1$, in blue, and $\mathbf{s}_2$, in orange, which we imagine represent the value of two stocks over a period of time. (This cell relies on some data loaded by the first cell in this activity.)

```
series_plot(s1, 'blue') + series_plot(s2, 'orange')
```

Even though one stock has a higher value than the other, the two appear to be related since they seem to rise and fall at roughly similar ways. We often say that they are *correlated*, and we would like to measure the degree to which they are correlated.

1. In order to compare the ways in which they rise and fall, we will first *demean* the time series; that is, for each time series, we will subtract its average value to obtain a new time series. There is a command, demean(s), that returns the demeaned time series of s. Use the Sage cell below to demean the series $\mathbf{s}_1$ and $\mathbf{s}_2$ and plot.

6.1. THE DOT PRODUCT

```
ds1 = demean(s1)
ds2 = demean(s2)
series_plot(ds1, 'blue') + series_plot(ds2, 'orange')
```

2. If the demeaned series are $\tilde{\mathbf{s}}_1$ and $\tilde{\mathbf{s}}_2$, then the correlation between $\mathbf{s}_1$ and $\mathbf{s}_2$ is defined to be

$$\text{corr}(\mathbf{s}_1, \mathbf{s}_2) = \frac{\tilde{\mathbf{s}}_1 \cdot \tilde{\mathbf{s}}_2}{|\tilde{\mathbf{s}}_1| \, |\tilde{\mathbf{s}}_2|}.$$

Given the geometric interpretation of the dot product, the correlation equals the cosine of the angle between the demeaned time series, and therefore $\text{corr}(\mathbf{s}_1, \mathbf{s}_2)$ is between -1 and 1.

Find the correlation between $\mathbf{s}_1$ and $\mathbf{s}_2$.

3. Suppose that two time series are such that their demeaned time series are scalar multiples of one another, as in Figure 6.1.9

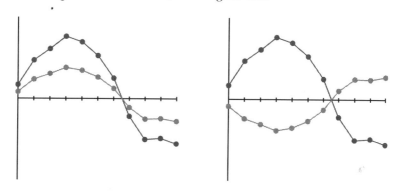

Figure 6.1.9 On the left, the demeaned time series are positive scalar multiples of one another. On the right, they are negative scalar multiples.

For instance, suppose we have time series $\mathbf{t}_1$ and $\mathbf{t}_2$ whose demeaned time series $\tilde{\mathbf{t}}_1$ and $\tilde{\mathbf{t}}_2$ are positive scalar multiples of one another. What is the angle between the demeaned vectors? What does this say about the correlation $\text{corr}(\mathbf{t}_1, \mathbf{t}_2)$?

4. Suppose the demeaned time series $\tilde{\mathbf{t}}_1$ and $\tilde{\mathbf{t}}_2$ are negative scalar multiples of one another, what is the angle between the demeaned vectors? What does this say about the correlation $\text{corr}(\mathbf{t}_1, \mathbf{t}_2)$?

5. Use the Sage cell below to plot the time series $\mathbf{s}_1$ and $\mathbf{s}_3$ and find their correlation.

```
series_plot(s1, 'blue') + series_plot(s3, 'orange')
```

6. Use the Sage cell below to plot the time series $\mathbf{s}_1$ and $\mathbf{s}_4$ and find their correlation.

```
series_plot(s1, 'blue') + series_plot(s4, 'orange')
```

6.1.2 k-means clustering

A typical problem in data science is to find some underlying patterns in a dataset. Suppose, for instance, that we have the set of 177 data points plotted in Figure 6.1.10. Notice that the points are not scattered around haphazardly; instead, they seem to form clusters. Our goal here is to develop a strategy for detecting the clusters.

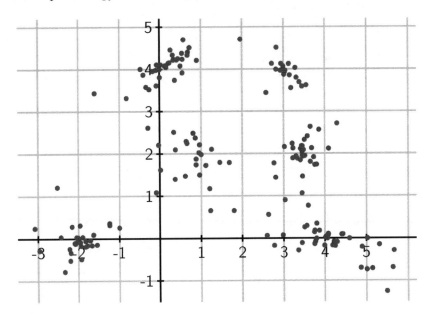

Figure 6.1.10 A set of 177 data points.

To see how this could be useful, suppose we have medical data describing a group of patients, some of whom have been diagnosed with a specific condition, such as diabetes. Perhaps we have a record of age, weight, blood sugar, cholesterol, and other attributes for each patient. It could be that the data points for the group diagnosed as having the condition form a cluster that is somewhat distinct from the rest of the data. Suppose that we are able to identify that cluster and that we are then presented with a new patient that has not been tested for the condition. If the attributes for that patient place them in that cluster, we might identify them as being at risk for the condition and prioritize them for appropriate screenings.

If there are many attributes for each patient, the data may be high-dimensional and not easily visualized. We would therefore like to develop an algorithm that separates the data points into clusters without human intervention. We call the result a *clustering*.

The next activity introduces a technique, called k-means clustering, that helps us find clusterings. To do so, we will view the data points as vectors so that the distance between two data points equals the length of the vector joining them. That is, if two points are represented by the vectors $\mathbf{v}$ and $\mathbf{w}$, then the distance between the points is $|\mathbf{v} - \mathbf{w}|$.

Activity 6.1.4. To begin, we identify the *centroid*, or the average, of a set of vectors

6.1. THE DOT PRODUCT

$\mathbf{v}_1, \mathbf{v}_2, \ldots, \mathbf{v}_n$ as
$$\frac{1}{n}(\mathbf{v}_1 + \mathbf{v}_2 + \ldots + \mathbf{v}_n).$$

a. Find the centroid of the vectors
$$\mathbf{v}_1 = \begin{bmatrix} 1 \\ 1 \end{bmatrix}, \mathbf{v}_2 = \begin{bmatrix} 4 \\ 1 \end{bmatrix}, \mathbf{v}_3 = \begin{bmatrix} 4 \\ 4 \end{bmatrix}.$$

and sketch the vectors and the centroid using Figure 6.1.11. You may wish to simply plot the points represented by the tips of the vectors rather than drawing the vectors themselves.

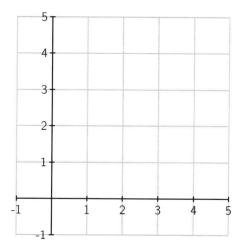

Figure 6.1.11 The vectors $\mathbf{v}_1$, $\mathbf{v}_2$, $\mathbf{v}_3$ and their centroid.

Notice that the centroid lies in the center of the points defined by the vectors.

b. Now we'll illustrate an algorithm that forms clusterings. To begin, consider the following points, represented as vectors,
$$\mathbf{v}_1 = \begin{bmatrix} -2 \\ 1 \end{bmatrix}, \mathbf{v}_2 = \begin{bmatrix} 1 \\ 1 \end{bmatrix}, \mathbf{v}_3 = \begin{bmatrix} 1 \\ 2 \end{bmatrix}, \mathbf{v}_4 = \begin{bmatrix} 3 \\ 2 \end{bmatrix},$$
which are shown in Figure 6.1.12.

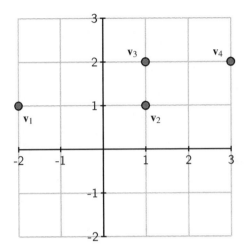

Figure 6.1.12 We will group this set of four points into two clusters.

Suppose that we would like to group these points into $k = 2$ clusters. (Later on, we'll see how to choose an appropriate value for k, the number of clusters.) We begin by choosing two points c_1 and c_2 at random and declaring them to be the "centers"' of the two clusters.

For example, suppose we randomly choose $c_1 = \mathbf{v}_2$ and $c_2 = \mathbf{v}_3$ as the center of two clusters. The cluster centered on $c_1 = \mathbf{v}_2$ will be the set of points that are closer to $c_1 = \mathbf{v}_2$ than to $c_2 = \mathbf{v}_3$. Determine which of the four data points are in this cluster, which we denote by C_1, and circle them in Figure 6.1.12.

c. The second cluster will consist of the data points that are closer to $c_2 = \mathbf{v}_3$ than $c_1 = \mathbf{v}_2$. Determine which of the four points are in this cluster, which we denote by C_2, and circle them in Figure 6.1.12.

d. We now have a clustering with two clusters, but we will try to improve upon it in the following way. First, find the centroids of the two clusters; that is, redefine c_1 to be the centroid of cluster C_1 and c_2 to be the centroid of C_2. Find those centroids and indicate them in Figure 6.1.13

6.1. THE DOT PRODUCT

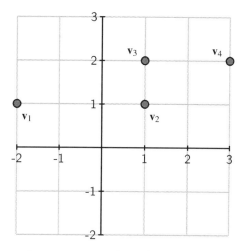

Figure 6.1.13 Indicate the new centroids and clusters.

Now update the cluster C_1 to be the set of points closer to c_1 than c_2. Update the cluster C_2 in a similar way and indicate the clusters in Figure 6.1.13.

e. Let's perform this last step again. That is, update the centroids c_1 and c_2 from the new clusters and then update the clusters C_1 and C_2. Indicate your centroids and clusters in Figure 6.1.14.

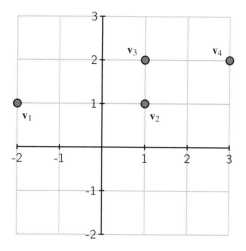

Figure 6.1.14 Indicate the new centroids and clusters.

Notice that this last step produces the same set of clusters so there is no point in repeating it. We declare this to be our final clustering.

This activity demonstrates our algorithm for finding a clustering. We first choose a value k and seek to break the data points into k clusters. The algorithm proceeds in the following way:

- Choose k points $c_1, c_2, \ldots, c_k$ at random from our dataset.

- Construct the cluster C_1 as the set of data points closest to c_1, C_2 as the set of data points closest to c_2, and so forth.
- Repeat the following until the clusters no longer change:
    - Find the centroids $c_1, c_2, \ldots, c_k$ of the current clusters.
    - Update the clusters $C_1, C_2, \ldots, C_k$.

The clusterings we find depend on the initial random choice of points $c_1, c_2, \ldots, c_k$. For instance, in the previous activity, we arrived, with the initial choice $c_1 = \mathbf{v}_2$ and $c_2 = \mathbf{v}_3$, at the clustering:

$$\begin{aligned} C_1 &= \{\mathbf{v}_1\} \\ C_2 &= \{\mathbf{v}_2, \mathbf{v}_3, \mathbf{v}_4\}. \end{aligned}$$

If we instead choose the initial points to be $c_1 = \mathbf{v}_3$ and $c_2 = \mathbf{v}_4$, we eventually find the clustering:

$$\begin{aligned} C_1 &= \{\mathbf{v}_1, \mathbf{v}_2, \mathbf{v}_3\} \\ C_2 &= \{\mathbf{v}_4\}. \end{aligned}$$

Is there a way that we can determine which clustering is the better of the two? It seems like a better clustering will be one for which the points in a cluster are, on average, closer to the centroid of their cluster. If we have a clustering, we therefore define a function, called the *objective*, which measures the average of the square of the distance from each point to the centroid of the cluster to which that point belongs. A clustering with a smaller objective will have clusters more tightly centered around their centroids, which should result in a better clustering.

For example, when we obtain the clustering:

$$\begin{aligned} C_1 &= \{\mathbf{v}_1, \mathbf{v}_2, \mathbf{v}_3\} \\ C_2 &= \{\mathbf{v}_4\}. \end{aligned}$$

with centroids $c_1 = \begin{bmatrix} 0 \\ 4/3 \end{bmatrix}$ and $c_2 = \mathbf{v}_4 = \begin{bmatrix} 3 \\ 2 \end{bmatrix}$, we find the objective to be

$$\frac{1}{4}\left(|\mathbf{v}_1 - c_1|^2 + |\mathbf{v}_2 - c_1|^2 + |\mathbf{v}_3 - c_1|^2 + |\mathbf{v}_4 - c_2|^2\right) = \frac{5}{3}.$$

Activity 6.1.5. We'll now use the objective to compare clusterings and to choose an appropriate value of k.

a. In the previous activity, one initial choice of c_1 and c_2 led to the clustering:

$$\begin{aligned} C_1 &= \{\mathbf{v}_1\} \\ C_2 &= \{\mathbf{v}_2, \mathbf{v}_3, \mathbf{v}_4\} \end{aligned}$$

with centroids $c_1 = \mathbf{v}_1$ and $c_2 = \begin{bmatrix} 5/3 \\ 5/3 \end{bmatrix}$. Find the objective of this clustering.

b. We have now seen two clusterings and computed their objectives. Recall that

6.1. THE DOT PRODUCT

our dataset is shown in Figure 6.1.12. Which of the two clusterings feels like the better fit? How is this fit reflected in the values of the objectives?

c. Evaluating the following cell will load and display a dataset consisting of 177 data points. This dataset has the name data.

```
url='https://raw.githubusercontent.com/davidaustinm/'
url+='ula_modules/master/k_means.py'
sage.repl.load.load(url, globals())
list_plot(data, color='blue', size=20, aspect_ratio=1)
```

Given this plot of the data, what would seem like a reasonable number of clusters?

d. In the following cell, you may choose a value of k and then run the algorithm to determine and display a clustering and its objective. If you run the algorithm a few times with the same value of k, you will likely see different clusterings having different objectives. This is natural since our algorithm starts by making a random choice of points $c_1, c_2, \ldots, c_k$, and a different choices may lead to different clusterings. Choose a value of k and run the algorithm a few times. Notice that clusterings having lower objectives seem to fit the data better. Repeat this experiment with a few different values of k.

```
k = 2    # you may change the value of k here
clusters, centroids, objective = kmeans(data, k)
print('Objective =', objective)
plotclusters(clusters, centroids)
```

e. For a given value of k, our strategy is to run the algorithm several times and choose the clustering with the smallest objective. After choosing a value of k, the following cell will run the algorithm 10 times and display the clustering having the smallest objective.

```
k = 2    # you may change the value of k here
clusters, centroids, objective = minimalobjective(data, k)
print('Objective =', objective)
plotclusters(clusters, centroids)
```

For each value of k between 2 and 9, find the clustering having the smallest objective and plot your findings in Figure 6.1.15.

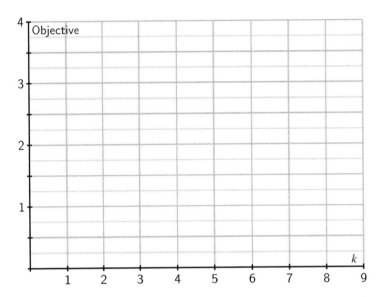

Figure 6.1.15 Construct a plot of the minimal objective as it depends on the choice of k.

This plot is called an *elbow plot* due to its shape. Notice how the objective decreases sharply when k is small and then flattens out. This leads to a location, called the elbow, where the objective transitions from being sharply decreasing to relatively flat. This means that increasing k beyond the elbow does not significantly decrease the objective, which makes the elbow a good choice for k.

Where does the elbow occur in your plot above? How does this compare to the best value of k that you estimated by simply looking at the data in Item c.

Of course, we could increase k until each data point is its own cluster. However, this defeats the point of the technique, which is to group together nearby data points in the hope that they share common features, thus providing insight into the structure of the data.

We have now seen how our algorithm and the objective identify a reasonable value for k, the number of the clusters, and produce a good clustering having k clusters. Notice that we don't claim to have found the best clustering as the true test of any clustering will be in how it helps us understand the dataset and helps us make predictions about any new data that we may encounter.

6.1.3 Summary

This section introduced the dot product and the ability to investigate geometric relationships between vectors.

6.1. THE DOT PRODUCT

- The dot product of two vectors $\mathbf{v}$ and $\mathbf{w}$ satisfies these properties:

$$\mathbf{v} \cdot \mathbf{v} = |\mathbf{v}|^2$$
$$\mathbf{v} \cdot \mathbf{w} = |\mathbf{v}||\mathbf{w}|\cos\theta$$

  where θ is the angle between $\mathbf{v}$ and $\mathbf{w}$.

- The vectors $\mathbf{v}$ and $\mathbf{w}$ are orthogonal when $\mathbf{v} \cdot \mathbf{w} = 0$.

- We explored some applications of the dot product to the similarity of vectors, correlation of time series, and k-means clustering.

6.1.4 Exercises

1. Consider the vectors

$$\mathbf{v} = \begin{bmatrix} 2 \\ 0 \\ 3 \\ -2 \end{bmatrix}, \quad \mathbf{w} = \begin{bmatrix} 1 \\ -3 \\ 4 \\ 1 \end{bmatrix}.$$

   a. Find the lengths of the vectors, $|\mathbf{v}|$ and $|\mathbf{w}|$.

   b. Find the dot product $\mathbf{v} \cdot \mathbf{w}$ and use it to find the angle θ between $\mathbf{v}$ and $\mathbf{w}$.

2. Consider the three vectors

$$\mathbf{u} = \begin{bmatrix} 1 \\ -2 \\ 2 \end{bmatrix}, \quad \mathbf{v} = \begin{bmatrix} 1 \\ 1 \\ 1 \end{bmatrix}, \quad \mathbf{w} = \begin{bmatrix} 0 \\ 2 \\ -1 \end{bmatrix}.$$

   a. Find the dot products $\mathbf{u} \cdot \mathbf{u}$, $\mathbf{u} \cdot \mathbf{v}$, and $\mathbf{u} \cdot \mathbf{w}$.

   b. Use the dot products you just found to evaluate:

   1. $|\mathbf{u}|$.
   2. $(-5\mathbf{u}) \cdot \mathbf{v}$.
   3. $\mathbf{u} \cdot (-3\mathbf{v} + 2\mathbf{w})$.
   4. $\left|\frac{1}{|\mathbf{u}|}\mathbf{u}\right|$.

   c. For what value of k is $\mathbf{u}$ orthogonal to $k\mathbf{v} + 5\mathbf{w}$?

3. Suppose that $\mathbf{v}$ and $\mathbf{w}$ are vectors where

$$\mathbf{v} \cdot \mathbf{v} = 4, \quad \mathbf{w} \cdot \mathbf{w} = 20, \quad \mathbf{v} \cdot \mathbf{w} = 8.$$

   a. What is $|\mathbf{v}|$?

   b. What is the angle between $\mathbf{v}$ and $\mathbf{w}$?

   c. Suppose that t is a scalar. Find the value of t for which $\mathbf{v}$ is orthogonal to $\mathbf{w} + t\mathbf{v}$?

4. Suppose that $\mathbf{v} = 3\mathbf{w}$.

   a. What is the relationship between $\mathbf{v} \cdot \mathbf{v}$ and $\mathbf{w} \cdot \mathbf{w}$?

   b. What is the relationship between $|\mathbf{v}|$ and $|\mathbf{w}|$?

   c. If $\mathbf{v} = s\mathbf{w}$ for some scalar s, what is the relationship between $\mathbf{v} \cdot \mathbf{v}$ and $\mathbf{w} \cdot \mathbf{w}$? What is the relationship between $|\mathbf{v}|$ and $|\mathbf{w}|$?

   d. Suppose that $\mathbf{v} = \begin{bmatrix} 3 \\ -2 \\ 2 \end{bmatrix}$. Find a scalar s so that $s\mathbf{v}$ has length 1.

5. Given vectors $\mathbf{v}$ and $\mathbf{w}$, explain why
$$|\mathbf{v} + \mathbf{w}|^2 + |\mathbf{v} - \mathbf{w}|^2 = 2|\mathbf{v}|^2 + 2|\mathbf{w}|^2.$$
Sketch two vectors $\mathbf{v}$ and $\mathbf{w}$ and explain why this fact is called the *parallelogram law*.

6. Consider the vectors
$$\mathbf{v}_1 = \begin{bmatrix} 2 \\ 0 \\ 4 \end{bmatrix}, \quad \mathbf{v}_2 = \begin{bmatrix} -1 \\ 2 \\ -4 \end{bmatrix}.$$
and a general vector $\mathbf{x} = \begin{bmatrix} x \\ y \\ z \end{bmatrix}$.

   a. Write an equation in terms of x, y, and z that describes all the vectors $\mathbf{x}$ orthogonal to $\mathbf{v}_1$.

   b. Write a linear system that describes all the vectors $\mathbf{x}$ orthogonal to both $\mathbf{v}_1$ and $\mathbf{v}_2$.

   c. Write the solution set to this linear system in parametric form. What type of geometric object does this solution set represent? Indicate with a rough sketch why this makes sense.

   d. Give a parametric description of all vectors orthogonal to $\mathbf{v}_1$. What type of geometric object does this represent? Indicate with a rough sketch why this makes sense.

7. Explain your responses to these questions.

   a. Suppose that $\mathbf{v}$ is orthogonal to both $\mathbf{w}_1$ and $\mathbf{w}_2$. Can you guarantee that $\mathbf{v}$ is also orthogonal to any linear combination $c_1\mathbf{w}_1 + c_2\mathbf{w}_2$?

   b. Suppose that $\mathbf{v}$ is orthogonal to itself. What can you say about $\mathbf{v}$?

8. Suppose that $\mathbf{v}_1$, $\mathbf{v}_2$, and $\mathbf{v}_3$ form a basis for $\mathbb{R}^3$ and that each vector is orthogonal to the other two. Suppose also that $\mathbf{v}$ is another vector in $\mathbb{R}^3$.

   a. Explain why $\mathbf{v} = c_1\mathbf{v}_1 + c_2\mathbf{v}_2 + c_3\mathbf{v}_3$ for some scalars c_1, c_2, and c_3.

   b. Beginning with the expression
$$\mathbf{v} \cdot \mathbf{v}_1 = (c_1\mathbf{v}_1 + c_2\mathbf{v}_2 + c_3\mathbf{v}_3) \cdot \mathbf{v}_1,$$

6.1. THE DOT PRODUCT

apply the distributive property of dot products to explain why

$$c_1 = \frac{\mathbf{v} \cdot \mathbf{v}_1}{\mathbf{v}_1 \cdot \mathbf{v}_1}.$$

Find similar expressions for c_2 and c_3.

c. Verify that

$$\mathbf{v}_1 = \begin{bmatrix} 1 \\ 2 \\ 1 \end{bmatrix}, \quad \mathbf{v}_2 = \begin{bmatrix} 1 \\ -1 \\ 1 \end{bmatrix}, \quad \mathbf{v}_3 = \begin{bmatrix} 1 \\ 0 \\ -1 \end{bmatrix}$$

form a basis for $\mathbb{R}^3$ and that each vector is orthogonal to the other two. Use what you've discovered in this problem to write the vector $\mathbf{v} = \begin{bmatrix} 3 \\ 5 \\ -1 \end{bmatrix}$ as a linear combination of $\mathbf{v}_1$, $\mathbf{v}_2$, and $\mathbf{v}_3$.

9. Suppose that $\mathbf{v}_1$, $\mathbf{v}_2$, and $\mathbf{v}_3$ are three nonzero vectors that are pairwise orthogonal; that is, each vector is orthogonal to the other two.

   a. Explain why $\mathbf{v}_3$ cannot be a linear combination of $\mathbf{v}_1$ and $\mathbf{v}_2$.

   b. Explain why this set of three vectors is linearly independent.

10. In the next chapter, we will consider certain $n \times n$ matrices A and define a function

    $$q(\mathbf{x}) = \mathbf{x} \cdot (A\mathbf{x}),$$

    where $\mathbf{x}$ is a vector in $\mathbb{R}^n$.

    a. Suppose that $A = \begin{bmatrix} 1 & 2 \\ 2 & 1 \end{bmatrix}$ and $\mathbf{x} = \begin{bmatrix} 2 \\ 1 \end{bmatrix}$. Evaluate $q(\mathbf{x}) = \mathbf{x} \cdot (A\mathbf{x})$.

    b. For a general vector $\mathbf{x} = \begin{bmatrix} x \\ y \end{bmatrix}$, evaluate $q(\mathbf{x}) = \mathbf{x} \cdot (A\mathbf{x})$ as an expression involving x and y.

    c. Suppose that $\mathbf{v}$ is an eigenvector of a matrix A with associated eigenvalue λ and that $\mathbf{v}$ has length 1. What is the value of the function $q(\mathbf{v})$?

11. Back in Section 1.1, we saw that equations of the form $Ax + By = C$ represent lines in the plane. In this exercise, we will see how this expression arises geometrically.

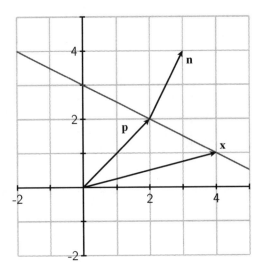

Figure 6.1.16 A line, a point **p** on the line, and a vector **n** perpendicular to the line.

a. Find the slope and vertical intercept of the line shown in Figure 6.1.16. Then write an equation for the line in the form $y = mx + b$.

b. Suppose that **p** is a point on the line, that **n** is a vector perpendicular to the line, and that $\mathbf{x} = \begin{bmatrix} x \\ y \end{bmatrix}$ is a general point on the line. Sketch the vector $\mathbf{x} - \mathbf{p}$ and describe the angle between this vector and the vector **n**.

c. What is the value of the dot product $\mathbf{n} \cdot (\mathbf{x} - \mathbf{p})$?

d. Explain why the equation of the line can be written in the form $\mathbf{n} \cdot \mathbf{x} = \mathbf{n} \cdot \mathbf{p}$.

e. Identify the vectors **p** and **n** for the line illustrated in Figure 6.1.16 and use them to write the equation of the line in terms of x and y. Verify that this expression is algebraically equivalent to the equation $y = mx + b$ that you earlier found for this line.

f. Explain why any line in the plane can be described by an equation having the form $Ax + By = C$. What is the significance of the vector $\begin{bmatrix} A \\ B \end{bmatrix}$?

6.2 Orthogonal complements and the matrix transpose

We've now seen how the dot product enables us to determine the angle between two vectors and, more specifically, when two vectors are orthogonal. Moving forward, we will explore how the orthogonality condition simplifies many common tasks, such as expressing a vector as a linear combination of a given set of vectors.

This section introduces the notion of an orthogonal complement, the set of vectors each of which is orthogonal to a prescribed subspace. We'll also find a way to describe dot products using matrix products, which allows us to study orthogonality using many of the tools for understanding linear systems that we developed earlier.

Preview Activity 6.2.1.

a. Sketch the vector $\mathbf{v} = \begin{bmatrix} -1 \\ 2 \end{bmatrix}$ on Figure 6.2.1 and one vector that is orthogonal to it.

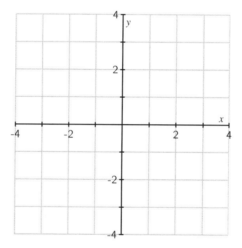

Figure 6.2.1 Sketch the vector $\mathbf{v}$ and one vector orthogonal to it.

b. If a vector $\mathbf{x}$ is orthogonal to $\mathbf{v}$, what do we know about the dot product $\mathbf{v} \cdot \mathbf{x}$?

c. If we write $\mathbf{x} = \begin{bmatrix} x \\ y \end{bmatrix}$, use the dot product to write an equation for the vectors orthogonal to $\mathbf{v}$ in terms of x and y.

d. Use this equation to sketch the set of all vectors orthogonal to $\mathbf{v}$ in Figure 6.2.1.

e. Section 3.5 introduced the column space Col(A) and null space Nul(A) of a matrix A. If A is a matrix, what is the meaning of the null space Nul(A)?

f. What is the meaning of the column space Col(A)?

6.2.1 Orthogonal complements

The preview activity presented us with a vector **v** and led us through the process of describing all the vectors orthogonal to **v**. Notice that the set of scalar multiples of **v** describes a line L, a 1-dimensional subspace of $\mathbb{R}^2$. We then described a second line consisting of all the vectors orthogonal to **v**. Notice that every vector on this line is orthogonal to every vector on the line L. We call this new line the *orthogonal complement* of L and denote it by $L^\perp$. The lines L and $L^\perp$ are illustrated on the left of Figure 6.2.2.

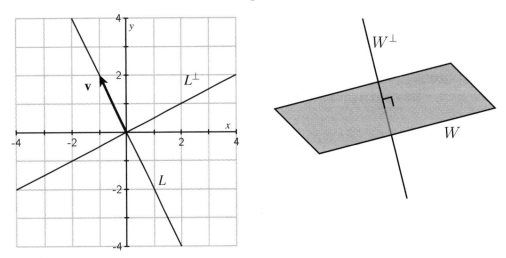

Figure 6.2.2 On the left is a line L and its orthogonal complement $L^\perp$. On the right is a plane W and its orthogonal complement $W^\perp$ in $\mathbb{R}^3$.

The next definition places this example into a more general context.

Definition 6.2.3 Given a subspace W of $\mathbb{R}^m$, the **orthogonal complement** of W is the set of vectors in $\mathbb{R}^m$ each of which is orthogonal to every vector in W. We denote the orthogonal complement by $W^\perp$.

A typical example appears on the right of Figure 6.2.2. Here we see a plane W, a two-dimensional subspace of $\mathbb{R}^3$, and its orthogonal complement $W^\perp$, which is a line in $\mathbb{R}^3$.

As the next activity demonstrates, the orthogonal complement of a subspace W is itself a subspace of $\mathbb{R}^m$.

Activity 6.2.2. Suppose that $\mathbf{w}_1 = \begin{bmatrix} 1 \\ 0 \\ -2 \end{bmatrix}$ and $\mathbf{w}_2 = \begin{bmatrix} 1 \\ 1 \\ -1 \end{bmatrix}$ form a basis for W, a two-dimensional subspace of $\mathbb{R}^3$. We will find a description of the orthogonal complement $W^\perp$.

a. Suppose that the vector **x** is orthogonal to $\mathbf{w}_1$. If we write $\mathbf{x} = \begin{bmatrix} x_1 \\ x_2 \\ x_3 \end{bmatrix}$, use the

6.2. ORTHOGONAL COMPLEMENTS AND THE MATRIX TRANSPOSE

fact that $\mathbf{w}_1 \cdot \mathbf{x} = 0$ to write a linear equation for x_1, x_2, and x_3.

b. Suppose that $\mathbf{x}$ is also orthogonal to $\mathbf{w}_2$. In the same way, write a linear equation for x_1, x_2, and x_3 that arises from the fact that $\mathbf{w}_2 \cdot \mathbf{x} = 0$.

c. If $\mathbf{x}$ is orthogonal to both $\mathbf{w}_1$ and $\mathbf{w}_2$, these two equations give us a linear system $B\mathbf{x} = \mathbf{0}$ for some matrix B. Identify the matrix B and write a parametric description of the solution space to the equation $B\mathbf{x} = \mathbf{0}$.

d. Since $\mathbf{w}_1$ and $\mathbf{w}_2$ form a basis for the two-dimensional subspace W, any vector $\mathbf{w}$ in W can be written as a linear combination

$$\mathbf{w} = c_1\mathbf{w}_1 + c_2\mathbf{w}_2.$$

If $\mathbf{x}$ is orthogonal to both $\mathbf{w}_1$ and $\mathbf{w}_2$, use the distributive property of dot products to explain why $\mathbf{x}$ is orthogonal to $\mathbf{w}$.

e. Give a basis for the orthogonal complement $W^\perp$ and state the dimension $\dim W^\perp$.

f. Describe $(W^\perp)^\perp$, the orthogonal complement of $W^\perp$.

Example 6.2.4 If L is the line defined by $\mathbf{v} = \begin{bmatrix} 1 \\ -2 \\ 3 \end{bmatrix}$ in $\mathbb{R}^3$, we will describe the orthogonal complement $L^\perp$, the set of vectors orthogonal to L.

If $\mathbf{x}$ is orthogonal to L, it must be orthogonal to $\mathbf{v}$ so we have

$$\mathbf{v} \cdot \mathbf{x} = x_1 - 2x_2 + 3x_3 = 0.$$

We can describe the solutions to this equation parametrically as

$$\mathbf{x} = \begin{bmatrix} x_1 \\ x_2 \\ x_3 \end{bmatrix} = \begin{bmatrix} 2x_2 - 3x_3 \\ x_2 \\ x_3 \end{bmatrix} = x_2 \begin{bmatrix} 2 \\ 1 \\ 0 \end{bmatrix} + x_3 \begin{bmatrix} -3 \\ 0 \\ 1 \end{bmatrix}.$$

Therefore, the orthogonal complement $L^\perp$ is a plane, a two-dimensional subspace of $\mathbb{R}^3$, spanned by the vectors $\begin{bmatrix} 2 \\ 1 \\ 0 \end{bmatrix}$ and $\begin{bmatrix} -3 \\ 0 \\ 1 \end{bmatrix}$.

Example 6.2.5 Suppose that W is the 2-dimensional subspace of $\mathbb{R}^5$ with basis

$$\mathbf{w}_1 = \begin{bmatrix} -1 \\ -2 \\ 2 \\ 3 \\ -4 \end{bmatrix}, \quad \mathbf{w}_2 = \begin{bmatrix} 2 \\ 4 \\ 2 \\ 0 \\ 2 \end{bmatrix}.$$

We will give a description of the orthogonal complement $W^\perp$.

If **x** is in $W^\perp$, we know that **x** is orthogonal to both $\mathbf{w}_1$ and $\mathbf{w}_2$. Therefore,

$$\mathbf{w}_1 \cdot \mathbf{x} = -x_1 - 2x_2 + 2x_3 + 3x_4 - 4x_5 = 0$$
$$\mathbf{w}_2 \cdot \mathbf{x} = 2x_1 + 4x_2 + 2x_3 + 0x_4 + 2x_5 = 0$$

In other words, $B\mathbf{x} = \mathbf{0}$ where

$$B = \begin{bmatrix} -1 & -2 & 2 & 3 & -4 \\ 2 & 4 & 2 & 0 & 2 \end{bmatrix} \sim \begin{bmatrix} 1 & 2 & 0 & -1 & 2 \\ 0 & 0 & 1 & 1 & -1 \end{bmatrix}.$$

The solutions may be described parametrically as

$$\mathbf{x} = \begin{bmatrix} x_1 \\ x_2 \\ x_3 \\ x_4 \\ x_5 \end{bmatrix} = x_2 \begin{bmatrix} -2 \\ 1 \\ 0 \\ 0 \\ 0 \end{bmatrix} + x_4 \begin{bmatrix} 1 \\ 0 \\ -1 \\ 1 \\ 0 \end{bmatrix} + x_5 \begin{bmatrix} -2 \\ 0 \\ 1 \\ 0 \\ 1 \end{bmatrix}.$$

The distributive property of dot products implies that any vector that is orthogonal to both $\mathbf{w}_1$ and $\mathbf{w}_2$ is also orthogonal to any linear combination of $\mathbf{w}_1$ and $\mathbf{w}_2$ since

$$(c_1 \mathbf{w}_1 + c_2 \mathbf{w}_2) \cdot \mathbf{x} = c_1 \mathbf{w}_1 \cdot \mathbf{x} + c_2 \mathbf{w}_2 \cdot \mathbf{x} = 0.$$

Therefore, $W^\perp$ is a 3-dimensional subspace of $\mathbb{R}^5$ with basis

$$\mathbf{v}_1 = \begin{bmatrix} -2 \\ 1 \\ 0 \\ 0 \\ 0 \end{bmatrix}, \quad \mathbf{v}_2 = \begin{bmatrix} 1 \\ 0 \\ -1 \\ 1 \\ 0 \end{bmatrix}, \quad \mathbf{v}_3 = \begin{bmatrix} -2 \\ 0 \\ 1 \\ 0 \\ 1 \end{bmatrix}.$$

One may check that the vectors $\mathbf{v}_1$, $\mathbf{v}_2$, and $\mathbf{v}_3$ are each orthogonal to both $\mathbf{w}_1$ and $\mathbf{w}_2$.

6.2.2 The matrix transpose

The previous activity and examples show how we can describe the orthogonal complement of a subspace as the solution set of a particular linear system. We will make this connection more explicit by defining a new matrix operation called the *transpose*.

Definition 6.2.6 The **transpose** of the $m \times n$ matrix A is the $n \times m$ matrix A^T whose rows are the columns of A.

Example 6.2.7 If $A = \begin{bmatrix} 4 & -3 & 0 & 5 \\ -1 & 2 & 1 & 3 \end{bmatrix}$, then $A^T = \begin{bmatrix} 4 & -1 \\ -3 & 2 \\ 0 & 1 \\ 5 & 3 \end{bmatrix}$

6.2. ORTHOGONAL COMPLEMENTS AND THE MATRIX TRANSPOSE

Activity 6.2.3. This activity illustrates how multiplying a vector by A^T is related to computing dot products with the columns of A. You'll develop a better understanding of this relationship if you compute the dot products and matrix products in this activity without using technology.

a. If $B = \begin{bmatrix} 3 & 4 \\ -1 & 2 \\ 0 & -2 \end{bmatrix}$, write the matrix B^T.

b. Suppose that
$$\mathbf{v}_1 = \begin{bmatrix} 2 \\ 0 \\ -2 \end{bmatrix}, \quad \mathbf{v}_2 = \begin{bmatrix} 1 \\ 1 \\ 2 \end{bmatrix}, \quad \mathbf{w} = \begin{bmatrix} -2 \\ 2 \\ 3 \end{bmatrix}.$$

Find the dot products $\mathbf{v}_1 \cdot \mathbf{w}$ and $\mathbf{v}_2 \cdot \mathbf{w}$.

c. Now write the matrix $A = \begin{bmatrix} \mathbf{v}_1 & \mathbf{v}_2 \end{bmatrix}$ and its transpose A^T. Find the product $A^T \mathbf{w}$ and describe how this product computes both dot products $\mathbf{v}_1 \cdot \mathbf{w}$ and $\mathbf{v}_2 \cdot \mathbf{w}$.

d. Suppose that $\mathbf{x}$ is a vector that is orthogonal to both $\mathbf{v}_1$ and $\mathbf{v}_2$. What does this say about the dot products $\mathbf{v}_1 \cdot \mathbf{x}$ and $\mathbf{v}_2 \cdot \mathbf{x}$? What does this say about the product $A^T \mathbf{x}$?

e. Use the matrix A^T to give a parametric description of all the vectors $\mathbf{x}$ that are orthogonal to $\mathbf{v}_1$ and $\mathbf{v}_2$.

f. Remember that $\text{Nul}(A^T)$, the null space of A^T, is the solution set of the equation $A^T \mathbf{x} = \mathbf{0}$. If $\mathbf{x}$ is a vector in $\text{Nul}(A^T)$, explain why $\mathbf{x}$ must be orthogonal to both $\mathbf{v}_1$ and $\mathbf{v}_2$.

g. Remember that $\text{Col}(A)$, the column space of A, is the set of linear combinations of the columns of A. Therefore, any vector in $\text{Col}(A)$ can be written as $c_1 \mathbf{v}_1 + c_2 \mathbf{v}_2$. If $\mathbf{x}$ is a vector in $\text{Nul}(A^T)$, explain why $\mathbf{x}$ is orthogonal to every vector in $\text{Col}(A)$.

The previous activity demonstrates an important connection between the matrix transpose and dot products. More specifically, the components of the product $A^T \mathbf{x}$ are simply the dot products of the columns of A with $\mathbf{x}$. We will make frequent use of this observation so let's record it as a proposition.

Proposition 6.2.8 *If A is the matrix whose columns are $\mathbf{v}_1, \mathbf{v}_2, \ldots, \mathbf{v}_n$, then*

$$A^T \mathbf{x} = \begin{bmatrix} \mathbf{v}_1 \cdot \mathbf{x} \\ \mathbf{v}_2 \cdot \mathbf{x} \\ \vdots \\ \mathbf{v}_n \cdot \mathbf{x} \end{bmatrix}$$

Example 6.2.9 Suppose that W is a subspace of $\mathbb{R}^4$ having basis

$$\mathbf{w}_1 = \begin{bmatrix} 1 \\ 0 \\ 2 \\ 1 \end{bmatrix}, \quad \mathbf{w}_2 = \begin{bmatrix} 2 \\ 1 \\ 3 \\ 4 \end{bmatrix},$$

and that we wish to describe the orthogonal complement $W^\perp$.

If A is the matrix $A = \begin{bmatrix} \mathbf{w}_1 & \mathbf{w}_2 \end{bmatrix}$ and $\mathbf{x}$ is in $W^\perp$, we have

$$A^T \mathbf{x} = \begin{bmatrix} \mathbf{w}_1 \cdot \mathbf{x} \\ \mathbf{w}_2 \cdot \mathbf{x} \end{bmatrix} = \begin{bmatrix} 0 \\ 0 \end{bmatrix}.$$

Describing vectors $\mathbf{x}$ that are orthogonal to both $\mathbf{w}_1$ and $\mathbf{w}_2$ is therefore equivalent to the more familiar task of describing the solution set $A^T \mathbf{x} = \mathbf{0}$. To do so, we find the reduced row echelon form of A^T and write the solution set parametrically as

$$\mathbf{x} = x_3 \begin{bmatrix} -2 \\ 1 \\ 1 \\ 0 \end{bmatrix} + x_4 \begin{bmatrix} -1 \\ -2 \\ 0 \\ 1 \end{bmatrix}.$$

Once again, the distributive property of dot products tells us that such a vector is also orthogonal to any linear combination of $\mathbf{w}_1$ and $\mathbf{w}_2$ so this solution set is, in fact, the orthogonal complement $W^\perp$. Indeed, we see that the vectors

$$\mathbf{v}_1 = \begin{bmatrix} -2 \\ 1 \\ 1 \\ 0 \end{bmatrix}, \quad \mathbf{v}_2 = \begin{bmatrix} -1 \\ -2 \\ 0 \\ 1 \end{bmatrix}$$

form a basis for $W^\perp$, which is a two-dimensional subspace of $\mathbb{R}^4$.

To place this example in a slightly more general context, note that $\mathbf{w}_1$ and $\mathbf{w}_2$, the columns of A, form a basis of W. Since $\text{Col}(A)$, the column space of A is the subspace of linear combinations of the columns of A, we have $W = \text{Col}(A)$.

This example also shows that the orthogonal complement $W^\perp = \text{Col}(A)^\perp$ is described by the solution set of $A^T \mathbf{x} = \mathbf{0}$. This solution set is what we have called $\text{Nul}(A^T)$, the null space of A^T. In this way, we see the following proposition, which is visually represented in Figure 6.2.11.

Proposition 6.2.10 *For any matrix A, the orthogonal complement of $\text{Col}(A)$ is $\text{Nul}(A^T)$; that is,*

$$\text{Col}(A)^\perp = \text{Nul}(A^T).$$

6.2. ORTHOGONAL COMPLEMENTS AND THE MATRIX TRANSPOSE

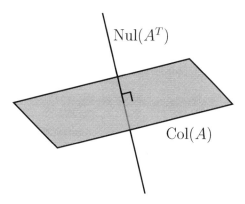

Figure 6.2.11 The orthogonal complement of the column space of A is the null space of A^T.

6.2.3 Properties of the matrix transpose

The transpose is a simple algebraic operation performed on a matrix. The next activity explores some of its properties.

Activity 6.2.4. In Sage, the transpose of a matrix A is given by A.T. Define the matrices

$$A = \begin{bmatrix} 1 & 0 & -3 \\ 2 & -2 & 1 \end{bmatrix}, \quad B = \begin{bmatrix} 3 & -4 & 1 \\ 0 & 1 & 2 \end{bmatrix}, \quad C = \begin{bmatrix} 1 & 0 & -3 \\ 2 & -2 & 1 \\ 3 & 2 & 0 \end{bmatrix}.$$

a. Evaluate $(A + B)^T$ and $A^T + B^T$. What do you notice about the relationship between these two matrices?

b. What happens if you transpose a matrix twice; that is, what is $(A^T)^T$?

c. Find $\det(C)$ and $\det(C^T)$. What do you notice about the relationship between these determinants?

d.   1. Find the product AC and its transpose $(AC)^T$.
   2. Is it possible to compute the product $A^T C^T$? Explain why or why not.
   3. Find the product $C^T A^T$ and compare it to $(AC)^T$. What do you notice about the relationship between these two matrices?

e. What is the transpose of the identity matrix I?

f. If a square matrix D is invertible, explain why you can guarantee that D^T is invertible and why $(D^T)^{-1} = (D^{-1})^T$.

In spite of the fact that we are looking at some specific examples, this activity demonstrates the following general properties of the transpose, which may be verified with a little effort.

> **Properties of the transpose.**
>
> Here are some properties of the matrix transpose, expressed in terms of general matrices A, B, and C. We assume that C is a square matrix.
>
> - If $A + B$ is defined, then $(A + B)^T = A^T + B^T$.
> - $(sA)^T = sA^T$.
> - $(A^T)^T = A$.
> - $\det(C) = \det(C^T)$.
> - If AB is defined, then $(AB)^T = B^T A^T$. Notice that the order of the multiplication is reversed.
> - $(C^T)^{-1} = (C^{-1})^T$.

There is one final property we wish to record though we will wait until Section 7.4 to explain why it is true.

Proposition 6.2.12 *For any matrix A, we have*

$$\operatorname{rank}(A) = \operatorname{rank}(A^T).$$

This proposition is important because it implies a relationship between the dimensions of a subspace and its orthogonal complement. For instance, if A is an $m \times n$ matrix, we saw in Section 3.5 that $\dim \operatorname{Col}(A) = \operatorname{rank}(A)$ and $\dim \operatorname{Nul}(A) = n - \operatorname{rank}(A)$.

Now suppose that W is an n-dimensional subspace of $\mathbb{R}^m$ with basis $\mathbf{w}_1, \mathbf{w}_2, \ldots, \mathbf{w}_n$. If we form the $m \times n$ matrix $A = \begin{bmatrix} \mathbf{w}_1 & \mathbf{w}_2 & \ldots & \mathbf{w}_n \end{bmatrix}$, then $\operatorname{Col}(A) = W$ so that

$$\operatorname{rank}(A) = \dim \operatorname{Col}(A) = \dim W = n.$$

The transpose A^T is an $n \times m$ matrix having $\operatorname{rank}(A^T) = \operatorname{rank}(A) = n$. Since $W^\perp = \operatorname{Nul}(A^T)$, we have

$$\dim W^\perp = \dim \operatorname{Nul}(A^T) = m - \operatorname{rank}(A^T) = m - n = m - \dim W.$$

This explains the following proposition.

Proposition 6.2.13 *If W is a subspace of $\mathbb{R}^m$, then*

$$\dim W + \dim W^\perp = m.$$

Example 6.2.14 In Example 6.2.4, we constructed the orthogonal complement of a line in $\mathbb{R}^3$. The dimension of the orthogonal complement should be $3 - 1 = 2$, which explains why we found the orthogonal complement to be a plane.

Example 6.2.15 In Example 6.2.5, we looked at W, a 2-dimensional subspace of $\mathbb{R}^5$ and found its orthogonal complement $W^\perp$ to be a $5 - 2 = 3$-dimensional subspace of $\mathbb{R}^5$.

6.2. ORTHOGONAL COMPLEMENTS AND THE MATRIX TRANSPOSE

Activity 6.2.5.

a. Suppose that W is a 5-dimensional subspace of $\mathbb{R}^9$ and that A is a matrix whose columns form a basis for W; that is, $\text{Col}(A) = W$.

1. What is the shape of A?
2. What is the rank of A?
3. What is the shape of A^T?
4. What is the rank of A^T?
5. What is $\dim \text{Nul}(A^T)$?
6. What is $\dim W^\perp$?
7. How are the dimensions of W and $W^\perp$ related?

b. Suppose that W is a subspace of $\mathbb{R}^4$ having basis

$$\mathbf{w}_1 = \begin{bmatrix} 1 \\ 0 \\ 2 \\ -1 \end{bmatrix}, \quad \mathbf{w}_2 = \begin{bmatrix} -1 \\ 2 \\ -6 \\ 3 \end{bmatrix}.$$

1. Find the dimensions $\dim W$ and $\dim W^\perp$.
2. Find a basis for $W^\perp$. It may be helpful to know that the Sage command `A.right_kernel()` produces a basis for $\text{Nul}(A)$.

3. Verify that each of the basis vectors you found for $W^\perp$ are orthogonal to the basis vectors for W.

6.2.4 Summary

This section introduced the matrix transpose, its connection to dot products, and its use in describing the orthogonal complement of a subspace.

- The columns of the matrix A are the rows of the matrix transpose A^T.
- The components of the product $A^T\mathbf{x}$ are the dot products of $\mathbf{x}$ with the columns of A.
- The orthogonal complement of the column space of A equals the null space of A^T; that is, $\text{Col}(A)^\perp = \text{Nul}(A^T)$.
- If W is a subspace of $\mathbb{R}^p$, then

$$\dim W + \dim W^\perp = p.$$

6.2.5 Exercises

1. Suppose that W is a subspace of $\mathbb{R}^4$ with basis

$$\mathbf{w}_1 = \begin{bmatrix} -2 \\ 2 \\ 2 \\ -4 \end{bmatrix}, \quad \mathbf{w}_2 = \begin{bmatrix} -2 \\ 3 \\ 5 \\ -5 \end{bmatrix}.$$

   a. What are the dimensions $\dim W$ and $\dim W^\perp$?

   b. Find a basis for $W^\perp$.

   c. Verify that each of the basis vectors for $W^\perp$ are orthogonal to $\mathbf{w}_1$ and $\mathbf{w}_2$.

2. Consider the matrix $A = \begin{bmatrix} -1 & -2 & -2 \\ 1 & 3 & 4 \\ 2 & 1 & -2 \end{bmatrix}$.

   a. Find $\text{rank}(A)$ and a basis for $\text{Col}(A)$.

   b. Determine the dimension of $\text{Col}(A)^\perp$ and find a basis for it.

3. Suppose that W is the subspace of $\mathbb{R}^4$ defined as the solution set of the equation

$$x_1 - 3x_2 + 5x_3 - 2x_4 = 0.$$

   a. What are the dimensions $\dim W$ and $\dim W^\perp$?

   b. Find a basis for W.

   c. Find a basis for $W^\perp$.

   d. In general, how can you easily find a basis for $W^\perp$ when W is defined by

$$Ax_1 + Bx_2 + Cx_3 + Dx_4 = 0?$$

4. Determine whether the following statements are true or false and explain your reasoning.

   a. If $A = \begin{bmatrix} 2 & 1 \\ 1 & 1 \\ -3 & 1 \end{bmatrix}$, then $\mathbf{x} = \begin{bmatrix} 4 \\ -5 \\ 1 \end{bmatrix}$ is in $\text{Col}(A)^\perp$.

   b. If A is a 2×3 matrix and B is a 3×4 matrix, then $(AB)^T = A^T B^T$ is a 4×2 matrix.

   c. If the columns of A are $\mathbf{v}_1, \mathbf{v}_2,$ and $\mathbf{v}_3$ and $A^T \mathbf{x} = \begin{bmatrix} 2 \\ 0 \\ 1 \end{bmatrix}$, then $\mathbf{x}$ is orthogonal to $\mathbf{v}_2$.

   d. If A is a 4×4 matrix with $\text{rank}(A) = 3$, then $\text{Col}(A)^\perp$ is a line in $\mathbb{R}^4$.

   e. If A is a 5×7 matrix with $\text{rank}(A) = 5$, then $\text{rank}(A^T) = 7$.

6.2. ORTHOGONAL COMPLEMENTS AND THE MATRIX TRANSPOSE

5. Apply properties of matrix operations to simplify the following expressions.
   a. $A^T(BA^T)^{-1}$
   b. $(A+B)^T(A+B)$
   c. $[A(A+B)^T]^T$
   d. $(A+2I)^T$

6. A symmetric matrix A is one for which $A = A^T$.
   a. Explain why a symmetric matrix must be square.
   b. If A and B are general matrices and D is a square diagonal matrix, which of the following matrices can you guarantee are symmetric?
      1. D
      2. BAB^{-1}
      3. AA^T.
      4. BDB^T

7. If A is a square matrix, remember that the characteristic polynomial of A is $\det(A - \lambda I)$ and that the roots of the characteristic polynomial are the eigenvalues of A.
   a. Explain why A and A^T have the same characteristic polynomial.
   b. Explain why A and A^T have the same set of eigenvalues.
   c. Suppose that A is diagonalizable with diagonalization $A = PDP^{-1}$. Explain why A^T is diagonalizable and find a diagonalization.

8. This exercise introduces a version of the Pythagorean theorem that we'll use later.
   a. Suppose that $\mathbf{v}$ and $\mathbf{w}$ are orthogonal to one another. Use the dot product to explain why
   $$|\mathbf{v}+\mathbf{w}|^2 = |\mathbf{v}|^2 + |\mathbf{w}|^2.$$
   b. Suppose that W is a subspace of $\mathbb{R}^m$ and that $\mathbf{z}$ is a vector in $\mathbb{R}^m$ for which
   $$\mathbf{z} = \mathbf{x} + \mathbf{y},$$
   where $\mathbf{x}$ is in W and $\mathbf{y}$ is in $W^\perp$. Explain why
   $$|\mathbf{z}|^2 = |\mathbf{x}|^2 + |\mathbf{y}|^2,$$
   which is an expression of the Pythagorean theorem.

9. In the next chapter, symmetric matrices---that is, matrices for which $A = A^T$---play an important role. It turns out that eigenvectors of a symmetric matrix that are associated to different eigenvalues are orthogonal. We will explain this fact in this exercise.
   a. Viewing a vector as a matrix having one column, we may write $\mathbf{x} \cdot \mathbf{y} = \mathbf{x}^T \mathbf{y}$. If A is a matrix, explain why $\mathbf{x} \cdot (A\mathbf{y}) = (A^T \mathbf{x}) \cdot \mathbf{y}$.

b. We have seen that the matrix $A = \begin{bmatrix} 1 & 2 \\ 2 & 1 \end{bmatrix}$ has eigenvectors $\mathbf{v}_1 = \begin{bmatrix} 1 \\ 1 \end{bmatrix}$, with associated eigenvalue $\lambda_1 = 3$, and $\mathbf{v}_2 = \begin{bmatrix} 1 \\ -1 \end{bmatrix}$, with associated eigenvalue $\lambda_2 = -1$. Verify that A is symmetric and that $\mathbf{v}_1$ and $\mathbf{v}_2$ are orthogonal.

c. Suppose that A is a general symmetric matrix and that $\mathbf{v}_1$ is an eigenvector associated to eigenvalue λ_1 and that $\mathbf{v}_2$ is an eigenvector associated to a different eigenvalue λ_2. Beginning with $\mathbf{v}_1 \cdot (A\mathbf{v}_2)$, apply the identity from the first part of this exercise to explain why $\mathbf{v}_1$ and $\mathbf{v}_2$ are orthogonal.

10. Given an $m \times n$ matrix A, the *row space* of A is the column space of A^T; that is, $\text{Row}(A) = \text{Col}(A^T)$.

    a. Suppose that A is a 7×15 matrix. For what p is $\text{Row}(A)$ a subspace of $\mathbb{R}^p$?

    b. How can Proposition 6.2.10 help us describe $\text{Row}(A)^\perp$?

    c. Suppose that $A = \begin{bmatrix} -1 & -2 & 2 & 1 \\ 2 & 4 & -1 & 5 \\ 1 & 2 & 0 & 3 \end{bmatrix}$. Find bases for $\text{Row}(A)$ and $\text{Row}(A)^\perp$.

6.3 Orthogonal bases and projections

We know that a linear system $A\mathbf{x} = \mathbf{b}$ is inconsistent when $\mathbf{b}$ is not in $\operatorname{Col}(A)$, the column space of A. Later in this chapter, we'll develop a strategy for dealing with inconsistent systems by finding $\widehat{\mathbf{b}}$, the vector in $\operatorname{Col}(A)$ that minimizes the distance to $\mathbf{b}$. The equation $A\mathbf{x} = \widehat{\mathbf{b}}$ is therefore consistent and its solution set can provide us with useful information about the original system $A\mathbf{x} = \mathbf{b}$.

In this section and the next, we'll develop some techniques that enable us to find $\widehat{\mathbf{b}}$, the vector in a given subspace W that is closest to a given vector $\mathbf{b}$.

Preview Activity 6.3.1. For this activity, it will be helpful to recall the distributive property of dot products:

$$\mathbf{v} \cdot (c_1 \mathbf{w}_1 + c_2 \mathbf{w}_2) = c_1 \mathbf{v} \cdot \mathbf{w}_1 + c_2 \mathbf{v} \cdot \mathbf{w}_2.$$

We'll work with the basis of $\mathbb{R}^2$ formed by the vectors

$$\mathbf{w}_1 = \begin{bmatrix} 1 \\ 2 \end{bmatrix}, \quad \mathbf{w}_2 = \begin{bmatrix} -2 \\ 1 \end{bmatrix}.$$

a. Verify that the vectors $\mathbf{w}_1$ and $\mathbf{w}_2$ are orthogonal.

b. Suppose that $\mathbf{b} = \begin{bmatrix} 7 \\ 4 \end{bmatrix}$ and find the dot products $\mathbf{w}_1 \cdot \mathbf{b}$ and $\mathbf{w}_2 \cdot \mathbf{b}$.

c. We would like to express $\mathbf{b}$ as a linear combination of $\mathbf{w}_1$ and $\mathbf{w}_2$, which means that we need to find weights c_1 and c_2 such that

$$\mathbf{b} = c_1 \mathbf{w}_1 + c_2 \mathbf{w}_2.$$

To find the weight c_1, dot both sides of this expression with $\mathbf{w}_1$:

$$\mathbf{b} \cdot \mathbf{w}_1 = (c_1 \mathbf{w}_1 + c_2 \mathbf{w}_2) \cdot \mathbf{w}_1,$$

and apply the distributive property.

d. In a similar fashion, find the weight c_2.

e. Verify that $\mathbf{b} = c_1 \mathbf{w}_1 + c_2 \mathbf{w}_2$ using the weights you have found.

We frequently ask to write a given vector as a linear combination of given basis vectors. In the past, we have done this by solving a linear system. The preview activity illustrates how this task can be simplified when the basis vectors are orthogonal to each other. We'll explore this and other uses of orthogonal bases in this section.

6.3.1 Orthogonal sets

The preview activity dealt with a basis of $\mathbb{R}^2$ formed by two orthogonal vectors. More generally, we will consider a set of orthogonal vectors, as described in the next definition.

Definition 6.3.1 By an *orthogonal set* of vectors, we mean a set of nonzero vectors each of which is orthogonal to the others.

Example 6.3.2 The 3-dimensional vectors

$$\mathbf{w}_1 = \begin{bmatrix} 1 \\ -1 \\ 1 \end{bmatrix}, \quad \mathbf{w}_2 = \begin{bmatrix} 1 \\ 1 \\ 0 \end{bmatrix}, \quad \mathbf{w}_3 = \begin{bmatrix} 1 \\ -1 \\ -2 \end{bmatrix}.$$

form an orthogonal set, which can be verified by computing

$$\begin{aligned} \mathbf{w}_1 \cdot \mathbf{w}_2 &= 0 \\ \mathbf{w}_1 \cdot \mathbf{w}_3 &= 0 \\ \mathbf{w}_2 \cdot \mathbf{w}_3 &= 0. \end{aligned}$$

Notice that this set of vectors forms a basis for $\mathbb{R}^3$.

Example 6.3.3 The vectors

$$\mathbf{w}_1 = \begin{bmatrix} 1 \\ 1 \\ 1 \\ 1 \end{bmatrix}, \quad \mathbf{w}_2 = \begin{bmatrix} 1 \\ 1 \\ -1 \\ -1 \end{bmatrix}, \quad \mathbf{w}_3 = \begin{bmatrix} 1 \\ -1 \\ 1 \\ -1 \end{bmatrix}$$

form an orthogonal set of 4-dimensional vectors. Since there are only three vectors, this set does not form a basis for $\mathbb{R}^4$. It does, however, form a basis for a 3-dimensional subspace W of $\mathbb{R}^4$.

Suppose that a vector $\mathbf{b}$ is a linear combination of an orthogonal set of vectors $\mathbf{w}_1, \mathbf{w}_2, \ldots, \mathbf{w}_n$; that is, suppose that

$$c_1 \mathbf{w}_1 + c_2 \mathbf{w}_2 + \cdots + c_n \mathbf{w}_n = \mathbf{b}.$$

Just as in the preview activity, we can find the weight c_1 by dotting both sides with $\mathbf{w}_1$ and applying the distributive property of dot products:

$$\begin{aligned} (c_1 \mathbf{w}_1 + c_2 \mathbf{w}_2 + \cdots + c_n \mathbf{w}_n) \cdot \mathbf{w}_1 &= \mathbf{b} \cdot \mathbf{w}_1 \\ c_1 \mathbf{w}_1 \cdot \mathbf{w}_1 + c_2 \mathbf{w}_2 \cdot \mathbf{w}_1 + \cdots + c_n \mathbf{w}_n \cdot \mathbf{w}_1 &= \mathbf{b} \cdot \mathbf{w}_1 \\ c_1 \mathbf{w}_1 \cdot \mathbf{w}_1 &= \mathbf{b} \cdot \mathbf{w}_1 \\ c_1 &= \frac{\mathbf{b} \cdot \mathbf{w}_1}{\mathbf{w}_1 \cdot \mathbf{w}_1}. \end{aligned}$$

Notice how the presence of an orthogonal set causes most of the terms in the sum to vanish. In the same way, we find that

$$c_i = \frac{\mathbf{b} \cdot \mathbf{w}_i}{\mathbf{w}_i \cdot \mathbf{w}_i}.$$

6.3. ORTHOGONAL BASES AND PROJECTIONS

so that
$$\mathbf{b} = \frac{\mathbf{b}\cdot\mathbf{w}_1}{\mathbf{w}_1\cdot\mathbf{w}_1}\mathbf{w}_1 + \frac{\mathbf{b}\cdot\mathbf{w}_2}{\mathbf{w}_2\cdot\mathbf{w}_2}\mathbf{w}_2 + \cdots + \frac{\mathbf{b}\cdot\mathbf{w}_n}{\mathbf{w}_n\cdot\mathbf{w}_n}\mathbf{w}_n.$$

We'll record this fact in the following proposition.

Proposition 6.3.4 *If a vector $\mathbf{b}$ is a linear combination of an orthogonal set of vectors $\mathbf{w}_1, \mathbf{w}_2, \ldots, \mathbf{w}_n$, then*
$$\mathbf{b} = \frac{\mathbf{b}\cdot\mathbf{w}_1}{\mathbf{w}_1\cdot\mathbf{w}_1}\mathbf{w}_1 + \frac{\mathbf{b}\cdot\mathbf{w}_2}{\mathbf{w}_2\cdot\mathbf{w}_2}\mathbf{w}_2 + \cdots + \frac{\mathbf{b}\cdot\mathbf{w}_n}{\mathbf{w}_n\cdot\mathbf{w}_n}\mathbf{w}_n.$$

Using this proposition, we can see that an orthogonal set of vectors must be linearly independent. Suppose, for instance, that $\mathbf{w}_1, \mathbf{w}_2, \ldots, \mathbf{w}_n$ is a set of nonzero orthogonal vectors and that one of the vectors is a linear combination of the others, say,
$$\mathbf{w}_3 = c_1\mathbf{w}_1 + c_2\mathbf{w}_2.$$

We therefore know that
$$\mathbf{w}_3 = \frac{\mathbf{w}_3\cdot\mathbf{w}_1}{\mathbf{w}_1\cdot\mathbf{w}_1}\mathbf{w}_1 + \frac{\mathbf{w}_3\cdot\mathbf{w}_2}{\mathbf{w}_2\cdot\mathbf{w}_1}\mathbf{w}_2 = \mathbf{0},$$

which cannot happen since we know that $\mathbf{w}_3$ is nonzero. This tells us that

Proposition 6.3.5 *An orthogonal set of vectors $\mathbf{w}_1, \mathbf{w}_2, \ldots, \mathbf{w}_n$ is linearly independent.*

If the vectors in an orthogonal set have dimension m, they form a linearly independent set in $\mathbb{R}^m$ and are therefore a basis for the subspace $W = \text{Span}\{\mathbf{w}_1, \mathbf{w}_2, \ldots, \mathbf{w}_n\}$. If there are m vectors in the orthogonal set, they form a basis for $\mathbb{R}^m$.

Activity 6.3.2. Consider the vectors

$$\mathbf{w}_1 = \begin{bmatrix} 1 \\ -1 \\ 1 \end{bmatrix}, \quad \mathbf{w}_2 = \begin{bmatrix} 1 \\ 1 \\ 0 \end{bmatrix}, \quad \mathbf{w}_3 = \begin{bmatrix} 1 \\ -1 \\ -2 \end{bmatrix}.$$

a. Verify that this set forms an orthogonal set of 3-dimensional vectors.

b. Explain why we know that this set of vectors forms a basis for $\mathbb{R}^3$.

c. Suppose that $\mathbf{b} = \begin{bmatrix} 2 \\ 4 \\ -4 \end{bmatrix}$. Find the weights c_1, c_2, and c_3 that express $\mathbf{b}$ as a linear combination $\mathbf{b} = c_1\mathbf{w}_1 + c_2\mathbf{w}_2 + c_3\mathbf{w}_3$ using Proposition 6.3.4.

d. If we multiply a vector $\mathbf{v}$ by a positive scalar s, the length of $\mathbf{v}$ is also multiplied by s; that is, $|s\mathbf{v}| = s\,|\mathbf{v}|$.

Using this observation, find a vector $\mathbf{u}_1$ that is parallel to $\mathbf{w}_1$ and has length 1. Such vectors are called *unit vectors*.

e. Similarly, find a unit vector $\mathbf{u}_2$ that is parallel to $\mathbf{w}_2$ and a unit vector $\mathbf{u}_3$ that is parallel to $\mathbf{w}_3$.

f. Construct the matrix $Q = \begin{bmatrix} \mathbf{u}_1 & \mathbf{u}_2 & \mathbf{u}_3 \end{bmatrix}$ and find the product $Q^T Q$. Use Proposition 6.2.8 to explain your result.

This activity introduces an important way of modifying an orthogonal set so that the vectors in the set have unit length. Recall that we may multiply any nonzero vector $\mathbf{w}$ by a scalar so that the new vector has length 1. For instance, we know that if s is a positive scalar, then $|s\mathbf{w}| = s\,|\mathbf{w}|$. To obtain a vector $\mathbf{u}$ having unit length, we want

$$|\mathbf{u}| = |s\mathbf{w}| = s\,|\mathbf{w}| = 1$$

so that $s = 1/|\mathbf{w}|$. Therefore,

$$\mathbf{u} = \frac{1}{|\mathbf{w}|}\mathbf{w}$$

becomes a unit vector parallel to $\mathbf{w}$.

Orthogonal sets in which the vectors have unit length are called *orthonormal* and are especially convenient.

Definition 6.3.6 An *orthonormal* set is an orthogonal set of vectors each of which has unit length.

Example 6.3.7 The vectors

$$\mathbf{u}_1 = \begin{bmatrix} 1/\sqrt{2} \\ 1/\sqrt{2} \end{bmatrix}, \qquad \mathbf{u}_2 = \begin{bmatrix} -1/\sqrt{2} \\ 1/\sqrt{2} \end{bmatrix}$$

are an orthonormal set of vectors in $\mathbb{R}^2$ and form an orthonormal basis for $\mathbb{R}^2$.

If we form the matrix

$$Q = \begin{bmatrix} \mathbf{u}_1 & \mathbf{u}_2 \end{bmatrix} = \begin{bmatrix} 1/\sqrt{2} & -1/\sqrt{2} \\ 1/\sqrt{2} & 1/\sqrt{2} \end{bmatrix},$$

we find that $Q^T Q = I$ since Proposition 6.2.8 tells us that

$$Q^T Q = \begin{bmatrix} \mathbf{u}_1 \cdot \mathbf{u}_1 & \mathbf{u}_1 \cdot \mathbf{u}_2 \\ \mathbf{u}_2 \cdot \mathbf{u}_1 & \mathbf{u}_2 \cdot \mathbf{u}_2 \end{bmatrix} = \begin{bmatrix} 1 & 0 \\ 0 & 1 \end{bmatrix}$$

The previous activity and example illustrate the next proposition.

Proposition 6.3.8 *If the columns of the $m \times n$ matrix Q form an orthonormal set, then $Q^T Q = I_n$, the $n \times n$ identity matrix.*

6.3.2 Orthogonal projections

We now turn to an important problem that will appear in many forms in the rest of our explorations. Suppose, as shown in Figure 6.3.9, that we have a subspace W of $\mathbb{R}^m$ and a vector $\mathbf{b}$ that is not in that subspace. We would like to find the vector $\widehat{\mathbf{b}}$ in W that is closest to $\mathbf{b}$, meaning the distance between $\widehat{\mathbf{b}}$ and $\mathbf{b}$ is as small as possible.

6.3. ORTHOGONAL BASES AND PROJECTIONS

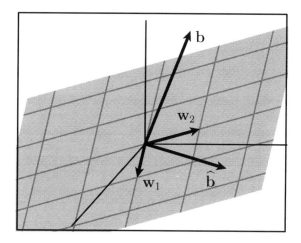

Figure 6.3.9 Given a plane in $\mathbb{R}^3$ and a vector **b** not in the plane, we wish to find the vector $\widehat{\mathbf{b}}$ in the plane that is closest to **b**.

To get started, let's consider a simpler problem where we have a line L in $\mathbb{R}^2$, defined by the vector **w**, and another vector **b** that is not on the line, as shown on the left of Figure 6.3.10. We wish to find $\widehat{\mathbf{b}}$, the vector on the line that is closest to **b**, as illustrated in the right of Figure 6.3.10.

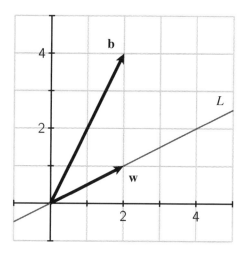 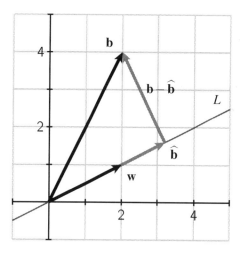

Figure 6.3.10 Given a line L and a vector **b**, we seek the vector $\widehat{\mathbf{b}}$ on L that is closest to **b**.

To find $\widehat{\mathbf{b}}$, we require that $\mathbf{b} - \widehat{\mathbf{b}}$ be orthogonal to L. For instance, if **y** is another vector on the line, as shown in Figure 6.3.11, then the Pythagorean theorem implies that

$$|\mathbf{b} - \mathbf{y}|^2 = |\mathbf{b} - \widehat{\mathbf{b}}|^2 + |\widehat{\mathbf{b}} - \mathbf{y}|^2$$

which means that $|\mathbf{b} - \mathbf{y}| \geq |\mathbf{b} - \widehat{\mathbf{b}}|$. Therefore, $\widehat{\mathbf{b}}$ is closer to **b** than any other vector on the line L.

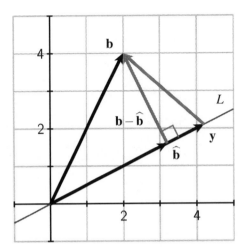

Figure 6.3.11 The vector $\widehat{\mathbf{b}}$ is closer to $\mathbf{b}$ than $\mathbf{y}$ because $\mathbf{b} - \widehat{\mathbf{b}}$ is orthogonal to L.

Definition 6.3.12 Given a vector $\mathbf{b}$ in $\mathbb{R}^m$ and a subspace W of $\mathbb{R}^m$, the *orthogonal projection* of $\mathbf{b}$ onto W is the vector $\widehat{\mathbf{b}}$ in W that is closest to $\mathbf{b}$. It is characterized by the property that $\mathbf{b} - \widehat{\mathbf{b}}$ is orthogonal to W.

Activity 6.3.3. This activity demonstrates how to determine the orthogonal projection of a vector onto a subspace of $\mathbb{R}^m$.

a. Let's begin by considering a line L, defined by the vector $\mathbf{w} = \begin{bmatrix} 2 \\ 1 \end{bmatrix}$, and a vector $\mathbf{b} = \begin{bmatrix} 2 \\ 4 \end{bmatrix}$ not on L, as illustrated in Figure 6.3.13.

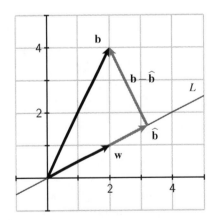 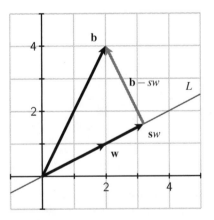

Figure 6.3.13 Finding the orthogonal projection of $\mathbf{b}$ onto the line defined by $\mathbf{w}$.

1. To find $\widehat{\mathbf{b}}$, first notice that $\widehat{\mathbf{b}} = s\mathbf{w}$ for some scalar s. Since $\mathbf{b} - \widehat{\mathbf{b}} = \mathbf{b} - s\mathbf{w}$ is

6.3. ORTHOGONAL BASES AND PROJECTIONS

orthogonal to $\mathbf{w}$, what do we know about the dot product

$$(\mathbf{b} - s\mathbf{w}) \cdot \mathbf{w}?$$

2. Apply the distributive property of dot products to find the scalar s. What is the vector $\widehat{\mathbf{b}}$, the orthogonal projection of $\mathbf{b}$ onto L?

3. More generally, explain why the orthogonal projection of $\mathbf{b}$ onto the line defined by $\mathbf{w}$ is

$$\widehat{\mathbf{b}} = \frac{\mathbf{b} \cdot \mathbf{w}}{\mathbf{w} \cdot \mathbf{w}} \mathbf{w}.$$

b. The same ideas apply more generally. Suppose we have an orthogonal set of vectors $\mathbf{w}_1 = \begin{bmatrix} 2 \\ 2 \\ -1 \end{bmatrix}$ and $\mathbf{w}_2 = \begin{bmatrix} 1 \\ 0 \\ 2 \end{bmatrix}$ that define a plane W in $\mathbb{R}^3$. If $\mathbf{b} = \begin{bmatrix} 3 \\ 9 \\ 6 \end{bmatrix}$ another vector in $\mathbb{R}^3$, we seek the vector $\widehat{\mathbf{b}}$ on the plane W closest to $\mathbf{b}$. As before, the vector $\mathbf{b} - \widehat{\mathbf{b}}$ will be orthogonal to W, as illustrated in Figure 6.3.14.

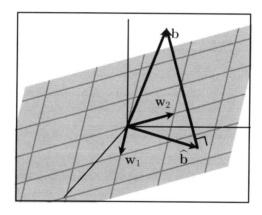

Figure 6.3.14 Given a plane W defined by the orthogonal vectors $\mathbf{w}_1$ and $\mathbf{w}_2$ and another vector $\mathbf{b}$, we seek the vector $\widehat{\mathbf{b}}$ on W closest to $\mathbf{b}$.

1. The vector $\mathbf{b} - \widehat{\mathbf{b}}$ is orthogonal to W. What does this say about the dot products: $(\mathbf{b} - \widehat{\mathbf{b}}) \cdot \mathbf{w}_1$ and $(\mathbf{b} - \widehat{\mathbf{b}}) \cdot \mathbf{w}_2$?

2. Since $\widehat{\mathbf{b}}$ is in the plane W, we can write it as a linear combination $\widehat{\mathbf{b}} = c_1 \mathbf{w}_1 + c_2 \mathbf{w}_2$. Then

$$\mathbf{b} - \widehat{\mathbf{b}} = \mathbf{b} - (c_1 \mathbf{w}_1 + c_2 \mathbf{w}_2).$$

Find the weight c_1 by dotting $\mathbf{b} - \widehat{\mathbf{b}}$ with $\mathbf{w}_1$ and applying the distributive property of dot products. Similarly, find the weight c_2.

3. What is the vector $\widehat{\mathbf{b}}$, the orthogonal projection of $\mathbf{b}$ onto the plane W?

c. Suppose that W is a subspace of $\mathbb{R}^m$ with orthogonal basis $\mathbf{w}_1, \mathbf{w}_2, \ldots, \mathbf{w}_n$ and that $\mathbf{b}$ is a vector in $\mathbb{R}^m$. Explain why the orthogonal projection of $\mathbf{b}$ onto W is

the vector
$$\widehat{\mathbf{b}} = \frac{\mathbf{b} \cdot \mathbf{w}_1}{\mathbf{w}_1 \cdot \mathbf{w}_1} \mathbf{w}_1 + \frac{\mathbf{b} \cdot \mathbf{w}_2}{\mathbf{w}_2 \cdot \mathbf{w}_2} \mathbf{w}_2 + \cdots + \frac{\mathbf{b} \cdot \mathbf{w}_n}{\mathbf{w}_n \cdot \mathbf{w}_n} \mathbf{w}_n.$$

d. Suppose that $\mathbf{u}_1, \mathbf{u}_2, \ldots, \mathbf{u}_n$ is an *orthonormal* basis for W; that is, the vectors are orthogonal to one another and have unit length. Explain why the orthogonal projection is
$$\widehat{\mathbf{b}} = (\mathbf{b} \cdot \mathbf{u}_1) \mathbf{u}_1 + (\mathbf{b} \cdot \mathbf{u}_2) \mathbf{u}_2 + \cdots + (\mathbf{b} \cdot \mathbf{u}_n) \mathbf{u}_n.$$

e. If $Q = \begin{bmatrix} \mathbf{u}_1 & \mathbf{u}_2 & \cdots & \mathbf{u}_n \end{bmatrix}$ is the matrix whose columns are an orthonormal basis of W, use Proposition 6.2.8 to explain why $\widehat{\mathbf{b}} = QQ^T \mathbf{b}$.

In all the cases considered in the activity, we are looking for $\widehat{\mathbf{b}}$, the vector in a subspace W closest to a vector $\mathbf{b}$, which is found by requiring that $\mathbf{b} - \widehat{\mathbf{b}}$ be orthogonal to W. This means that $(\mathbf{b} - \widehat{\mathbf{b}}) \cdot \mathbf{w} = 0$ for any vector $\mathbf{w}$ in W.

If we have an orthogonal basis $\mathbf{w}_1, \mathbf{w}_2, \ldots, \mathbf{w}_n$ for W, then $\widehat{\mathbf{b}} = c_1 \mathbf{w}_1 + c_w \mathbf{w}_2 + \cdots + c_n \mathbf{w}_n$. Therefore,

$$(\mathbf{b} - \widehat{\mathbf{b}}) \cdot \mathbf{w}_i = 0$$
$$\mathbf{b} \cdot \mathbf{w}_i = \widehat{\mathbf{b}} \cdot \mathbf{w}_i$$
$$\mathbf{b} \cdot \mathbf{w}_i = (c_1 \mathbf{w}_1 + c_2 \mathbf{w}_2 + \cdots + c_n \mathbf{w}_n) \cdot \mathbf{w}_i$$
$$\mathbf{b} \cdot \mathbf{w}_i = c_i \mathbf{w}_i \cdot \mathbf{w}_i$$
$$c_i = \frac{\mathbf{b} \cdot \mathbf{w}_i}{\mathbf{w}_i \cdot \mathbf{w}_i}.$$

This leads to the projection formula:

Proposition 6.3.15 Projection formula. *If W is a subspace of $\mathbb{R}^m$ having an orthogonal basis $\mathbf{w}_1, \mathbf{w}_2, \ldots, \mathbf{w}_n$ and $\mathbf{b}$ is a vector in $\mathbb{R}^m$, then the orthogonal projection of $\mathbf{b}$ onto W is*

$$\widehat{\mathbf{b}} = \frac{\mathbf{b} \cdot \mathbf{w}_1}{\mathbf{w}_1 \cdot \mathbf{w}_1} \mathbf{w}_1 + \frac{\mathbf{b} \cdot \mathbf{w}_2}{\mathbf{w}_2 \cdot \mathbf{w}_2} \mathbf{w}_2 + \cdots + \frac{\mathbf{b} \cdot \mathbf{w}_n}{\mathbf{w}_n \cdot \mathbf{w}_n} \mathbf{w}_n.$$

Caution.
Remember that the projection formula given in Proposition 6.3.15 applies only when the basis $\mathbf{w}_1, \mathbf{w}_2, \cdots, \mathbf{w}_n$ of W is *orthogonal*.

If we have an orthonormal basis $\mathbf{u}_1, \mathbf{u}_2, \ldots, \mathbf{u}_n$ for W, the projection formula simplifies to
$$\widehat{\mathbf{b}} = (\mathbf{b} \cdot \mathbf{u}_1) \mathbf{u}_1 + (\mathbf{b} \cdot \mathbf{u}_2) \mathbf{u}_2 + \cdots + (\mathbf{b} \cdot \mathbf{u}_n) \mathbf{u}_n.$$

If we then form the matrix
$$Q = \begin{bmatrix} \mathbf{u}_1 & \mathbf{u}_2 & \cdots & \mathbf{u}_n \end{bmatrix},$$
this expression may be succintly written
$$\widehat{\mathbf{b}} = (\mathbf{b} \cdot \mathbf{u}_1) \mathbf{u}_1 + (\mathbf{b} \cdot \mathbf{u}_2) \mathbf{u}_2 + \cdots + (\mathbf{b} \cdot \mathbf{u}_n) \mathbf{u}_n$$

6.3. ORTHOGONAL BASES AND PROJECTIONS

$$= \begin{bmatrix} \mathbf{u}_1 & \mathbf{u}_2 & \ldots & \mathbf{u}_n \end{bmatrix} \begin{bmatrix} \mathbf{u}_1 \cdot \mathbf{b} \\ \mathbf{u}_2 \cdot \mathbf{b} \\ \vdots \\ \mathbf{u}_n \cdot \mathbf{b} \end{bmatrix}$$

$$= QQ^T \mathbf{b}$$

This leads to the following proposition.

Proposition 6.3.16 *If $\mathbf{u}_1, \mathbf{u}_2, \ldots, \mathbf{u}_n$ is an orthonormal basis for a subspace W of $\mathbb{R}^m$, then the matrix transformation that projects vectors in $\mathbb{R}^m$ orthogonally onto W is represented by the matrix QQ^T where*

$$Q = \begin{bmatrix} \mathbf{u}_1 & \mathbf{u}_2 & \ldots & \mathbf{u}_n \end{bmatrix}.$$

Example 6.3.17 In the previous activity, we looked at the plane W defined by the two orthogonal vectors

$$\mathbf{w}_1 = \begin{bmatrix} 2 \\ 2 \\ -1 \end{bmatrix}, \quad \mathbf{w}_2 = \begin{bmatrix} 1 \\ 0 \\ 2 \end{bmatrix}.$$

We can form an orthonormal basis by scalar multiplying these vectors to have unit length:

$$\mathbf{u}_1 = \frac{1}{3} \begin{bmatrix} 2 \\ 2 \\ -1 \end{bmatrix} = \begin{bmatrix} 2/3 \\ 2/3 \\ -1/3 \end{bmatrix}, \quad \mathbf{u}_2 = \frac{1}{\sqrt{5}} \begin{bmatrix} 1 \\ 0 \\ 2 \end{bmatrix} = \begin{bmatrix} 1/\sqrt{5} \\ 0 \\ 2/\sqrt{5} \end{bmatrix}.$$

Using these vectors, we form the matrix

$$Q = \begin{bmatrix} 2/3 & 1/\sqrt{5} \\ 2/3 & 0 \\ -1/3 & 2/\sqrt{5} \end{bmatrix}.$$

The projection onto the plane W is then given by the matrix

$$QQ^T = \begin{bmatrix} 2/3 & 1/\sqrt{5} \\ 2/3 & 0 \\ -1/3 & 2/\sqrt{5} \end{bmatrix} \begin{bmatrix} 2/3 & 2/3 & -1/3 \\ 1/\sqrt{5} & 0 & 2/\sqrt{5} \end{bmatrix} = \begin{bmatrix} 29/45 & 4/9 & 8/45 \\ 4/9 & 4/9 & -2/9 \\ 8/45 & -2/9 & 41/45 \end{bmatrix}.$$

Let's check that this works by considering the vector $\mathbf{b} = \begin{bmatrix} 1 \\ 0 \\ 0 \end{bmatrix}$ and finding $\widehat{\mathbf{b}}$, its orthogonal projection onto the plane W. In terms of the original basis $\mathbf{w}_1$ and $\mathbf{w}_2$, the projection formula from Proposition 6.3.15 tells us that

$$\widehat{\mathbf{b}} = \frac{\mathbf{b} \cdot \mathbf{w}_1}{\mathbf{w}_1 \cdot \mathbf{w}_1} \mathbf{w}_1 + \frac{\mathbf{b} \cdot \mathbf{w}_2}{\mathbf{w}_2 \cdot \mathbf{w}_2} \mathbf{w}_2 = \begin{bmatrix} 29/45 \\ 4/9 \\ 8/45 \end{bmatrix}$$

Alternatively, we use the matrix QQ^T, as in Proposition 6.3.16, to find that

$$\widehat{\mathbf{b}} = QQ^T\mathbf{b} = \begin{bmatrix} 29/45 & 4/9 & 8/45 \\ 4/9 & 4/9 & -2/9 \\ 8/45 & -2/9 & 41/45 \end{bmatrix} \begin{bmatrix} 1 \\ 0 \\ 0 \end{bmatrix} = \begin{bmatrix} 29/45 \\ 4/9 \\ 8/45 \end{bmatrix}.$$

Activity 6.3.4.

a. Suppose that L is the line in $\mathbb{R}^3$ defined by the vector $\mathbf{w} = \begin{bmatrix} 1 \\ 2 \\ -2 \end{bmatrix}$.

1. Find an orthonormal basis $\mathbf{u}$ for L.
2. Construct the matrix $Q = \begin{bmatrix} \mathbf{u} \end{bmatrix}$ and use it to construct the matrix P that projects vectors orthogonally onto L.
3. Use your matrix to find $\widehat{\mathbf{b}}$, the orthogonal projection of $\mathbf{b} = \begin{bmatrix} 1 \\ 1 \\ 1 \end{bmatrix}$ onto L.
4. Find rank(P) and explain its geometric significance.

b. The vectors

$$\mathbf{w}_1 = \begin{bmatrix} 1 \\ 1 \\ 1 \\ 1 \end{bmatrix}, \quad \mathbf{w}_2 = \begin{bmatrix} 0 \\ 1 \\ 1 \\ -2 \end{bmatrix}$$

form an orthogonal basis of W, a two-dimensional subspace of $\mathbb{R}^4$.

1. Use the projection formula from Proposition 6.3.15 to find $\widehat{\mathbf{b}}$, the orthogonal projection of $\mathbf{b} = \begin{bmatrix} 9 \\ 2 \\ -2 \\ 3 \end{bmatrix}$ onto W.

2. Find an orthonormal basis $\mathbf{u}_1$ and $\mathbf{u}_2$ for W and use it to construct the matrix P that projects vectors orthogonally onto W. Check that $P\mathbf{b} = \widehat{\mathbf{b}}$, the orthogonal projection you found in the previous part of this activity.
3. Find rank(P) and explain its geometric significance.
4. Find a basis for $W^\perp$.
5. Find a vector $\mathbf{b}^\perp$ in $W^\perp$ such that

$$\mathbf{b} = \widehat{\mathbf{b}} + \mathbf{b}^\perp.$$

6. If Q is the matrix whose columns are $\mathbf{u}_1$ and $\mathbf{u}_2$, find the product Q^TQ and explain your result.

6.3. ORTHOGONAL BASES AND PROJECTIONS

This activity demonstrates one issue of note. We found $\widehat{\mathbf{b}}$, the orthogonal projection of $\mathbf{b}$ onto W, by requiring that $\mathbf{b} - \widehat{\mathbf{b}}$ be orthogonal to W. In other words, $\mathbf{b} - \widehat{\mathbf{b}}$ is a vector in the orthogonal complement $W^\perp$, which we may denote $\mathbf{b}^\perp$. This explains the following proposition, which is illustrated in Figure 6.3.19

Proposition 6.3.18 *If W is a subspace of $\mathbb{R}^n$ with orthogonal complement $W^\perp$, then any n-dimensional vector $\mathbf{b}$ can be uniquely written as*
$$\mathbf{b} = \widehat{\mathbf{b}} + \mathbf{b}^\perp$$
where $\widehat{\mathbf{b}}$ is in W and $\mathbf{b}^\perp$ is in $W^\perp$. The vector $\widehat{\mathbf{b}}$ is the orthogonal projection of $\mathbf{b}$ onto W and $\mathbf{b}^\perp$ is the orthogonal projection of $\mathbf{b}$ onto $W^\perp$.

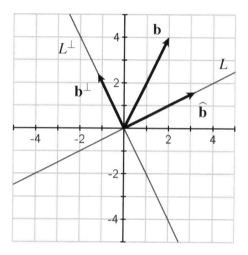

Figure 6.3.19 A vector $\mathbf{b}$ along with $\widehat{\mathbf{b}}$, its orthogonal projection onto the line L, and $\mathbf{b}^\perp$, its orthogonal projection onto the orthogonal complement $L^\perp$.

Let's summarize what we've found. If Q is a matrix whose columns $\mathbf{u}_1, \mathbf{u}_2, \ldots, \mathbf{u}_n$ form an orthonormal set in $\mathbb{R}^m$, then

- $Q^T Q = I_n$, the $n \times n$ identity matrix, because this product computes the dot products between the columns of Q.

- QQ^T is the matrix the projects vectors orthogonally onto W, the subspace of $\mathbb{R}^m$ spanned by $\mathbf{u}_1, \ldots, \mathbf{u}_n$.

As we've said before, matrix multiplication depends on the order in which we multiply the matrices, and we see this clearly here.

Because $Q^T Q = I$, there is a temptation to say that Q is invertible. This is usually not the case, however. Remember that an invertible matrix must be a square matrix, and the matrix Q will only be square if $n = m$. In this case, there are m vectors in the orthonormal set so the subspace W spanned by the vectors $\mathbf{u}_1, \mathbf{u}_2, \ldots, \mathbf{u}_m$ is $\mathbb{R}^m$. If $\mathbf{b}$ is a vector in $\mathbb{R}^m$, then $\widehat{\mathbf{b}} = QQ^T\mathbf{b}$ is the orthogonal projection of $\mathbf{b}$ onto $\mathbb{R}^m$. In other words, $QQ^T\mathbf{b}$ is the closest vector in $\mathbb{R}^m$ to $\mathbf{b}$, and this closest vector must be $\mathbf{b}$ itself. Therefore, $QQ^T\mathbf{b} = \mathbf{b}$, which means that $QQ^T = I$. In this case, Q is an invertible matrix.

Example 6.3.20 Consider the orthonormal set of vectors

$$\mathbf{u}_1 = \begin{bmatrix} 1/\sqrt{3} \\ -1/\sqrt{3} \\ 1/\sqrt{3} \end{bmatrix}, \quad \mathbf{u}_2 = \begin{bmatrix} 1/\sqrt{2} \\ 1/\sqrt{2} \\ 0 \end{bmatrix}$$

and the matrix they define

$$Q = \begin{bmatrix} 1/\sqrt{3} & 1/\sqrt{2} \\ -1/\sqrt{3} & 1/\sqrt{2} \\ 1/\sqrt{3} & 0 \end{bmatrix}.$$

In this case, $\mathbf{u}_1$ and $\mathbf{u}_2$ span a plane, a 2-dimensional subspace of $\mathbb{R}^3$. We know that $Q^TQ = I_2$ and QQ^T projects vectors orthogonally onto the plane. However, Q is not a square matrix so it cannot be invertible.

Example 6.3.21 Now consider the orthonormal set of vectors

$$\mathbf{u}_1 = \begin{bmatrix} 1/\sqrt{3} \\ -1/\sqrt{3} \\ 1/\sqrt{3} \end{bmatrix}, \quad \mathbf{u}_2 = \begin{bmatrix} 1/\sqrt{2} \\ 1/\sqrt{2} \\ 0 \end{bmatrix}, \quad \mathbf{u}_3 = \begin{bmatrix} 1/\sqrt{6} \\ -1/\sqrt{6} \\ -2/\sqrt{6} \end{bmatrix}$$

and the matrix they define

$$Q = \begin{bmatrix} 1/\sqrt{3} & 1/\sqrt{2} & 1/\sqrt{6} \\ -1/\sqrt{3} & 1/\sqrt{2} & -1/\sqrt{6} \\ 1/\sqrt{3} & 0 & -2/\sqrt{6} \end{bmatrix}.$$

Here, $\mathbf{u}_1$, $\mathbf{u}_2$, and $\mathbf{u}_3$ form a basis for $\mathbb{R}^3$ so that both $Q^TQ = I_3$ and $QQ^T = I_3$. Therefore, Q is a square matrix and is invertible.

Moreover, since $Q^TQ = I$, we see that $Q^{-1} = Q^T$ so finding the inverse of Q is as simple as writing its transpose. Matrices with this property are very special and will play an important role in our upcoming work. We will therefore give them a special name.

Definition 6.3.22 A square $m \times m$ matrix Q whose columns form an orthonormal basis for $\mathbb{R}^m$ is called *orthogonal*.

This terminology can be a little confusing. We call a basis orthogonal if the basis vectors are orthogonal to one another. However, a matrix is orthogonal if the columns are orthogonal to one another and have unit length. It pays to keep this in mind when reading statements about orthogonal bases and orthogonal matrices. In the meantime, we record the following proposition.

Proposition 6.3.23 *An orthogonal matrix Q is invertible and its inverse $Q^{-1} = Q^T$.*

6.3.3 Summary

This section introduced orthogonal sets and the projection formula that allows us to project vectors orthogonally onto a subspace.

6.3. ORTHOGONAL BASES AND PROJECTIONS

- Given an orthogonal set $\mathbf{w}_1, \mathbf{w}_2, \ldots, \mathbf{w}_n$ that spans an n-dimensional subspace W of $\mathbb{R}^m$, the orthogonal projection of $\mathbf{b}$ onto W is the vector in W closest to $\mathbf{b}$ and may be written as
$$\widehat{\mathbf{b}} = \frac{\mathbf{b} \cdot \mathbf{w}_1}{\mathbf{w}_1 \cdot \mathbf{w}_1} \mathbf{w}_1 + \frac{\mathbf{b} \cdot \mathbf{w}_2}{\mathbf{w}_2 \cdot \mathbf{w}_2} \mathbf{w}_2 + \cdots + \frac{\mathbf{b} \cdot \mathbf{w}_n}{\mathbf{w}_n \cdot \mathbf{w}_n} \mathbf{w}_n.$$

- If $\mathbf{u}_1, \mathbf{u}_2, \ldots, \mathbf{u}_n$ is an orthonormal basis of W and Q is the matrix whose columns are $\mathbf{u}_i$, then the matrix $P = QQ^T$ projects vectors orthogonally onto W.

- If the columns of Q form an orthonormal basis for an n-dimensional subspace of $\mathbb{R}^m$, then $Q^T Q = I_n$.

- An orthogonal matrix Q is a square matrix whose columns form an orthonormal basis. In this case, $QQ^T = Q^T Q = I$ so that $Q^{-1} = Q^T$.

6.3.4 Exercises

1. Suppose that
$$\mathbf{w}_1 = \begin{bmatrix} 1 \\ 1 \\ 1 \end{bmatrix}, \quad \mathbf{w}_2 = \begin{bmatrix} 1 \\ -2 \\ 1 \end{bmatrix}.$$

    a. Verify that $\mathbf{w}_1$ and $\mathbf{w}_2$ form an orthogonal basis for a plane W in $\mathbb{R}^3$.

    b. Use Proposition 6.3.15 to find $\widehat{\mathbf{b}}$, the orthogonal projection of $\mathbf{b} = \begin{bmatrix} 2 \\ 1 \\ -1 \end{bmatrix}$ onto W.

    c. Find an orthonormal basis $\mathbf{u}_1, \mathbf{u}_2$ for W.

    d. Find the matrix P representing the matrix transformation that projects vectors in $\mathbb{R}^3$ orthogonally onto W. Verify that $\widehat{\mathbf{b}} = P\mathbf{b}$.

    e. Determine rank(P) and explain its geometric significance.

2. Consider the vectors
$$\mathbf{w}_1 = \begin{bmatrix} 1 \\ 1 \\ 1 \end{bmatrix}, \quad \mathbf{w}_2 = \begin{bmatrix} -1 \\ 0 \\ 1 \end{bmatrix}, \quad \mathbf{w}_3 = \begin{bmatrix} 1 \\ -2 \\ 1 \end{bmatrix}.$$

    a. Explain why these vectors form an orthogonal basis for $\mathbb{R}^3$.

    b. Suppose that $A = \begin{bmatrix} \mathbf{w}_1 & \mathbf{w}_2 & \mathbf{w}_3 \end{bmatrix}$ and evaluate the product $A^T A$. Why is this product a diagonal matrix and what is the significance of the diagonal entries?

    c. Express the vector $\mathbf{b} = \begin{bmatrix} -3 \\ -6 \\ 3 \end{bmatrix}$ as a linear combination of $\mathbf{w}_1, \mathbf{w}_2$, and $\mathbf{w}_3$.

    d. Multiply the vectors $\mathbf{w}_1, \mathbf{w}_2, \mathbf{w}_3$ by appropriate scalars to find an orthonormal basis $\mathbf{u}_1, \mathbf{u}_2, \mathbf{u}_3$ of $\mathbb{R}^3$.

e. If $Q = \begin{bmatrix} \mathbf{u}_1 & \mathbf{u}_2 & \mathbf{u}_3 \end{bmatrix}$, find the matrix product QQ^T and explain the result.

3. Suppose that
$$\mathbf{w}_1 = \begin{bmatrix} 1 \\ 1 \\ 0 \\ -1 \end{bmatrix}, \quad \mathbf{w}_2 = \begin{bmatrix} 1 \\ 0 \\ 1 \\ 1 \end{bmatrix}$$
form an orthogonal basis for a subspace W of $\mathbb{R}^4$.

   a. Find $\widehat{\mathbf{b}}$, the orthogonal projection of $\mathbf{b} = \begin{bmatrix} 2 \\ -1 \\ -6 \\ 7 \end{bmatrix}$ onto W.

   b. Find the vector $\mathbf{b}^\perp$ in $W^\perp$ such that $\mathbf{b} = \widehat{\mathbf{b}} + \mathbf{b}^\perp$.

   c. Find a basis for $W^\perp$. and express $\mathbf{b}^\perp$ as a linear combination of the basis vectors.

4. Consider the vectors
$$\mathbf{w}_1 = \begin{bmatrix} 1 \\ 1 \\ 0 \\ 0 \end{bmatrix}, \quad \mathbf{w}_2 = \begin{bmatrix} 0 \\ 0 \\ 1 \\ 1 \end{bmatrix}, \quad \mathbf{b} = \begin{bmatrix} 2 \\ -4 \\ 1 \\ 3 \end{bmatrix}.$$

   a. If L is the line defined by the vector $\mathbf{w}_1$, find the vector in L closest to $\mathbf{b}$. Call this vector $\widehat{\mathbf{b}}_1$.

   b. If W is the subspace spanned by $\mathbf{w}_1$ and $\mathbf{w}_2$, find the vector in W closest to $\mathbf{b}$. Call this vector $\widehat{\mathbf{b}}_2$.

   c. Determine whether $\widehat{\mathbf{b}}_1$ or $\widehat{\mathbf{b}}_2$ is closer to $\mathbf{b}$ and explain why.

5. Suppose that $\mathbf{w} = \begin{bmatrix} 2 \\ -1 \\ 2 \end{bmatrix}$ defines a line L in $\mathbb{R}^3$.

   a. Find the orthogonal projections of the vectors $\begin{bmatrix} 1 \\ 0 \\ 0 \end{bmatrix}, \begin{bmatrix} 0 \\ 1 \\ 0 \end{bmatrix}, \begin{bmatrix} 0 \\ 0 \\ 1 \end{bmatrix}$ onto L.

   b. Find the matrix $P = \frac{1}{|\mathbf{w}|^2}\mathbf{w}\mathbf{w}^T$.

   c. Use Proposition 2.5.6 to explain why the columns of P are related to the orthogonal projections you found in the first part of this exericse.

6. Suppose that
$$\mathbf{v}_1 = \begin{bmatrix} 1 \\ 0 \\ 3 \end{bmatrix}, \quad \mathbf{v}_2 = \begin{bmatrix} 2 \\ 2 \\ 2 \end{bmatrix}$$

6.3. ORTHOGONAL BASES AND PROJECTIONS

form the basis for a plane W in $\mathbb{R}^3$.

a. Find a basis for the line that is the orthogonal complement $W^\perp$.

b. Given the vector $\mathbf{b} = \begin{bmatrix} 6 \\ -6 \\ 2 \end{bmatrix}$, find $\mathbf{y}$, the orthogonal projection of $\mathbf{b}$ onto the line $W^\perp$.

c. Explain why the vector $\mathbf{z} = \mathbf{b} - \mathbf{y}$ must be in W and write $\mathbf{z}$ as a linear combination of $\mathbf{v}_1$ and $\mathbf{v}_2$.

7. Determine whether the following statements are true or false and explain your thinking.

    a. If the columns of Q form an orthonormal basis for a subspace W and $\mathbf{w}$ is a vector in W, then $QQ^T\mathbf{w} = \mathbf{w}$.

    b. An orthogonal set of vectors in $\mathbb{R}^8$ can have no more than 8 vectors.

    c. If Q is a 7×5 matrix whose columns are orthonormal, then $QQ^T = I_7$.

    d. If Q is a 7×5 matrix whose columns are orthonormal, then $Q^T Q = I_5$.

    e. If the orthogonal projection of $\mathbf{b}$ onto a subspace W satisfies $\widehat{\mathbf{b}} = \mathbf{0}$, then $\mathbf{b}$ is in $W^\perp$.

8. Suppose that Q is an orthogonal matrix.

    a. Remembering that $\mathbf{v} \cdot \mathbf{w} = \mathbf{v}^T \mathbf{w}$, explain why
    $$Q\mathbf{x} \cdot (Q\mathbf{y}) = \mathbf{x} \cdot \mathbf{y}.$$

    b. Explain why $|Q\mathbf{x}| = |\mathbf{x}|$.

    This means that the length of a vector is unchanged after multiplying by an orthogonal matrix.

    c. If λ is a real eigenvalue of Q, explain why $\lambda = \pm 1$.

9. Explain why the following statements are true.

    a. If Q is an orthogonal matrix, then $\det Q = \pm 1$.

    b. If Q is a 8×4 matrix whose columns are orthonormal, then QQ^T is an 8×8 matrix whose rank is 4.

    c. If $\widehat{\mathbf{b}}$ is the orthogonal projection of $\mathbf{b}$ onto a subspace W, then $\mathbf{b} - \widehat{\mathbf{b}}$ is the orthogonal projection of $\mathbf{b}$ onto $W^\perp$.

10. This exercise is about 2×2 orthogonal matrices.

    a. In Section 2.6, we saw that the matrix $\begin{bmatrix} \cos\theta & -\sin\theta \\ \sin\theta & \cos\theta \end{bmatrix}$ represents a rotation by an angle θ. Explain why this matrix is an orthogonal matrix.

    b. We also saw that the matrix $\begin{bmatrix} \cos\theta & \sin\theta \\ \sin\theta & -\cos\theta \end{bmatrix}$ represents a reflection in a line. Explain why this matrix is an orthogonal matrix.

c. Suppose that $\mathbf{u}_1 = \begin{bmatrix} \cos \theta \\ \sin \theta \end{bmatrix}$ is a 2-dimensional unit vector. Use a sketch to indicate all the possible vectors $\mathbf{u}_2$ such that $\mathbf{u}_1$ and $\mathbf{u}_2$ form an orthonormal basis of $\mathbb{R}^2$.

d. Explain why every 2×2 orthogonal matrix is either a rotation or a reflection.

6.4 Finding orthogonal bases

The last section demonstrated the value of working with orthogonal, and especially orthonormal, sets. If we have an orthogonal basis $\mathbf{w}_1, \mathbf{w}_2, \ldots, \mathbf{w}_n$ for a subspace W, the Projection Formula 6.3.15 tells us that the orthogonal projection of a vector $\mathbf{b}$ onto W is

$$\widehat{\mathbf{b}} = \frac{\mathbf{b} \cdot \mathbf{w}_1}{\mathbf{w}_1 \cdot \mathbf{w}_1} \mathbf{w}_1 + \frac{\mathbf{b} \cdot \mathbf{w}_2}{\mathbf{w}_2 \cdot \mathbf{w}_2} \mathbf{w}_2 + \cdots + \frac{\mathbf{b} \cdot \mathbf{w}_n}{\mathbf{w}_n \cdot \mathbf{w}_n} \mathbf{w}_n.$$

An orthonormal basis $\mathbf{u}_1, \mathbf{u}_2, \ldots, \mathbf{u}_n$ is even more convenient: after forming the matrix $Q = \begin{bmatrix} \mathbf{u}_1 & \mathbf{u}_2 & \cdots & \mathbf{u}_n \end{bmatrix}$, we have $\widehat{\mathbf{b}} = QQ^T\mathbf{b}$.

In the examples we've seen so far, however, orthogonal bases were given to us. What we need now is a way to form orthogonal bases. In this section, we'll explore an algorithm that begins with a basis for a subspace and creates an orthogonal basis. Once we have an orthogonal basis, we can scale each of the vectors appropriately to produce an orthonormal basis.

Preview Activity 6.4.1. Suppose we have a basis for $\mathbb{R}^2$ consisting of the vectors

$$\mathbf{v}_1 = \begin{bmatrix} 1 \\ 1 \end{bmatrix}, \quad \mathbf{v}_2 = \begin{bmatrix} 0 \\ 2 \end{bmatrix}$$

as shown in Figure 6.4.1. Notice that this basis is not orthogonal.

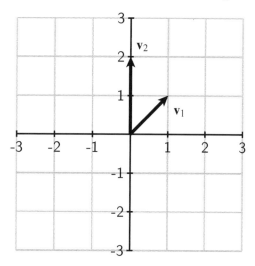

Figure 6.4.1 A basis for $\mathbb{R}^2$.

a. Find the vector $\widehat{\mathbf{v}}_2$ that is the orthogonal projection of $\mathbf{v}_2$ onto the line defined by $\mathbf{v}_1$.

b. Explain why $\mathbf{v}_2 - \widehat{\mathbf{v}}_2$ is orthogonal to $\mathbf{v}_1$.

c. Define the new vectors $\mathbf{w}_1 = \mathbf{v}_1$ and $\mathbf{w}_2 = \mathbf{v}_2 - \widehat{\mathbf{v}}_2$ and sketch them in Figure 6.4.2. Explain why $\mathbf{w}_1$ and $\mathbf{w}_2$ define an orthogonal basis for $\mathbb{R}^2$.

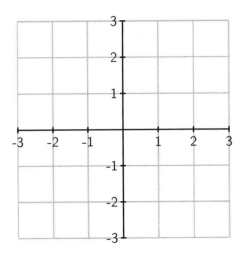

Figure 6.4.2 Sketch the new basis w_1 and w_2.

d. Write the vector $b = \begin{bmatrix} 8 \\ -10 \end{bmatrix}$ as a linear combination of w_1 and w_2.

e. Scale the vectors w_1 and w_2 to produce an orthonormal basis u_1 and u_2 for $\mathbb{R}^2$.

6.4.1 Gram-Schmidt orthogonalization

The preview activity illustrates the main idea behind an algorithm, known as *Gram-Schmidt orthogonalization*, that begins with a basis for some subspace of $\mathbb{R}^m$ and produces an orthogonal or orthonormal basis. The algorithm relies on our construction of the orthogonal projection. Remember that we formed the orthogonal projection $\widehat{b}$ of b onto a subspace W by requiring that $b - \widehat{b}$ is orthogonal to W as shown in Figure 6.4.3.

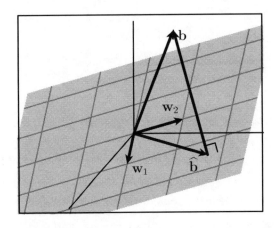

Figure 6.4.3 If $\widehat{b}$ is the orthogonal projection of b onto W, then $b - \widehat{b}$ is orthogonal to W.

6.4. FINDING ORTHOGONAL BASES

This observation guides our construction of an orthogonal basis for it allows us to create a vector that is orthogonal to a given subspace. Let's see how the Gram-Schmidt algorithm works.

Activity 6.4.2. Suppose that W is a three-dimensional subspace of $\mathbb{R}^4$ with basis:

$$\mathbf{v}_1 = \begin{bmatrix} 1 \\ 1 \\ 1 \\ 1 \end{bmatrix}, \quad \mathbf{v}_2 = \begin{bmatrix} 1 \\ 3 \\ 2 \\ 2 \end{bmatrix}, \quad \mathbf{v}_3 = \begin{bmatrix} 1 \\ -3 \\ -3 \\ -3 \end{bmatrix}.$$

We can see that this basis is not orthogonal by noting that $\mathbf{v}_1 \cdot \mathbf{v}_2 = 8$. Our goal is to create an orthogonal basis $\mathbf{w}_1, \mathbf{w}_2$, and $\mathbf{w}_3$ for W.

To begin, we declare that $\mathbf{w}_1 = \mathbf{v}_1$, and we call W_1 the line defined by $\mathbf{w}_1$.

a. Find the vector $\widehat{\mathbf{v}}_2$ that is the orthogonal projection of $\mathbf{v}_2$ onto W_1, the line defined by $\mathbf{w}_1$.

b. Form the vector $\mathbf{w}_2 = \mathbf{v}_2 - \widehat{\mathbf{v}}_2$ and verify that it is orthogonal to $\mathbf{w}_1$.

c. Explain why $\text{Span}\{\mathbf{w}_1, \mathbf{w}_2\} = \text{Span}\{\mathbf{v}_1, \mathbf{v}_2\}$ by showing that any linear combination of $\mathbf{v}_1$ and $\mathbf{v}_2$ can be written as a linear combination of $\mathbf{w}_1$ and $\mathbf{w}_2$ and vice versa.

d. The vectors $\mathbf{w}_1$ and $\mathbf{w}_2$ are an orthogonal basis for a two-dimensional subspace W_2 of $\mathbb{R}^4$. Find the vector $\widehat{\mathbf{v}}_3$ that is the orthogonal projection of $\mathbf{v}_3$ onto W_2.

e. Verify that $\mathbf{w}_3 = \mathbf{v}_3 - \widehat{\mathbf{v}}_3$ is orthogonal to both $\mathbf{w}_1$ and $\mathbf{w}_2$.

f. Explain why $\mathbf{w}_1, \mathbf{w}_2$, and $\mathbf{w}_3$ form an orthogonal basis for W.

g. Now find an orthonormal basis for W.

As this activity illustrates, Gram-Schmidt orthogonalization begins with a basis $\mathbf{v}_1 \mathbf{v}_2, \ldots, \mathbf{v}_n$ for a subspace W of $\mathbb{R}^m$ and creates an orthogonal basis for W. Let's work through a second example.

Example 6.4.4 Let's start with the basis

$$\mathbf{v}_1 = \begin{bmatrix} 2 \\ -1 \\ 2 \end{bmatrix}, \quad \mathbf{v}_2 = \begin{bmatrix} -3 \\ 3 \\ 0 \end{bmatrix}, \quad \mathbf{v}_3 = \begin{bmatrix} -2 \\ 7 \\ 1 \end{bmatrix},$$

which is a basis for $\mathbb{R}^3$.

To get started, we'll simply set $\mathbf{w}_1 = \mathbf{v}_1 = \begin{bmatrix} 2 \\ -1 \\ 2 \end{bmatrix}$. We construct $\mathbf{w}_2$ from $\mathbf{v}_2$ by subtracting its orthogonal projection onto W_1, the line defined by $\mathbf{w}_1$. This gives

$$\mathbf{w}_2 = \mathbf{v}_2 - \frac{\mathbf{v}_2 \cdot \mathbf{w}_1}{\mathbf{w}_1 \cdot \mathbf{w}_1} \mathbf{w}_1 = \mathbf{v}_2 + \mathbf{w}_1 = \begin{bmatrix} -1 \\ 2 \\ 2 \end{bmatrix}.$$

Notice that we found $\mathbf{v}_2 = -\mathbf{w}_1 + \mathbf{w}_2$. Therefore, we can rewrite any linear combination of $\mathbf{v}_1$ and $\mathbf{v}_2$ as

$$c_1 \mathbf{v}_1 + c_2 \mathbf{v}_2 = c_1 \mathbf{w}_1 + c_2(-\mathbf{w}_1 + \mathbf{w}_2) = (c_1 - c_2)\mathbf{w}_1 + c_2 \mathbf{w}_2,$$

a linear combination of $\mathbf{w}_1$ and $\mathbf{w}_2$. This tells us that

$$W_2 = \text{Span}\{\mathbf{w}_1, \mathbf{w}_2\} = \text{Span}\{\mathbf{v}_1, \mathbf{v}_2\}.$$

In other words, $\mathbf{w}_1$ and $\mathbf{w}_2$ is a orthogonal basis for W_2, the 2-dimensional subspace that is the span of $\mathbf{v}_1$ and $\mathbf{v}_2$.

Finally, we form $\mathbf{w}_3$ from $\mathbf{v}_3$ by subtracting its orthogonal projection onto W_2:

$$\mathbf{w}_3 = \mathbf{v}_3 - \frac{\mathbf{v}_3 \cdot \mathbf{w}_1}{\mathbf{w}_1 \cdot \mathbf{w}_1} \mathbf{w}_1 - \frac{\mathbf{v}_3 \cdot \mathbf{w}_2}{\mathbf{w}_2 \cdot \mathbf{w}_2} \mathbf{w}_2 = \mathbf{v}_3 + \mathbf{w}_1 - 2\mathbf{w}_2 = \begin{bmatrix} 2 \\ 2 \\ -1 \end{bmatrix}.$$

We can now check that

$$\mathbf{w}_1 = \begin{bmatrix} 2 \\ -1 \\ 2 \end{bmatrix}, \quad \mathbf{w}_2 = \begin{bmatrix} -1 \\ 2 \\ 2 \end{bmatrix}, \quad \mathbf{w}_3 = \begin{bmatrix} 2 \\ 2 \\ -1 \end{bmatrix},$$

is an orthogonal set. Furthermore, we have, as before, $\text{Span}\{\mathbf{w}_1, \mathbf{w}_2, \mathbf{w}_3\} = \text{Span}\{\mathbf{v}_1, \mathbf{v}_2, \mathbf{v}_3\}$, which says that we have found a new orthogonal basis for $\mathbb{R}^3$.

To create an orthonormal basis, we form unit vectors parallel to each of the vectors in the orthogonal basis:

$$\mathbf{u}_1 = \begin{bmatrix} 2/3 \\ -1/3 \\ 2/3 \end{bmatrix}, \quad \mathbf{u}_2 = \begin{bmatrix} -1/3 \\ 2/3 \\ 2/3 \end{bmatrix}, \quad \mathbf{u}_3 = \begin{bmatrix} 2/3 \\ 2/3 \\ -1/3 \end{bmatrix}.$$

More generally, if we have a basis $\mathbf{v}_1, \mathbf{v}_2, \ldots, \mathbf{v}_n$ for a subspace W of $\mathbb{R}^m$, the Gram-Schmidt algorithm creates an orthogonal basis for W in the following way:

$$\mathbf{w}_1 = \mathbf{v}_1$$
$$\mathbf{w}_2 = \mathbf{v}_2 - \frac{\mathbf{v}_2 \cdot \mathbf{w}_1}{\mathbf{w}_1 \cdot \mathbf{w}_1} \mathbf{w}_1$$
$$\mathbf{w}_3 = \mathbf{v}_3 - \frac{\mathbf{v}_3 \cdot \mathbf{w}_1}{\mathbf{w}_1 \cdot \mathbf{w}_1} \mathbf{w}_1 - \frac{\mathbf{v}_3 \cdot \mathbf{w}_2}{\mathbf{w}_2 \cdot \mathbf{w}_2} \mathbf{w}_2$$

6.4. FINDING ORTHOGONAL BASES

$$\vdots$$

$$\mathbf{w}_n = \mathbf{v}_n - \frac{\mathbf{v}_n \cdot \mathbf{w}_1}{\mathbf{w}_1 \cdot \mathbf{w}_1}\mathbf{w}_1 - \frac{\mathbf{v}_n \cdot \mathbf{w}_2}{\mathbf{w}_2 \cdot \mathbf{w}_2}\mathbf{w}_2 - \ldots - \frac{\mathbf{v}_n \cdot \mathbf{w}_{n-1}}{\mathbf{w}_{n-1} \cdot \mathbf{w}_{n-1}}\mathbf{w}_{n-1}.$$

From here, we may form an orthonormal basis by constructing a unit vector parallel to each vector in the orthogonal basis: $\mathbf{u}_j = 1/|\mathbf{w}_j|\ \mathbf{w}_j$.

Activity 6.4.3. Sage can automate these computations for us. Before we begin, however, it will be helpful to understand how we can combine things using a list in Python. For instance, if the vectors v1, v2, and v3 form a basis for a subspace, we can bundle them together using square brackets: [v1, v2, v3]. Furthermore, we could assign this to a variable, such as basis = [v1, v2, v3].

Evaluating the following cell will load in some special commands.

```
url='https://raw.githubusercontent.com/davidaustinm/'
url+='ula_modules/master/orthogonality.py'
sage.repl.load.load(url, globals())
```

- There is a command to apply the projection formula: projection(b, basis) returns the orthogonal projection of b onto the subspace spanned by basis, which is a list of vectors.

- The command unit(w) returns a unit vector parallel to w.

- Given a collection of vectors, say, v1 and v2, we can form the matrix whose columns are v1 and v2 using matrix([v1, v2]).T. When given a list of vectors, Sage constructs a matrix whose *rows* are the given vectors. For this reason, we need to apply the transpose.

Let's now consider W, the subspace of $\mathbb{R}^5$ having basis

$$\mathbf{v}_1 = \begin{bmatrix} 14 \\ -6 \\ 8 \\ 2 \\ -6 \end{bmatrix}, \quad \mathbf{v}_2 = \begin{bmatrix} 5 \\ -3 \\ 4 \\ 3 \\ -7 \end{bmatrix}, \quad \mathbf{v}_3 = \begin{bmatrix} 2 \\ 3 \\ 0 \\ -2 \\ 1 \end{bmatrix}.$$

a. Apply the Gram-Schmidt algorithm to find an orthogonal basis $\mathbf{w}_1, \mathbf{w}_2$, and $\mathbf{w}_3$ for W.

b. Find $\widehat{\mathbf{b}}$, the orthogonal projection of $\mathbf{b} = \begin{bmatrix} -5 \\ 11 \\ 0 \\ -1 \\ 5 \end{bmatrix}$ onto W.

c. Explain why we know that $\widehat{\mathbf{b}}$ is a linear combination of the original vectors $\mathbf{v}_1$, $\mathbf{v}_2$, and $\mathbf{v}_3$ and then find weights so that

$$\widehat{\mathbf{b}} = c_1\mathbf{v}_1 + c_2\mathbf{v}_2 + c_3\mathbf{v}_3.$$

d. Find an orthonormal basis $\mathbf{u}_1$, $\mathbf{u}_2$, for $\mathbf{u}_3$ for W and form the matrix Q whose columns are these vectors.

e. Find the product Q^TQ and explain the result.

f. Find the matrix P that projects vectors orthogonally onto W and verify that $P\mathbf{b}$ gives $\widehat{\mathbf{b}}$, the orthogonal projection that you found earlier.

6.4.2 QR factorizations

Now that we've seen how the Gram-Schmidt algorithm forms an orthonormal basis for a given subspace, we will explore how the algorithm leads to an important matrix factorization known as the QR factorization.

Activity 6.4.4. Suppose that A is the 4×3 matrix whose columns are

$$\mathbf{v}_1 = \begin{bmatrix} 1 \\ 1 \\ 1 \\ 1 \end{bmatrix}, \quad \mathbf{v}_2 = \begin{bmatrix} 1 \\ 3 \\ 2 \\ 2 \end{bmatrix}, \quad \mathbf{v}_3 = \begin{bmatrix} 1 \\ -3 \\ -3 \\ -3 \end{bmatrix}.$$

These vectors form a basis for W, the subspace of $\mathbb{R}^4$ that we encountered in Activity 6.4.2. Since these vectors are the columns of A, we have $\text{Col}(A) = W$.

a. When we implemented Gram-Schmidt, we first found an orthogonal basis $\mathbf{w}_1$, $\mathbf{w}_2$, and $\mathbf{w}_3$ using

$$\mathbf{w}_1 = \mathbf{v}_1$$
$$\mathbf{w}_2 = \mathbf{v}_2 - \frac{\mathbf{v}_2 \cdot \mathbf{w}_1}{\mathbf{w}_1 \cdot \mathbf{w}_1}\mathbf{w}_1$$
$$\mathbf{w}_3 = \mathbf{v}_3 - \frac{\mathbf{v}_3 \cdot \mathbf{w}_1}{\mathbf{w}_1 \cdot \mathbf{w}_1}\mathbf{w}_1 - \frac{\mathbf{v}_3 \cdot \mathbf{w}_2}{\mathbf{w}_2 \cdot \mathbf{w}_2}\mathbf{w}_2.$$

Use these expressions to write $\mathbf{v}_1$, $\mathbf{v}_1$, and $\mathbf{v}_3$ as linear combinations of $\mathbf{w}_1$, $\mathbf{w}_2$, and $\mathbf{w}_3$.

b. We next normalized the orthogonal basis $\mathbf{w}_1$, $\mathbf{w}_2$, and $\mathbf{w}_3$ to obtain an orthonormal basis $\mathbf{u}_1$, $\mathbf{u}_2$, and $\mathbf{u}_3$.

Write the vectors $\mathbf{w}_i$ as scalar multiples of $\mathbf{u}_i$. Then use these expressions to write $\mathbf{v}_1$, $\mathbf{v}_1$, and $\mathbf{v}_3$ as linear combinations of $\mathbf{u}_1$, $\mathbf{u}_2$, and $\mathbf{u}_3$.

6.4. FINDING ORTHOGONAL BASES

c. Suppose that $Q = \begin{bmatrix} \mathbf{u}_1 & \mathbf{u}_2 & \mathbf{u}_3 \end{bmatrix}$. Use the result of the previous part to find a vector $\mathbf{r}_1$ so that $Q\mathbf{r}_1 = \mathbf{v}_1$.

d. Then find vectors $\mathbf{r}_2$ and $\mathbf{r}_3$ such that $Q\mathbf{r}_2 = \mathbf{v}_2$ and $Q\mathbf{r}_3 = \mathbf{v}_3$.

e. Construct the matrix $R = \begin{bmatrix} \mathbf{r}_1 & \mathbf{r}_2 & \mathbf{r}_3 \end{bmatrix}$. Remembering that $A = \begin{bmatrix} \mathbf{v}_1 & \mathbf{v}_2 & \mathbf{v}_3 \end{bmatrix}$, explain why $A = QR$.

f. What is special about the shape of R?

g. Suppose that A is a 10×6 matrix whose columns are linearly independent. This means that the columns of A form a basis for $W = \text{Col}(A)$, a 6-dimensional subspace of $\mathbb{R}^{10}$. Suppose that we apply Gram-Schmidt orthogonalization to create an orthonormal basis whose vectors form the columns of Q and that we write $A = QR$. What are the shape of Q and what the shape of R?

When the columns of a matrix A are linearly independent, they form a basis for $\text{Col}(A)$ so that we can perform the Gram-Schmidt algorithm. The previous activity shows how this leads to a factorization of A as the product of a matrix Q whose columns are an orthonormal basis for $\text{Col}(A)$ and an upper triangular matrix R.

Proposition 6.4.5 *QR factorization. If A is an $m \times n$ matrix whose columns are linearly independent, we may write $A = QR$ where Q is an $m \times n$ matrix whose columns form an orthonormal basis for $\text{Col}(A)$ and R is an $n \times n$ upper triangular matrix.*

Example 6.4.6 We'll consider the matrix $A = \begin{bmatrix} 2 & -3 & -2 \\ -1 & 3 & 7 \\ 2 & 0 & 1 \end{bmatrix}$ whose columns, which we'll denote $\mathbf{v}_1, \mathbf{v}_2$, and $\mathbf{v}_3$, are the basis of $\mathbb{R}^3$ that we considered in Example 6.4.4. There we found an orthogonal basis $\mathbf{w}_1, \mathbf{w}_2$, and $\mathbf{w}_3$ that satisfied

$$\mathbf{v}_1 = \mathbf{w}_1$$
$$\mathbf{v}_2 = -\mathbf{w}_1 + \mathbf{w}_2$$
$$\mathbf{v}_3 = -\mathbf{w}_1 + 2\mathbf{w}_2 + \mathbf{w}_3.$$

In terms of the resulting orthonormal basis $\mathbf{u}_1, \mathbf{u}_2$, and $\mathbf{u}_3$, we had

$$\mathbf{w}_1 = 3\mathbf{u}_1, \qquad \mathbf{w}_2 = 3\mathbf{u}_2, \qquad \mathbf{w}_3 = 3\mathbf{u}_3$$

so that

$$\mathbf{v}_1 = 3\mathbf{u}_1$$
$$\mathbf{v}_2 = -3\mathbf{u}_1 + 3\mathbf{u}_2$$
$$\mathbf{v}_3 = -3\mathbf{u}_1 + 6\mathbf{u}_2 + 3\mathbf{u}_3.$$

Therefore, if $Q = \begin{bmatrix} \mathbf{u}_1 & \mathbf{u}_2 & \mathbf{u}_3 \end{bmatrix}$, we have the QR factorization

$$A = Q \begin{bmatrix} 3 & -3 & -3 \\ 0 & 3 & 6 \\ 0 & 0 & 3 \end{bmatrix} = QR.$$

The value of the QR factorization will become clear in the next section where we use it to solve least-squares problems.

Activity 6.4.5. As before, we would like to use Sage to automate the process of finding and using the QR factorization of a matrix A. Evaluating the following cell provides a command QR(A) that returns the factorization, which may be stored using, for example, Q, R = QR(A).

```
url='https://raw.githubusercontent.com/davidaustinm/'
url+='ula_modules/master/orthogonality.py'
sage.repl.load.load(url, globals())
```

Suppose that A is the following matrix whose columns are linearly independent.

$$A = \begin{bmatrix} 1 & 0 & -3 \\ 0 & 2 & -1 \\ 1 & 0 & 1 \\ 1 & 3 & 5 \end{bmatrix}.$$

a. If $A = QR$, what is the shape of Q and R? What is special about the form of R?

b. Find the QR factorization using Q, R = QR(A) and verify that R has the predicted shape and that $A = QR$.

c. Find the matrix P that orthogonally projects vectors onto $\text{Col}(A)$.

d. Find $\widehat{\mathbf{b}}$, the orthogonal projection of $\mathbf{b} = \begin{bmatrix} 4 \\ -17 \\ -14 \\ 22 \end{bmatrix}$ onto $\text{Col}(A)$.

e. Explain why the equation $A\mathbf{x} = \widehat{\mathbf{b}}$ must be consistent and then find $\mathbf{x}$.

In fact, Sage provides its own version of the QR factorization that is a bit different than the way we've developed the factorization here. For this reason, we have provided our own version of the factorization.

6.4.3 Summary

This section explored the Gram-Schmidt orthogonalization algorithm and how it leads to the matrix factorization $A = QR$ when the columns of A are linearly independent.

- Beginning with a basis $\mathbf{v}_1, \mathbf{v}_2, \ldots, \mathbf{v}_n$ for a subspace W of $\mathbb{R}^m$, the vectors

$$\mathbf{w}_1 = \mathbf{v}_1$$

6.4. FINDING ORTHOGONAL BASES

$$\mathbf{w}_2 = \mathbf{v}_2 - \frac{\mathbf{v}_2 \cdot \mathbf{w}_1}{\mathbf{w}_1 \cdot \mathbf{w}_1}\mathbf{w}_1$$

$$\mathbf{w}_3 = \mathbf{v}_3 - \frac{\mathbf{v}_3 \cdot \mathbf{w}_1}{\mathbf{w}_1 \cdot \mathbf{w}_1}\mathbf{w}_1 - \frac{\mathbf{v}_3 \cdot \mathbf{w}_2}{\mathbf{w}_2 \cdot \mathbf{w}_2}\mathbf{w}_2$$

$$\vdots$$

$$\mathbf{w}_n = \mathbf{v}_n - \frac{\mathbf{v}_n \cdot \mathbf{w}_1}{\mathbf{w}_1 \cdot \mathbf{w}_1}\mathbf{w}_1 - \frac{\mathbf{v}_n \cdot \mathbf{w}_2}{\mathbf{w}_2 \cdot \mathbf{w}_2}\mathbf{w}_2 - \ldots - \frac{\mathbf{v}_n \cdot \mathbf{w}_{n-1}}{\mathbf{w}_{n-1} \cdot \mathbf{w}_{n-1}}\mathbf{w}_{n-1}$$

form an orthogonal basis for W.

- We may scale each vector $\mathbf{w}_i$ appropriately to obtain an orthonormal basis $\mathbf{u}_1, \mathbf{u}_2, \ldots, \mathbf{u}_n$.

- Expressing the Gram-Schmidt algorithm in matrix form shows that, if the columns of A are linearly independent, then we can write $A = QR$, where the columns of Q form an orthonormal basis for $\operatorname{Col}(A)$ and R is upper triangular.

6.4.4 Exercises

1. Suppose that a subspace W of $\mathbb{R}^3$ has a basis formed by

$$\mathbf{v}_1 = \begin{bmatrix} 1 \\ 1 \\ 1 \end{bmatrix}, \quad \mathbf{v}_2 = \begin{bmatrix} 1 \\ -2 \\ -2 \end{bmatrix}.$$

   a. Find an orthogonal basis for W.

   b. Find an orthonormal basis for W.

   c. Find the matrix P that projects vectors orthogonally onto W.

   d. Find the orthogonal projection of $\begin{bmatrix} 3 \\ 4 \\ -2 \end{bmatrix}$ onto W.

2. Find the QR factorization of $A = \begin{bmatrix} 4 & 7 \\ -2 & 4 \\ 4 & 4 \end{bmatrix}$.

3. Consider the basis of $\mathbb{R}^3$ given by the vectors

$$\mathbf{v}_1 = \begin{bmatrix} 2 \\ -2 \\ 2 \end{bmatrix}, \quad \mathbf{v}_2 = \begin{bmatrix} -1 \\ -3 \\ 1 \end{bmatrix}, \quad \mathbf{v}_3 = \begin{bmatrix} 2 \\ 0 \\ -5 \end{bmatrix}.$$

   a. Apply the Gram-Schmit orthogonalization algorithm to find an orthonormal basis $\mathbf{u}_1, \mathbf{u}_2, \mathbf{u}_3$ for $\mathbb{R}^3$.

   b. If A is the 3×3 whose columns are $\mathbf{v}_1, \mathbf{v}_2$, and $\mathbf{v}_3$, find the QR factorization of A.

c. Suppose that we want to solve the equation $A\mathbf{x} = \mathbf{b} = \begin{bmatrix} -9 \\ 1 \\ 7 \end{bmatrix}$, which we can rewrite as $QR\mathbf{x} = \mathbf{b}$.

1. If we set $\mathbf{y} = R\mathbf{x}$, the equation $QR\mathbf{x} = \mathbf{b}$ becomes $Q\mathbf{y} = \mathbf{b}$. Explain how to solve the equation $Q\mathbf{y} = \mathbf{b}$ in a computationally efficient manner.
2. Explain how to solve the equation $R\mathbf{x} = \mathbf{y}$ in a computationally efficient manner.
3. Find the solution $\mathbf{x}$ by first solving $Q\mathbf{y} = \mathbf{b}$ and then $R\mathbf{x} = \mathbf{y}$.

4. Consider the vectors

$$\mathbf{v}_1 = \begin{bmatrix} 1 \\ -1 \\ -1 \\ 1 \\ 1 \end{bmatrix}, \quad \mathbf{v}_2 = \begin{bmatrix} 2 \\ 1 \\ 4 \\ -4 \\ 2 \end{bmatrix}, \quad \mathbf{v}_3 = \begin{bmatrix} 5 \\ -4 \\ -3 \\ 7 \\ 1 \end{bmatrix}$$

and the subspace W of $\mathbb{R}^5$ that they span.

a. Find an orthonormal basis for W.

b. Find the 5×5 matrix that projects vectors orthogonally onto W.

c. Find $\widehat{\mathbf{b}}$, the orthogonal projection of $\mathbf{b} = \begin{bmatrix} -8 \\ 3 \\ -12 \\ 8 \\ -4 \end{bmatrix}$ onto W.

d. Express $\widehat{\mathbf{b}}$ as a linear combination of $\mathbf{v}_1$, $\mathbf{v}_2$, and $\mathbf{v}_3$.

5. Consider the set of vectors

$$\mathbf{v}_1 = \begin{bmatrix} 2 \\ 1 \\ 1 \end{bmatrix}, \quad \mathbf{v}_2 = \begin{bmatrix} 1 \\ 2 \\ 2 \end{bmatrix}, \quad \mathbf{v}_3 = \begin{bmatrix} 3 \\ 0 \\ 0 \end{bmatrix}.$$

a. What happens when we apply the Gram-Schmit orthogonalization algorithm?

b. Why does the algorithm fail to produce an orthogonal basis for $\mathbb{R}^3$?

6. Suppose that A is a matrix with linearly independent columns and having the factorization $A = QR$. Determine whether the following statements are true or false and explain your thinking.

a. It follows that $R = Q^T A$.

b. The matrix R is invertible.

c. The product Q^TQ projects vectors orthogonally onto Col(A).

d. The columns of Q are an orthogonal basis for Col(A).

e. The orthogonal complement Col(A)$^\perp$ = Nul(Q^T).

7. Suppose we have the QR factorization $A = QR$, where A is a 7×4 matrix.

    a. What is the shape of the product QQ^T? Explain the significance of this product.

    b. What is the shape of the product Q^TQ? Explain the significance of this product.

    c. What is the shape of the matrix R?

    d. If R is a diagonal matrix, what can you say about the columns of A?

8. Suppose we have the QR factorization $A = QR$ where the columns of A are $\mathbf{a}_1, \mathbf{a}_2, \ldots, \mathbf{a}_n$ and the columns of R are $\mathbf{r}_1, \mathbf{r}_2, \ldots, \mathbf{r}_n$.

    a. How can the matrix product A^TA be expressed in terms of dot products?

    b. How can the matrix product R^TR be expressed in terms of dot products?

    c. Explain why $A^TA = R^TR$.

    d. Explain why the dot products $\mathbf{a}_i \cdot \mathbf{a}_j = \mathbf{r}_i \cdot \mathbf{r}_j$.

6.5 Orthogonal least squares

Suppose we collect some data when performing an experiment and plot it as shown on the left of Figure 6.5.1. Notice that there is no line on which all the points lie; in fact, it would be surprising if there were since we can expect some uncertainty in the measurements recorded. There does, however, appear to be a line, as shown on the right, on which the points *almost* lie.

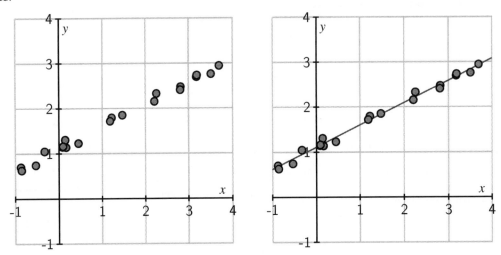

Figure 6.5.1 A collection of points and a line approximating the linear relationship implied by them.

In this section, we'll explore how the techniques developed in this chapter enable us to find the line that best approximates the data. More specifically, we'll see how the search for a line passing through the data points leads to an inconsistent system $A\mathbf{x} = \mathbf{b}$. Since we are unable to find a solution, we instead seek the vector $\mathbf{x}$ where $A\mathbf{x}$ is as close as possible to $\mathbf{b}$. Orthogonal projection gives us just the right tool for doing this.

Preview Activity 6.5.1.

a. Is there a solution to the equation $A\mathbf{x} = \mathbf{b}$ where A and $\mathbf{b}$ are such that

$$\begin{bmatrix} 1 & 2 \\ 2 & 5 \\ -1 & 0 \end{bmatrix} \mathbf{x} = \begin{bmatrix} 5 \\ -3 \\ -1 \end{bmatrix}.$$

b. We know that $\begin{bmatrix} 1 \\ 2 \\ -1 \end{bmatrix}$ and $\begin{bmatrix} 2 \\ 5 \\ 0 \end{bmatrix}$ form a basis for Col(A). Find an orthogonal basis for Col(A).

c. Find the orthogonal projection $\widehat{\mathbf{b}}$ of $\mathbf{b}$ onto Col(A).

6.5. ORTHOGONAL LEAST SQUARES

d. Explain why the equation $A\mathbf{x} = \widehat{\mathbf{b}}$ must be consistent and then find its solution.

6.5.1 A first example

When we've encountered inconsistent systems in the past, we've simply said there is no solution and moved on. The preview activity, however, shows how we can find approximate solutions to an inconsistent system: if there are no solutions to $A\mathbf{x} = \mathbf{b}$, we instead solve the consistent system $A\mathbf{x} = \widehat{\mathbf{b}}$, the orthogonal projection of $\mathbf{b}$ onto $\text{Col}(A)$. As we'll see, this solution is, in a specific sense, the best possible.

Activity 6.5.2. Suppose we have three data points $(1, 1)$, $(2, 1)$, and $(3, 3)$ and that we would like to find a line passing through them.

   a. Plot these three points in Figure 6.5.2. Are you able to draw a line that passes through all three points?

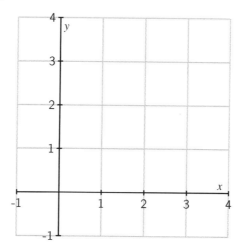

Figure 6.5.2 Plot the three data points here.

   b. Remember that the equation of a line can be written as $b + mx = y$ where m is the slope and b is the y-intercept. We will try to find b and m so that the three points lie on the line.

   The first data point $(1, 1)$ gives an equation for b and m. In particular, we know that when $x = 1$, then $y = 1$ so we have $b + m(1) = 1$ or $b + m = 1$. Use the other two data points to create a linear system describing m and b.

   c. We have obtained a linear system having three equations, one from each data point, for the two unknowns b and m. Identify a matrix A and vector $\mathbf{b}$ so that the system has the form $A\mathbf{x} = \mathbf{b}$, where $\mathbf{x} = \begin{bmatrix} b \\ m \end{bmatrix}$.

Notice that the unknown vector $\mathbf{x} = \begin{bmatrix} b \\ m \end{bmatrix}$ describes the line that we seek.

d. Is there a solution to this linear system? How does this question relate to your attempt to draw a line through the three points above?

e. Since this system is inconsistent, we know that **b** is not in the column space Col(A). Find an orthogonal basis for Col(A) and use it to find the orthogonal projection $\widehat{\mathbf{b}}$ of **b** onto Col(A).

f. Since $\widehat{\mathbf{b}}$ is in Col(A), the equation $A\mathbf{x} = \widehat{\mathbf{b}}$ is consistent. Find its solution $\mathbf{x} = \begin{bmatrix} b \\ m \end{bmatrix}$ and sketch the line $y = b + mx$ in Figure 6.5.2. We say that this is the line of best fit.

This activity illustrates the idea behind a technique known as *orthogonal least squares*, which we have been working toward throughout this chapter. If the data points are denoted as (x_i, y_i), we construct the matrix A and vector **b** as

$$A = \begin{bmatrix} 1 & x_1 \\ 1 & x_2 \\ 1 & x_3 \end{bmatrix}, \quad \mathbf{b} = \begin{bmatrix} y_1 \\ y_2 \\ y_3 \end{bmatrix}.$$

With the vector $\mathbf{x} = \begin{bmatrix} b \\ m \end{bmatrix}$ representing the line $b + mx = y$, we see that the equation $A\mathbf{x} = \mathbf{b}$ describes a line passing through all the data points. In our activity, it is visually apparent that there is no such line, which agrees with the fact that the equation $A\mathbf{x} = \mathbf{b}$ is inconsistent.

Remember that $\widehat{\mathbf{b}}$, the orthogonal projection of **b** onto Col(A), is the closest vector in Col(A) to **b**. Therefore, when we solve the equation $A\mathbf{x} = \widehat{\mathbf{b}}$, we are finding the vector **x** so that

$$A\mathbf{x} = \begin{bmatrix} b + mx_1 \\ b + mx_2 \\ b + mx_3 \end{bmatrix} \text{ is as close to } \mathbf{b} = \begin{bmatrix} y_1 \\ y_2 \\ y_3 \end{bmatrix} \text{ as possible.}$$

Let's think about what this means within the context of this problem.

The difference $\mathbf{b} - A\mathbf{x} = \begin{bmatrix} y_1 - (b + mx_1) \\ y_2 - (b + mx_2) \\ y_3 - (b + mx_3) \end{bmatrix}$ so that the square of the distance between $A\mathbf{x}$ and **b** is

$$|\mathbf{b} - A\mathbf{x}|^2 = (y_1 - (b + mx_1))^2 + (y_2 - (b + mx_2))^2 + (y_3 - (b + mx_3))^2.$$

Our approach finds the values for b and m that make this sum of squares as small as possible, which is why we call this a *least-squares* problem.

Drawing the line defined by the vector $\mathbf{x} = \begin{bmatrix} b \\ m \end{bmatrix}$, the quantity $y_i - (b + mx_i)$ reflects the

6.5. ORTHOGONAL LEAST SQUARES

vertical distance between the line and the data point (x_i, y_i), as shown in Figure 6.5.3. Seen in this way, the square of the distance $|\mathbf{b} - A\mathbf{x}|^2$ is a measure of how much the line defined by the vector $\mathbf{x}$ misses the data points. The solution to the least-squares problem is the line that misses the data points by the smallest amount possible.

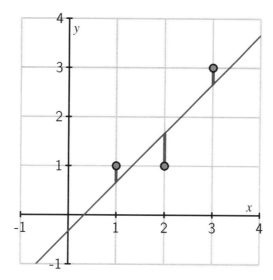

Figure 6.5.3 The solution of the least-squares problem and the vertical distances between the line and the data points.

6.5.2 Solving least-squares problems

Now that we've seen an example of what we're trying to accomplish, let's put this technique into a more general framework.

Given an inconsistent system $A\mathbf{x} = \mathbf{b}$, we seek the vector $\mathbf{x}$ that minimizes the distance from $A\mathbf{x}$ to $\mathbf{b}$. In other words, $\mathbf{x}$ satisfies $A\mathbf{x} = \widehat{\mathbf{b}}$, where $\widehat{\mathbf{b}}$ is the orthogonal projection of $\mathbf{b}$ onto the column space $\text{Col}(A)$. We know the equation $A\mathbf{x} = \widehat{\mathbf{b}}$ is consistent since $\widehat{\mathbf{b}}$ is in $\text{Col}(A)$, and we know there is only one solution if we assume that the columns of A are linearly independent.

We will usually denote the solution of $A\mathbf{x} = \widehat{\mathbf{b}}$ by $\widehat{\mathbf{x}}$ and call this vector the *least-squares approximate solution* of $A\mathbf{x} = \mathbf{b}$ to distinguish it from a (possibly non-existent) solution of $A\mathbf{x} = \mathbf{b}$.

There is an alternative method for finding $\widehat{\mathbf{x}}$ that does not involve first finding the orthogonal projection $\widehat{\mathbf{b}}$. Remember that $\widehat{\mathbf{b}}$ is defined by the fact that $\widehat{\mathbf{b}} - \mathbf{b}$ is orthogonal to $\text{Col}(A)$. In other words, $\widehat{\mathbf{b}} - \mathbf{b}$ is in the orthogonal complement $\text{Col}(A)^{\perp}$, which Proposition 6.2.10 tells us is the same as $\text{Nul}(A^T)$. Since $\widehat{\mathbf{b}} - \mathbf{b}$ is in $\text{Nul}(A^T)$, it follows that

$$A^T(\widehat{\mathbf{b}} - \mathbf{b}) = 0.$$

Because the least-squares approximate solution is the vector $\widehat{\mathbf{x}}$ such that $A\widehat{\mathbf{x}} = \widehat{\mathbf{b}}$, we can rearrange this equation to see that

$$A^T(A\widehat{\mathbf{x}} - \mathbf{b}) = 0$$
$$A^T A\widehat{\mathbf{x}} - A^T\mathbf{b} = 0$$
$$A^T A\widehat{\mathbf{x}} = A^T\mathbf{b}.$$

This equation is called the *normal equation*, and we have the following proposition.

Proposition 6.5.4 *If the columns of A are linearly independent, then there is a unique least-squares approximate solution $\widehat{\mathbf{x}}$ to the equation $A\mathbf{x} = \mathbf{b}$ given by the normal equation*

$$A^T A\widehat{\mathbf{x}} = A^T\mathbf{b}.$$

Example 6.5.5 Consider the equation

$$\begin{bmatrix} 2 & 1 \\ 2 & 0 \\ -1 & 3 \end{bmatrix} \mathbf{x} = \begin{bmatrix} 16 \\ -1 \\ 7 \end{bmatrix}$$

with matrix A and vector $\mathbf{b}$. Since this equation is inconsistent, we will find the least-squares approximate solution $\widehat{\mathbf{x}}$ by solving the normal equation $A^T A\widehat{\mathbf{x}} = A^T\mathbf{b}$, which has the form

$$A^T A\widehat{\mathbf{x}} = \begin{bmatrix} 9 & -1 \\ -1 & 10 \end{bmatrix} = \begin{bmatrix} 23 \\ 37 \end{bmatrix} = A^T\mathbf{b}$$

and the solution $\widehat{\mathbf{x}} = \begin{bmatrix} 3 \\ 4 \end{bmatrix}$.

Activity 6.5.3. The rate at which a cricket chirps is related to the outdoor temperature, as reflected in some experimental data that we'll study in this activity. The chirp rate C is expressed in chirps per second while the temperature T is in degrees Fahrenheit. Evaluate the following cell to load the data:

```
base='https://raw.githubusercontent.com/davidaustinm/'
url=base+'ula_modules/master/orthogonality.py'
sage.repl.load.load(url, globals())
url=base+'ula_modules/master/data/crickets.csv'
df = pd.read_csv(url)
data = [vector(row) for row in df.values]
chirps = vector(df['Chirps'])
temps = vector(df['Temperature'])
print(df)
list_plot(data, color='blue', size=40, xmin=12, xmax=22, ymin=60,
    ymax=100)
```

Evaluating this cell also provides:

- the vectors `chirps` and `temps` formed from the columns of the dataset.

- the command `onesvec(n)`, which creates an n-dimensional vector whose entries

6.5. ORTHOGONAL LEAST SQUARES

are all one.

- Remember that you can form a matrix whose columns are the vectors v1 and v2 with `matrix([v1, v2]).T`.

We would like to represent this relationship by a linear function

$$\beta_0 + \beta_1 C = T.$$

a. Use the first data point $(C_1, T_1) = (20.0, 88.6)$ to write an equation involving β_0 and β_1.

b. Suppose that we represent the unknowns using a vector $\mathbf{x} = \begin{bmatrix} \beta_0 \\ \beta_1 \end{bmatrix}$. Use the 15 data points to create the matrix A and vector $\mathbf{b}$ so that the linear system $A\mathbf{x} = \mathbf{b}$ describes the unknown vector $\mathbf{x}$.

c. Write the normal equations $A^T A \widehat{\mathbf{x}} = A^T \mathbf{b}$; that is, find the matrix $A^T A$ and the vector $A^T \mathbf{b}$.

d. Solve the normal equations to find $\widehat{\mathbf{x}}$, the least-squares approximate solution to the equation $A\mathbf{x} = \mathbf{b}$. Call your solution xhat since x has another meaning in Sage.

What are the values of β_0 and β_1 that you found?

e. If the chirp rate is 22 chirps per second, what is your prediction for the temperature?

You can plot the data and your line, assuming you called the solution xhat, using the cell below.

```
plot_model(xhat, data, domain=(12, 22))
```

This example demonstrates an approach, called *linear regression*, in which a collection of data is modeled using a linear function found by solving a least-squares problem. Once we have the linear function that best fits the data, we can make predictions about situations that we haven't encountered in the data.

If we're going to use our function to make predictions, it's natural to ask how much confidence we have in these predictions. This is a statistical question that leads to a rich and well-developed theory[1], which we won't explore in much detail here. However, there is one simple measure of how well our linear function fits the data that is known as the coefficient of determination and denoted by R^2.

We have seen that the square of the distance $|\mathbf{b} - A\mathbf{x}|^2$ measures the amount by which the

---

[1]For example, see Gareth James, Daniela Witten, Trevor Hastie, Robert Tibshirani. *An Introduction to Statistical Learning: with Applications in R*. Springer, 2013.

line fails to pass through the data points. When the line is close to the data points, we expect this number to be small. However, the size of this measure depends on the scale of the data. For instance, the two lines shown in Figure 6.5.6 seem to fit the data equally well, but $|\mathbf{b}-A\widehat{\mathbf{x}}|^2$ is 100 times larger on the right.

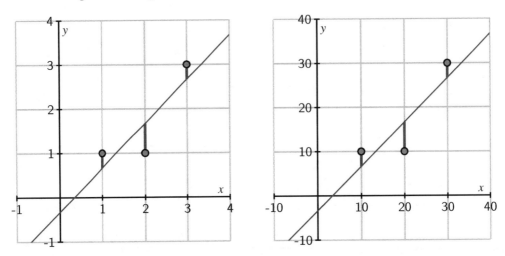

Figure 6.5.6 The lines appear to fit equally well in spite of the fact that $|\mathbf{b} - A\widehat{\mathbf{x}}|^2$ differs by a factor of 100.

The coefficient of determination R^2 is defined by normalizing $|\mathbf{b} - A\widehat{\mathbf{x}}|^2$ so that it is independent of the scale. Recall that we described how to demean a vector in Section 6.1: given a vector $\mathbf{v}$, we obtain $\widetilde{\mathbf{v}}$ by subtracting the average of the components from each component.

Definition 6.5.7 Coefficient of determination. The coefficient of determination is

$$R^2 = 1 - \frac{|\mathbf{b} - A\widehat{\mathbf{x}}|^2}{|\widetilde{\mathbf{b}}|^2},$$

where $\widetilde{\mathbf{b}}$ is the vector obtained by demeaning $\mathbf{b}$.

A more complete explanation of this definition relies on the concept of variance, which we explore in Exercise 6.5.6.12 and the next chapter. For the time being, it's enough to know that $0 \leq R^2 \leq 1$ and that the closer R^2 is to 1, the better the line fits the data. In our original example, illustrated in Figure 6.5.6, we find that $R^2 = 0.75$, and in our study of cricket chirp rates, we have $R^2 = 0.69$. However, assessing the confidence we have in predictions made by solving a least-squares problem can require considerable thought, and it would be naive to rely only on the value of R^2.

6.5.3 Using QR factorizations

As we've seen, the least-squares approximate solution $\widehat{\mathbf{x}}$ to $A\mathbf{x} = \mathbf{b}$ may be found by solving the normal equation $A^T A\widehat{\mathbf{x}} = A^T\mathbf{b}$, and this can be a practical strategy for some problems. However, this approach can be problematic as small rounding errors can accumulate and lead to inaccurate final results.

6.5. ORTHOGONAL LEAST SQUARES

As the next activity demonstrates, there is an alternate method for finding the least-squares approximate solution $\widehat{\mathbf{x}}$ using a QR factorization of the matrix A, and this method is preferable as it is numerically more reliable.

Activity 6.5.4.

a. Suppose we are interested in finding the least-squares approximate solution to the equation $A\mathbf{x} = \mathbf{b}$ and that we have the QR factorization $A = QR$. Explain why the least-squares approximate solution is given by solving

$$A\widehat{\mathbf{x}} = QQ^T\mathbf{b}$$

$$QR\widehat{\mathbf{x}} = QQ^T\mathbf{b}$$

b. Multiply both sides of the second expression by Q^T and explain why

$$R\widehat{\mathbf{x}} = Q^T\mathbf{b}.$$

Since R is upper triangular, this is a relatively simple equation to solve using back substitution, as we saw in Section 5.1. We will therefore write the least-squares approximate solution as

$$\widehat{\mathbf{x}} = R^{-1}Q^T\mathbf{b},$$

and put this to use in the following context.

c. Brozak's formula, which is used to calculate a person's body fat index BFI, is

$$BFI = 100\left(\frac{4.57}{\rho} - 4.142\right)$$

where ρ denotes a person's body density in grams per cubic centimeter. Obtaining an accurate measure of ρ is difficult, however, because it requires submerging the person in water and measuring the volume of water displaced. Instead, we will gather several other body measurements, which are more easily obtained, and use it to predict BFI.

For instance, suppose we take 10 patients and measure their weight w in pounds, height h in inches, abdomen a in centimeters, wrist circumference r in centimeters, neck circumference n in centimeters, and BFI. Evaluating the following cell loads and displays the data.

```
base='https://raw.githubusercontent.com/davidaustinm/'
url=base+'ula_modules/master/orthogonality.py'
sage.repl.load.load(url, globals())
url=base+'/ula_modules/master/data/bfi.csv'
df = pd.read_csv(url)
weight = vector(df['Weight'])
height = vector(df['Height'])
abdomen = vector(df['Abdomen'])
wrist = vector(df['Wrist'])
neck = vector(df['Neck'])
BFI = vector(df['BFI'])
print(df)
```

In addition, that cell provides:

(a) vectors weight, height, abdomen, wrist, neck, and BFI formed from the columns of the dataset.

(b) the command onesvec(n), which returns an n-dimensional vector whose entries are all one.

(c) the command QR(A) that returns the QR factorization of A as Q, R = QR(A).

(d) the command demean(v), which returns the demeaned vector $\widetilde{\mathbf{v}}$.

We would like to find the linear function

$$\beta_0 + \beta_1 w + \beta_2 h + \beta_3 a + \beta_4 r + \beta_5 n = BFI$$

that best fits the data.

Use the first data point to write an equation for the parameters $\beta_0, \beta_1, \ldots, \beta_5$.

d. Describe the linear system $A\mathbf{x} = \mathbf{b}$ for these parameters. More specifically, describe how the matrix A and the vector $\mathbf{b}$ are formed.

e. Construct the matrix A and find its QR factorization in the cell below.

f. Find the least-squares approximate solution $\widehat{\mathbf{x}}$ by solving the equation $R\widehat{\mathbf{x}} = Q^T\mathbf{b}$. You may want to use N(xhat) to display a decimal approximation of the vector. What are the parameters $\beta_0, \beta_1, \ldots, \beta_5$ that best fit the data?

g. Find the coefficient of determination R^2 for your parameters. What does this imply about the quality of the fit?

h. Suppose a person's measurements are: weight 190, height 70, abdomen 90, wrist 18, and neck 35. Estimate this person's BFI.

To summarize, we have seen that

6.5. ORTHOGONAL LEAST SQUARES

Proposition 6.5.8 *If the columns of A are linearly independent and we have the QR factorization $A = QR$, then the least-squares approximate solution $\widehat{x}$ to the equation $Ax = b$ is given by*

$$\widehat{x} = R^{-1}Q^T b.$$

6.5.4 Polynomial Regression

In the examples we've seen so far, we have fit a linear function to a dataset. Sometimes, however, a polynomial, such as a quadratic function, may be more appropriate. It turns out that the techniques we've developed in this section are still useful as the next activity demonstrates.

Activity 6.5.5.

a. Suppose that we have a small dataset containing the points $(0,2)$, $(1,1)$, $(2,3)$, and $(3,3)$, such as appear when the following cell is evaluated.

```
url='https://raw.githubusercontent.com/davidaustinm/'
url+='ula_modules/master/orthogonality.py'
sage.repl.load.load(url, globals())
data = [[0, 2], [1, 1], [2, 3], [3, 3]]
list_plot(data, color='blue', size=40)
```

In addition to loading and plotting the data, evaluating that cell provides the following commands:

- Q, R = QR(A) returns the QR factorization of A.
- demean(v) returns the demeaned vector $\widetilde{v}$.

Let's fit a quadratic function of the form

$$\beta_0 + \beta_1 x + \beta_2 x^2 = y$$

to this dataset.

Write four equations, one for each data point, that describe the coefficients β_0, β_1, and β_2.

b. Express these four equations as a linear system $Ax = b$ where $x = \begin{bmatrix} \beta_0 \\ \beta_1 \\ \beta_2 \end{bmatrix}$.

Find the QR factorization of A and use it to find the least-squares approximate solution $\widehat{x}$.

c. Use the parameters β_0, β_1, and β_2 that you found to write the quadratic function that fits the data. You can plot this function, along with the data, by entering your function in the place indicated below.

```
list_plot(data, color='blue', size=40) + plot( **your
        function here**,
 0, 3, color='red')
```

d. What is your predicted y value when $x = 1.5$?

e. Find the coefficient of determination R^2 for the quadratic function. What does this say about the quality of the fit?

f. Now fit a cubic polynomial of the form

$$\beta_0 + \beta_1 x + \beta_2 x^2 + \beta_3 x^3 = y$$

to this dataset.

g. Find the coefficient of determination R^2 for the cubic function. What does this say about the quality of the fit?

h. What do you notice when you plot the cubic function along with the data? How does this reflect the value of R^2 that you found?

```
list_plot(data, color='blue', size=40) + plot( **your
        function here**,
 0, 3, color='red')
```

The matrices A that you created in the last activity when fitting a quadratic and cubic function to a dataset have a special form. In particular, if the data points are labeled (x_i, y_i) and we seek a degree k polynomial, then

$$A = \begin{bmatrix} 1 & x_1 & x_1^2 & \cdots & x_1^k \\ 1 & x_2 & x_2^2 & \cdots & x_2^k \\ \vdots & \vdots & \vdots & \ddots & \vdots \\ 1 & x_m & x_m^2 & \cdots & x_m^k \end{bmatrix}.$$

This is called a *Vandermonde* matrix of degree k.

Activity 6.5.6. This activity explores a dataset describing Arctic sea ice and that comes from Sustainability Math.[2]

Evaluating the cell below will plot the extent of Arctic sea ice, in millions of square kilometers, during the twelve months of 2012.

6.5. ORTHOGONAL LEAST SQUARES

```
base='https://raw.githubusercontent.com/davidaustinm/'
url=base+'ula_modules/master/orthogonality.py'
sage.repl.load.load(url, globals())
url=base+'/ula_modules/master/data/sea_ice.csv'
df = pd.read_csv(url)
data = [vector([row[0], row[2]]) for row in df.values]
month = vector(df['Month'])
ice = vector(df['2012'])
print(df[['Month', '2012']])
list_plot(data, color='blue', size=40)
```

In addition, you have access to a few special variables and commands:

- `month` is the vector of month values and `ice` is the vector of sea ice values from the table above.
- `vandermonde(x, k)` constructs the Vandermonde matrix of degree k using the points in the vector x.
- `Q, R = QR(A)` provides the QR factorization of A.
- `demean(v)` returns the demeaned vector $\widetilde{\mathbf{v}}$.

a. Find the vector $\widehat{\mathbf{x}}$, the least-squares approximate solution to the linear system that results from fitting a degree 5 polynomial to the data.

b. If your result is stored in the variable xhat, you may plot the polynomial and the data together using the following cell.

```
plot_model(xhat, data)
```

c. Find the coefficient of determination R^2 for this polynomial fit.

d. Repeat these steps to fit a degree 8 polynomial to the data, plot the polynomial with the data, and find R^2.

e. Repeat one more time by fitting a degree 11 polynomial to the data, creating a plot, and finding R^2.

It's certainly true that higher degree polynomials fit the data better, as seen by the increasing values of R^2, but that's not always a good thing. For instance, when $k = 11$, you may notice that the graph of the polynomial wiggles a little more than we would expect. In this case, the polynomial is trying too hard to fit the data, which usually contains some uncertainty, especially if it's obtained from measurements. The error built in to the data is called *noise*, and its presence means that we shouldn't expect our polynomial to fit the data perfectly. When we choose a polynomial whose degree is too high, we give the noise too much weight in the model, which leads to some undesirable behavior, like the wiggles in the graph.

Fitting the data with a polynomial whose degree is too high is called *overfitting*, a phenomenon that can appear in many machine learning applications. Generally speaking, we would like to choose k large enough to capture the essential features of the data but not so large that we overfit and build the noise into the model. There are ways to determine the optimal value of k, but we won't pursue that here.

f. Choosing a reasonable value of k, estimate the extent of Arctic sea ice at month 6.5, roughly at the Summer Solstice.

6.5.5 Summary

This section introduced some types of least-squares problems and a framework for working with them.

- Given an inconsistent system $A\mathbf{x} = \mathbf{b}$, we find $\widehat{\mathbf{x}}$, the least-squares approximate solution, by requiring that $A\widehat{\mathbf{x}}$ be as possible to $\mathbf{b}$ as possible. In other words, $A\widehat{\mathbf{x}} = \widehat{\mathbf{b}}$ where $\widehat{\mathbf{b}}$ is the orthogonal projection of $\mathbf{b}$ onto Col(A).

- One way to find $\widehat{\mathbf{x}}$ is by solving the normal equations $A^T A \widehat{\mathbf{x}} = A^T \mathbf{b}$. This is not our preferred method since numerical problems can arise.

- A second way to find $\widehat{\mathbf{x}}$ uses a QR factorization of A. If $A = QR$, then $\widehat{\mathbf{x}} = R^{-1} Q^T \mathbf{b}$ and finding R^{-1} is computationally feasible since R is upper triangular.

- This technique may be applied widely and is useful for modeling data. We saw examples in this section where linear functions of several input variables and polynomials provided effective models for different datasets.

- A simple measure of the quality of the fit is the coefficient of determination R^2 though some additional thought should be given in real applications.

6.5.6 Exercises

Evaluating the following cell loads in some commands that will be helpful in the following exercises. In particular, there are commands:

- QR(A) that returns the QR factorization of A as Q, R = QR(A),

- onesvec(n) that returns the n-dimensional vector whose entries are all 1,

---
[2]sustainabilitymath.org

6.5. ORTHOGONAL LEAST SQUARES

- demean(v) that demeans the vector v,
- vandermonde(x, k) that returns the Vandermonde matrix of degree k formed from the components of the vector x, and
- plot_model(xhat, data) that plots the data and the model xhat.

```
url='https://raw.githubusercontent.com/davidaustinm/'
url+='ula_modules/master/orthogonality.py'
sage.repl.load.load(url, globals())
```

1. Suppose we write the linear system

$$\begin{bmatrix} 1 & -1 \\ 2 & -1 \\ -1 & 3 \end{bmatrix} \mathbf{x} = \begin{bmatrix} -8 \\ 5 \\ -10 \end{bmatrix}$$

as $A\mathbf{x} = \mathbf{b}$.

   a. Find an orthogonal basis for $\text{Col}(A)$.

   b. Find $\widehat{\mathbf{b}}$, the orthogonal projection of $\mathbf{b}$ onto $\text{Col}(A)$.

   c. Find a solution to the linear system $A\mathbf{x} = \widehat{\mathbf{b}}$.

2. Consider the data in Table 6.5.9.

   Table 6.5.9 A dataset with four points.

x	y
1	1
2	1
3	1
4	2

   a. Set up the linear system $A\mathbf{x} = \mathbf{b}$ that describes the line $b + mx = y$ passing through these points.

   b. Write the normal equations that describe the least-squares approximate solution to $A\mathbf{x} = \mathbf{b}$.

   c. Find the least-squares approximate solution $\widehat{\mathbf{x}}$ and plot the data and the resulting line.

   d. What is your predicted y-value when $x = 3.5$?

   e. Find the coefficient of determination R^2.

3. Consider the four points in Table 6.5.9.

a. Set up a linear system $A\mathbf{x} = \mathbf{b}$ that describes a quadratic function

$$\beta_0 + \beta_1 x + \beta_2 x^2 = y$$

passing through the points.

b. Use a QR factorization to find the least-squares approximate solution $\widehat{\mathbf{x}}$ and plot the data and the graph of the resulting quadratic function.

c. What is your predicted y-value when $x = 3.5$?

d. Find the coefficient of determination R^2.

4. Consider the data in Table 6.5.10.

Table 6.5.10 A simple dataset

x_1	x_2	y
1	1	4.2
1	2	3.3
2	1	5.9
2	2	5.1
3	2	7.5
3	3	6.3

a. Set up a linear system $A\mathbf{x} = \mathbf{b}$ that describes the relationship

$$\beta_0 + \beta_1 x_1 + \beta_2 x_2 = y.$$

b. Find the least-squares approximate solution $\widehat{\mathbf{x}}$.

c. What is your predicted y-value when $x_1 = 2.4$ and $x_2 = 2.9$?

d. Find the coefficient of determination R^2.

5. Determine whether the following statements are true or false and explain your thinking.

a. If $A\mathbf{x} = \mathbf{b}$ is consistent, then $\widehat{\mathbf{x}}$ is a solution to $A\mathbf{x} = \mathbf{b}$.

b. If $R^2 = 1$, then the least-squares approximate solution $\widehat{\mathbf{x}}$ is also a solution to the original equation $A\mathbf{x} = \mathbf{b}$.

c. Given the QR factorization $A = QR$, we have $A\widehat{\mathbf{x}} = Q^T Q \mathbf{b}$.

d. A QR factorization provides a method for finding the least-squares approximate solution to $A\mathbf{x} = \mathbf{b}$ that is more reliable than solving the normal equations.

e. A solution to $AA^T \mathbf{x} = A\mathbf{b}$ is the least-squares approximate solution to $A\mathbf{x} = \mathbf{b}$.

6. Explain your response to the following questions.

a. If $\widehat{\mathbf{x}} = \mathbf{0}$, what does this say about the vector $\mathbf{b}$?

6.5. ORTHOGONAL LEAST SQUARES

b. If the columns of A are orthonormal, how can you easily find the least-squares approximate solution to $A\mathbf{x} = \mathbf{b}$?

7. The following cell loads in some data showing the number of people in Bangladesh living without electricity over 27 years. It also defines vectors year, which records the years in the dataset, and people, which records the number of people.

```
base='https://raw.githubusercontent.com/davidaustinm/'
url=base+'ula_modules/master/orthogonality.py'
sage.repl.load.load(url, globals())
url=base+'ula_modules/master/data/bangladesh.csv'
df = pd.read_csv(url)
data = [vector(row) for row in df.values]
year = vector(df['Year'])
people = vector(df['People'])
print(df)
list_plot(data, size=40, color='blue')
```

a. Suppose we want to write
$$N = \beta_0 + \beta_1 t$$
where t is the year and N is the number of people. Construct the matrix A and vector $\mathbf{b}$ so that the linear system $A\mathbf{x} = \mathbf{b}$ describes the vector $\mathbf{x} = \begin{bmatrix} \beta_0 \\ \beta_1 \end{bmatrix}$.

b. Using a QR factorization of A, find the values of β_0 and β_1 in the least-squares approximate solution $\widehat{\mathbf{x}}$.

c. What is the coefficient of determination R^2 and what does this tell us about the quality of the approximation?

d. What is your prediction for the number of people living without electricity in 1985?

e. Estimate the year in which there will be no people living without electricity.

8. This problem concerns a dataset describing planets in our Solar system. For each planet, we have the length L of the semi-major axis, essentially the distance from the planet to the Sun in AU (astronomical units), and the period P, the length of time in years required to complete one orbit around the Sun.

We would like to model this data using the function $P = CL^r$ where C and r are parameters we need to determine. Since this isn't a linear function, we will transform this relationship by taking the natural logarithm of both sides to obtain
$$\ln(P) = \ln(C) + r \ln(L).$$

Evaluating the following cell loads the dataset and defines two vectors logaxis, whose components are $\ln(L)$, and logperiod, whose components are $\ln(P)$.

```
import numpy as np
base='https://raw.githubusercontent.com/davidaustinm/'
url=base+'ula_modules/master/orthogonality.py'
sage.repl.load.load(url, globals())
url=base+'ula_modules/master/data/planets.csv'
df = pd.read_csv(url, index_col=0)
logaxis = vector(np.log(df['Semi-major_axis']))
logperiod = vector(np.log(df['Period']))
print(df)
```

a. Construct the matrix A and vector $\mathbf{b}$ so that the solution to $A\mathbf{x} = \mathbf{b}$ is the vector
$$\mathbf{x} = \begin{bmatrix} \ln(C) \\ r \end{bmatrix}.$$

b. Find the least-squares approximate solution $\hat{\mathbf{x}}$. What does this give for the values of C and r?

c. Find the coefficient of determination R^2. What does this tell us about the quality of the approximation?

d. Suppose that the orbit of an asteroid has a semi-major axis whose length is $L = 4.0$ AU. Estimate the period P of the asteroid's orbit.

e. Halley's Comet has a period of $P = 75$ years. Estimate the length of its semi-major axis.

9. Evaluating the following cell loads a dataset describing the temperature in the Earth's atmosphere at various altitudes. There are also two vectors altitude, expressed in kilometers, and temperature, in degrees Celsius.

```
base='https://raw.githubusercontent.com/davidaustinm/'
url=base+'ula_modules/master/orthogonality.py'
sage.repl.load.load(url, globals())
url=base+'ula_modules/master/data/altitude-temps.csv'
df = pd.read_csv(url)
data = [vector(row) for row in df.values]
altitude = vector(df['Altitude'])
temperature = vector(df['Temperature'])
print(df)
list_plot(data, size=40, color='blue')
```

a. Describe how to form the matrix A and vector $\mathbf{b}$ so that the linear system $A\mathbf{x} = \mathbf{b}$ describes a degree k polynomial fitting the data.

b. After choosing a value of k, construct the matrix A and vector $\mathbf{b}$, and find the least-squares approximate solution $\hat{\mathbf{x}}$.

c. Plot the polynomial and data using plot_model(xhat, data).

d. Now examine what happens as you vary the degree of the polynomial k. Choose

6.5. ORTHOGONAL LEAST SQUARES

an appropriate value of k that seems to capture the most important features of the data while avoiding overfitting, and explain your choice.

e. Use your value of k to estimate the temperature at an altitude of 55 kilometers.

10. The following cell loads some data describing 1057 houses in a particular real estate market. For each house, we record the living area in square feet, the lot size in acres, the age in years, and the price in dollars. The cell also defines variables area, size, age, and price.

```
base='https://raw.githubusercontent.com/davidaustinm/'
url=base+'ula_modules/master/orthogonality.py'
sage.repl.load.load(url, globals())
url=base+'ula_modules/master/data/housing.csv'
df = pd.read_csv(url, index_col=0)
df = df.fillna(df.mean())
area = vector(df['Living.Area'])
size = vector(df['Lot.Size'])
age = vector(df['Age'])
price = vector(df['Price'])
df
```

We will use linear regression to predict the price of a house given its living area, lot size, and age:

$$\beta_0 + \beta_1 \text{ Living Area} + \beta_2 \text{ Lot Size} + \beta_3 \text{ Age} = \text{Price}.$$

a. Use a QR factorization to find the least-squares approximate solution $\widehat{x}$.

b. Discuss the significance of the signs of β_1, β_2, and β_3.

c. If two houses are identical except for differing in age by one year, how would you predict that their prices compare to each another?

d. Find the coefficient of determination R^2. What does this say about the quality of the fit?

e. Predict the price of a house whose living area is 2000 square feet, lot size is 1.5 acres, and age is 50 years.

11. We observed that if the columns of A are linearly independent, then there is a unique least-squares approximate solution to the equation $A\mathbf{x} = \mathbf{b}$ because the equation $A\widehat{\mathbf{x}} = \widehat{\mathbf{b}}$ has a unique solution. We also said that $\widehat{\mathbf{x}}$ is the unique solution to the normal equation $A^T A\widehat{\mathbf{x}} = A^T \mathbf{b}$ without explaining why this equation has a unique solution. This exercise offers an explanation.

Assuming that the columns of A are linearly independent, we would like to conclude that the equation $A^T A\widehat{\mathbf{x}} = A^T \mathbf{b}$ has a unique solution.

a. Suppose that $\mathbf{x}$ is a vector for which $A^T A\mathbf{x} = \mathbf{0}$. Explain why the following argument is valid and allows us to conclude that $A\mathbf{x} = \mathbf{0}$.

$$A^T A \mathbf{x} = \mathbf{0}$$
$$\mathbf{x} \cdot A^T A \mathbf{x} = \mathbf{x} \cdot \mathbf{0} = 0$$
$$(A\mathbf{x}) \cdot (A\mathbf{x}) = 0$$
$$|A\mathbf{x}|^2 = 0.$$

In other words, if $A^T A \mathbf{x} = \mathbf{0}$, we know that $A\mathbf{x} = \mathbf{0}$.

b. If the columns of A are linearly independent and $A\mathbf{x} = \mathbf{0}$, what do we know about the vector $\mathbf{x}$?

c. Explain why $A^T A \mathbf{x} = \mathbf{0}$ can only happen when $\mathbf{x} = \mathbf{0}$.

d. Assuming that the columns of A are linearly independent, explain why $A^T A \widehat{\mathbf{x}} = A^T \mathbf{b}$ has a unique solution.

12. This problem is about the meaning of the coefficient of determination R^2 and its connection to variance, a topic that appears in the next section. Throughout this problem, we consider the linear system $A\mathbf{x} = \mathbf{b}$ and the approximate least-squares solution $\widehat{\mathbf{x}}$, where $A\widehat{\mathbf{x}} = \widehat{\mathbf{b}}$. We suppose that A is an $m \times n$ matrix, and we will denote the m-dimensional

$$\text{vector } \mathbf{1} = \begin{bmatrix} 1 \\ 1 \\ \vdots \\ 1 \end{bmatrix}.$$

a. Explain why $\overline{\mathbf{b}}$, the mean of the components of $\mathbf{b}$, can be found as the dot product

$$\overline{\mathbf{b}} = \frac{1}{m} \mathbf{b} \cdot \mathbf{1}.$$

b. In the examples we have seen in this section, explain why $\mathbf{1}$ is in $\text{Col}(A)$.

c. If we write $\mathbf{b} = \widehat{\mathbf{b}} + \mathbf{b}^\perp$, explain why

$$\mathbf{b}^\perp \cdot \mathbf{1} = 0$$

and hence why the mean of the components of $\mathbf{b}^\perp$ is zero.

d. The variance of an m-dimensional vector $\mathbf{v}$ is $\text{Var}(\mathbf{v}) = \frac{1}{m} |\widetilde{\mathbf{v}}|^2$, where $\widetilde{\mathbf{v}}$ is the vector obtained by demeaning $\mathbf{v}$.

Explain why
$$\text{Var}(\mathbf{b}) = \text{Var}(\widehat{\mathbf{b}}) + \text{Var}(\mathbf{b}^\perp).$$

e. Explain why

$$\frac{|\mathbf{b} - A\widehat{\mathbf{x}}|^2}{|\widetilde{\mathbf{b}}|^2} = \frac{\text{Var}(\mathbf{b}^\perp)}{\text{Var}(\mathbf{b})}.$$

6.5. ORTHOGONAL LEAST SQUARES

and hence

$$R^2 = \frac{\text{Var}(\widehat{\mathbf{b}})}{\text{Var}(\mathbf{b})} = \frac{\text{Var}(A\widehat{\mathbf{x}})}{\text{Var}(\mathbf{b})}.$$

These expressions indicate why it is sometimes said that R^2 measures the "fraction of variance explained" by the function we are using to fit the data. As seen in the previous exercise, there may be other features that are not recorded in the dataset that influence the quantity we wish to predict.

f. Explain why $0 \leq R^2 \leq 1$.

CHAPTER 7

Singular value decompositions

Chapter 4 demonstrated several important uses for the theory of eigenvalues and eigenvectors. For example, knowing the eigenvalues and eigenvectors of a matrix A enabled us to make predictions about the long-term behavior of dynamical systems in which some initial state $\mathbf{x}_0$ evolves according to the rule $\mathbf{x}_{k+1} = A\mathbf{x}_k$.

We can't, however, apply this theory to every problem we might meet. First, eigenvectors only exist when the matrix A is square, and we have seen situations, such as the least-squares problems in Section 6.5, where the matrices we're interested in are not square. Second, even when A is square, there may not be a basis for $\mathbb{R}^m$ consisting of eigenvectors of A, an important condition we required for some of our work.

This chapter introduces singular value decompositions, whose singular values and singular vectors may be viewed as a generalization of eigenvalues and eigenvectors. In fact, we will see that every matrix, whether square or not, has a singular value decomposition and that knowing it gives us a great deal of insight into the matrix. It's been said that having a singular value decomposition is like looking at a matrix with X-ray vision as the decomposition reveals essential features of the matrix.

7.1 Symmetric matrices and variance

In this section, we will revisit the theory of eigenvalues and eigenvectors for the special class of matrices that are *symmetric*, meaning that the matrix equals its transpose. This understanding of symmetric matrices will enable us to form singular value decompositions later in the chapter. We'll also begin studying variance in this section as it provides an important context that motivates some of our later work.

To begin, remember that if A is a square matrix, we say that $\mathbf{v}$ is an eigenvector of A with associated eigenvalue λ if $A\mathbf{v} = \lambda\mathbf{v}$. In other words, for these special vectors, the operation of matrix multiplication simplifies to scalar multiplication.

Preview Activity 7.1.1. This preview activity reminds us how a basis of eigenvectors can be used to relate a square matrix to a diagonal one.

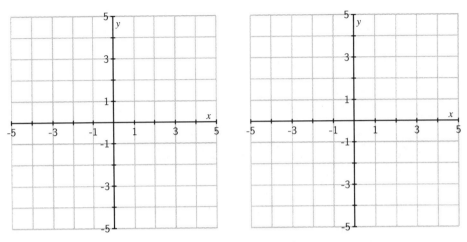

Figure 7.1.1 Use these plots to sketch the vectors requested in the preview activity.

a. Suppose that $D = \begin{bmatrix} 3 & 0 \\ 0 & -1 \end{bmatrix}$ and that $\mathbf{e}_1 = \begin{bmatrix} 1 \\ 0 \end{bmatrix}$ and $\mathbf{e}_2 = \begin{bmatrix} 0 \\ 1 \end{bmatrix}$.

  1. Sketch the vectors $\mathbf{e}_1$ and $D\mathbf{e}_1$ on the left side of Figure 7.1.1.
  2. Sketch the vectors $\mathbf{e}_2$ and $D\mathbf{e}_2$ on the left side of Figure 7.1.1.
  3. Sketch the vectors $\mathbf{e}_1 + 2\mathbf{e}_2$ and $D(\mathbf{e}_1 + 2\mathbf{e}_2)$ on the left side.
  4. Give a geometric description of the matrix transformation defined by D.

b. Now suppose we have vectors $\mathbf{v}_1 = \begin{bmatrix} 1 \\ 1 \end{bmatrix}$ and $\mathbf{v}_2 = \begin{bmatrix} -1 \\ 1 \end{bmatrix}$ and that A is a 2×2 matrix such that
$$A\mathbf{v}_1 = 3\mathbf{v}_1, \qquad A\mathbf{v}_2 = -\mathbf{v}_2.$$
That is, $\mathbf{v}_1$ and $\mathbf{v}_2$ are eigenvectors of A with associated eigenvalues 3 and -1.

  1. Sketch the vectors $\mathbf{v}_1$ and $A\mathbf{v}_1$ on the right side of Figure 7.1.1.
  2. Sketch the vectors $\mathbf{v}_2$ and $A\mathbf{v}_2$ on the right side of Figure 7.1.1.
  3. Sketch the vectors $\mathbf{v}_1 + 2\mathbf{v}_2$ and $A(\mathbf{v}_1 + 2\mathbf{v}_2)$ on the right side.
  4. Give a geometric description of the matrix transformation defined by A.

c. In what ways are the matrix transformations defined by D and A related to one another?

The preview activity asks us to compare the matrix transformations defined by two matrices, a diagonal matrix D and a matrix A whose eigenvectors are given to us. The transformation defined by D stretches horizontally by a factor of 3 and reflects in the horizontal axis, as shown in Figure 7.1.2

7.1. SYMMETRIC MATRICES AND VARIANCE

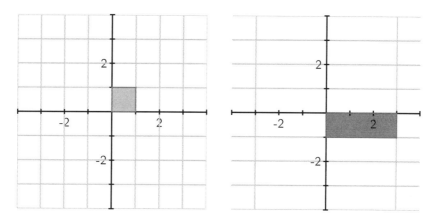

Figure 7.1.2 The matrix transformation defined by D.

By contrast, the transformation defined by A stretches the plane by a factor of 3 in the direction of $\mathbf{v}_1$ and reflects in the line defined by $\mathbf{v}_1$, as seen in Figure 7.1.3.

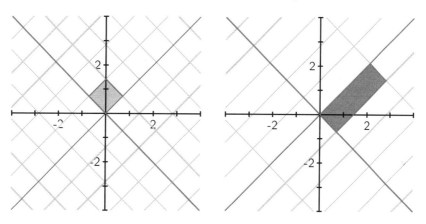

Figure 7.1.3 The matrix transformation defined by A.

In this way, we see that the matrix transformations defined by these two matrices are equivalent after a 45° rotation. This notion of equivalence is what we called *similarity* in Section 4.3. There we considered a square $m \times m$ matrix A that provided enough eigenvectors to form a basis of $\mathbb{R}^m$. For example, suppose we can construct a basis for $\mathbb{R}^m$ using eigenvectors $\mathbf{v}_1, \mathbf{v}_2, \ldots, \mathbf{v}_m$ having associated eigenvalues $\lambda_1, \lambda_2, \ldots, \lambda_m$. Forming the matrices,

$$P = \begin{bmatrix} \mathbf{v}_1 & \mathbf{v}_2 & \ldots & \mathbf{v}_m \end{bmatrix}, \qquad D = \begin{bmatrix} \lambda_1 & 0 & \ldots & 0 \\ 0 & \lambda_2 & \ldots & 0 \\ \vdots & \vdots & \ddots & \vdots \\ 0 & 0 & \ldots & \lambda_m \end{bmatrix},$$

enables us to write $A = PDP^{-1}$. This is what it means for A to be diagonalizable.

For the example in the preview activity, we are led to form

$$P = \begin{bmatrix} 1 & -1 \\ 1 & 1 \end{bmatrix}, \qquad D = \begin{bmatrix} 3 & 0 \\ 0 & -1 \end{bmatrix}$$

which tells us that $A = PDP^{-1} = \begin{bmatrix} 1 & 2 \\ 2 & 1 \end{bmatrix}$.

Notice that the matrix A has eigenvectors $\mathbf{v}_1$ and $\mathbf{v}_2$ that not only form a basis for $\mathbb{R}^2$ but, in fact, form an orthogonal basis for $\mathbb{R}^2$. Given the prominent role played by orthogonal bases in the last chapter, we would like to understand what conditions on a matrix enable us to form an orthogonal basis of eigenvectors.

7.1.1 Symmetric matrices and orthogonal diagonalization

Let's begin by looking at some examples in the next activity.

Activity 7.1.2. Remember that the Sage command A.right_eigenmatrix() attempts to find a basis for $\mathbb{R}^m$ consisting of eigenvectors of A. In particular, the assignment D, P = A.right_eigenmatrix() provides a diagonal matrix D constructed from the eigenvalues of A with the columns of P containing the associated eigenvectors.

   a. For each of the following matrices, determine whether there is a basis for $\mathbb{R}^2$ consisting of eigenvectors of that matrix. When there is such a basis, form the matrices P and D and verify that the matrix equals PDP^{-1}.

   1. $\begin{bmatrix} 3 & -4 \\ 4 & 3 \end{bmatrix}$.

   2. $\begin{bmatrix} 1 & 1 \\ -1 & 3 \end{bmatrix}$.

   3. $\begin{bmatrix} 1 & 0 \\ -1 & 2 \end{bmatrix}$.

   4. $\begin{bmatrix} 9 & 2 \\ 2 & 6 \end{bmatrix}$.

   b. For which of these examples is it possible to form an orthogonal basis for $\mathbb{R}^2$ consisting of eigenvectors?

   c. For any such matrix A, find an orthonormal basis of eigenvectors and explain why $A = QDQ^{-1}$ where Q is an orthogonal matrix.

   d. Finally, explain why $A = QDQ^T$ in this case.

   e. When $A = QDQ^T$, what is the relationship between A and A^T?

The examples in this activity illustrate a range of possibilities. First, a matrix may have complex eigenvalues, in which case it will not be diagonalizable. Second, even if all the eigenvalues are real, there may not be a basis of eigenvalues if the dimension of one of the eigenspaces is less than the algebraic multiplicity of the associated eigenvalue.

7.1. SYMMETRIC MATRICES AND VARIANCE

We are interested in matrices for which there is an orthogonal basis of eigenvectors. When this happens, we can create an orthonormal basis of eigenvectors by scaling each eigenvector in the basis so that its length is 1. Putting these orthonormal vectors into a matrix Q produces an orthogonal matrix, which means that $Q^T = Q^{-1}$. We then have

$$A = QDQ^{-1} = QDQ^T.$$

In this case, we say that A is *orthogonally diagonalizable*.

Definition 7.1.4 If there is an orthonormal basis of $\mathbb{R}^n$ consisting of eigenvectors of the matrix A, we say that A is *orthogonally diagonalizable*. In particular, we can write $A = QDQ^T$ where Q is an orthogonal matrix.

When A is orthogonally diagonalizable, notice that

$$A^T = (QDQ^T)^T = (Q^T)^T D^T Q^T = QDQ^T = A.$$

That is, when A is orthogonally diagonalizable, $A = A^T$ and we say that A is *symmetric*.

Definition 7.1.5 A *symmetric* matrix A is one for which $A = A^T$.

Example 7.1.6 Consider the matrix $A = \begin{bmatrix} -2 & 36 \\ 36 & -23 \end{bmatrix}$, which has eigenvectors $\mathbf{v}_1 = \begin{bmatrix} 4 \\ 3 \end{bmatrix}$, with associated eigenvalue $\lambda_1 = 25$, and $\mathbf{v}_2 = \begin{bmatrix} 3 \\ -4 \end{bmatrix}$, with associated eigenvalue $\lambda_2 = -50$. Notice that $\mathbf{v}_1$ and $\mathbf{v}_2$ are orthogonal so we can form an orthonormal basis of eigenvectors:

$$\mathbf{u}_1 = \begin{bmatrix} 4/5 \\ 3/5 \end{bmatrix}, \quad \mathbf{u}_1 = \begin{bmatrix} 3/5 \\ -4/5 \end{bmatrix}.$$

In this way, we construct the matrices

$$Q = \begin{bmatrix} 4/5 & 3/5 \\ 3/5 & -4/5 \end{bmatrix}, \quad D = \begin{bmatrix} 25 & 0 \\ 0 & -50 \end{bmatrix}$$

and note that $A = QDQ^T$.

Notice also that, as expected, A is symmetric; that is, $A = A^T$.

Example 7.1.7 If $A = \begin{bmatrix} 1 & 2 \\ 2 & 1 \end{bmatrix}$, then there is an orthogonal basis of eigenvectors $\mathbf{v}_1 = \begin{bmatrix} 1 \\ 1 \end{bmatrix}$ and $\mathbf{v}_2 = \begin{bmatrix} -1 \\ 1 \end{bmatrix}$ with eigenvalues $\lambda_1 = 3$ and $\lambda_2 = -1$. Using these eigenvectors, we form the orthogonal matrix Q consisting of eigenvectors and the diagonal matrix D, where

$$Q = \begin{bmatrix} 1/\sqrt{2} & -1/\sqrt{2} \\ 1/\sqrt{2} & 1/\sqrt{2} \end{bmatrix}, \quad D = \begin{bmatrix} 3 & 0 \\ 0 & -1 \end{bmatrix}.$$

Then we have $A = QDQ^T$.

Notice that the matrix transformation represented by Q is a $45°$ rotation while that represented by $Q^T = Q^{-1}$ is a $-45°$ rotation. Therefore, if we multiply a vector $\mathbf{x}$ by A, we can decompose the multiplication as

$$Ax = Q(D(Q^T x)).$$

That is, we first rotate **x** by −45°, then apply the diagonal matrix D, which stretches and reflects, and finally rotate by 45°. We may visualize this factorization as in Figure 7.1.8.

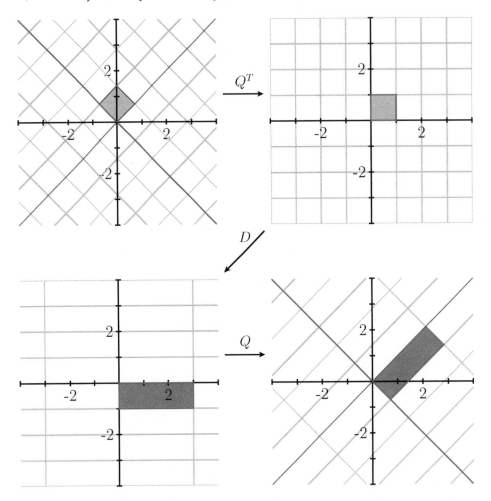

Figure 7.1.8 The transformation defined by $A = QDQ^T$ can be interpreted as a sequence of geometric transformations: Q^T rotates by −45°, D stretches and reflects, and Q rotates by 45°.

In fact, a similar picture holds any time the matrix A is orthogonally diagonalizable.

We have seen that a matrix that is orthogonally diagonalizable must be symmetric. In fact, it turns out that any symmetric matrix is orthogonally diagonalizable. We record this fact in the next theorem.

Theorem 7.1.9 The Spectral Theorem. *The matrix A is orthogonally diagonalizable if and only if A is symmetric.*

7.1. SYMMETRIC MATRICES AND VARIANCE

Activity 7.1.3. Each of the following matrices is symmetric so the Spectral Theorem tells us that each is orthogonally diagonalizable. The point of this activity is to find an orthogonal diagonalization for each matrix.

To begin, find a basis for each eigenspace. Use this basis to find an orthogonal basis for each eigenspace and put these bases together to find an orthogonal basis for $\mathbb{R}^m$ consisting of eigenvectors. Use this basis to write an orthogonal diagonalization of the matrix.

a. $\begin{bmatrix} 0 & 2 \\ 2 & 3 \end{bmatrix}$.

b. $\begin{bmatrix} 4 & -2 & 14 \\ -2 & 19 & -16 \\ 14 & -16 & 13 \end{bmatrix}$.

c. $\begin{bmatrix} 5 & 4 & 2 \\ 4 & 5 & 2 \\ 2 & 2 & 2 \end{bmatrix}$.

d. Consider the matrix $A = B^T B$ where $B = \begin{bmatrix} 0 & 1 & 2 \\ 2 & 0 & 1 \end{bmatrix}$. Explain how we know that A is symmetric and then find an orthogonal diagonalization of A.

As the examples in Activity 7.1.3 illustrate, the Spectral Theorem implies a number of things. Namely, if A is a symmetric $m \times m$ matrix, then

- the eigenvalues of A are real.
- there is a basis of $\mathbb{R}^m$ consisting of eigenvectors.
- two eigenvectors that are associated to different eigenvalues are orthogonal.

We won't justify the first two facts here since that would take us rather far afield. However, it will be helpful to explain the third fact. To begin, notice the following:

$$\mathbf{v} \cdot (A\mathbf{w}) = \mathbf{v}^T A \mathbf{w} = (A^T \mathbf{v})^T \mathbf{w} = (A^T \mathbf{v}) \cdot \mathbf{w}.$$

This is a useful fact that we'll employ quite a bit in the future so let's summarize it in the following proposition.

Proposition 7.1.10 *For any matrix A, we have*

$$\mathbf{v} \cdot (A\mathbf{w}) = (A^T \mathbf{v}) \cdot \mathbf{w}.$$

In particular, if A is symmetric, then

$$\mathbf{v} \cdot (A\mathbf{w}) = (A\mathbf{v}) \cdot \mathbf{w}.$$

Example 7.1.11 Suppose a symmetric matrix A has eigenvectors $\mathbf{v}_1$, with associated eigenvalue $\lambda_1 = 3$, and $\mathbf{v}_2$, with associated eigenvalue $\lambda_2 = 10$. Notice that

$$(A\mathbf{v}_1) \cdot \mathbf{v}_2 = 3\mathbf{v}_1 \cdot \mathbf{v}_2$$
$$\mathbf{v}_1 \cdot (A\mathbf{v}_2) = 10\mathbf{v}_1 \cdot \mathbf{v}_2.$$

Since $(A\mathbf{v}_1) \cdot \mathbf{v}_2 = \mathbf{v}_1 \cdot (A\mathbf{v}_2)$ by Proposition 7.1.10, we have

$$3\mathbf{v}_1 \cdot \mathbf{v}_2 = 10\mathbf{v}_1 \cdot \mathbf{v}_2,$$

which can only happen if $\mathbf{v}_1 \cdot \mathbf{v}_2 = 0$. Therefore, $\mathbf{v}_1$ and $\mathbf{v}_2$ are orthogonal.

More generally, the same argument shows that two eigenvectors of a symmetric matrix associated to distinct eigenvalues are orthogonal.

7.1.2 Variance

Many of the ideas we'll encounter in this chapter, such as orthogonal diagonalizations, can be applied to the study of data. In fact, it can be useful to understand these applications because they provide an important context in which mathematical ideas have a more concrete meaning and their motivation appears more clearly. For that reason, we will now introduce the statistical concept of variance as a way to gain insight into the significance of orthogonal diagonalizations.

Given a set of data points, their variance measures how spread out the points are. The next activity looks at some examples.

Activity 7.1.4. We'll begin with a set of three data points

$$\mathbf{d}_1 = \begin{bmatrix} 1 \\ 1 \end{bmatrix}, \quad \mathbf{d}_2 = \begin{bmatrix} 2 \\ 1 \end{bmatrix}, \quad \mathbf{d}_3 = \begin{bmatrix} 3 \\ 4 \end{bmatrix}.$$

a. Find the centroid, or mean, $\overline{\mathbf{d}} = \frac{1}{N}\sum_j \mathbf{d}_j$. Then plot the data points and their centroid in Figure 7.1.12.

7.1. SYMMETRIC MATRICES AND VARIANCE

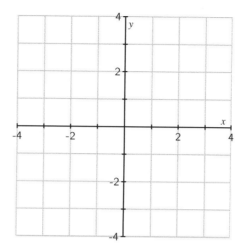

Figure 7.1.12 Plot the data points and their centroid here.

b. Notice that the centroid lies in the center of the data so the spread of the data will be measured by how far away the points are from the centroid. To simplify our calculations, find the demeaned data points

$$\widetilde{\mathbf{d}}_j = \mathbf{d}_j - \overline{\mathbf{d}}$$

and plot them in Figure 7.1.13.

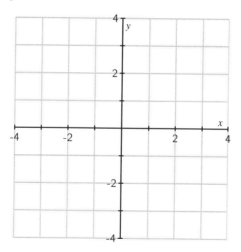

Figure 7.1.13 Plot the demeaned data points $\widetilde{\mathbf{d}}_j$ here.

c. Now that the data has been demeaned, we will define the total variance as the average of the squares of the distances from the origin; that is, the total variance is

$$V = \frac{1}{N} \sum_j |\widetilde{\mathbf{d}}_j|^2.$$

Find the total variance V for our set of three points.

d. Now plot the projections of the demeaned data onto the x and y axes using Figure 7.1.14 and find the variances V_x and V_y of the projected points.

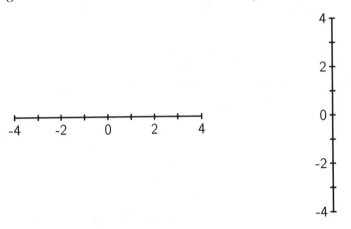

Figure 7.1.14 Plot the projections of the demeaned data onto the x and y axes.

e. Which of the variances, V_x and V_y, is larger and how does the plot of the projected points explain your response?

f. What do you notice about the relationship between V, V_x, and V_y? How does the Pythagorean theorem explain this relationship?

g. Plot the projections of the demeaned data points onto the lines defined by vectors $\mathbf{v}_1 = \begin{bmatrix} 1 \\ 1 \end{bmatrix}$ and $\mathbf{v}_2 = \begin{bmatrix} -1 \\ 1 \end{bmatrix}$ using Figure 7.1.15 and find the variances $V_{\mathbf{v}_1}$ and $V_{\mathbf{v}_2}$ of these projected points.

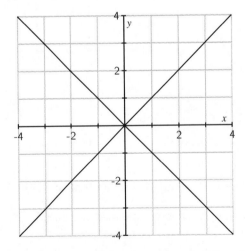

Figure 7.1.15 Plot the projections of the deameaned data onto the lines defined by $\mathbf{v}_1$ and $\mathbf{v}_2$.

7.1. SYMMETRIC MATRICES AND VARIANCE

h. What is the relationship between the total variance V and V_{v_1} and V_{v_2}? How does the Pythagorean theorem explain your response?

Notice that variance enjoys an additivity property. Consider, for instance, the situation where our data points are two-dimensional and suppose that the demeaned points are $\widetilde{\mathbf{d}}_j = \begin{bmatrix} \widetilde{x}_j \\ \widetilde{y}_j \end{bmatrix}$. We have

$$|\widetilde{\mathbf{d}}_j|^2 = \widetilde{x}_j^2 + \widetilde{y}_j^2.$$

If we take the average over all data points, we find that the total variance V is the sum of the variances in the x and y directions:

$$\frac{1}{N}\sum_j |\widetilde{\mathbf{d}}_j|^2 = \frac{1}{N}\sum_j \widetilde{x}_j^2 + \frac{1}{N}\sum_j \widetilde{y}_j^2$$

$$V = V_x + V_y.$$

More generally, suppose that we have an orthonormal basis $\mathbf{u}_1$ and $\mathbf{u}_2$. If we project the demeaned points onto the line defined by $\mathbf{u}_1$, we obtain the points $(\widetilde{\mathbf{d}}_j \cdot \mathbf{u}_1)\mathbf{u}_1$ so that

$$V_{\mathbf{u}_1} = \frac{1}{N}\sum_j |(\widetilde{\mathbf{d}}_j \cdot \mathbf{u}_1)\mathbf{u}_1|^2 = \frac{1}{N}\sum_j (\widetilde{\mathbf{d}}_j \cdot \mathbf{u}_1)^2.$$

For each of our demeaned data points, the Projection Formula tells us that

$$\widetilde{\mathbf{d}}_j = (\widetilde{\mathbf{d}}_j \cdot \mathbf{u}_1)\mathbf{u}_1 + (\widetilde{\mathbf{d}}_j \cdot \mathbf{u}_2)\mathbf{u}_2.$$

We then have

$$|\widetilde{\mathbf{d}}_j|^2 = \widetilde{\mathbf{d}}_j \cdot \widetilde{\mathbf{d}}_j = (\widetilde{\mathbf{d}}_j \cdot \mathbf{u}_1)^2 + (\widetilde{\mathbf{d}}_j \cdot \mathbf{u}_2)^2$$

since $\mathbf{u}_1 \cdot \mathbf{u}_2 = 0$. When we average over all the data points, we find that the total variance V is the sum of the variances in the $\mathbf{u}_1$ and $\mathbf{u}_2$ directions. This leads to the following proposition, in which this observation is expressed more generally.

Proposition 7.1.16 Additivity of Variance. *If W is a subspace with orthonormal basis $\mathbf{u}_1, \mathbf{u}_2, \ldots, \mathbf{u}_n$, then the variance of the points projected onto W is the sum of the variances in the $\mathbf{u}_j$ directions:*

$$V_W = V_{\mathbf{u}_1} + V_{\mathbf{u}_2} + \ldots + V_{\mathbf{u}_n}.$$

The next activity demonstrates a more efficient way to find the variance $V_\mathbf{u}$ in a particular direction and connects our discussion of variance with symmetric matrices.

Activity 7.1.5. Let's return to the dataset from the previous activity in which we have demeaned data points:

$$\widetilde{\mathbf{d}}_1 = \begin{bmatrix} -1 \\ -1 \end{bmatrix}, \quad \widetilde{\mathbf{d}}_2 = \begin{bmatrix} 0 \\ -1 \end{bmatrix}, \quad \widetilde{\mathbf{d}}_3 = \begin{bmatrix} 1 \\ 2 \end{bmatrix}.$$

Our goal is to compute the variance $V_\mathbf{u}$ in the direction defined by a unit vector $\mathbf{u}$.

To begin, form the demeaned data matrix

$$A = \begin{bmatrix} \tilde{\mathbf{d}}_1 & \tilde{\mathbf{d}}_2 & \tilde{\mathbf{d}}_3 \end{bmatrix}$$

and suppose that $\mathbf{u}$ is a unit vector.

a. Write the vector $A^T \mathbf{u}$ in terms of the dot products $\tilde{\mathbf{d}}_j \cdot \mathbf{u}$.

b. Explain why $V_{\mathbf{u}} = \frac{1}{3}|A^T \mathbf{u}|^2$.

c. Apply Proposition 7.1.10 to explain why

$$V_{\mathbf{u}} = \frac{1}{3}|A^T \mathbf{u}|^2 = \frac{1}{3}(A^T \mathbf{u}) \cdot (A^T \mathbf{u}) = \mathbf{u}^T \left(\frac{1}{3}AA^T\right) \mathbf{u} = \mathbf{u} \cdot \left(\frac{1}{3}AA^T\right) \mathbf{u} =$$

d. In general, the matrix $C = \frac{1}{N} AA^T$ is called the *covariance* matrix of the dataset, and it is useful because the variance $V_{\mathbf{u}} = \mathbf{u} \cdot (C\mathbf{u})$, as we have just seen. Find the matrix C for our dataset with three points.

e. Use the covariance matrix to find the variance $V_{\mathbf{u}_1}$ when $\mathbf{u}_1 = \begin{bmatrix} 1/\sqrt{5} \\ 2/\sqrt{5} \end{bmatrix}$.

f. Use the covariance matrix to find the variance $V_{\mathbf{u}_2}$ when $\mathbf{u}_2 = \begin{bmatrix} -2/\sqrt{5} \\ 1/\sqrt{5} \end{bmatrix}$. Since $\mathbf{u}_1$ and $\mathbf{u}_2$ are orthogonal, verify that the sum of $V_{\mathbf{u}_1}$ and $V_{\mathbf{u}_2}$ gives the total variance.

g. Explain why the covariance matrix C is a symmetric matrix.

This activity introduced the covariance matrix of a dataset, which is defined to be $C = \frac{1}{N} AA^T$ where A is the matrix of demeaned data points. Notice that

$$C^T = \frac{1}{N}(AA^T)^T = \frac{1}{N} AA^T = C,$$

which tells us that C is symmetric. In particular, we know that it is orthogonally diagonalizable, an observation that will play an important role in the future.

This activity also demonstrates the significance of the covariance matrix, which is recorded in the following proposition.

Proposition 7.1.17 *If C is the covariance matrix associated to a demeaned dataset and $\mathbf{u}$ is a unit vector, then the variance of the demeaned points projected onto the line defined by $\mathbf{u}$ is*

$$V_{\mathbf{u}} = \mathbf{u} \cdot C\mathbf{u}.$$

Our goal in the future will be to find directions $\mathbf{u}$ where the variance is as large as possible and directions where it is as small as possible. The next activity demonstrates why this is useful.

7.1. SYMMETRIC MATRICES AND VARIANCE

Activity 7.1.6.

a. Evaluating the following Sage cell loads a dataset consisting of 100 demeaned data points and provides a plot of them. It also provides the demeaned data matrix A.

```
import pandas as pd
url='https://raw.githubusercontent.com/davidaustinm/'
url+='ula_modules/master/data/variance-data.csv'
df=pd.read_csv(url, header=None)
data=[vector(row) for row in df.values]
A=matrix(data).T
list_plot(data, size=20, color='blue', aspect_ratio=1)
```

What is the shape of the covariance matrix C? Find C and verify your response.

b. By visually inspecting the data, determine which is larger, V_x or V_y. Then compute both of these quantities to verify your response.

c. What is the total variance V?

d. In approximately what direction is the variance greatest? Choose a reasonable vector $\mathbf{u}$ that points in approximately that direction and find $V_\mathbf{u}$.

e. In approximately what direction is the variance smallest? Choose a reasonable vector $\mathbf{w}$ that points in approximately that direction and find $V_\mathbf{w}$.

f. How are the directions $\mathbf{u}$ and $\mathbf{w}$ in the last two parts of this problem related to one another? Why does this relationship hold?

This activity illustrates how variance can identify a line along which the data are concentrated. When the data primarily lie along a line defined by a vector $\mathbf{u}_1$, then the variance in that direction will be large while the variance in an orthogonal direction $\mathbf{u}_2$ will be small.

Remember that variance is additive, according to Proposition 7.1.16, so that if $\mathbf{u}_1$ and $\mathbf{u}_2$ are orthogonal unit vectors, then the total variance is

$$V = V_{\mathbf{u}_1} + V_{\mathbf{u}_2}.$$

Therefore, if we choose $\mathbf{u}_1$ to be the direction where $V_{\mathbf{u}_1}$ is a maximum, then $V_{\mathbf{u}_2}$ will be a minimum.

In the next section, we will use an orthogonal diagonalization of the covariance matrix C to find the directions having the greatest and smallest variances. In this way, we will be able to determine when data are concentrated along a line or subspace.

7.1.3 Summary

This section explored both symmetric matrices and variance. In particular, we saw that

- A matrix A is orthogonally diagonalizable if there is an orthonormal basis of eigenvectors. In particular, we can write $A = QDQ^T$, where D is a diagonal matrix of eigenvalues and Q is an orthogonal matrix of eigenvectors.

- The Spectral Theorem tells us that a matrix A is orthogonally diagonalizable if and only if it is symmetric; that is, $A = A^T$.

- The variance of a dataset can be computed using the covariance matrix $C = \frac{1}{N} AA^T$, where A is the matrix of demeaned data points. In particular, the variance of the demeaned data points projected onto the line defined by the unit vector $\mathbf{u}$ is $V_\mathbf{u} = \mathbf{u} \cdot C\mathbf{u}$.

- Variance is additive so that if W is a subspace with orthonormal basis $\mathbf{u}_1, \mathbf{u}_2, \ldots, \mathbf{u}_n$, then
$$V_W = V_{\mathbf{u}_1} + V_{\mathbf{u}_2} + \ldots + V_{\mathbf{u}_n}.$$

7.1.4 Exercises

1. For each of the following matrices, find the eigenvalues and a basis for each eigenspace. Determine whether the matrix is diagonalizable and, if so, find a diagonalization. Determine whether the matrix is orthogonally diagonalizable and, if so, find an orthogonal diagonalization.

    a. $\begin{bmatrix} 5 & 1 \\ -1 & 3 \end{bmatrix}$

    b. $\begin{bmatrix} 0 & 1 \\ 1 & 0 \end{bmatrix}$

    c. $\begin{bmatrix} 1 & 0 & 0 \\ 2 & -2 & 0 \\ 0 & 1 & 4 \end{bmatrix}$

    d. $\begin{bmatrix} 2 & 5 & -4 \\ 5 & -7 & 5 \\ -4 & 5 & 2 \end{bmatrix}$

2. Consider the matrix $A = \begin{bmatrix} 1 & 2 & 2 \\ 2 & 1 & 2 \\ 2 & 2 & 1 \end{bmatrix}$ whose eigenvalues are $\lambda_1 = 5$, $\lambda_2 = -1$, and $\lambda_3 = -1$.

    a. Explain why A is orthogonally diagonalizable.

    b. Find an orthonormal basis for the eigenspace E_5.

    c. Find a basis for the eigenspace E_{-1}.

7.1. SYMMETRIC MATRICES AND VARIANCE

   d. Now find an orthonormal basis for E_{-1}.

   e. Find matrices D and Q such that $A = QDQ^T$.

3. Find an orthogonal diagonalization, if one exists, for the following matrices.

   a. $\begin{bmatrix} 11 & 4 & 12 \\ 4 & -3 & -16 \\ 12 & -16 & 1 \end{bmatrix}$.

   b. $\begin{bmatrix} 1 & 0 & 2 \\ 0 & 1 & 2 \\ -2 & -2 & 1 \end{bmatrix}$.

   c. $\begin{bmatrix} 9 & 3 & 3 & 3 \\ 3 & 9 & 3 & 3 \\ 3 & 3 & 9 & 3 \\ 3 & 3 & 3 & 9 \end{bmatrix}$.

4. Suppose that A is an $m \times n$ matrix and that $B = A^T A$.

   a. Explain why B is orthogonally diagonalizable.

   b. Explain why $\mathbf{v} \cdot (B\mathbf{v}) = |A\mathbf{v}|^2$.

   c. Suppose that $\mathbf{u}$ is an eigenvector of B with associated eigenvalue λ and that $\mathbf{u}$ has unit length. Explain why $\lambda = |A\mathbf{u}|^2$.

   d. Explain why the eigenvalues of B are nonnegative.

   e. If C is the covariance matrix associated to a demeaned dataset, explain why the eigenvalues of C are nonnegative.

5. Suppose that you have the data points

$$(2,0), (2,3), (4,1), (3,2), (4,4).$$

   a. Find the demeaned data points.

   b. Find the total variance V of the dataset.

   c. Find the variance in the direction $\mathbf{e}_1 = \begin{bmatrix} 1 \\ 0 \end{bmatrix}$ and the variance in the direction $\mathbf{e}_2 = \begin{bmatrix} 0 \\ 1 \end{bmatrix}$.

   d. Project the demeaned data points onto the line defined by $\mathbf{v}_1 = \begin{bmatrix} 2 \\ 1 \end{bmatrix}$ and find the variance of these projected points.

e. Project the demeaned data points onto the line defined by $v_2 = \begin{bmatrix} 1 \\ -2 \end{bmatrix}$ and find the variance of these projected points.

f. How and why are the results of from the last two parts related to the total variance?

6. Suppose you have six 2-dimensional data points arranged in the matrix

$$\begin{bmatrix} 2 & 0 & 4 & 4 & 5 & 3 \\ 1 & 0 & 3 & 5 & 4 & 5 \end{bmatrix}.$$

a. Find the matrix A of demeaned data points and plot the points in Figure 7.1.18.

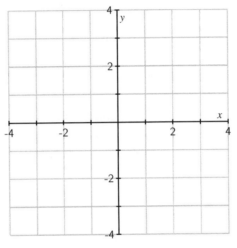

Figure 7.1.18 A plot for the demeaned data points.

b. Construct the covariance matrix C and explain why you know that it is orthogonally diagonalizable.

c. Find an orthogonal diagonalization of C.

d. Sketch the lines corresponding to the two eigenvectors on the plot above.

e. Find the variances in the directions of the eigenvectors.

7. Suppose that C is the covariance matrix of a demeaned dataset.

a. Suppose that u is an eigenvector of C with associated eigenvalue λ and that u has unit length. Explain why $V_u = \lambda$.

b. Suppose that the covariance matrix of a demeaned dataset can be written as $C = QDQ^T$ where

$$Q = \begin{bmatrix} u_1 & u_2 \end{bmatrix}, \qquad D = \begin{bmatrix} 10 & 0 \\ 0 & 0 \end{bmatrix}.$$

What is V_{u_2}? What does this tell you about the demeaned data?

c. Explain why the total variance of a dataset equals the sum of the eigenvalues of

7.1. SYMMETRIC MATRICES AND VARIANCE

the covariance matrix.

8. Determine whether the following statements are true or false and explain your thinking.

   a. If A is an invertible, orthogonally diagonalizable matrix, then so is A^{-1}.

   b. If $\lambda = 2 + i$ is an eigenvalue of A, then A cannot be orthogonally diagonalizable.

   c. If there is a basis for $\mathbb{R}^m$ consisting of eigenvectors of A, then A is orthogonally diagonalizable.

   d. If $\mathbf{u}$ and $\mathbf{v}$ are eigenvectors of a symmetric matrix associated to eigenvalues -2 and 3, then $\mathbf{u} \cdot \mathbf{v} = 0$.

   e. If A is a square matrix, then $\mathbf{u} \cdot (A\mathbf{v}) = (A\mathbf{u}) \cdot \mathbf{v}$.

9. Suppose that A is a noninvertible, symmetric 3×3 matrix having eigenvectors

$$\mathbf{v}_1 = \begin{bmatrix} 2 \\ -1 \\ 2 \end{bmatrix}, \quad \mathbf{v}_2 = \begin{bmatrix} 1 \\ 4 \\ 1 \end{bmatrix}$$

and associated eigenvalues $\lambda_1 = 20$ and $\lambda_2 = -4$. Find matrices Q and D such that $A = QDQ^T$.

10. Suppose that W is a plane in $\mathbb{R}^3$ and that P is the 3×3 matrix that projects vectors orthogonally onto W.

    a. Explain why P is orthogonally diagonalizable.

    b. What are the eigenvalues of P?

    c. Explain the relationship between the eigenvectors of P and the plane W.

7.2 Quadratic forms

With our understanding of symmetric matrices and variance in hand, we'll now explore how to determine the directions in which the variance of a dataset is as large as possible and where it is as small as possible. This is part of a much larger story involving a type of function, called a *quadratic form*, that we'll introduce here.

Preview Activity 7.2.1. Let's begin by looking at an example. Suppose we have three data points that form the demeaned data matrix

$$A = \begin{bmatrix} 2 & 1 & -3 \\ 1 & 2 & -3 \end{bmatrix}$$

a. Plot the demeaned data points in Figure 7.2.1. In which direction does the variance appear to be largest and in which does it appear to be smallest?

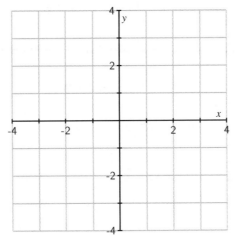

Figure 7.2.1 Use this coordinate grid to plot the demeaned data points.

b. Construct the covariance matrix C and determine the variance in the direction of $\begin{bmatrix} 1 \\ 1 \end{bmatrix}$ and the variance in the direction of $\begin{bmatrix} -1 \\ 1 \end{bmatrix}$.

c. What is the total variance of this dataset?

d. Generally speaking, if C is the covariance matrix of a dataset and $\mathbf{u}$ is an eigenvector of C having unit length and with associated eigenvalue λ, what is $V_{\mathbf{u}}$?

7.2. QUADRATIC FORMS

7.2.1 Quadratic forms

Given a matrix A of N demeaned data points, the symmetric covariance matrix $C = \frac{1}{N}AA^T$ determines the variance in a particular direction

$$V_{\mathbf{u}} = \mathbf{u} \cdot (C\mathbf{u}),$$

where $\mathbf{u}$ is a unit vector defining the direction.

More generally, a symmetric $m \times m$ matrix A defines a function $q : \mathbb{R}^m \to \mathbb{R}$ by

$$q(\mathbf{x}) = \mathbf{x} \cdot (A\mathbf{x}).$$

Notice that this expression is similar to the one we use to find the variance $V_{\mathbf{u}}$ in terms of the covariance matrix C. The only difference is that we allow $\mathbf{x}$ to be any vector rather than requiring it to be a unit vector.

Example 7.2.2 Suppose that $A = \begin{bmatrix} 1 & 2 \\ 2 & 1 \end{bmatrix}$. If we write $\mathbf{x} = \begin{bmatrix} x_1 \\ x_2 \end{bmatrix}$, then we have

$$q\left(\begin{bmatrix} x_1 \\ x_2 \end{bmatrix}\right) = \begin{bmatrix} x_1 \\ x_2 \end{bmatrix} \cdot \left(\begin{bmatrix} 1 & 2 \\ 2 & 1 \end{bmatrix} \begin{bmatrix} x_1 \\ x_2 \end{bmatrix}\right)$$

$$= \begin{bmatrix} x_1 \\ x_2 \end{bmatrix} \cdot \begin{bmatrix} x_1 + 2x_2 \\ 2x_1 + x_2 \end{bmatrix}$$

$$= x_1^2 + 2x_1 x_2 + 2x_1 x_2 + x_2^2$$

$$= x_1^2 + 4x_1 x_2 + x_2^2.$$

We may evaluate the quadratic form using some input vectors:

$$q\left(\begin{bmatrix} 1 \\ 0 \end{bmatrix}\right) = 1, \quad q\left(\begin{bmatrix} 1 \\ 1 \end{bmatrix}\right) = 6, \quad q\left(\begin{bmatrix} 2 \\ 4 \end{bmatrix}\right) = 52.$$

Notice that the value of the quadratic form is a scalar.

Definition 7.2.3 If A is a symmetric $m \times m$ matrix, the *quadratic form* defined by A is the function $q_A(\mathbf{x}) = \mathbf{x} \cdot (A\mathbf{x})$.

> **Activity 7.2.2.** Let's look at some more examples of quadratic forms.
>
> a. Consider the symmetric matrix $D = \begin{bmatrix} 3 & 0 \\ 0 & -1 \end{bmatrix}$. Write the quadratic form $q_D(\mathbf{x})$ defined by D in terms of the components of $\mathbf{x} = \begin{bmatrix} x_1 \\ x_2 \end{bmatrix}$. What is the value of $q_D\left(\begin{bmatrix} 2 \\ -4 \end{bmatrix}\right)$?
>
> b. Given the symmetric matrix $A = \begin{bmatrix} 2 & 5 \\ 5 & -3 \end{bmatrix}$, write the quadratic form $q_A(\mathbf{x})$ de-

fined by A and evaluate $q_A\left(\begin{bmatrix} 2 \\ -1 \end{bmatrix}\right)$.

c. Suppose that $q\left(\begin{bmatrix} x_1 \\ x_2 \end{bmatrix}\right) = 3x_1^2 - 4x_1x_2 + 4x_2^2$. Find a symmetric matrix A such that q is the quadratic form defined by A.

d. Suppose that q is a quadratic form and that $q(\mathbf{x}) = 3$. What is $q(2\mathbf{x})$? $q(-\mathbf{x})$? $q(10\mathbf{x})$?

e. Suppose that A is a symmetric matrix and $q_A(\mathbf{x})$ is the quadratic form defined by A. Suppose that $\mathbf{x}$ is an eigenvector of A with associated eigenvalue -4 and with length 7. What is $q_A(\mathbf{x})$?

Linear algebra is principally about things that are linear. However, quadratic forms, as the name implies, have a distinctly non-linear character. First, if $A = \begin{bmatrix} a & b \\ b & c \end{bmatrix}$, is a symmetric matrix, then the associated quadratic form is

$$q_A\left(\begin{bmatrix} x_1 \\ x_2 \end{bmatrix}\right) = ax_1^2 + 2bx_1x_2 + cx_2^2.$$

Notice how the variables x_1 and x_2 are multiplied together, which tells us this isn't a linear function.

This expression assumes an especially simple form when D is a diagonal matrix. In particular, if $D = \begin{bmatrix} a & 0 \\ 0 & c \end{bmatrix}$, then $q_D\left(\begin{bmatrix} x_1 \\ x_2 \end{bmatrix}\right) = ax_1^2 + cx_2^2$. This is special because there is no cross-term involving x_1x_2.

Remember that matrix transformations have the property that $T(s\mathbf{x}) = sT(\mathbf{x})$. Quadratic forms behave differently:

$$q_A(s\mathbf{x}) = (s\mathbf{x}) \cdot (A(s\mathbf{x})) = s^2 \mathbf{x} \cdot (A\mathbf{x}) = s^2 q_A(\mathbf{x}).$$

For instance, when we multiply $\mathbf{x}$ by the scalar 2, then $q_A(2\mathbf{x}) = 4q_A(\mathbf{x})$. Also, notice that $q_A(-\mathbf{x}) = q_A(\mathbf{x})$ since the scalar is squared.

Finally, evaluating a quadratic form on an eigenvector has a particularly simple form. Suppose that $\mathbf{x}$ is an eigenvector of A with associated eigenvalue λ. We then have

$$q_A(\mathbf{x}) = \mathbf{x} \cdot (A\mathbf{x}) = \lambda \mathbf{x} \cdot \mathbf{x} = \lambda |\mathbf{x}|^2.$$

Let's now return to our motivating question: in which direction $\mathbf{u}$ is the variance $V_\mathbf{u} = \mathbf{u} \cdot (C\mathbf{u})$ of a dataset as large as possible and in which is it as small as possible. Remembering that the vector $\mathbf{u}$ is a unit vector, we can now state a more general form of this question: If $q_A(\mathbf{x})$ is a quadratic form, for which unit vectors $\mathbf{u}$ is $q_A(\mathbf{u}) = \mathbf{u} \cdot (A\mathbf{u})$ as large as possible and for which is it as small as possible? Since a unit vector specifies a direction, we will often ask for the directions in which the quadratic form $q(\mathbf{x})$ is at its maximum or minimum value.

7.2. QUADRATIC FORMS

Activity 7.2.3. We can gain some intuition about this problem by graphing the quadratic form and paying particular attention to the unit vectors.

a. Evaluating the following cell defines the matrix $D = \begin{bmatrix} 3 & 0 \\ 0 & -1 \end{bmatrix}$ and displays the graph of the associated quadratic form $q_D(\mathbf{x})$. In addition, the points corresponding to vectors $\mathbf{u}$ with unit length are displayed as a curve.

```
url='https://raw.githubusercontent.com/davidaustinm/'
url+='ula_modules/master/quad_plot.py'
sage.repl.load.load(url, globals())

## We define our matrix here
A = matrix(2, 2, [3, 0, 0, -1])

quad_plot(A)
```

Notice that the matrix D is diagonal. In which directions does the quadratic form have its maximum and minimum values?

b. Write the quadratic form q_D associated to D. What is the value of $q_D\left(\begin{bmatrix} 1 \\ 0 \end{bmatrix}\right)$? What is the value of $q_D\left(\begin{bmatrix} 0 \\ 1 \end{bmatrix}\right)$?

c. Consider a unit vector $\mathbf{u} = \begin{bmatrix} u_1 \\ u_2 \end{bmatrix}$ so that $u_1^2 + u_2^2 = 1$, an expression we can rewrite as $u_1^2 = 1 - u_2^2$. Write the quadratic form $q_D(\mathbf{u})$ and replace u_1^2 by $1 - u_2^2$. Now explain why the maximum of $q_D(\mathbf{u})$ is 3. In which direction does the maximum occur? Does this agree with what you observed from the graph above?

d. Write the quadratic form $q_D(\mathbf{u})$ and replace u_2^2 by $1 - u_1^2$. What is the minimum value of $q_D(\mathbf{u})$ and in which direction does the minimum occur?

e. Use the previous Sage cell to change the matrix to $A = \begin{bmatrix} 1 & 2 \\ 2 & 1 \end{bmatrix}$ and display the graph of the quadratic form $q_A(\mathbf{x}) = \mathbf{x} \cdot (A\mathbf{x})$. Determine the directions in which the maximum and minimum occur?

f. Remember that $A = \begin{bmatrix} 1 & 2 \\ 2 & 1 \end{bmatrix}$ is symmetric so that $A = QDQ^T$ where D is the diagonal matrix above and Q is the orthogonal matrix that rotates vectors by 45°. Notice that

$$q_A(\mathbf{u}) = \mathbf{u} \cdot (A\mathbf{u}) = \mathbf{u} \cdot (QDQ^T\mathbf{u}) = (Q^T\mathbf{u}) \cdot (DQ^T\mathbf{u}) = q_D(\mathbf{v})$$

where $\mathbf{v} = Q^T\mathbf{u}$. That is, we have $q_A(\mathbf{u}) = q_D(\mathbf{v})$.

Explain why $\mathbf{v} = Q^T\mathbf{u}$ is also a unit vector; that is, explain why

$$|\mathbf{v}|^2 = |Q^T\mathbf{u}|^2 = (Q^T\mathbf{u}) \cdot (Q^T\mathbf{u}) = 1.$$

g. Using the fact that $q_A(\mathbf{u}) = q_D(\mathbf{v})$, explain how we now know the maximum value of $q_A(\mathbf{u})$ is 3 and determine the direction in which it occurs. Also, determine the minimum value of $q_A(\mathbf{u})$ and determine the direction in which it occurs.

This activity demonstrates how the eigenvalues of A determine the maximum and minimum values of the quadratic form $q_A(\mathbf{u})$ when evaluated on unit vectors and how the associated eigenvectors determine the directions in which the maximum and minimum values occur. Let's look at another example so that this connection is clear.

Example 7.2.4 Consider the symmetric matrix $A = \begin{bmatrix} -7 & -6 \\ -6 & 2 \end{bmatrix}$. Because A is symmetric, we know that it can be orthogonally diagonalized. In fact, we have $A = QDQ^T$ where

$$D = \begin{bmatrix} 5 & 0 \\ 0 & -10 \end{bmatrix}, \quad Q = \begin{bmatrix} 1/\sqrt{5} & 2/\sqrt{5} \\ -2/\sqrt{5} & 1/\sqrt{5} \end{bmatrix}.$$

From this diagonalization, we know that $\lambda_1 = 5$ is the largest eigenvalue of A with associated eigenvector $\mathbf{u}_1 = \begin{bmatrix} 1/\sqrt{5} \\ -2/\sqrt{5} \end{bmatrix}$ and that $\lambda_2 = -10$ is the smallest eigenvalue with associated eigenvector $\mathbf{u}_2 = \begin{bmatrix} 2/\sqrt{5} \\ 1/\sqrt{5} \end{bmatrix}$.

Let's first study the quadratic form $q_D(\mathbf{u}) = 5u_1^2 - 10u_2^2$ because the absence of the cross-term makes it comparatively simple. Remembering that $\mathbf{u}$ is a unit vector, we have $u_1^2 + u_2^2 = 1$, which means that $u_1^2 = 1 - u_2^2$. Therefore,

$$q_D(\mathbf{u}) = 5u_1^2 - 10u_2^2 = 5(1 - u_2^2) - 10u_2^2 = 5 - 15u_2^2.$$

This tells us that $q_D(\mathbf{u})$ has a maximum value of 5, which occurs when $u_2 = 0$ or in the direction $\begin{bmatrix} 1 \\ 0 \end{bmatrix}$.

In the same way, rewriting $u_2^2 = 1 - u_1^2$ allows us to conclude that the minimum value of $q_D(\mathbf{u})$ is -10, which occurs in the direction $\begin{bmatrix} 0 \\ 1 \end{bmatrix}$.

Let's now return to the matrix A whose quadratic form q_A is related to q_D because $A = QDQ^T$. In particular, we have

$$q_A(\mathbf{u}) = \mathbf{u} \cdot (A\mathbf{u}) = \mathbf{u} \cdot (QDQ^T\mathbf{u}) = (Q^T\mathbf{u}) \cdot (DQ^T\mathbf{u}) = \mathbf{v} \cdot (D\mathbf{v}) = q_D(\mathbf{v}).$$

In other words, we have $q_A(\mathbf{u}) = q_D(\mathbf{v})$ where $\mathbf{v} = Q^T\mathbf{u}$. This is quite useful because it allows us to relate the values of q_A to those of q_D, which we already understand quite well.

Now it turns out that $\mathbf{v}$ is also a unit vector because

$$|\mathbf{v}|^2 = \mathbf{v} \cdot \mathbf{v} = (Q^T\mathbf{u}) \cdot (Q^T\mathbf{u}) = \mathbf{u} \cdot (QQ^T\mathbf{u}) = \mathbf{u} \cdot \mathbf{u} = |\mathbf{u}|^2 = 1.$$

Therefore, the maximum value of $q_A(\mathbf{u})$ is the same as $q_D(\mathbf{v})$, which we know to be 5 and which occurs in the direction $\mathbf{v} = \begin{bmatrix} 1 \\ 0 \end{bmatrix}$. This means that the maximum value of $q_A(\mathbf{u})$ is

7.2. QUADRATIC FORMS

also 5 and that this occurs in the direction $\mathbf{u} = Q\mathbf{v} = Q\begin{bmatrix} 1 \\ 0 \end{bmatrix} = \begin{bmatrix} 1/\sqrt{5} \\ -2/\sqrt{5} \end{bmatrix}$. We now know that the maximum value of $q_A(\mathbf{u})$ is the largest eigenvalue $\lambda_1 = 5$ and that this maximum value occurs in the direction of an associated eigenvector.

In the same way, we see that the minimum value of $q_A(\mathbf{u})$ is the smallest eigenvalue $\lambda_2 = -10$ and that this minimum occurs in the direction of $\mathbf{u} = Q\begin{bmatrix} 0 \\ 1 \end{bmatrix} = \begin{bmatrix} 2/\sqrt{5} \\ 1/\sqrt{5} \end{bmatrix}$, an associated eigenvector.

More generally, we have

Proposition 7.2.5 *Suppose that A is a symmetric matrix, that we list its eigenvalues in decreasing order $\lambda_1 \geq \lambda_2 \ldots \geq \lambda_m$, and that $\mathbf{u}_1, \mathbf{u}_2, \ldots, \mathbf{u}_m$ is a basis of associated eigenvectors. The maximum value of $q_A(\mathbf{u})$ among all unit vectors $\mathbf{u}$ is λ_1, which occurs in the direction $\mathbf{u}_1$. Similarly, the minimum value of $q_A(\mathbf{u})$ is λ_m, which occurs in the direction $\mathbf{u}_m$.*

Example 7.2.6 Suppose that A is the symmetric matrix $A = \begin{bmatrix} 0 & 6 & 3 \\ 6 & 3 & 6 \\ 0 & 6 & 6 \end{bmatrix}$, which may be orthogonally diagonalized as $A = QDQ^T$ where

$$D = \begin{bmatrix} 12 & 0 & 0 \\ 0 & 3 & 0 \\ 0 & 0 & -6 \end{bmatrix}, \quad Q = \begin{bmatrix} 1/3 & 2/3 & 2/3 \\ 2/3 & 1/3 & -2/3 \\ 2/3 & -2/3 & 1/3 \end{bmatrix}.$$

We see that the maximum value of $q_A(\mathbf{u})$ is 12, which occurs in the direction $\begin{bmatrix} 1/3 \\ 2/3 \\ 2/3 \end{bmatrix}$, and the minimum value is -6, which occurs in the direction $\begin{bmatrix} 2/3 \\ -2/3 \\ 1/3 \end{bmatrix}$.

Example 7.2.7 Suppose we have the matrix of demeaned data points $A = \begin{bmatrix} 2 & 1 & -3 \\ 1 & 2 & -3 \end{bmatrix}$ that we considered in Preview Activity 7.2.1. The data points are shown in Figure 7.2.8.

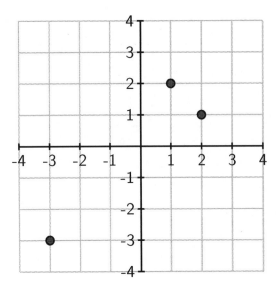

Figure 7.2.8 The set of demeaned data points from Preview Activity 7.2.1.

Constructing the covariance matrix $C = \frac{1}{3} AA^T$ gives $C = \begin{bmatrix} 14/3 & 13/3 \\ 13/3 & 14/3 \end{bmatrix}$, which has eigenvalues $\lambda_1 = 9$, with associated eigenvector $\begin{bmatrix} 1/\sqrt{2} \\ 1/\sqrt{2} \end{bmatrix}$, and $\lambda_2 = 1/3$, with associated eigenvector $\begin{bmatrix} -1/\sqrt{2} \\ 1/\sqrt{2} \end{bmatrix}$.

Remember that the variance in a direction $\mathbf{u}$ is $V_{\mathbf{u}} = \mathbf{u} \cdot (C\mathbf{u}) = q_C(\mathbf{u})$. Therefore, the variance attains a maximum value of 9 in the direction $\begin{bmatrix} 1/\sqrt{2} \\ 1/\sqrt{2} \end{bmatrix}$ and a minimum value of $1/3$ in the direction $\begin{bmatrix} -1/\sqrt{2} \\ 1/\sqrt{2} \end{bmatrix}$. Figure 7.2.9 shows the data projected onto the lines defined by these vectors.

7.2. QUADRATIC FORMS

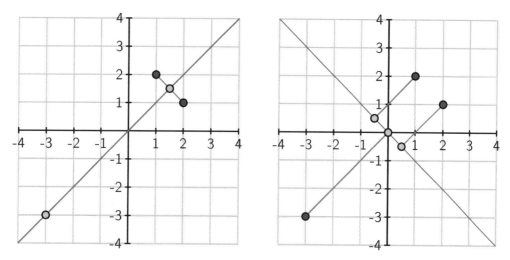

Figure 7.2.9 The demeaned data from Preview Activity 7.2.1 is shown projected onto the lines of maximal and minimal variance.

Remember that variance is additive, as stated in Proposition 7.1.16, which tells us that the total variance is $V = 9 + 1/3 = 28/3$.

We've been focused on finding the directions in which a quadratic form attains its maximum and minimum values, but there's another important observation to make after this activity. Recall how we used the fact that a symmetric matrix is orthogonally diagonalizable: if $A = QDQ^T$, then $q_A(\mathbf{u}) = q_D(\mathbf{v})$ where $\mathbf{v} = Q^T\mathbf{u}$.

More generally, if we define $\mathbf{y} = Q^T\mathbf{x}$, we have

$$q_A(\mathbf{x}) = \mathbf{x} \cdot (A\mathbf{x}) = \mathbf{x} \cdot (QDQ^T\mathbf{x}) = (Q^T\mathbf{x}) \cdot (DQ^T\mathbf{x}) = \mathbf{y} \cdot (D\mathbf{y}) = q_D(\mathbf{y})$$

Remembering that the quadratic form associated to a diagonal form has no cross terms, we obtain

$$q_A(\mathbf{x}) = q_D(\mathbf{y}) = \lambda_1 y_1^2 + \lambda_2 y_2^2 + \ldots + \lambda_m y_m^2.$$

In other words, after a change of coordinates, the quadratic form q_A can be written without cross terms. This is known as the Principle Axes Theorem.

Theorem 7.2.10 Principle Axes Theorem. *If A is a symmetric $m \times m$ matrix with eigenvalues $\lambda_1, \lambda_2, \ldots, \lambda_m$, then the quadratic form q_A can be written, after an orthogonal change of coordinates $\mathbf{y} = Q^T\mathbf{x}$, as*

$$q_A(\mathbf{x}) = \lambda_1 y_1^2 + \lambda_2 y_2^2 + \ldots + \lambda_m y_m^2.$$

We will put this to use in the next section.

7.2.2 Definite symmetric matrices

While our questions about variance provide some motivation for exploring quadratic forms, these functions appear in a variety of other contexts so it's worth spending some more time with them. For example, quadratic forms appear in multivariable calculus when describing the behavior of a function of several variables near a critical point and in physics when describing the kinetic energy of a rigid body.

The following definition will be important in this section.

Definition 7.2.11 A symmetric matrix A is called *positive definite* if its associated quadratic form satisfies $q_A(\mathbf{x}) > 0$ for any nonzero vector $\mathbf{x}$. If $q_A(\mathbf{x}) \geq 0$ for all nonzero vectors $\mathbf{x}$, we say that A is *positive semidefinite*.

Likewise, we say that A is *negative definite* if $q_A(\mathbf{x}) < 0$ for all nonzero vectors $\mathbf{x}$.

Finally, A is called *indefinite* if $q_A(\mathbf{x}) > 0$ for some $\mathbf{x}$ and $q_A(\mathbf{x}) < 0$ for others.

> **Activity 7.2.4.** This activity explores the relationship between the eigenvalues of a symmetric matrix and its definiteness.
>
> a. Consider the diagonal matrix $D = \begin{bmatrix} 4 & 0 \\ 0 & 2 \end{bmatrix}$ and write its quadratic form $q_D(\mathbf{x})$ in terms of the components of $\mathbf{x} = \begin{bmatrix} x_1 \\ x_2 \end{bmatrix}$. How does this help you decide whether D is positive definite or not?
>
> b. Now consider $D = \begin{bmatrix} 4 & 0 \\ 0 & 0 \end{bmatrix}$ and write its quadratic form $q_D(\mathbf{x})$ in terms of x_1 and x_2. What can you say about the definiteness of D?
>
> c. If D is a diagonal matrix, what condition on the diagonal entries guarantee that D is
>
>   1. positive definite?
>   2. positive semidefinite?
>   3. negative definite?
>   4. negative semidefinite?
>   5. indefinite?
>
> d. Suppose that A is a symmetric matrix with eigenvalues 4 and 2 so that $A = QDQ^T$ where $D = \begin{bmatrix} 4 & 0 \\ 0 & 2 \end{bmatrix}$. If $\mathbf{y} = Q^T\mathbf{x}$, then we have $q_A(\mathbf{x}) = q_D(\mathbf{y})$. Explain why this tells us that A is positive definite.
>
> e. Suppose that A is a symmetric matrix with eigenvalues 4 and 0. What can you say about the definiteness of A in this case?
>
> f. What condition on the eigenvalues of a symmetric matrix A guarantees that A is
>
>   1. positive definite?
>   2. positive semidefinite?
>   3. negative definite?
>   4. negative semidefinite?
>   5. indefinite?

7.2. QUADRATIC FORMS

As seen in this activity, it is straightforward to determine the definiteness of a diagonal matrix. For instance, if $D = \begin{bmatrix} 7 & 0 \\ 0 & 5 \end{bmatrix}$, then

$$q_D(\mathbf{x}) = 7x_1^2 + 5x_2^2.$$

This shows that $q_D(\mathbf{x}) > 0$ when either x_1 or x_2 is not zero so we conclude that D is positive definite. In the same way, we see that D is positive semidefinite if all the diagonal entries are nonnegative.

Understanding this behavior for diagonal matrices enables us to understand more general symmetric matrices. As we saw previously, the quadratic form for a symmetric matrix $A = QDQ^T$ agrees with the quadratic form for the diagonal matrix D after a change of coordinates. In particular,

$$q_A(\mathbf{x}) = q_D(\mathbf{y})$$

where $\mathbf{y} = Q^T\mathbf{x}$. Now the diagonal entries of D are the eigenvalues of A from which we conclude that $q_A(\mathbf{x}) > 0$ if all the eigenvalues of A are positive. Likewise, $q_A(\mathbf{x}) \geq 0$ if all the eigenvalues are nonnegative.

Proposition 7.2.12 *A symmetric matrix is positive definite if all its eigenvalues are positive. It is positive semidefinite if all its eigenvalues are nonnegative.*

Likewise, a symmetric matrix is indefinite if some eigenvalues are positive and some are negative.

We will now apply what we've learned about quadratic forms to study the nature of critical points in multivariable calculus. The rest of this section assumes that the reader is familiar with ideas from multivariable calculus and can be skipped by others.

First, suppose that $f(x, y)$ is a differentiable function. We will use f_x and f_y to denote the partial derivatives of f with respect to x and y. Similarly, f_{xx}, f_{xy}, f_{yx} and f_{yy} denote the second partial derivatives. You may recall that the mixed partials, f_{xy} and f_{yx} are equal under a mild assumption on the function f. A typical question in calculus is to determine where this function has its maximum and minimum values.

Any local maximum or minimum of f appears at a critical point (x_0, y_0) where

$$f_x(x_0, y_0) = 0, \qquad f_y(x_0, y_0) = 0.$$

Near a critical point, the quadratic approximation of f tells us that

$$f(x, y) \approx f(x_0, y_0) + \frac{1}{2}f_{xx}(x_0, y_0)(x - x_0)^2$$
$$+ f_{xy}(x_0, y_0)(x - x_0)(y - y_0) + \frac{1}{2}f_{yy}(x_0, y_0)(y - y_0)^2.$$

Activity 7.2.5. Let's explore how our understanding of quadratic forms helps us determine the behavior of a function f near a critical point.

a. Consider the function $f(x, y) = 2x^3 - 6xy + 3y^2$. Find the partial derivatives f_x and f_y and use these expressions to determine the critical points of f.

b. Evaluate the second partial derivatives f_{xx}, f_{xy}, and f_{yy}.

c. Let's first consider the critical point $(1, 1)$. Use the quadratic approximation as written above to find an expression approximating f near the critical point.

d. Using the vector $\mathbf{w} = \begin{bmatrix} x - 1 \\ y - 1 \end{bmatrix}$, rewrite your approximation as

$$f(x, y) \approx f(1, 1) + q_A(\mathbf{w})$$

for some matrix A. What is the matrix A in this case?

e. Find the eigenvalues of A. What can you conclude about the definiteness of A?

f. Recall that (x_0, y_0) is a local minimum for f if $f(x, y) > f(x_0, y_0)$ for nearby points (x, y). Explain why our understanding of the eigenvalues of A shows that $(1, 1)$ is a local minimum for f.

```
x, y = var('x', 'y')
plot3d(2*x^3 - 6*x*y + 3*y^2, (x, 0.75,1.25), (y,0.75,1.25))
```

Near a critical point (x_0, y_0) of a function $f(x, y)$, we can write

$$f(x, y) \approx f(x_0, y_0) + q_A(\mathbf{w})$$

where $\mathbf{w} = \begin{bmatrix} x - x_0 \\ y - y_0 \end{bmatrix}$ and $A = \frac{1}{2}\begin{bmatrix} f_{xx}(x_0, y_0) & f_{xy}(x_0, y_0) \\ f_{yx}(x_0, y_0) & f_{yy}(x_0, y_0) \end{bmatrix}$. If A is positive definite, then $q_A(\mathbf{w}) > 0$, which tells us that

$$f(x, y) \approx f(x_0, y_0) + q_A(\mathbf{w}) > f(x_0, y_0)$$

and that the critical point (x_0, y_0) is therefore a local minimum.

The matrix

$$H = \begin{bmatrix} f_{xx}(x_0, y_0) & f_{xy}(x_0, y_0) \\ f_{yx}(x_0, y_0) & f_{yy}(x_0, y_0) \end{bmatrix}$$

is called the *Hessian* of f, and we see now that the eigenvalues of this symmetric matrix determine the nature of the critical point (x_0, y_0). In particular, if the eigenvalues are both positive, then q_H is positive definite, and the critical point is a local minimum.

This observation leads to the Second Derivative Test for multivariable functions.

Proposition 7.2.13 Second Derivative Test. *The nature of a critical point of a multivariable function is determined by the Hessian H of the function at the critical point. If*

- *H has all positive eigenvalues, the critical point is a local minimum.*

- *H has all negative eigenvalues, the critical point is a local maximum.*

- *H has both positive and negative eigenvalues, the critical point is neither a local maximum nor minimum.*

Most multivariable calculus texts assume that the reader is not familiar with linear algebra and so write the second derivative test for functions of two variables in terms of $D = \det(H)$. If

7.2. QUADRATIC FORMS

- $D > 0$ and $f_{xx}(x_0, y_0)) > 0$, then (x_0, y_0) is a local minimum.
- $D > 0$ and $f_{xx}(x_0, y_0)) < 0$, then (x_0, y_0) is a local maximum.
- $D < 0$, then (x_0, y_0) is neither a local maximum nor minimum.

The conditions in this version of the second derivative test are simply algebraic criteria that tell us about the definiteness of the Hessian matrix H.

7.2.3 Summary

This section explored quadratic forms, functions that are defined by symmetric matrices.

- If A is a symmetric matrix, then the quadratic form defined by A is the function $q_A(\mathbf{x}) = \mathbf{x} \cdot (A\mathbf{x})$. Quadratic forms appear when studying the variance of a dataset. If C is the covariance matrix, then the variance in the direction defined by a unit vector $\mathbf{u}$ is $q_C(\mathbf{u}) = \mathbf{u} \cdot (C\mathbf{u}) = V_\mathbf{u}$.

  Similarly, quadratic forms appear in multivariable calculus when analyzing the behavior of a function of several variables near a critical point.

- If λ_1 is the largest eigenvalue of a symmetric matrix A and λ_m the smallest, then the maximum value of $q_A(\mathbf{u})$ among unit vectors $\mathbf{u}$, is λ_1, and this maximum value occurs in the direction of $\mathbf{u}_1$, a unit eigenvector associated to λ_1.

  Similarly, the minimum value of $q_A(\mathbf{u})$ is λ_m, which appears in the direction of $\mathbf{u}_m$, an eigenvector associated to λ_m.

- A symmetric matrix is positive definite if its eigenvalues are all positive, positive semi-definite if its eigenvalues are all nonnegative, and indefinite if it has both positive and negative eigenvalues.

- If the Hessian H of a multivariable function f is positive definite at a critical point, then the critical point is a local minimum. Likewise, if the Hessian is negative definite, the critical point is a local maximum.

7.2.4 Exercises

1. Suppose that $A = \begin{bmatrix} 4 & 2 \\ 2 & 7 \end{bmatrix}$.

   a. Find an orthogonal diagonalization of A.

   b. Evaluate the quadratic form $q_A\left(\begin{bmatrix} 1 \\ 1 \end{bmatrix}\right)$.

   c. Find the unit vector $\mathbf{u}$ for which $q_A(\mathbf{u})$ is as large as possible. What is the value of $q_A(\mathbf{u})$ in this direction?

   d. Find the unit vector $\mathbf{u}$ for which $q_A(\mathbf{u})$ is as small as possible. What is the value of $q_A(\mathbf{u})$ in this direction?

2. Consider the quadratic form

$$q\left(\begin{bmatrix} x_1 \\ x_2 \end{bmatrix}\right) = 3x_1^2 - 4x_1x_2 + 6x_2^2.$$

   a. Find a matrix A such that $q(\mathbf{x}) = \mathbf{x}^T A \mathbf{x}$.

   b. Find the maximum and minimum values of $q(\mathbf{u})$ among all unit vectors $\mathbf{u}$ and describe the directions in which they occur.

3. Suppose that A is a demeaned data matrix:

$$A = \begin{bmatrix} 1 & -2 & 0 & 1 \\ 1 & -1 & -1 & 1 \end{bmatrix}.$$

   a. Find the covariance matrix C.

   b. What is the variance of the data projected onto the line defined by $\mathbf{u} = \begin{bmatrix} 1/\sqrt{2} \\ 1/\sqrt{2} \end{bmatrix}$.

   c. What is the total variance?

   d. In which direction is the variance greatest and what is the variance in this direction?

4. Consider the matrix $A = \begin{bmatrix} 4 & -3 & -3 \\ -3 & 4 & -3 \\ -3 & -3 & 4 \end{bmatrix}$.

   a. Find Q and D such that $A = QDQ^T$.

   b. Find the maximum and minimum values of $q(\mathbf{u}) = \mathbf{x}^T A \mathbf{x}$ among all unit vectors $\mathbf{u}$.

   c. Describe the direction in which the minimum value occurs. What can you say about the direction in which the maximum occurs?

5. Consider the matrix $B = \begin{bmatrix} -2 & 1 \\ 4 & -2 \\ 2 & -1 \end{bmatrix}$.

   a. Find the matrix A so that $q\left(\begin{bmatrix} x_1 \\ x_2 \end{bmatrix}\right) = |B\mathbf{x}|^2 = q_A(\mathbf{x})$.

   b. Find the maximum and minimum values of $q(\mathbf{u})$ among all unit vectors $\mathbf{u}$ and describe the directions in which they occur.

   c. What does the minimum value of $q(\mathbf{u})$ tell you about the matrix B?

6. Consider the quadratic form

$$q\left(\begin{bmatrix} x_1 \\ x_2 \\ x_3 \end{bmatrix}\right) = 7x_1^2 + 4x_2^2 + 7x_3^2 - 2x_1x_2 - 4x_1x_3 - 2x_2x_3.$$

a. What can you say about the definiteness of the matrix A that defines the quadratic form?

b. Find a matrix Q so that the change of coordinates $\mathbf{y} = Q^T \mathbf{x}$ transforms the quadratic form into one that has no cross terms. Write the quadratic form in terms of $\mathbf{y}$.

c. What are the maximum and minimum values for $q(\mathbf{u})$ among all unit vectors $\mathbf{u}$?

7. Explain why the following statements are true.

   a. Given any matrix B, the matrix $B^T B$ is a symmetric, positive semidefinite matrix.

   b. If both A and B are symmetric, positive definite matrices, then $A + B$ is a symmetric, positive definite matrix.

   c. If A is a symmetric, invertible, positive definite matrix, then A^{-1} is also.

8. Determine whether the following statements are true or false and explain your reasoning.

   a. If A is an indefinite matrix, we can't know whether it is positive definite or not.

   b. If the smallest eigenvalue of A is 3, then A is positive definite.

   c. If C is the covariance matrix associated with a dataset, then C is positive semidefinite.

   d. If A is a symmetric 2×2 matrix and the maximum and minimum values of $q_A(\mathbf{u})$ occur at $\begin{bmatrix} 1 \\ 0 \end{bmatrix}$ and $\begin{bmatrix} 0 \\ 1 \end{bmatrix}$, then A is diagonal.

   e. If A is negative definite and Q is an orthogonal matrix with $B = QAQ^T$, then B is negative definite.

9. Determine the critical points for each of the following functions. At each critical point, determine the Hessian H, describe the definiteness of H, and determine whether the critical point is a local maximum or minimum.

   a. $f(x, y) = xy + \frac{2}{x} + \frac{2}{y}$.

   b. $f(x, y) = x^4 + y^4 - 4xy$.

10. Consider the function $f(x, y, z) = x^4 + y^4 + z^4 - 4xyz$.

    a. Show that f has a critical point at $(-1, 1, -1)$ and construct the Hessian H at that point.

    b. Find the eigenvalues of H. Is this a definite matrix of some kind?

    c. What does this imply about whether $(-1, 1, -1)$ is a local maximum or minimum?

7.3 Principal Component Analysis

We are sometimes presented with a dataset having many data points that live in a high dimensional space. For instance, we looked at a dataset describing body fat index (BFI) in Activity 6.5.4 where each data point is six-dimensional. Developing an intuitive understanding of the data is hampered by the fact that it cannot be visualized.

This section explores a technique called *principal component analysis*, which enables us to reduce the dimension of a dataset so that it may be visualized or studied in a way so that interesting features more readily stand out. Our previous work with variance and the orthogonal diagonalization of symmetric matrices provides the key ideas.

> **Preview Activity 7.3.1.** We will begin by recalling our earlier discussion of variance. Suppose we have a dataset that leads to the covariance matrix
>
> $$C = \begin{bmatrix} 7 & -4 \\ -4 & 13 \end{bmatrix}.$$
>
> a. Suppose that $\mathbf{u}$ is a unit eigenvector of C with eigenvalue λ. What is the variance $V_\mathbf{u}$ in the $\mathbf{u}$ direction?
>
> b. Find an orthogonal diagonalization of C.
>
> c. What is the total variance?
>
> d. In which direction is the variance greatest and what is the variance in this direction? If we project the data onto this line, how much variance is lost?
>
> e. In which direction is the variance smallest and how is this direction related to the direction of maximum variance?

Here are some ideas we've seen previously that will be particularly useful for us in this section. Remember that the covariance matrix of a dataset is $C = \frac{1}{N}AA^T$ where A is the matrix of N demeaned data points.

- When $\mathbf{u}$ is a unit vector, the variance of the demeaned data after projecting onto the line defined by $\mathbf{u}$ is given by the quadratic form $V_\mathbf{u} = \mathbf{u} \cdot (C\mathbf{u})$.

- In particular, if $\mathbf{u}$ is a unit eigenvector of C with associated eigenvalue λ, then $V_\mathbf{u} = \lambda$.

- Moreover, variance is additive, as we recorded in Proposition 7.1.16: if W is a subspace having an orthonormal basis $\mathbf{u}_1, \mathbf{u}_2, \ldots, \mathbf{u}_n$, then the variance

$$V_W = V_{\mathbf{u}_1} + V_{\mathbf{u}_2} + \ldots + V_{\mathbf{u}_n}.$$

7.3.1 Principal Component Analysis

Let's begin by looking at an example that illustrates the central theme of this technique.

Activity 7.3.2. Suppose that we work with a dataset having 100 five-dimensional data points. The demeaned data matrix A is therefore 5×100 and leads to the covariance matrix $C = \frac{1}{100} AA^T$, which is a 5×5 matrix. Because C is symmetric, the Spectral Theorem tells us it is orthogonally diagonalizable so suppose that $C = QDQ^T$ where

$$Q = \begin{bmatrix} \mathbf{u}_1 & \mathbf{u}_2 & \mathbf{u}_3 & \mathbf{u}_4 & \mathbf{u}_5 \end{bmatrix}, \quad D = \begin{bmatrix} 13 & 0 & 0 & 0 & 0 \\ 0 & 10 & 0 & 0 & 0 \\ 0 & 0 & 2 & 0 & 0 \\ 0 & 0 & 0 & 0 & 0 \\ 0 & 0 & 0 & 0 & 0 \end{bmatrix}.$$

a. What is $V_{\mathbf{u}_2}$, the variance in the $\mathbf{u}_2$ direction?

b. Find the variance of the data projected onto the line defined by $\mathbf{u}_4$. What does this say about the data?

c. What is the total variance of the data?

d. Consider the 2-dimensional subspace spanned by $\mathbf{u}_1$ and $\mathbf{u}_2$. If we project the data onto this subspace, what fraction of the total variance is represented by the variance of the projected data?

e. How does this question change if we project onto the 3-dimensional subspace spanned by $\mathbf{u}_1$, $\mathbf{u}_2$, and $\mathbf{u}_3$?

f. What does this tell us about the data?

This activity demonstrates how the eigenvalues of the covariance matrix can tell us when data are clustered around, or even wholly contained within, a smaller dimensional subspace. In particular, the original data is 5-dimensional, but we see that it actually lies in a 3-dimensional subspace of $\mathbb{R}^5$. Later in this section, we'll see how to use this observation to work with the data as if it were three-dimensional, an idea known as *dimensional reduction*.

The eigenvectors $\mathbf{u}_j$ of the covariance matrix are called *principal components*, and we will order them so that their associated eigenvalues decrease. Generally speaking, we hope that the first few principal components retain most of the variance, as the example in the activity demonstrates. In that example, we have the sequence of subspaces

- W_1, the 1-dimensional subspace spanned by $\mathbf{u}_1$, which retains $13/25 = 52\%$ of the total variance,

- W_2, the 2-dimensional subspace spanned by $\mathbf{u}_1$ and $\mathbf{u}_2$, which retains $23/25 = 92\%$ of the variance, and

- W_3, the 3-dimensional subspace spanned by $\mathbf{u}_1$, $\mathbf{u}_2$, and $\mathbf{u}_3$, which retains all of the variance.

Notice how we retain more of the total variance as we increase the dimension of the subspace onto which the data are projected. Eventually, projecting the data onto W_3 retains all the variance, which tells us the data must lie in W_3, a smaller dimensional subspace of $\mathbb{R}^5$.

In fact, these subspaces are the best possible. We know that the first principal component $\mathbf{u}_1$ is the eigenvector of C associated to the largest eigenvalue. This means that the variance is as large as possible in the $\mathbf{u}_1$ direction. In other words, projecting onto any other line will retain a smaller amount of variance. Similarly, projecting onto any other 2-dimensional subspace besides W_2 will retain less variance than projecting onto W_2. The principal components have the wonderful ability to pick out the best possible subspaces to retain as much variance as possible.

Of course, this is a contrived example. Typically, the presence of noise in a dataset means that we do not expect all the points to be wholly contained in a smaller dimensional subspace. In fact, the 2-dimensional subspace W_2 retains 92% of the variance. Depending on the situation, we may want to write off the remaining 8% of the variance as noise in exchange for the convenience of working with a smaller dimensional subspace. As we'll see later, we will seek a balance using a number of principal components large enough to retain most of the variance but small enough to be easy to work with.

Activity 7.3.3. We will work here with a dataset having 100 3-dimensional demeaned data points. Evaluating the following cell will plot those data points and define the demeaned data matrix A whose shape is 3×100.

```
url='https://raw.githubusercontent.com/davidaustinm/'
url+='ula_modules/master/pca_demo.py'
sage.repl.load.load(url, globals())
```

Notice that the data appears to cluster around a plane though it does not seem to be wholly contained within that plane.

a. Use the matrix A to construct the covariance matrix C. Then determine the variance in the direction of $\mathbf{u} = \begin{bmatrix} 1/3 \\ 2/3 \\ 2/3 \end{bmatrix}$?

b. Find the eigenvalues of C and determine the total variance.

Notice that Sage does not necessarily sort the eigenvalues in decreasing order.

c. Use the `right_eigenmatrix()` command to find the eigenvectors of C. Remembering that the Sage command `B.column(1)` retrieves the vector represented by the second column of B, define vectors u1, u2, and u3 representing the three principal components in order of decreasing eigenvalues. How can you check if these vectors are an orthonormal basis for $\mathbb{R}^3$?

7.3. PRINCIPAL COMPONENT ANALYSIS

d. What fraction of the total variance is retained by projecting the data onto W_1, the subspace spanned by $\mathbf{u}_1$? What fraction of the total variance is retained by projecting onto W_2, the subspace spanned by $\mathbf{u}_1$ and $\mathbf{u}_2$? What fraction of the total variance do we lose by projecting onto W_2?

e. If we project a data point $\mathbf{x}$ onto W_2, the Projection Formula tells us we obtain

$$\widehat{\mathbf{x}} = (\mathbf{u}_1 \cdot \mathbf{x})\mathbf{u}_1 + (\mathbf{u}_2 \cdot \mathbf{x})\mathbf{u}_2.$$

Rather than viewing the projected data in $\mathbb{R}^3$, we will record the coordinates of $\widehat{\mathbf{x}}$ in the basis defined by $\mathbf{u}_1$ and $\mathbf{u}_2$; that is, we will record the coordinates

$$\begin{bmatrix} \mathbf{u}_1 \cdot \mathbf{x} \\ \mathbf{u}_2 \cdot \mathbf{x} \end{bmatrix}.$$

Construct the matrix Q so that $Q^T\mathbf{x} = \begin{bmatrix} \mathbf{u}_1 \cdot \mathbf{x} \\ \mathbf{u}_2 \cdot \mathbf{x} \end{bmatrix}$.

f. Since each column of A represents a data point, the matrix $Q^T A$ represents the coordinates of the projected data points. Evaluating the following cell will plot those projected data points.

```
pca_plot(Q.T*A)
```

Notice how this plot enables us to view the data as if it were two-dimensional. Why is this plot wider than it is tall?

This example is a more realistic illustration of principal component analysis. The plot of the 3-dimensional data appears to show that the data lies close to a plane, and the principal components will identify this plane. Starting with the 3×100 matrix of demeaned data A, we construct the covariance matrix $C = \frac{1}{100} AA^T$ and study its eigenvalues. Notice that the first two principal components account for more than 98% of the variance, which means we can expect the points to lie close to W_2, the two-dimensional subspace spanned by $\mathbf{u}_1$ and $\mathbf{u}_2$.

Since W_2 is a subspace of $\mathbb{R}^3$, projecting the data points onto W_2 gives a list of 100 points in $\mathbb{R}^3$. In order to visualize them more easily, we instead consider the coordinates of the projections in the basis defined by $\mathbf{u}_1$ and $\mathbf{u}_2$. For instance, we know that the projection of a data point $\mathbf{x}$ is

$$\widehat{\mathbf{x}} = (\mathbf{u}_1 \cdot \mathbf{x})\mathbf{u}_1 + (\mathbf{u}_2 \cdot \mathbf{x})\mathbf{u}_2,$$

which is a three-dimensional vector. Instead, we can record the coordinates $\begin{bmatrix} \mathbf{u}_1 \cdot \mathbf{x} \\ \mathbf{u}_2 \cdot \mathbf{x} \end{bmatrix}$ and plot them in the two-dimensional coordinate plane, as illustrated in Figure 7.3.1.

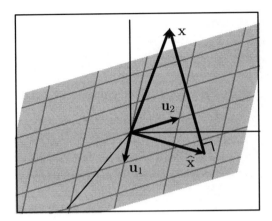 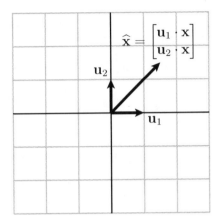

Figure 7.3.1 The projection $\widehat{\mathbf{x}}$ of a data point $\mathbf{x}$ onto W_2 is a three-dimensional vector, which may be represented by the two coordinates describing this vector as a linear combination of $\mathbf{u}_1$ and $\mathbf{u}_2$.

If we form the matrix $Q = \begin{bmatrix} \mathbf{u}_1 & \mathbf{u}_2 \end{bmatrix}$, then we have

$$Q^T \mathbf{x} = \begin{bmatrix} \mathbf{u}_1 \cdot \mathbf{x} \\ \mathbf{u}_2 \cdot \mathbf{x} \end{bmatrix}.$$

This means that the columns of $Q^T A$ represent the coordinates of the projected points, which may now be plotted in the plane.

In this plot, the first coordinate, represented by the horizontal coordinate, represents the projection of a data point onto the line defined by $\mathbf{u}_1$ while the second coordinate represents the projection onto the line defined by $\mathbf{u}_2$. Since $\mathbf{u}_1$ is the first principal component, the variance in the $\mathbf{u}_1$ direction is greater than the variance in the $\mathbf{u}_2$ direction. For this reason, the plot will be more spread out in the horizontal direction than in the vertical.

7.3.2 Using Principal Component Analysis

Now that we've explored the ideas behind principal component analysis, we will look at a few examples that illustrate its use.

Activity 7.3.4. The next cell will load a dataset describing the average consumption of various food groups for citizens in each of the four nations of the United Kingdom. The units for each entry are grams per person per week.

```
import pandas as pd
url='https://raw.githubusercontent.com/davidaustinm/'
url+='ula_modules/master/data/uk-diet.csv'
df = pd.read_csv(url, index_col=0)
data_mean = vector(df.T.mean())
A = matrix([vector(row) for row in (df.T-df.T.mean()).values]).T
df
```

7.3. PRINCIPAL COMPONENT ANALYSIS

We will view this as a dataset consisting of four points in $\mathbb{R}^{17}$. As such, it is impossible to visualize and studying the numbers themselves doesn't lead to much insight.

In addition to loading the data, evaluating the cell above created a vector data_mean, which is the mean of the four data points, and A, the 17×4 matrix of demeaned data.

a. What is the average consumption of Beverages across the four nations?

b. Find the covariance matrix C and its eigenvalues. Because there are four points in $\mathbb{R}^{17}$ whose mean is zero, there are only three nonzero eigenvalues.

c. For what percentage of the total variance does the first principal component account?

d. Find the first principal component u_1 and project the four demeaned data points onto the line defined by u_1. Plot those points on Figure 7.3.2

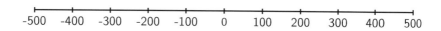

Figure 7.3.2 A plot of the demeaned data projected onto the first principal component.

e. For what percentage of the total variance do the first two principal components account?

f. Find the coordinates of the demeaned data points projected onto W_2, the two-dimensional subspace of $\mathbb{R}^{17}$ spanned by the first two principal components.

Plot these coordinates in Figure 7.3.3.

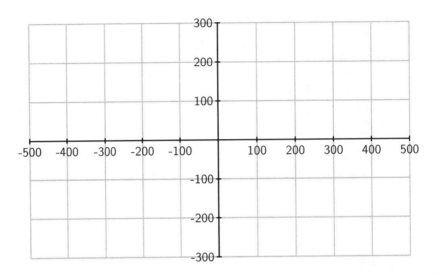

Figure 7.3.3 The coordinates of the demeaned data points projected onto the first two principal components.

g. What information do these plots reveal that is not clear from consideration of the original data points?

h. Study the first principal component u_1 and find the first component of u_1, which corresponds to the dietary category Alcoholic Drinks. (To do this, you may wish to use N(u1, digits=2) for a result that's easier to read.) If a data point lies on the far right side of the plot in Figure 7.3.3, what does it mean about that nation's consumption of Alcoholic Drinks?

This activity demonstrates how principal component analysis enables us to extract information from a dataset that may not be easily obtained otherwise. As in our previous example, we see that the data points lie quite close to a two-dimensional subspace of $\mathbb{R}^{17}$. In fact, W_2, the subspace spanned by the first two principal components, accounts for more than 96% of the variance. More importantly, when we project the data onto W_2, it becomes apparent that Northern Ireland is fundamentally different from the other three nations.

With some additional thought, we can determine more specific ways in which Northern Ireland is different. On the 2-dimensional plot, Northern Ireland lies far to the right compared to the other three nations. Since the data has been demeaned, the origin $(0,0)$ in this plot corresponds to the average of the four nations. The coordinates of the point representing Northern Ireland are about $(477, 59)$, meaning that the projected data point differs from the mean by about $477u_1 + 59u_2$.

Let's just focus on the contribution from u_1. We see that the ninth component of u_1, the one that describes Fresh Fruit, is about -0.63. This means that the ninth component of $477u_1$ differs from the mean by about $477(-0.63) = -300$ grams per person per week. So roughly speaking, people in Northern Ireland are eating about 300 fewer grams of Fresh Fruit than the average across the four nations. This is borne out by looking at the original data, which show that the consumption of Fresh Fruit in Northern Ireland is significantly less than the

7.3. PRINCIPAL COMPONENT ANALYSIS

other nations. Examing the other components of $\mathbf{u}_1$ shows other ways in which Northern Ireland differs from the other three nations.

> **Activity 7.3.5.** In this activity, we'll look at a well-known dataset[1] that describes 150 irises representing three species of iris: Iris setosa, Iris versicolor, and Iris virginica. For each flower, the length and width of its sepal and the length and width of its petal, all in centimeters, are recorded.
>
>
>
> **Figure 7.3.4** One of the three species, iris versicolor, represented in the dataset showing three shorter petals and three longer sepals. (Source: Wikipedia[2], License: GNU Free Documetation License[3])
>
> Evaluating the following cell will load the dataset, which consists of 150 points in $\mathbb{R}^4$. In addition, we have a vector data_mean, a four-dimensional vector holding the mean of the data points, and A, the 4×150 demeaned data matrix.
>
> ```
> url='https://raw.githubusercontent.com/davidaustinm/'
> url+='ula_modules/master/pca_iris.py'
> sage.repl.load.load(url, globals())
> df.T
> ```
>
> Since the data is four-dimensional, we are not able to visualize it. Of course, we could forget about two of the measurements and plot the 150 points represented by their, say, sepal length and sepal width.
>
> ```
> sepal_plot()
> ```
>
> a. What is the mean sepal width?
>
> b. Find the covariance matrix C and its eigenvalues.

c. Find the fraction of variance for which the first two principal components account.

d. Construct the first two principal components $\mathbf{u}_1$ and $\mathbf{u}_2$ along with the matrix Q whose columns are $\mathbf{u}_1$ and $\mathbf{u}_2$.

e. As we have seen, the columns of the matrix $Q^T A$ hold the coordinates of the demeaned data points after projecting onto W_2, the subspace spanned by the first two principal components. Evaluating the following cell shows a plot of these coordinates.

```
pca_plot(Q.T*A)
```

Suppose we have a flower whose coordinates in this plane are $(-2.5, -0.75)$. To what species does this iris most likely belong? Find an estimate of the sepal length, sepal width, petal length, and petal width for this flower.

f. Suppose you have an iris, but you only know that its sepal length is 5.65 cm and its sepal width is 2.75 cm. Knowing only these two measurements, determine the coordinates (c_1, c_2) in the plane where this iris lies. To what species does this iris most likely belong? Now estimate the petal length and petal width of this iris.

g. Suppose you find another iris whose sepal width is 3.2 cm and whose petal width is 2.2 cm. Find the coordinates (c_1, c_2) of this iris and determine the species to which it most likely belongs. Also, estimate the sepal length and the petal length.

7.3.3 Summary

This section has explored principal component analysis as a technique to reduce the dimension of a dataset. From the demeaned data matrix A, we form the covariance matrix $C = \frac{1}{N} A A^T$, where N is the number of data points.

- The eigenvectors $\mathbf{u}_1, \mathbf{u}_2, \ldots \mathbf{u}_m$, of C are called the principal components. We arrange them so that their corresponding eigenvalues are in decreasing order.

- If W_n is the subspace spanned by the first n principal components, then the variance of the demeaned data projected onto W_n is the sum of the first n eigenvalues of C. No other n-dimensional subspace retains more variance when the data is projected onto it.

---
[1] archive.ics.uci.edu
[2] gvsu.edu/s/21D
[3] gvsu.edu/s/21E

- If Q is the matrix whose columns are the first n principal components, then the columns of $Q^T A$ hold the coordinates, expressed in the basis $\mathbf{u}_1, \ldots, \mathbf{u}_n$, of the data once projected onto W_n.

- Our goal is to use a number of principal components that is large enough to retain most of the variance in the dataset but small enough to be manageable.

7.3.4 Exercises

1. Suppose that
$$Q = \begin{bmatrix} -1/\sqrt{2} & 1/\sqrt{2} \\ 1/\sqrt{2} & 1/\sqrt{2} \end{bmatrix}, \quad D_1 = \begin{bmatrix} 75 & 0 \\ 0 & 74 \end{bmatrix}, \quad D_2 = \begin{bmatrix} 100 & 0 \\ 0 & 1 \end{bmatrix}$$
and that we have two datasets, one whose covariance matrix is $C_1 = QD_1Q^T$ and one whose covariance matrix is $C_2 = QD_2Q^T$. For each dataset, find

   a. the total variance.

   b. the fraction of variance represented by the first principal component.

   c. a verbal description of how the demeaned data points appear when plotted in the plane.

2. Suppose that a dataset has mean $\begin{bmatrix} 13 \\ 5 \\ 7 \end{bmatrix}$ and that its associated covariance matrix is
$$C = \begin{bmatrix} 275 & -206 & 251 \\ -206 & 320 & -206 \\ 251 & -206 & 275 \end{bmatrix}.$$

   a. What fraction of the variance is represented by the first two principal components?

   b. If $\begin{bmatrix} 30 \\ -3 \\ 26 \end{bmatrix}$ is one of the data points, find the coordinates when the demeaned point is projected into the plane defined by the first two principal components.

   c. If a projected data point has coordinates $\begin{bmatrix} 12 \\ -25 \end{bmatrix}$, find an estimate for the original data point.

3. Evaluating the following cell loads a 2×100 demeaned data matrix A.

   ```
   url='https://raw.githubusercontent.com/davidaustinm/'
   url+='ula_modules/master/pca_ex.py'
   sage.repl.load.load(url, globals())
   ```

a. Find the principal components $\mathbf{u}_1$ and $\mathbf{u}_2$ and the variance in the direction of each principal component.

b. What is the total variance?

c. What can you conclude about this dataset?

4. Determine whether the following statements are true or false and explain your thinking.

   a. If the eigenvalues of the covariance matrix are λ_1, λ_2, and λ_3, then λ_3 is the variance of the demeaned data points when projected on the third principal component $\mathbf{u}_3$.

   b. Principal component analysis always allows us to construct a smaller dimensional representation of a dataset without losing any information.

   c. If the eigenvalues of the covariance matrix are 56, 32, and 0, then the demeaned data points all lie on a line in $\mathbb{R}^3$.

5. In Activity 7.3.5, we looked at a dataset consisting of four measurements of 150 irises. These measurements are sepal length, sepal width, petal length, and petal width.

   a. Find the first principal component $\mathbf{u}_1$ and describe the meaning of its four components. Which component is most significant? What can you say about the relative importance of the four measurements?

   b. When the dataset is plotted in the plane defined by $\mathbf{u}_1$ and $\mathbf{u}_2$, the specimens from the species iris-setosa lie on the left side of the plot. What does this tell us about how iris-setosa differs from the other two species in the four measurements?

   c. In general, which species is closest to the "average iris"?

6. This problem explores a dataset describing 333 penguins. There are three species, Adelie, Chinstrap, and Gentoo, as illustrated on the left of Figure 7.3.5, as well as both male and female penguins in the dataset.

Figure 7.3.5 Artwork by @allison_horst[4]

Evaluating the next cell will load and display the data. The meaning of the culmen length and width is contained in the illustration on the right of Figure 7.3.5.

7.3. PRINCIPAL COMPONENT ANALYSIS

```
url='https://raw.githubusercontent.com/davidaustinm/'
url+='ula_modules/master/pca_penguins.py'
sage.repl.load.load(url, globals())
df.T
```

This dataset is a bit different from others that we've looked at because the scale of the measurements is significantly different. For instance, the measurements for the body mass are roughly 100 times as large as those for the culmen length. For this reason, we will standardize the data by first demeaning it, as usual, and then rescaling each measurement by the reciprocal of its standard deviation. The result is stored in the 4×333 matrix A.

a. Find the covariance matrix and its eigenvalues.

b. What fraction of the total variance is explained by the first two principal components?

c. Construct the 2×333 matrix B whose columns are the coordinates of the demeaned data points projected onto the first two principal components. The following cell will create the plot.

```
pca_plot(B)
```

d. Examine the components of the first two principal component vectors. How does the body mass of Gentoo penguins compare to that of the other two species?

e. What seems to be generally true about the culmen measurements for a Chinstrap penguin compared to a Adelie?

f. You can plot just the males or females using the following cell.

```
pca_plot(B, sex='female')
```

What seems to be generally true about the body mass measurements for a male Gentoo compared to a female Gentoo?

[4]gvsu.edu/s/21G

7.4 Singular Value Decompositions

The Spectral Theorem has motivated the past few sections. In particular, we applied the fact that symmetric matrices can be orthogonally diagonalized to simplify quadratic forms, which enabled us to use principal component analysis to reduce the dimension of a dataset.

But what can we do with matrices that are not symmetric or even square? For instance, the following matrices are not diagonalizable, much less orthogonally so:

$$\begin{bmatrix} 2 & 1 \\ 0 & 2 \end{bmatrix}, \quad \begin{bmatrix} 1 & 1 & 0 \\ -1 & 0 & 1 \end{bmatrix}.$$

In this section, we will develop a description of matrices called the *singular value decomposition* that is, in many ways, analogous to an orthogonal diagonalization. For example, we have seen that any symmetric matrix can be written in the form QDQ^T where Q is an orthogonal matrix and D is diagonal. A singular value decomposition will have the form $U\Sigma V^T$ where U and V are orthogonal and Σ is diagonal. Most notably, we will see that *every* matrix has a singular value decomposition whether it's symmetric or not.

Preview Activity 7.4.1. Let's review orthogonal diagonalizations and quadratic forms as our understanding of singular value decompositions will rely on them.

a. Suppose that A is any matrix. Explain why the matrix $G = A^T A$ is symmetric.

b. Suppose that $A = \begin{bmatrix} 1 & 2 \\ -2 & -1 \end{bmatrix}$. Find the matrix $G = A^T A$ and write out the quadratic form $q_G\left(\begin{bmatrix} x_1 \\ x_2 \end{bmatrix}\right)$ as a function of x_1 and x_2.

c. What is the maximum value of $q_G(\mathbf{x})$ and in which direction does it occur?

d. What is the minimum value of $q_G(\mathbf{x})$ and in which direction does it occur?

e. What is the geometric relationship between the directions in which the maximum and minimum values occur?

7.4.1 Finding singular value decompositions

We will begin by explaining what a singular value decomposition is and how we can find one for a given matrix A.

Recall how the orthogonal diagonalization of a symmetric matrix is formed: if A is symmetric, we write $A = QDQ^T$ where the diagonal entries of D are the eigenvalues of A and the columns of Q are the associated eigenvectors. Moreover, the eigenvalues are related to the maximum and minimum values of the associated quadratic form $q_A(\mathbf{u})$ among all unit vectors.

7.4. SINGULAR VALUE DECOMPOSITIONS

A general matrix, particularly a matrix that is not square, may not have eigenvalues and eigenvectors, but we can discover analogous features, called *singular values* and *singular vectors*, by studying a function somewhat similar to a quadratic form. More specifically, any matrix A defines a function

$$l_A(\mathbf{x}) = |A\mathbf{x}|,$$

which measures the length of $A\mathbf{x}$. For example, the diagonal matrix $D = \begin{bmatrix} 3 & 0 \\ 0 & -2 \end{bmatrix}$ gives the function $l_D(\mathbf{x}) = \sqrt{9x_1^2 + 4x_2^2}$. The presence of the square root means that this function is not a quadratic form. We can, however, define the singular values and vectors by looking for the maximum and minimum of this function $l_A(\mathbf{u})$ among all unit vectors $\mathbf{u}$.

While $l_A(\mathbf{x})$ is not itself a quadratic form, it becomes one if we square it:

$$(l_A(\mathbf{x}))^2 = |A\mathbf{x}|^2 = (A\mathbf{x}) \cdot (A\mathbf{x}) = \mathbf{x} \cdot (A^T A \mathbf{x}) = q_{A^T A}(\mathbf{x}).$$

We call $G = A^T A$, the *Gram matrix* associated to A and note that

$$l_A(\mathbf{x}) = \sqrt{q_G(\mathbf{x})}.$$

This is important in the next activity, which introduces singular values and singular vectors.

Activity 7.4.2. The following interactive figure will help us explore singular values and vectors geometrically before we begin a more algebraic approach.

There is an interactive diagram, available at gvsu.edu/s/0YE, that accompanies this activity.

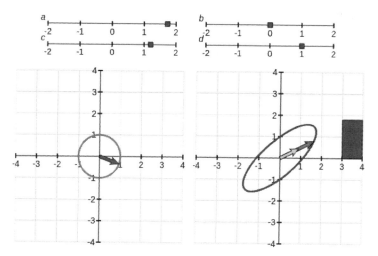

Figure 7.4.1 Singular values, right singular vectors and left singular vectors

Select the matrix $A = \begin{bmatrix} 1 & 2 \\ -2 & -1 \end{bmatrix}$. As we vary the vector $\mathbf{x}$, we see the vector $A\mathbf{x}$ on the right in gray while the height of the blue bar to the right tells us $l_A(\mathbf{x}) = |A\mathbf{x}|$.

a. The first *singular value* σ_1 is the maximum value of $l_A(\mathbf{x})$ and an associated *right singular vector* $\mathbf{v}_1$ is a unit vector describing a direction in which this maximum occurs.

Use the diagram to find the first singular value σ_1 and an associated right singular vector $\mathbf{v}_1$.

b. The second singular value σ_2 is the minimum value of $l_A(\mathbf{x})$ and an associated right singular vector $\mathbf{v}_2$ is a unit vector describing a direction in which this minimum occurs.

Use the diagram to find the second singular value σ_2 and an associated right singular vector $\mathbf{v}_2$.

c. Here's how we can find the right singular values and vectors without using the diagram. Remember that $l_A(\mathbf{x}) = \sqrt{q_G(\mathbf{x})}$ where $G = A^T A$ is the Gram matrix associated to A. Since G is symmetric, it is orthogonally diagonalizable. Find G and an orthogonal diagonalization of it.

What is the maximum value of the quadratic form $q_G(\mathbf{x})$ among all unit vectors and in which direction does it occur? What is the minimum value of $q_G(\mathbf{x})$ and in which direction does it occur?

d. Because $l_A(\mathbf{x}) = \sqrt{q_G(\mathbf{x})}$, the first singular value σ_1 will be the square root of the maximum value of $q_G(\mathbf{x})$ and σ_2 the square root of the minimum. Verify that the singular values that you found from the diagram are the square roots of the maximum and minimum values of $q_G(\mathbf{x})$.

e. Verify that the right singular vectors $\mathbf{v}_1$ and $\mathbf{v}_2$ that you found from the diagram are the directions in which the maximum and minimum values occur.

f. Finally, we introduce the *left singular vectors* $\mathbf{u}_1$ and $\mathbf{u}_2$ by requiring that $A\mathbf{v}_1 = \sigma_1 \mathbf{u}_1$ and $A\mathbf{v}_2 = \sigma_2 \mathbf{u}_2$. Find the two left singular vectors.

g. Form the matrices

$$U = \begin{bmatrix} \mathbf{u}_1 & \mathbf{u}_2 \end{bmatrix}, \quad \Sigma = \begin{bmatrix} \sigma_1 & 0 \\ 0 & \sigma_2 \end{bmatrix}, \quad V = \begin{bmatrix} \mathbf{v}_1 & \mathbf{v}_2 \end{bmatrix}$$

and explain why $AV = U\Sigma$.

h. Finally, explain why $A = U\Sigma V^T$ and verify that this relationship holds for this specific example.

As this activity shows, the singular values of A are the maximum and minimum values of $l_A(\mathbf{x}) = |A\mathbf{x}|$ among all unit vectors and the right singular vectors $\mathbf{v}_1$ and $\mathbf{v}_2$ are the directions in which they occur. The key to finding the singular values and vectors is to utilize the Gram matrix G and its associated quadratic form $q_G(\mathbf{x})$. We will illustrate with some more examples.

7.4. SINGULAR VALUE DECOMPOSITIONS

Example 7.4.2 We will find a singular value decomposition of the matrix $A = \begin{bmatrix} 1 & 2 \\ -1 & 2 \end{bmatrix}$. Notice that this matrix is not symmetric so it cannot be orthogonally diagonalized.

We begin by constructing the Gram matrix $G = A^T A = \begin{bmatrix} 2 & 0 \\ 0 & 8 \end{bmatrix}$. Since G is symmetric, it can be orthogonally diagonalized with

$$D = \begin{bmatrix} 8 & 0 \\ 0 & 2 \end{bmatrix}, \quad Q = \begin{bmatrix} 0 & 1 \\ 1 & 0 \end{bmatrix}.$$

We now know that the maximum value of the quadratic form $q_G(\mathbf{x})$ is 8, which occurs in the direction $\begin{bmatrix} 0 \\ 1 \end{bmatrix}$. Since $l_A(\mathbf{x}) = \sqrt{q_G(\mathbf{x})}$, this tells us that the maximum value of $l_A(\mathbf{x})$, the first singular value, is $\sigma_1 = \sqrt{8}$ and that this occurs in the direction of the first right singular vector $\mathbf{v}_1 = \begin{bmatrix} 0 \\ 1 \end{bmatrix}$.

In the same way, we also know that the second singular value $\sigma_2 = \sqrt{2}$ with associated right singular vector $\mathbf{v}_2 = \begin{bmatrix} 1 \\ 0 \end{bmatrix}$.

The first left singular vector $\mathbf{u}_1$ is defined by $A\mathbf{v}_1 = \begin{bmatrix} 2 \\ 2 \end{bmatrix} = \sigma_1 \mathbf{u}_1$. Because $\sigma_1 = \sqrt{8}$, we have $\mathbf{u}_1 = \begin{bmatrix} 1/\sqrt{2} \\ 1/\sqrt{2} \end{bmatrix}$. Notice that $\mathbf{u}_1$ is a unit vector because $\sigma_1 = |A\mathbf{v}_1|$.

In the same way, the second left singular vector is defined by $A\mathbf{v}_2 = \begin{bmatrix} 1 \\ -1 \end{bmatrix} = \sigma_2 \mathbf{u}_2$, which gives us $\mathbf{u}_2 = \begin{bmatrix} 1/\sqrt{2} \\ -1/\sqrt{2} \end{bmatrix}$.

We then construct

$$U = \begin{bmatrix} \mathbf{u}_1 & \mathbf{u}_2 \end{bmatrix} = \begin{bmatrix} 1/\sqrt{2} & 1/\sqrt{2} \\ 1/\sqrt{2} & -1/\sqrt{2} \end{bmatrix}$$

$$\Sigma = \begin{bmatrix} \sigma_1 & 0 \\ 0 & \sigma_2 \end{bmatrix} = \begin{bmatrix} \sqrt{8} & 0 \\ 0 & \sqrt{2} \end{bmatrix}$$

$$V = \begin{bmatrix} \mathbf{v}_1 & \mathbf{v}_2 \end{bmatrix} = \begin{bmatrix} 0 & 1 \\ 1 & 0 \end{bmatrix}$$

We now have $AV = U\Sigma$ because

$$AV = \begin{bmatrix} A\mathbf{v}_1 & A\mathbf{v}_2 \end{bmatrix} = \begin{bmatrix} \sigma_1 \mathbf{u}_1 & \sigma_2 \mathbf{u}_2 \end{bmatrix} = \Sigma U.$$

Because the right singular vectors, the columns of V, are eigenvectors of the symmetric matrix G, they form an orthonormal basis, which means that V is orthogonal. Therefore, we have $(AV)V^T = A = U\Sigma V^T$. This gives the singular value decomposition

$$A = \begin{bmatrix} 1 & 2 \\ -1 & 2 \end{bmatrix} = \begin{bmatrix} 1/\sqrt{2} & 1/\sqrt{2} \\ 1/\sqrt{2} & -1/\sqrt{2} \end{bmatrix} \begin{bmatrix} \sqrt{8} & 0 \\ 0 & \sqrt{2} \end{bmatrix} \begin{bmatrix} 0 & 1 \\ 1 & 0 \end{bmatrix}^T = U\Sigma V^T.$$

To summarize, we find a singular value decomposition of a matrix A in the following way:

- Construct the Gram matrix $G = A^T A$ and find an orthogonal diagonalization to obtain eigenvalues λ_i and an orthonormal basis of eigenvectors.

- The singular values of A are the squares roots of eigenvalues λ_i of G; that is, $\sigma_i = \sqrt{\lambda_i}$. By convention, the singular values are listed in decreasing order: $\sigma_1 \geq \sigma_2 \geq \ldots$. The right singular vectors $\mathbf{v}_i$ are the associated eigenvectors of G.

- The left singular vectors $\mathbf{u}_i$ are found by $A\mathbf{v}_i = \sigma_i \mathbf{u}_i$. Because $\sigma_i = |A\mathbf{v}_i|$, we know that $\mathbf{u}_i$ will be a unit vector.

   In fact, the left singular vectors will also form an orthonormal basis. To see this, suppose that the associated singular values are nonzero. We then have:

$$\sigma_i \sigma_j (\mathbf{u}_i \cdot \mathbf{u}_j) = (\sigma_i \mathbf{u}_i) \cdot (\sigma_j \mathbf{u}_j) = (A\mathbf{v}_i) \cdot (A\mathbf{v}_j)$$
$$= \mathbf{v}_i \cdot (A^T A \mathbf{v}_j)$$
$$= \mathbf{v}_i \cdot (G\mathbf{v}_j) = \lambda_j \mathbf{v}_i \cdot \mathbf{v}_j = 0$$

since the right singular vectors are orthogonal.

Example 7.4.3 Let's find a singular value decomposition for the symmetric matrix $A = \begin{bmatrix} 1 & 2 \\ 2 & 1 \end{bmatrix}$. The associated Gram matrix is

$$G = A^T A = \begin{bmatrix} 5 & 4 \\ 4 & 5 \end{bmatrix},$$

which has an orthogonal diagonalization with

$$D = \begin{bmatrix} 9 & 0 \\ 0 & 1 \end{bmatrix}, \qquad Q = \begin{bmatrix} 1/\sqrt{2} & 1/\sqrt{2} \\ 1/\sqrt{2} & -1/\sqrt{2} \end{bmatrix}.$$

This gives singular values and vectors

$$\sigma_1 = 3, \qquad \mathbf{v}_1 = \begin{bmatrix} 1/\sqrt{2} \\ 1/\sqrt{2} \end{bmatrix}, \qquad \mathbf{u}_1 = \begin{bmatrix} 1/\sqrt{2} \\ 1/\sqrt{2} \end{bmatrix}$$

$$\sigma_2 = 1, \qquad \mathbf{v}_2 = \begin{bmatrix} 1/\sqrt{2} \\ -1/\sqrt{2} \end{bmatrix}, \qquad \mathbf{u}_2 = \begin{bmatrix} -1/\sqrt{2} \\ 1/\sqrt{2} \end{bmatrix}$$

and the singular value decomposition $A = U\Sigma V^T$ where

$$U = \begin{bmatrix} 1/\sqrt{2} & -1/\sqrt{2} \\ 1/\sqrt{2} & 1/\sqrt{2} \end{bmatrix}, \qquad \Sigma = \begin{bmatrix} 3 & 0 \\ 0 & 1 \end{bmatrix}, \qquad V = \begin{bmatrix} 1/\sqrt{2} & 1/\sqrt{2} \\ 1/\sqrt{2} & -1/\sqrt{2} \end{bmatrix}.$$

This example is special because A is symmetric. With a little thought, it's possible to relate this singular value decomposition to an orthogonal diagonalization of A using the fact that $G = A^T A = A^2$.

Activity 7.4.3. In this activity, we will construct the singular value decomposition of $A = \begin{bmatrix} 1 & 0 & -1 \\ 1 & 1 & 1 \end{bmatrix}$. Notice that this matrix is not square so there are no eigenvalues and eigenvectors associated to it.

a. Construct the Gram matrix $G = A^T A$ and find an orthogonal diagonalization of it.

b. Identify the singular values of A and the right singular vectors $\mathbf{v}_1, \mathbf{v}_2$, and $\mathbf{v}_3$. What is the dimension of these vectors? How many nonzero singular values are there?

c. Find the left singular vectors $\mathbf{u}_1$ and $\mathbf{u}_2$ using the fact that $A\mathbf{v}_i = \sigma_i \mathbf{u}_i$. What is the dimension of these vectors? What happens if you try to find a third left singular vector $\mathbf{u}_3$ in this way?

d. As before, form the orthogonal matrices U and V from the left and right singular vectors. What are the shapes of U and V? How do these shapes relate to the number of rows and columns of A?

e. Now form Σ so that it has the same shape as A:

$$\Sigma = \begin{bmatrix} \sigma_1 & 0 & 0 \\ 0 & \sigma_2 & 0 \end{bmatrix}$$

and verify that $A = U\Sigma V^T$.

f. How can you use this singular value decomposition of $A = U\Sigma V^T$ to easily find a singular value decomposition of $A^T = \begin{bmatrix} 1 & 1 \\ 0 & 1 \\ -1 & 1 \end{bmatrix}$?

Example 7.4.4 We will find a singular value decomposition of the matrix $A = \begin{bmatrix} 2 & -2 & 1 \\ -4 & -8 & -8 \end{bmatrix}$.

Finding an orthogonal diagonalization of $G = A^T A$ gives

$$D = \begin{bmatrix} 144 & 0 & 0 \\ 0 & 9 & 0 \\ 0 & 0 & 0 \end{bmatrix}, \quad Q = \begin{bmatrix} 1/3 & 2/3 & 2/3 \\ 2/3 & -2/3 & 1/3 \\ 2/3 & 1/3 & -2/3 \end{bmatrix},$$

which gives singular values $\sigma_1 = \sqrt{144} = 12$, $\sigma_2 = \sqrt{9} = 3$, and $\sigma_3 = 0$. The right singular vectors $\mathbf{v}_i$ appear as the columns of Q so that $V = Q$.

We now find

$$A\mathbf{v}_1 = \begin{bmatrix} 0 \\ -12 \end{bmatrix} = 12\mathbf{u}_1, \qquad \mathbf{u}_1 = \begin{bmatrix} 0 \\ -1 \end{bmatrix}$$

$$A\mathbf{v}_2 = \begin{bmatrix} 3 \\ 0 \end{bmatrix} = 3\mathbf{u}_1, \qquad \mathbf{u}_1 = \begin{bmatrix} 1 \\ 0 \end{bmatrix}$$

$$A\mathbf{v}_3 = \begin{bmatrix} 0 \\ 0 \end{bmatrix}$$

Notice that it's not possible to find a third left singular vector since $A\mathbf{v}_3 = \mathbf{0}$. We therefore form the matrices

$$U = \begin{bmatrix} 0 & 1 \\ -1 & 0 \end{bmatrix}, \qquad \Sigma = \begin{bmatrix} 12 & 0 & 0 \\ 0 & 3 & 0 \end{bmatrix}, \qquad V = \begin{bmatrix} 1/3 & 2/3 & 2/3 \\ 2/3 & -2/3 & 1/3 \\ 2/3 & 1/3 & -2/3 \end{bmatrix},$$

which gives the singular value decomposition $A = U\Sigma V^T$.

Notice that U is a 2×2 orthogonal matrix because A has two rows, and V is a 3×3 orthogonal matrix because A has three columns.

As we'll see in the next section, some additional work may be needed to construct the left singular vectors $\mathbf{u}_j$ if more of the singular values are zero, but we won't worry about that now. For the time being, let's record our work in the following theorem.

Theorem 7.4.5 The singular value decomposition. *An $m \times n$ matrix A may be written as $A = U\Sigma V^T$ where U is an orthogonal $m \times m$ matrix, V is an orthogonal $n \times n$ matrix, and Σ is an $m \times n$ matrix whose entries are zero except for the singular values of A which appear in decreasing order on the diagonal.*

Notice that a singular value decomposition of A gives us a singular value decomposition of A^T. More specifically, if $A = U\Sigma V^T$, then

$$A^T = (U\Sigma V^T)^T = V\Sigma^T U^T.$$

Proposition 7.4.6 *If $A = U\Sigma V^T$, then $A^T = V\Sigma^T U^T$. In other words, A and A^T share the same singular values, and the left singular vectors of A are the right singular vectors of A^T and vice-versa.*

As we said earlier, a singular value decomposition should be thought of a generalization of an orthogonal diagonalization. For instance, the Spectral Theorem tells us that a symmetric matrix can be written as QDQ^T. Many matrices, however, are not symmetric and so they are not orthogonally diagonalizable. However, every matrix has a singular value decomposition $U\Sigma V^T$. The price of this generalization is that we usually have two sets of singular vectors that form the orthogonal matrices U and V whereas a symmetric matrix has a single set of eignevectors that form the orthogonal matrix Q.

7.4.2 The structure of singular value decompositions

Now that we have an understanding of what a singular value decomposition is and how to construct it, let's explore the ways in which a singular value decomposition reveals the

7.4. SINGULAR VALUE DECOMPOSITIONS

underlying structure of the matrix. As we'll see, the matrices U and V in a singular value decomposition provide convenient bases for some important subspaces, such as the column and null spaces of the matrix. This observation will provide the key to some of our uses of these decompositions in the next section.

Activity 7.4.4. Let's suppose that a matrix A has a singular value decomposition $A = U\Sigma V^T$ where

$$U = \begin{bmatrix} \mathbf{u}_1 & \mathbf{u}_2 & \mathbf{u}_3 & \mathbf{u}_4 \end{bmatrix}, \quad \Sigma = \begin{bmatrix} 20 & 0 & 0 \\ 0 & 5 & 0 \\ 0 & 0 & 0 \\ 0 & 0 & 0 \end{bmatrix}, \quad V = \begin{bmatrix} \mathbf{v}_1 & \mathbf{v}_2 & \mathbf{v}_3 \end{bmatrix}.$$

a. What is the shape of A; that is, how many rows and columns does A have?

b. Suppose we write a three-dimensional vector $\mathbf{x}$ as a linear combination of right singular vectors:

$$\mathbf{x} = c_1 \mathbf{v}_1 + c_2 \mathbf{v}_2 + c_3 \mathbf{v}_3.$$

We would like to find an expression for $A\mathbf{x}$.

To begin, $V^T \mathbf{x} = \begin{bmatrix} \mathbf{v}_1 \cdot \mathbf{x} \\ \mathbf{v}_2 \cdot \mathbf{x} \\ \mathbf{v}_3 \cdot \mathbf{x} \end{bmatrix} = \begin{bmatrix} c_1 \\ c_2 \\ c_3 \end{bmatrix}.$

Now $\Sigma V^T \mathbf{x} = \begin{bmatrix} 20 & 0 & 0 \\ 0 & 5 & 0 \\ 0 & 0 & 0 \\ 0 & 0 & 0 \end{bmatrix} \begin{bmatrix} c_1 \\ c_2 \\ c_3 \end{bmatrix} = \begin{bmatrix} 20c_1 \\ 5c_2 \\ 0 \\ 0 \end{bmatrix}.$

And finally, $A\mathbf{x} = U\Sigma V^T \mathbf{x} = \begin{bmatrix} \mathbf{u}_1 & \mathbf{u}_2 & \mathbf{u}_3 & \mathbf{u}_4 \end{bmatrix} \begin{bmatrix} 20c_1 \\ 5c_2 \\ 0 \\ 0 \end{bmatrix} = 20c_1 \mathbf{u}_1 + 5c_2 \mathbf{u}_2.$

To summarize, we have $A\mathbf{x} = 20c_1 \mathbf{u}_1 + 5c_2 \mathbf{u}_2$.

What condition on c_1, c_2, and c_3 must be satisfied if $\mathbf{x}$ is a solution to the equation $A\mathbf{x} = 40\mathbf{u}_1 + 20\mathbf{u}_2$? Is there a unique solution or infinitely many?

c. Remembering that $\mathbf{u}_1$ and $\mathbf{u}_2$ are linearly independent, what condition on c_1, c_2, and c_3 must be satisfied if $A\mathbf{x} = \mathbf{0}$?

d. How do the right singular vectors $\mathbf{v}_i$ provide a basis for $\text{Nul}(A)$, the subspace of solutions to the equation $A\mathbf{x} = \mathbf{0}$?

e. Remember that $\mathbf{b}$ is in $\text{Col}(A)$ if the equation $A\mathbf{x} = \mathbf{b}$ is consistent, which means that

$$A\mathbf{x} = 20c_1 \mathbf{u}_1 + 5c_2 \mathbf{u}_2 = \mathbf{b}$$

for some coefficients c_1 and c_2. How do the left singular vectors $\mathbf{u}_i$ provide an orthonormal basis for $\text{Col}(A)$?

f. Remember that rank(A) is the dimension of the column space. What is rank(A) and how do the number of nonzero singular values determine rank(A)?

This activity shows how a singular value decomposition of a matrix encodes important information about its null and column spaces. More specifically, the left and right singular vectors provide orthonormal bases for Nul(A) and Col(A). This is one of the reasons that singular value decompositions are so useful.

Example 7.4.7 Suppose we have a singular value decomposition $A = U\Sigma V^T$ where $\Sigma = \begin{bmatrix} \sigma_1 & 0 & 0 & 0 & 0 \\ 0 & \sigma_2 & 0 & 0 & 0 \\ 0 & 0 & \sigma_3 & 0 & 0 \\ 0 & 0 & 0 & 0 & 0 \end{bmatrix}$. This means that A has four rows and five columns just as Σ does.

As in the activity, if $x = c_1 v_1 + c_2 v_2 + \ldots + c_5 v_5$, we have

$$A x = \sigma_1 c_1 u_1 + \sigma_2 c_2 u_2 + \sigma_3 c_3 u_3.$$

If b is in Col(A), then b must have the form

$$b = \sigma_1 c_1 u_1 + \sigma_2 c_2 u_2 + \sigma_3 c_3 u_3,$$

which says that b is a linear combination of u_1, u_2, and u_3. These three vectors therefore form a basis for Col(A). In fact, since they are columns in the orthogonal matrix U, they form an orthonormal basis for Col(A).

Remembering that rank(A) = dim Col(A), we see that rank(A) = 3, which results from the three nonzero singular values. In general, the rank r of a matrix A equals the number of nonzero singular values, and $u_1, u_2, \ldots, u_r$ form an orthonormal basis for Col(A).

Moreover, if $x = c_1 v_1 + c_2 v_2 + \ldots + c_5 v_5$ satisfies $Ax = 0$, then

$$Ax = \sigma_1 c_1 u_1 + \sigma_2 c_2 u_2 + \sigma_3 c_3 u_3 = 0,$$

which implies that $c_1 = 0$, $c_2 = 0$, and $c_3 = 0$. Therefore, $x = c_4 v_4 + c_5 v_5$ so v_4 and v_5 form an orthonormal basis for Nul(A).

More generally, if A is an $m \times n$ matrix and if rank(A) = r, the last $n - r$ right singular vectors form an orthonormal basis for Nul(A).

Generally speaking, if the rank of an $m \times n$ matrix A is r, then there are r nonzero singular values and Σ has the form

$$\begin{bmatrix} \sigma_1 & \ldots & 0 & \ldots & 0 \\ 0 & \ldots & 0 & \ldots & 0 \\ 0 & \ldots & \sigma_r & \ldots & 0 \\ 0 & \ldots & 0 & \ldots & 0 \\ \vdots & \vdots & \vdots & \ddots & \vdots \\ 0 & \ldots & 0 & \ldots & 0 \end{bmatrix},$$

7.4. SINGULAR VALUE DECOMPOSITIONS

The first r columns of U form an orthonormal basis for Col(A):

$$U = \begin{bmatrix} \underbrace{\mathbf{u}_1 \; \ldots \; \mathbf{u}_r}_{\text{Col}(A)} & \mathbf{u}_{r+1} \; \ldots \; \mathbf{u}_m \end{bmatrix}$$

and the last $n - r$ columns of V form an orthonormal basis for Nul(A):

$$V = \begin{bmatrix} \mathbf{v}_1 \; \ldots \; \mathbf{v}_r & \underbrace{\mathbf{v}_{r+1} \; \ldots \; \mathbf{v}_n}_{\text{Nul}(A)} \end{bmatrix}$$

Remember that Proposition 7.4.6 says that A and its transpose A^T share the same singular values. Since the rank of a matrix equals its number of nonzero singular values, this means that rank(A) = rank(A^T), a fact that we cited back in Section 6.2.

Proposition 7.4.8 *For any matrix A,*

$$\text{rank}(A) = \text{rank}(A^T).$$

If we have a singular value decomposition of an $m \times n$ matrix $A = U\Sigma V^T$, Proposition 7.4.6 also tells us that the left singular vectors of A are the right singular vectors of A^T. Therefore, U is the $m \times m$ matrix whose columns are the right singular vectors of A^T. This means that the last $m - r$ vectors form an orthonormal basis for Nul(A^T). Therefore, the columns of U provide orthonormal bases for Col(A) and Nul(A^T):

$$U = \begin{bmatrix} \underbrace{\mathbf{u}_1 \; \ldots \; \mathbf{u}_r}_{\text{Col}(A)} & \underbrace{\mathbf{u}_{r+1} \; \ldots \; \mathbf{u}_m}_{\text{Nul}(A^T)} \end{bmatrix}.$$

This reflects the familiar fact that Nul(A^T) is the orthogonal complement of Col(A).

In the same way, V is the $n \times n$ matrix whose columns are the left singular vectors of A^T, which means that the first r vectors form an orthonormal basis for Col(A^T). Because the columns of A^T are the rows of A, this subspace is sometimes called the *row space* of A and denoted Row(A). While we have yet to have an occasion to use Row(A), there are times when it is important to have an orthonormal basis for it, and a singular value decomposition provides just that. To summarize, the columns of V provide orthonormal bases for Col(A^T) and Nul(A):

$$V = \begin{bmatrix} \underbrace{\mathbf{v}_1 \; \ldots \; \mathbf{v}_r}_{\text{Col}(A^T)} & \underbrace{\mathbf{v}_{r+1} \; \ldots \; \mathbf{v}_m}_{\text{Nul}(A)} \end{bmatrix}$$

Considered altogether, the subspaces Col(A), Nul(A), Col(A^T), and Nul(A^T) are called the *four fundamental subspaces* associated to A. In addition to telling us the rank of a matrix, a singular value decomposition gives us orthonormal bases for all four fundamental subspaces.

Theorem 7.4.9 *Suppose A is an $m \times n$ matrix having a singular value decomposition $A = U\Sigma V^T$. Then*

- *$r = \text{rank}(A)$ is the number of nonzero singular values.*
- *The columns $\mathbf{u}_1, \mathbf{u}_2, \ldots, \mathbf{u}_r$ form an orthonormal basis for $\text{Col}(A)$.*
- *The columns $\mathbf{u}_{r+1}, \ldots, \mathbf{u}_m$ form an orthonormal basis for $\text{Nul}(A^T)$.*
- *The columns $\mathbf{v}_1, \mathbf{v}_2, \ldots, \mathbf{v}_r$ form an orthonormal basis for $\text{Col}(A^T)$.*
- *The columns $\mathbf{v}_{r+1}, \ldots, \mathbf{v}_n$ form an orthonormal basis for $\text{Nul}(A)$.*

When we previously outlined a procedure for finding a singular decomposition of an $m \times n$ matrix A, we found the left singular vectors $\mathbf{u}_j$ using the expression $A\mathbf{v}_j = \sigma_j \mathbf{u}_j$. This produces left singular vectors $\mathbf{u}_1, \mathbf{u}_2, \ldots, \mathbf{u}_r$, where $r = \text{rank}(A)$. If $r < m$, however, we still need to find the left singular vectors $\mathbf{u}_{r+1}, \ldots, \mathbf{u}_m$. Theorem 7.4.9 tells us how to do that: because those vectors form an orthonormal basis for $\text{Nul}(A^T)$, we can find them by solving $A^T\mathbf{x} = \mathbf{0}$ to obtain a basis for $\text{Nul}(A^T)$ and applying the Gram-Schmidt algorithm.

We won't worry about this issue too much, however, as we will frequently use software to find singular value decompositions for us.

7.4.3 Reduced singular value decompositions

As we'll see in the next section, there are times when it is helpful to express a singular value decomposition in a slightly different form.

Activity 7.4.5. Suppose we have a singular value decomposition $A = U\Sigma V^T$ where

$$U = \begin{bmatrix} \mathbf{u}_1 & \mathbf{u}_2 & \mathbf{u}_3 & \mathbf{u}_4 \end{bmatrix}, \quad \Sigma = \begin{bmatrix} 18 & 0 & 0 \\ 0 & 4 & 0 \\ 0 & 0 & 0 \\ 0 & 0 & 0 \end{bmatrix}, \quad V = \begin{bmatrix} \mathbf{v}_1 & \mathbf{v}_2 & \mathbf{v}_3 \end{bmatrix}.$$

a. What is the shape of A? What is $\text{rank}(A)$?

b. Identify bases for $\text{Col}(A)$ and $\text{Col}(A^T)$.

c. Explain why

$$U\Sigma = \begin{bmatrix} \mathbf{u}_1 & \mathbf{u}_2 \end{bmatrix} \begin{bmatrix} 18 & 0 & 0 \\ 0 & 4 & 0 \end{bmatrix}.$$

d. Explain why

$$\begin{bmatrix} 18 & 0 & 0 \\ 0 & 4 & 0 \end{bmatrix} V^T = \begin{bmatrix} 18 & 0 \\ 0 & 4 \end{bmatrix} \begin{bmatrix} \mathbf{v}_1 & \mathbf{v}_2 \end{bmatrix}^T.$$

e. If $A = U\Sigma V^T$, explain why $A = U_r \Sigma_r V_r^T$ where the columns of U_r are an orthonormal basis for $\text{Col}(A)$, Σ_r is a square, diagonal, invertible matrix, and the columns of V_r form an orthonormal basis for $\text{Col}(A^T)$.

7.4. SINGULAR VALUE DECOMPOSITIONS

We call this a *reduced singular value decomposition*.

Proposition 7.4.10 Reduced singular value decomposition. *If A is an $m \times n$ matrix having rank r, then $A = U_r \Sigma_r V_r^T$ where*

- U_r is an $m \times r$ matrix whose columns form an orthonormal basis for $\text{Col}(A)$,

- $\Sigma_r = \begin{bmatrix} \sigma_1 & 0 & \cdots & 0 \\ 0 & \sigma_2 & \cdots & 0 \\ \vdots & \vdots & \ddots & \vdots \\ 0 & 0 & 0 & \sigma_r \end{bmatrix}$ is an $r \times r$ diagonal, invertible matrix, and

- V_r is an $n \times r$ matrix whose columns form an orthonormal basis for $\text{Col}(A^T)$.

Example 7.4.11 In Example 7.4.4, we found the singular value decomposition

$$A = \begin{bmatrix} 2 & -2 & 1 \\ -4 & -8 & -8 \end{bmatrix} = \begin{bmatrix} 0 & 1 \\ -1 & 0 \end{bmatrix} \begin{bmatrix} 12 & 0 & 0 \\ 0 & 3 & 0 \end{bmatrix} \begin{bmatrix} 1/3 & 2/3 & 2/3 \\ 2/3 & -2/3 & 1/3 \\ 2/3 & 1/3 & -2/3 \end{bmatrix}^T.$$

Since there are two nonzero singular values, $\text{rank}(A) = 2$ so that the reduced singular value decomposition is

$$A = \begin{bmatrix} 2 & -2 & 1 \\ -4 & -8 & -8 \end{bmatrix} = \begin{bmatrix} 0 & 1 \\ -1 & 0 \end{bmatrix} \begin{bmatrix} 12 & 0 \\ 0 & 3 \end{bmatrix} \begin{bmatrix} 1/3 & 2/3 \\ 2/3 & -2/3 \\ 2/3 & 1/3 \end{bmatrix}^T.$$

7.4.4 Summary

This section has explored singular value decompositions, how to find them, and how they organize important information about a matrix.

- A singular value decomposition of a matrix A is a factorization where $A = U\Sigma V^T$. The matrix Σ has the same shape as A, and its only nonzero entries are the singular values of A, which appear in decreasing order on the diagonal. The matrices U and V are orthogonal and contain the left and right singular vectors, respectively, as their columns.

- To find a singular value decomposition of a matrix, we construct the Gram matrix $G = A^T A$, which is symmetric. The singular values of A are the square roots of the eigenvalues of G, and the right singular vectors $\mathbf{v}_j$ are the associated eigenvectors of G. The left singular vectors $\mathbf{u}_j$ are determined from the relationship $A\mathbf{v}_j = \sigma_j \mathbf{u}_j$.

- A singular value decomposition reveals fundamental information about a matrix. For instance, the number of nonzero singular values is the rank r of the matrix. The first r left singular vectors form an orthonormal basis for $\text{Col}(A)$ with the remaining left singular vectors forming an orthonormal basis of $\text{Nul}(A^T)$. The first r right singular vectors form an orthonormal basis for $\text{Col}(A^T)$ while the remaining right singular vectors form an orthonormal basis of $\text{Nul}(A)$.

- If A is a rank r matrix, we can write a reduced singular value decomposition as $A = U_r \Sigma_r V_r^T$ where the columns of U_r form an orthonormal basis for $\text{Col}(A)$, the columns of V_r form an orthonormal basis for $\text{Col}(A^T)$, and Σ_r is an $r \times r$ diagonal, invertible matrix.

7.4.5 Exercises

1. Consider the matrix $A = \begin{bmatrix} 1 & 2 & 1 \\ 0 & -1 & 2 \end{bmatrix}$.

   a. Find the Gram matrix $G = A^T A$ and use it to find the singular values and right singular vectors of A.

   b. Find the left singular vectors.

   c. Form the matrices U, Σ, and V and verify that $A = U \Sigma V^T$.

   d. What is $\text{rank}(A)$ and what does this say about $\text{Col}(A)$?

   e. Determine an orthonormal basis for $\text{Nul}(A)$.

2. Find singular value decompositions for the following matrices:

   a. $\begin{bmatrix} 0 & 0 \\ 0 & -8 \end{bmatrix}$.

   b. $\begin{bmatrix} 2 & 3 \\ 0 & 2 \end{bmatrix}$.

   c. $\begin{bmatrix} 4 & 0 & 0 \\ 0 & 0 & 2 \end{bmatrix}$.

   d. $\begin{bmatrix} 4 & 0 \\ 0 & 0 \\ 0 & 2 \end{bmatrix}$.

3. Consider the matrix $A = \begin{bmatrix} 2 & 1 \\ 1 & 2 \end{bmatrix}$.

   a. Find a singular value decomposition of A and verify that it is also an orthogonal diagonalization of A.

   b. If A is a symmetric, positive semidefinite matrix, explain why a singular value decomposition of A is an orthogonal diagonalization of A.

7.4. SINGULAR VALUE DECOMPOSITIONS

4. Suppose that the matrix A has the singular value decomposition

$$\begin{bmatrix} -0.46 & 0.52 & 0.46 & 0.55 \\ -0.82 & 0.00 & -0.14 & -0.55 \\ -0.04 & 0.44 & -0.85 & 0.28 \\ -0.34 & -0.73 & -0.18 & 0.55 \end{bmatrix} \begin{bmatrix} 6.2 & 0.0 & 0.0 \\ 0.0 & 4.1 & 0.0 \\ 0.0 & 0.0 & 0.0 \\ 0.0 & 0.0 & 0.0 \end{bmatrix} \begin{bmatrix} -0.74 & 0.62 & -0.24 \\ 0.28 & 0.62 & 0.73 \\ -0.61 & -0.48 & 0.64 \end{bmatrix}.$$

a. What are the dimensions of A?

b. What is rank(A)?

c. Find orthonormal bases for Col(A), Nul(A), Col(A^T), and Nul(A^T).

d. Find the orthogonal projection of $\mathbf{b} = \begin{bmatrix} 1 \\ 0 \\ 2 \\ -1 \end{bmatrix}$ onto Col(A).

5. Consider the matrix $A = \begin{bmatrix} 1 & 0 & -1 \\ 2 & 2 & 0 \\ -1 & 1 & 2 \end{bmatrix}$.

a. Construct the Gram matrix G and use it to find the singular values and right singular vectors $\mathbf{v}_1$, $\mathbf{v}_2$, and $\mathbf{v}_3$ of A. What are the matrices Σ and V in a singular value decomposition?

b. What is rank(A)?

c. Find as many left singular vectors $\mathbf{u}_j$ as you can using the relationship $A\mathbf{v}_j = \sigma_j \mathbf{u}_j$.

d. Find an orthonormal basis for Nul(A^T) and use it to construct the matrix U so that $A = U\Sigma V^T$.

e. State an orthonormal basis for Nul(A) and an orthonormal basis for Col(A).

6. Consider the matrix $B = \begin{bmatrix} 1 & 0 \\ 2 & -1 \\ 1 & 2 \end{bmatrix}$ and notice that $B = A^T$ where A is the matrix in Exercise 7.4.5.1.

a. Use your result from Exercise 7.4.5.1 to find a singular value decomposition of $B = U\Sigma V^T$.

b. What is rank(B)? Determine a basis for Col(B) and Col(B)$^\perp$.

c. Suppose that $\mathbf{b} = \begin{bmatrix} -3 \\ 4 \\ 7 \end{bmatrix}$. Use the bases you found in the previous part of this exericse to write $\mathbf{b} = \widehat{\mathbf{b}} + \mathbf{b}^\perp$, where $\widehat{\mathbf{b}}$ is in Col(B) and $\mathbf{b}^\perp$ is in Col(B)$^\perp$.

   d. Find the least-squares approximate solution to the equation $B\mathbf{x} = \mathbf{b}$.

7. Suppose that A is a square $m \times m$ matrix with singular value decomposition $A = U\Sigma V^T$.

   a. If A is invertible, find a singular value decomposition of A^{-1}.

   b. What condition on the singular values must hold for A to be invertible?

   c. How are the singular values of A and the singular values of A^{-1} related to one another?

   d. How are the right and left singular vectors of A related to the right and left singular vectors of A^{-1}?

8.
   a. If Q is an orthogonal matrix, remember that $Q^T Q = I$. Explain why $\det Q = \pm 1$.

   b. If $A = U\Sigma V^T$ is a singular value decomposition of a square matrix A, explain why $|\det A|$ is the product of the singular values of A.

   c. What does this say about the singular values of A if A is invertible?

9. If A is a matrix and $G = A^T A$ its Gram matrix, remember that

   $$\mathbf{x} \cdot (G\mathbf{x}) = \mathbf{x} \cdot (A^T A\mathbf{x}) = (A\mathbf{x}) \cdot (A\mathbf{x}) = |A\mathbf{x}|^2.$$

   a. For a general matrix A, explain why the eigenvalues of G are nonnegative.

   b. Given a symmetric matrix A having an eigenvalue λ, explain why λ^2 is an eigenvalue of G.

   c. If A is symmetric, explain why the singular values of A equal the absolute value of its eigenvalues: $\sigma_j = |\lambda_j|$.

10. Determine whether the following statements are true or false and explain your reasoning.

    a. If $A = U\Sigma V^T$ is a singular value decomposition of A, then $G = V(\Sigma^T \Sigma)V^T$ is an orthogonal diagonalization of its Gram matrix.

    b. If $A = U\Sigma V^T$ is a singular value decomposition of a rank 2 matrix A, then $\mathbf{v}_1$ and $\mathbf{v}_2$ form an orthonormal basis for the column space Col(A).

    c. If A is a symmetric matrix, then its set of singular values is the same as its set of eigenvalues.

    d. If A is a 10×7 matrix and $\sigma_7 = 4$, then the columns of A are linearly independent.

    e. The Gram matrix is always orthogonally diagonalizable.

11. Suppose that $A = U\Sigma V^T$ is a singular value decomposition of the $m \times n$ matrix A. If

7.4. SINGULAR VALUE DECOMPOSITIONS

$\sigma_1, \ldots, \sigma_r$ are the nonzero singular values, the general form of the matrix Σ is

$$\Sigma = \begin{bmatrix} \sigma_1 & \cdots & 0 & \cdots & 0 \\ 0 & \cdots & 0 & \cdots & 0 \\ 0 & \cdots & \sigma_r & \cdots & 0 \\ 0 & \cdots & 0 & \cdots & 0 \\ 0 & \vdots & 0 & \vdots & 0 \\ 0 & \cdots & 0 & \cdots & 0 \end{bmatrix}.$$

a. If you know that the columns of A are linearly independent, what more can you say about the form of Σ?

b. If you know that the columns of A span $\mathbb{R}^m$, what more can you say about the form of Σ?

c. If you know that the columns of A are linearly independent and span $\mathbb{R}^m$, what more can you say about the form of Σ?

7.5 Using Singular Value Decompositions

We've now seen what singular value decompositions are, how to construct them, and how they provide important information about a matrix such as orthonormal bases for the four fundamental subspaces. This puts us in a good position to begin using singular value decompositions to solve a wide variety of problems.

Given the fact that singular value decompositions so immediately convey fundamental data about a matrix, it seems natural that some of our previous work can be reinterpreted in terms of singular value decompositions. Therefore, we'll take some time in this section to revisit some familiar issues, such as least-squares problems and principal component analysis, while also looking at some new applications.

Preview Activity 7.5.1. Suppose that $A = U\Sigma V^T$ where

$$\Sigma = \begin{bmatrix} 13 & 0 & 0 & 0 \\ 0 & 8 & 0 & 0 \\ 0 & 0 & 2 & 0 \\ 0 & 0 & 0 & 0 \\ 0 & 0 & 0 & 0 \end{bmatrix},$$

vectors u_j form the columns of U, and vectors v_j form the columns of V.

a. What are the shapes of the matrices A, U, and V?

b. What is the rank of A?

c. Describe how to find an orthonormal basis for Col(A).

d. Describe how to find an orthonormal basis for Nul(A).

e. If the columns of Q form an orthonormal basis for Col(A), what is Q^TQ?

f. How would you form a matrix that projects vectors orthogonally onto Col(A)?

7.5.1 Least-squares problems

Least-squares problems, which we explored in Section 6.5, arise when we are confronted with an inconsistent linear system $Ax = b$. Since there is no solution to the system, we instead find the vector x minimizing the distance between b and Ax. That is, we find the vector $\widehat{x}$, the least-squares approximate solution, by solving $A\widehat{x} = \widehat{b}$ where $\widehat{b}$ is the orthogonal projection of b onto the column space of A.

If we have a singular value decomposition $A = U\Sigma V^T$, then the number of nonzero singular values r tells us the rank of A, and the first r columns of U form an orthonormal basis for Col(A). This basis may be used to project vectors onto Col(A) and hence to solve least-squares problems.

7.5. USING SINGULAR VALUE DECOMPOSITIONS

Before exploring this connection further, we will introduce Sage as a tool for automating the construction of singular value decompositions. One new feature is that we need to declare our matrix to consist of floating point entries. We do this by including RDF inside the matrix definition, as illustrated in the following cell.

```
A = matrix(RDF, 3, 2, [1,0,-1,1,1,1])
U, Sigma, V = A.SVD()
print(U)
print('---------')
print(Sigma)
print('---------')
print(V)
```

Activity 7.5.2. Consider the equation $A\mathbf{x} = \mathbf{b}$ where

$$\begin{bmatrix} 1 & 0 \\ 1 & 1 \\ 1 & 2 \end{bmatrix} \mathbf{x} = \begin{bmatrix} -1 \\ 3 \\ 6 \end{bmatrix}$$

a. Find a singular value decomposition for A using the Sage cell below. What are singular values of A?

b. What is r, the rank of A? How can we identify an orthonormal basis for Col(A)?

c. Form the reduced singular value decomposition $U_r \Sigma_r V_r^T$ by constructing: the matrix U_r, consisting of the first r columns of U; the matrix V_r, consisting of the first r columns of V; and Σ_r, a square $r \times r$ diagonal matrix. Verify that $A = U_r \Sigma_r V_r^T$.

You may find it convenient to remember that if B is a matrix defined in Sage, then B.matrix_from_columns(list) and B.matrix_from_rows(list) can be used to extract columns or rows from B. For instance, B.matrix_from_rows([0,1,2]) provides a matrix formed from the first three rows of B.

d. How does the reduced singular value decomposition provide a matrix whose columns are an orthonormal basis for Col(A)?

e. Explain why a least-squares approximate solution $\widehat{\mathbf{x}}$ satisfies

$$A\widehat{\mathbf{x}} = U_r U_r^T \mathbf{b}.$$

f. What is the product $V_r^T V_r$ and why does it have this form?

g. Explain why

$$\widehat{\mathbf{x}} = V_r \Sigma_r^{-1} U_r^T \mathbf{b}$$

is the least-squares approximate solution, and use this expression to find $\widehat{\mathbf{x}}$.

This activity demonstrates the power of a singular value decomposition to find a least-squares approximate solution for an equation $A\mathbf{x} = \mathbf{b}$. Because it immediately provides an orthonormal basis for Col(A), something that we've had to construct using the Gram-Schmidt process in the past, we can easily project $\mathbf{b}$ onto Col(A), which results in a simple expression for $\widehat{\mathbf{x}}$.

Proposition 7.5.1 *If $A = U_r \Sigma_r V_r^T$ is a reduced singular value decomposition of A, then a least-squares approximate solution to $A\mathbf{x} = \mathbf{b}$ is given by*

$$\widehat{\mathbf{x}} = V_r \Sigma_r^{-1} U_r^T \mathbf{b}.$$

If the columns of A are linearly independent, then the equation $A\widehat{\mathbf{x}} = \mathbf{b}$ has only one solution so there is a unique least-squares approximate solution $\widehat{\mathbf{x}}$. Otherwise, the expression in Proposition 7.5.1 produces the solution to $A\widehat{\mathbf{x}} = \mathbf{b}$ having the shortest length.

The matrix $A^+ = V_r \Sigma_r^{-1} U_r^T$ is known as the *Moore-Penrose psuedoinverse* of A. When A is invertible, $A^{-1} = A^+$.

7.5.2 Rank k approximations

If we have a singular value decomposition for a matrix A, we can form a sequence of matrices A_k that approximate A with increasing accuracy. This may feel familiar to calculus students who have seen the way in which a function $f(x)$ can be approximated by a linear function, a quadratic function, and so forth with increasing accuracy.

We'll begin with a singular value decomposition of a rank r matrix A so that $A = U\Sigma V^T$. To create the approximating matrix A_k, we keep the first k singular values and set the others to zero. For instance, if $\Sigma = \begin{bmatrix} 22 & 0 & 0 & 0 & 0 \\ 0 & 14 & 0 & 0 & 0 \\ 0 & 0 & 3 & 0 & 0 \\ 0 & 0 & 0 & 0 & 0 \end{bmatrix}$, we can form matrices

$$\Sigma^{(1)} = \begin{bmatrix} 22 & 0 & 0 & 0 & 0 \\ 0 & 0 & 0 & 0 & 0 \\ 0 & 0 & 0 & 0 & 0 \\ 0 & 0 & 0 & 0 & 0 \end{bmatrix}, \quad \Sigma^{(2)} = \begin{bmatrix} 22 & 0 & 0 & 0 & 0 \\ 0 & 14 & 0 & 0 & 0 \\ 0 & 0 & 0 & 0 & 0 \\ 0 & 0 & 0 & 0 & 0 \end{bmatrix}$$

and define $A_1 = U\Sigma^{(1)} V^T$ and $A_2 = U\Sigma^{(2)} V^T$. Because A_k has k nonzero singular values, we know that rank$(A_k) = k$. In fact, there is a sense in which A_k is the closest matrix to A among all rank k matrices.

Activity 7.5.3. Let's consider a matrix $A = U\Sigma V^T$ where

$$U = \begin{bmatrix} \frac{1}{2} & \frac{1}{2} & \frac{1}{2} & \frac{1}{2} \\ \frac{1}{2} & \frac{1}{2} & -\frac{1}{2} & -\frac{1}{2} \\ \frac{1}{2} & -\frac{1}{2} & \frac{1}{2} & -\frac{1}{2} \\ \frac{1}{2} & -\frac{1}{2} & -\frac{1}{2} & \frac{1}{2} \end{bmatrix}, \quad \Sigma = \begin{bmatrix} 500 & 0 & 0 & 0 \\ 0 & 100 & 0 & 0 \\ 0 & 0 & 20 & 0 \\ 0 & 0 & 0 & 4 \end{bmatrix}$$

7.5. USING SINGULAR VALUE DECOMPOSITIONS

$$V = \begin{bmatrix} \frac{1}{2} & \frac{1}{2} & \frac{1}{2} & \frac{1}{2} \\ \frac{1}{2} & -\frac{1}{2} & -\frac{1}{2} & \frac{1}{2} \\ -\frac{1}{2} & -\frac{1}{2} & \frac{1}{2} & \frac{1}{2} \\ -\frac{1}{2} & \frac{1}{2} & -\frac{1}{2} & \frac{1}{2} \end{bmatrix}$$

Evaluating the following cell will create the matrices U, V, and Sigma. Notice how the `diagonal_matrix` command provides a convenient way to form the diagonal matrix Σ.

```
h = 1/2
U = matrix(4,4,[h,h,h,h,  h,h,-h,-h,  h,-h,h,-h,  h,-h,-h,h])
V = matrix(4,4,[h,h,h,h,  h,-h,-h,h,  -h,-h,h,h,  -h,h,-h,h])
Sigma = diagonal_matrix([500, 100, 20, 4])
```

a. Form the matrix $A = U\Sigma V^T$. What is rank(A)?

b. Now form the approximating matrix $A_1 = U\Sigma^{(1)} V^T$. What is rank(A_1)?

c. Find the error in the approximation $A \approx A_1$ by finding $A - A_1$.

d. Now find $A_2 = U\Sigma^{(2)} V^T$ and the error $A - A_2$. What is rank(A_2)?

e. Find $A_3 = U\Sigma^{(3)} V^T$ and the error $A - A_3$. What is rank(A_3)?

f. What would happen if we were to compute A_4?

g. What do you notice about the error $A - A_k$ as k increases?

In this activity, the approximating matrix A_k has rank k because its singular value decomposition has k nonzero singular values. We then saw how the difference between A and the approximations A_k decreases as k increases, which means that the sequence A_k forms better approximations as k increases.

Another way to represent A_k is with a reduced singular value decomposition so that $A_k = U_k \Sigma_k V_k^T$ where

$$U_k = \begin{bmatrix} \mathbf{u}_1 & \cdots & \mathbf{u}_k \end{bmatrix}, \quad \Sigma_k = \begin{bmatrix} \sigma_1 & 0 & \cdots & 0 \\ 0 & \sigma_2 & \cdots & 0 \\ \vdots & \vdots & \ddots & \vdots \\ 0 & 0 & \cdots & \sigma_k \end{bmatrix}, \quad V_k = \begin{bmatrix} \mathbf{v}_1 & \cdots & \mathbf{v}_k \end{bmatrix}.$$

Notice that the rank 1 matrix A_1 then has the form $A_1 = \mathbf{u}_1 \begin{bmatrix} \sigma_1 \end{bmatrix} \mathbf{v}_1^T = \sigma_1 \mathbf{u}_1 \mathbf{v}_1^T$ and that we can similarly write:

$$A \approx A_1 = \sigma_1 \mathbf{u}_1 \mathbf{v}_1^T$$

$$A \approx A_2 = \sigma_1 \mathbf{u}_1 \mathbf{v}_1^T + \sigma_2 \mathbf{u}_2 \mathbf{v}_2^T$$
$$A \approx A_3 = \sigma_1 \mathbf{u}_1 \mathbf{v}_1^T + \sigma_2 \mathbf{u}_2 \mathbf{v}_2^T + \sigma_3 \mathbf{u}_3 \mathbf{v}_3^T$$
$$\vdots$$
$$A = A_r = \sigma_1 \mathbf{u}_1 \mathbf{v}_1^T + \sigma_2 \mathbf{u}_2 \mathbf{v}_2^T + \sigma_3 \mathbf{u}_3 \mathbf{v}_3^T + \ldots + \sigma_r \mathbf{u}_r \mathbf{v}_r^T.$$

Given two vectors $\mathbf{u}$ and $\mathbf{v}$, the matrix $\mathbf{u}\,\mathbf{v}^T$ is called the *outer product* of $\mathbf{u}$ and $\mathbf{v}$. (The dot product $\mathbf{u} \cdot \mathbf{v} = \mathbf{u}^T \mathbf{v}$ is sometimes called the *inner product*.) An outer product will always be a rank 1 matrix so we see above how A_k is obtained by adding together k rank 1 matrices, each of which gets us one step closer to the original matrix A.

7.5.3 Principal component analysis

In Section 7.3, we explored principal component analysis as a technique to reduce the dimension of a dataset. In particular, we constructed the covariance matrix C from a demeaned data matrix and saw that the eigenvalues and eigenvectors of C tell us about the variance of the dataset in different directions. We referred to the eigenvectors of C as *principal components* and found that projecting the data onto a subspace defined by the first few principal components frequently gave us a way to visualize the dataset. As we added more principal components, we retained more information about the original dataset. This feels similar to the rank k approximations we have just seen so let's explore the connection.

Suppose that we have a dataset with N points, that A represents the demeaned data matrix, that $A = U\Sigma V^T$ is a singular value decomposition, and that the singular values are A are denoted as σ_i. It follows that the covariance matrix

$$C = \frac{1}{N} A A^T = \frac{1}{N}(U\Sigma V^T)(U\Sigma V^T)^T = U\left(\frac{1}{N}\Sigma\Sigma^T\right)U^T.$$

Notice that $\frac{1}{N}\Sigma\Sigma^T$ is a diagonal matrix whose diagonal entries are $\frac{1}{N}\sigma_i^2$. Therefore, it follows that

$$C = U\left(\frac{1}{N}\Sigma\Sigma^T\right)U^T$$

is an orthogonal diagonalization of C showing that

- the principal components of the dataset, which are the eigenvectors of C, are given by the columns of U. In other words, the left singular vectors of A are the principal components of the dataset.

- the variance in the direction of a principal component is the associated eigenvalue of C and therefore
$$V_{\mathbf{u}_i} = \frac{1}{N}\sigma_i^2.$$

Activity 7.5.4. Let's revisit the iris dataset that we studied in Section 7.3. Remember that there are four measurements given for each of 150 irises and that each iris belongs

7.5. USING SINGULAR VALUE DECOMPOSITIONS

to one of three species.

Evaluating the following cell will load the dataset and define the demeaned data matrix A whose shape is 4×150.

```
url='https://raw.githubusercontent.com/davidaustinm/'
url+='ula_modules/master/pca_iris.py'
sage.repl.load.load(url, globals())
df.T
```

a. Find the singular values of A using the command A.singular_values() and use them to determine the variance $V_{\mathbf{u}_j}$ in the direction of each of the four principal components. What is the fraction of variance retained by the first two principal components?

b. We will now write the matrix $\Gamma = \Sigma V^T$ so that $A = U\Gamma$. Suppose that a demeaned data point, say, the 100th column of A, is written as a linear combination of principal components:

$$\mathbf{x} = c_1\mathbf{u}_1 + c_2\mathbf{u}_2 + c_3\mathbf{u}_3 + c_4\mathbf{u}_4.$$

Explain why $\begin{bmatrix} c_1 \\ c_2 \\ c_3 \\ c_4 \end{bmatrix}$, the vector of coordinates of $\mathbf{x}$ in the basis of principal components, appears as 100th column of Γ.

c. Suppose that we now project this demeaned data point $\mathbf{x}$ orthogonally onto the subspace spanned by the first two principal components $\mathbf{u}_1$ and $\mathbf{u}_2$. What are the coordinates of the projected point in this basis and how can we find them in the matrix Γ?

d. Alternatively, consider the approximation $A_2 = U_2 \Sigma_2 V_2^T$ of the demeaned data matrix A. Explain why the 100th column of A_2 represents the projection of $\mathbf{x}$ onto the two-dimensional subspace spanned by the first two principal components, $\mathbf{u}_1$ and $\mathbf{u}_2$. Then explain why the coefficients in that projection, $c_1\mathbf{u}_1 + c_2\mathbf{u}_2$, form the two-dimensional vector $\begin{bmatrix} c_1 \\ c_2 \end{bmatrix}$ that is the 100th column of $\Gamma_2 = \Sigma_2 V_2^T$.

e. Now we've seen that the columns of $\Gamma_2 = \Sigma_2 V_2^T$ form the coordinates of the demeaned data points projected on to the two-dimensional subspace spanned by $\mathbf{u}_1$ and $\mathbf{u}_2$. In the cell below, find a singular value decomposition of A and use it to form the matrix Gamma2. When you evaluate this cell, you will see a plot of the projected demeaned data plots, similar to the one we created in Section 7.3.

```
# Form the SVD of A and use it to form Gamma2

Gamma2 = 

# The following will plot the projected demeaned data points
data = Gamma2.columns()
(list_plot(data[:50], color='blue', aspect_ratio=1) +
 list_plot(data[50:100], color='orange') +
 list_plot(data[100:], color='green'))
```

In our first encounter with principal component analysis, we began with a demeaned data matrix A, formed the covariance matrix C, and used the eigenvalues and eigenvectors of C to project the demeaned data onto a smaller dimensional subspace. In this section, we have seen that a singular value decomposition of A provides a more direct route: the left singular vectors of A form the principal components and the approximating matrix A_k represents the data points projected onto the subspace spanned by the first k principal components. The coordinates of a projected demeaned data point are given by the columns of $\Gamma_k = \Sigma_k V_k^T$.

7.5.4 Image compressing and denoising

In addition to principal component analysis, the approximations A_k of a matrix A obtained from a singular value decomposition can be used in image processing. Remember that we studied the JPEG compression algorithm, whose foundation is the change of basis defined by the Discrete Cosine Transform, in Section 3.3. We will now see how a singular value decomposition provides another tool for both compressing images and removing noise in them.

Activity 7.5.5. Evaluating the following cell loads some data that we'll use in this activity. To begin, it defines and displays a 25×15 matrix A.

```
url='https://raw.githubusercontent.com/davidaustinm/'
url+='ula_modules/master/svd_compress.py'
sage.repl.load.load(url, globals())
print(A)
```

    a. If we interpret 0 as black and 1 as white, this matrix represents an image as shown below.

```
display_matrix(A)
```

We will explore how the singular value decomposition helps us to compress this image.

        1. By inspecting the image represented by A, identify a basis for Col(A) and

7.5. USING SINGULAR VALUE DECOMPOSITIONS

determine rank(A).

2. The following cell plots the singular values of A. Explain how this plot verifies that the rank is what you found in the previous part.

   ```
   plot_sv(A)
   ```

3. There is a command approximate(A, k) that creates the approximation A_k. Use the cell below to define k and look at the images represented by the first few approximations. What is the smallest value of k for which $A = A_k$?

   ```
   k = 
   display_matrix(approximate(A, k))
   ```

4. Now we can see how the singular value decomposition allows us to compress images. Since this is a 25×15 matrix, we need $25 \cdot 15 = 375$ numbers to represent the image. However, we can also reconstruct the image using a small number of singular values and vectors:

   $$A = A_k = \sigma_1 \mathbf{u}_1 \mathbf{v}_1^T + \sigma_2 \mathbf{u}_2 \mathbf{v}_2^T + \ldots + \sigma_k \mathbf{u}_k \mathbf{v}_k^T.$$

   What are the dimensions of the singular vectors $\mathbf{u}_i$ and $\mathbf{v}_i$? Between the singular vectors and singular values, how many numbers do we need to reconstruct A_k for the smallest k for which $A = A_k$? This is the compressed size of the image.

5. The *compression ratio* is the ratio of the uncompressed size to the compressed size. What compression ratio does this represent?

b. Next we'll explore an example based on a photograph.

1. Consider the following image consisting of an array of 316×310 pixels stored in the matrix A.

   ```
   A = matrix(RDF, image)
   display_image(A)
   ```

   Plot the singular values of A.

   ```
   plot_sv(A)
   ```

2. Use the cell below to study the approximations A_k for $k = 1, 10, 20, 50, 100$.

   ```
   k = 1
   display_image(approximate(A, k))
   ```

   Notice how the approximating image A_k more closely approximates the original image A as k increases.

   What is the compression ratio when $k = 50$? What is the compression ratio when $k = 100$? Notice how a higher compression ratio leads to a lower quality reconstruction of the image.

c. A second, related application of the singular value decomposition to image processing is called *denoising*. For example, consider the image represented by the matrix A below.

```
A = matrix(RDF, noise.values)
display_matrix(A)
```

This image is similar to the image of the letter "O" we first studied in this activity, but there are splotchy regions in the background that result, perhaps, from scanning the image. We think of the splotchy regions as noise, and our goal is to improve the quality of the image by reducing the noise.

1. Plot the singular values below. How are the singular values of this matrix similar to those represented by the clean image that we considered earlier and how are they different?

```
plot_sv(A)
```

2. There is a natural point where the singular values dramatically decrease so it makes sense to think of the noise as being formed by the small singular values. To denoise the image, we will therefore replace A by its approximation A_k, where k is the point at which the singular values drop off. This has the effect of setting the small singular values to zero and hence eliminating the noise. Choose an appropriate value of k below and notice that the new image appears to be somewhat cleaned up as a result of removing the noise.

```
k = 
display_matrix(approximate(A, k))
```

Several examples illustrating how the singular value decomposition compresses images are available at this page from Tim Baumann.[1]

7.5.5 Analyzing Supreme Court cases

As we've seen, a singular value decomposition concentrates the most important features of a matrix into the first singular values and singular vectors. We will now use this observation to extract meaning from a large dataset giving the voting records of Supreme Court justices. A similar analysis appears in the paper A pattern analysis of the second Rehnquist U.S. Supreme Court[2] by Lawrence Sirovich.

The makeup of the Supreme Court was unusually stable during a period from 1994-2005 when it was led by Chief Justice William Rehnquist. This is sometimes called the *second Rehnquist court*. The justices during this period were:

- William Rehnquist

---
[1] timbaumann.info/projects.html
[2] gvsu.edu/s/21F

7.5. USING SINGULAR VALUE DECOMPOSITIONS

- Antonin Scalia
- Clarence Thomas
- Anthony Kennedy
- Sandra Day O'Connor
- John Paul Stevens
- David Souter
- Ruth Bader Ginsburg
- Stephen Breyer

During this time, there were 911 cases in which all nine judges voted. We would like to understand patterns in their voting.

Activity 7.5.6. Evaluating the following cell loads and displays a dataset describing the votes of each justice in these 911 cases. More specifically, an entry of +1 means that the justice represented by the row voted with the majority in the case represented by the column. An entry of -1 means that justice was in the minority. This information is also stored in the 9×911 matrix A.

```
url='https://raw.githubusercontent.com/davidaustinm/'
url+='ula_modules/master/svd_supreme.py'
sage.repl.load.load(url, globals())
A = matrix(RDF, cases.values)
cases
```

The justices are listed, very roughly, in order from more conservative to more progressive.

In this activity, it will be helpful to visualize the entries in various matrices and vectors. The next cell displays the first 50 columns of the matrix A with white representing an entry of +1, red representing -1, and black representing 0.

```
display_matrix(A.matrix_from_columns(range(50)))
```

  a. Plot the singular values of A below. Describe the significance of this plot, including the relative contributions from the singular values σ_k as k increases.

```
plot_sv(A)
```

  b. Form the singular value decomposition $A = U\Sigma V^T$ and the matrix of coefficients Γ so that $A = U\Gamma$.

```
```

  c. We will now study a particular case, the second case which appears as the column of A indexed by 1. There is a command `display_column(A, k)` that provides a visual display of the kth column of a matrix A. Describe the justices' votes in the second case.

d. Also, display the first left singular vector $\mathbf{u}_1$, the column of U indexed by 0, and the column of Γ holding the coefficients that express the second case as a linear combination of left singular vectors.

What does this tell us about how the second case is constructed as a linear combination of left singular vectors? What is the significance of the first left singular vector $\mathbf{u}_1$?

e. Let's now study the 48th case, which is represented by the column of A indexed by 47. Describe the voting pattern in this case.

f. Display the second left singular vector $\mathbf{u}_2$ and the vector of coefficients that express the 48th case as a linear combination of left singular vectors.

Describe how this case is constructed as a linear combination of singular vectors. What is the significance of the second left singular vector $\mathbf{u}_2$?

g. The data in Table 7.5.2 describes the number of cases decided by each possible vote count.

Table 7.5.2 Number of cases by vote count

Vote count	# of cases
9-0	405
8-1	89
7-2	111
6-3	118
5-4	188

How do the singular vectors $\mathbf{u}_1$ and $\mathbf{u}_2$ reflect this data? Would you characterize the court as leaning toward the conservatives or progressives? Use these singular vectors to explain your response.

h. Cases decided by a 5-4 vote are often the most impactful as they represent a sharp divide among the justices and, often, society at large. For that reason, we will now focus on the 5-4 decisions. Evaluating the next cell forms the 9×188 matrix B consisting of 5-4 decisions.

```
B = matrix(RDF, fivefour.values)
display_matrix(B.matrix_from_columns(range(50)))
```

7.5. USING SINGULAR VALUE DECOMPOSITIONS

Form the singular value decomposition of $B = U\Sigma V^T$ along with the matrix Γ of coefficients so that $B = U\Gamma$ and display the first left singular vector $\mathbf{u}_1$. Study how the 7th case, indexed by 6, is constructed as a linear combination of left singular vectors.

What does this singular vector tell us about the make up of the court and whether it leans towards the conservatives or progressives?

i. Display the second left singular vector $\mathbf{u}_2$ and study how the 6th case, indexed by 5, is constructed as a linear combination of left singular vectors.

What does $\mathbf{u}_2$ tell us about the relative importance of the justices' voting records?

j. By a *swing vote*, we mean a justice who is less inclined to vote with a particular bloc of justices but instead swings from one bloc to another with the potential to sway close decisions. What do the singular vectors $\mathbf{u}_1$ and $\mathbf{u}_2$ tell us about the presence of voting blocs on the court and the presence of a swing vote? Which justice represents the swing vote?

7.5.6 Summary

This section has demonstrated some uses of the singular value decomposition. Because the singular values appear in decreasing order, the decomposition has the effect of concentrating the most important features of the matrix into the first singular values and singular vectors.

- Because the first left singular vectors form an orthonormal basis for Col(A), a singular value decomposition provides a convenient way to project vectors onto Col(A) and therefore to solve least-squares problems.

- A singular value decomposition of a rank r matrix A leads to a series of approximations A_k of A where

$$A \approx A_1 = \sigma_1 \mathbf{u}_1 \mathbf{v}_1^T$$
$$A \approx A_2 = \sigma_1 \mathbf{u}_1 \mathbf{v}_1^T + \sigma_2 \mathbf{u}_2 \mathbf{v}_2^T$$
$$A \approx A_3 = \sigma_1 \mathbf{u}_1 \mathbf{v}_1^T + \sigma_2 \mathbf{u}_2 \mathbf{v}_2^T + \sigma_3 \mathbf{u}_3 \mathbf{v}_3^T$$
$$\vdots$$
$$A = A_r = \sigma_1 \mathbf{u}_1 \mathbf{v}_1^T + \sigma_2 \mathbf{u}_2 \mathbf{v}_2^T + \sigma_3 \mathbf{u}_3 \mathbf{v}_3^T + \ldots + \sigma_r \mathbf{u}_r \mathbf{v}_r^T$$

In each case, A_k is the rank k matrix that is closest to A.

- If A is a demeaned data matrix, the left singular vectors give the principal components of A, and the variance in the direction of a principal component can be simply expressed in terms of the corresponding singular value.

- The singular value decomposition has many applications. In this section, we looked at how the decomposition is used in image processing through the techniques of compression and denoising.

- Because the first few left singular vectors contain the most important features of a matrix, we can use a singular value decomposition to extract meaning from a large dataset as we did when analyzing the voting patterns of the second Rehnquist court.

7.5.7 Exercises

1. Suppose that
$$A = \begin{bmatrix} 2.1 & -1.9 & 0.1 & 3.7 \\ -1.5 & 2.7 & 0.9 & -0.6 \\ -0.4 & 2.8 & -1.5 & 4.2 \\ -0.4 & 2.4 & 1.9 & -1.8 \end{bmatrix}.$$

   a. Find the singular values of A. What is rank(A)?

   b. Find the sequence of matrices A_1, A_2, A_3, and A_4 where A_k is the rank k approximation of A.

2. Suppose we would like to find the best quadratic function
$$\beta_0 + \beta_1 x + \beta_2 x^2 = y$$
   fitting the points
$$(0, 1), (1, 0), (2, 1.5), (3, 4), (4, 8).$$

   a. Set up a linear system $A\mathbf{x} = \mathbf{b}$ describing the coefficients $\mathbf{x} = \begin{bmatrix} \beta_0 \\ \beta_1 \\ \beta_2 \end{bmatrix}$.

   b. Find the singular value decomposition of A.

   c. Use the singular value decomposition to find the least-squares approximate solution $\widehat{\mathbf{x}}$.

3. Remember that the outer product of two vectors $\mathbf{u}$ and $\mathbf{v}$ is the matrix $\mathbf{u}\mathbf{v}^T$.

   a. Suppose that $\mathbf{u} = \begin{bmatrix} 2 \\ -3 \end{bmatrix}$ and $\mathbf{v} = \begin{bmatrix} 2 \\ 0 \\ 1 \end{bmatrix}$. Evaluate the outer product $\mathbf{u}\mathbf{v}^T$. To get a clearer sense of how this works, perform this operation without using technology. How is each of the columns of $\mathbf{u}\mathbf{v}^T$ related to $\mathbf{u}$?

   b. Suppose $\mathbf{u}$ and $\mathbf{v}$ are general vectors. What is rank($\mathbf{u}\mathbf{v}^T$) and what is a basis for its column space Col($\mathbf{u}\mathbf{v}^T$)?

7.5. USING SINGULAR VALUE DECOMPOSITIONS

c. Suppose that **u** is a unit vector. What is the effect of multiplying a vector by the matrix **u** **u**^{T}?

4. Evaluating the following cell loads in a dataset recording some features of 1057 houses. Notice how the lot size varies over a relatively small range compared to the other features. For this reason, in addition to demeaning the data, we'll scale each feature by dividing by its standard deviation so that the range of values is similar for each feature. The matrix A holds the result.

```
import pandas as pd
url='https://raw.githubusercontent.com/davidaustinm/'
url+='ula_modules/master/data/housing.csv'
df = pd.read_csv(url, index_col=0)
df = df.fillna(df.mean())
std = (df-df.mean())/df.std()
A = matrix(std.values).T
df.T
```

a. Find the singular values of A and use them to determine the variance in the direction of the principal components.

b. For what fraction of the variance do the first two principal components account?

c. Find a singular value decomposition of A and construct the 2×1057 matrix B whose entries are the coordinates of the demeaned data points projected on to the two-dimensional subspace spanned by the first two principal components. You can plot the projected data points using `list_plot(B.columns())`.

d. Study the entries in the first two principal components $\mathbf{u}_1$ and $\mathbf{u}_2$. Would a more expensive house lie on the left, right, top, or bottom of the plot you constructed?

e. In what ways does a house that lies on the far left of the plot you constructed differ from an average house? In what ways does a house that lies near the top of the plot you constructed differ from an average house?

5. Let's revisit the voting records of justices on the second Rehnquist court. Evaluating the following cell will load the voting records of the justices in the 188 cases decided by a 5-4 vote and store them in the matrix A.

```
url='https://raw.githubusercontent.com/davidaustinm/'
url+='ula_modules/master/svd_supreme.py'
sage.repl.load.load(url, globals())
A = matrix(RDF, fivefour.values)
v = vector(188*[1])
fivefour
```

a. The cell above also defined the 188-dimensional vector **v** whose entries are all 1. What does the product $A\mathbf{v}$ represent? Use the following cell to evaluate this product.

    b. How does the product $A\mathbf{v}$ tell us which justice voted in the majority most frequently? What does this say about the presence of a swing vote on the court?

    c. How does this product tell us whether we should characterize this court as leaning conservative or progressive?

    d. How does this product tell us about the presence of a second swing vote on the court?

    e. Study the left singular vector $\mathbf{u}_3$ and describe how it reinforces the fact that there was a second swing vote. Who was this second swing vote?

6. The following cell loads a dataset that describes the percentages with which justices on the second Rehnquist court agreed with one another. For instance, the entry in the first row and second column is 72.78, which means that Justices Rehnquist and Scalia agreed with each other in 72.78% of the cases.

```
url='https://raw.githubusercontent.com/davidaustinm/'
url+='ula_modules/master/svd_supreme.py'
sage.repl.load.load(url, globals())
A = 1/100*matrix(RDF, agreement.values)
agreement
```

    a. Examine the matrix A. What special structure does this matrix have and why should we expect it to have this structure?

    b. Plot the singular values of A below. For what value of k would the approximation A_k be a reasonable approximation of A?

```
plot_sv(A)
```

    c. Find a singular value decomposition $A = U\Sigma V^T$ and examine the matrices U and V using, for instance, n(U, 3). What do you notice about the relationship between U and V and why should we expect this relationship to hold?

    d. The command approximate(A, k) will form the approximating matrix A_k. Study the matrix A_1 using the display_matrix command. Which justice or justices seem to be most agreeable, that is, most likely to agree with other justices? Which justice is least agreeable?

    e. Examine the difference $A_2 - A_1$ and describe how this tells us about the presence of voting blocs and swing votes on the court.

7.5. USING SINGULAR VALUE DECOMPOSITIONS

7. Suppose that $A = U_r \Sigma_r V_r^T$ is a reduced singular value decomposition of the $m \times n$ matrix A. The matrix $A^+ = V_r \Sigma_r^{-1} U_r^T$ is called the *Moore-Penrose inverse* of A.

   a. Explain why A^+ is an $n \times m$ matrix.

   b. If A is an invertible, square matrix, explain why $A^+ = A^{-1}$.

   c. Explain why $AA^+ \mathbf{b} = \widehat{\mathbf{b}}$, the orthogonal projection of $\mathbf{b}$ onto $\text{Col}(A)$.

   d. Explain why $A^+ A \mathbf{x} = \widehat{\mathbf{x}}$, the orthogonal projection of $\mathbf{x}$ onto $\text{Col}(A^T)$.

8. In Subsection 5.1.1, we saw how some linear algebraic computations are sensitive to round off error made by a computer. A singular value decomposition can help us understand when this situation can occur.

   For instance, consider the matrices

   $$A = \begin{bmatrix} 1.0001 & 1 \\ 1 & 1 \end{bmatrix}, \quad B = \begin{bmatrix} 1 & 1 \\ 1 & 1 \end{bmatrix}.$$

   The entries in these matrices are quite close to one another, but A is invertible while B is not. It seems like A is *almost* singular. In fact, we can measure how close a matrix is to being singular by forming the *condition number*, σ_1/σ_n, the ratio of the largest to smallest singular value. If A were singular, the condition number would be undefined because the singular value $\sigma_n = 0$. Therefore, we will think of matrices with large condition numbers as being close to singular.

   a. Define the matrix A and find a singular value decomposition. What is the condition number of A?

   b. Define the left singular vectors $\mathbf{u}_1$ and $\mathbf{u}_2$. Compare the results $A^{-1}\mathbf{b}$ when

      1. $\mathbf{b} = \mathbf{u}_1 + \mathbf{u}_2$.
      2. $\mathbf{b} = 2\mathbf{u}_1 + \mathbf{u}_2$.

      Notice how a small change in the vector $\mathbf{b}$ leads to a small change in $A^{-1}\mathbf{b}$.

   c. Now compare the results $A^{-1}\mathbf{b}$ when

      1. $\mathbf{b} = \mathbf{u}_1 + \mathbf{u}_2$.
      2. $\mathbf{b} = \mathbf{u}_1 + 2\mathbf{u}_2$.

      Notice now how a small change in $\mathbf{b}$ leads to a large change in $A^{-1}\mathbf{b}$.

   d. Previously, we saw that, if we write $\mathbf{x}$ in terms of left singular vectors $\mathbf{x} = c_1 \mathbf{v}_1 + c_2 \mathbf{v}_2$, then we have
      $$\mathbf{b} = A\mathbf{x} = c_1 \sigma_1 \mathbf{u}_1 + c_2 \sigma_2 \mathbf{u}_2.$$
      If we write $\mathbf{b} = d_1 \mathbf{u}_1 + d_2 \mathbf{u}_2$, explain why $A^{-1}\mathbf{b}$ is sensitive to small changes in d_2.

   Generally speaking, a square matrix A with a large condition number will demonstrate this type of behavior so that the computation of A^{-1} is likely to be affected by round off error. We call such a matrix *ill-conditioned*.

APPENDIX A

Sage Reference

We have introduced a number of Sage commands throughout the text, and the most important ones are summarized here in a single place.

Accessing Sage In addition to the Sage cellls included throughout the book, there are a number of ways to access Sage.

  a. There is a freely available Sage cell at `sagecell.sagemath.org`.

  b. You can save your Sage work by creating an account at `cocalc.com` and working in a Sage worksheet.

  c. There is a page of Sage cells at `gvsu.edu/s/0Ng`. The results obtained from evaluating one cell are available in other cells on that page. However, you will lose any work once the page is reloaded.

Creating matrices   There are a couple of ways to create matrices. For instance, the matrix

$$\begin{bmatrix} -2 & 3 & 0 & 4 \\ 1 & -2 & 1 & -3 \\ 0 & 2 & 3 & 0 \end{bmatrix}$$

can be created in either of the two following ways.

a.
```
matrix(3, 4, [-2, 3, 0, 4,
              1,-2, 1,-3,
              0, 2, 3, 0])
```

b.
```
matrix([ [-2, 3, 0, 4],
         [ 1,-2, 1,-3],
         [ 0, 2, 3, 0] ])
```

Be aware that Sage can treat mathematically equivalent matrices in different ways depending on how they are entered. For instance, the matrix

```
matrix([ [1, 2],
         [2, 1] ])
```

has integer entries while

```
matrix([ [1.0, 2.0],
         [2.0, 1.0] ])
```

has floating point entries.
If you would like the entries to be considered as floating point numbers, you can include RDF in the definition of the matrix.

```
matrix(RDF, [ [1, 2],
              [2, 1] ])
```

Special matrices   The 4 × 4 identity matrix can be created with

                  `identity_matrix(4)`

A diagonal matrix can be created from a list of its diagonal entries. For instance,

                  `diagonal_matrix([3,-4,2])`

Reduced row echelon form   The reduced row echelon form of a matrix can be obtained using the `rref()` function. For instance,

```
A = matrix([ [1,2], [2,1] ])
A.rref()
```

Vectors   A vector is defined by listing its components.

                  `v = vector([3,-1,2])`

Addition   The + operator performs vector and matrix addition.

```
v = vector([2,1])
w = vector([-3,2])
print(v+w)
```

```
A = matrix([[2,-3],[1,2]])
B = matrix([[-4,1],[3,-1]])
print(A+B)
```

Multiplication The * operator performs scalar multiplication of vectors and matrices.

```
v = vector([2,1])
print(3*v)
A = matrix([[2,1],[-3,2]])
print(3*A)
```

Similarly, the * is used for matrix-vector and matrix-matrix multiplication.

```
A = matrix([[2,-3],[1,2]])
v = vector([2,1])
print(A*v)
B = matrix([[-4,1],[3,-1]])
print(A*B)
```

Operations on vectors

a. The length of a vector v is found using v.norm().

b. The dot product of two vectors v and w is v*w.

Operations on matrices

a. The transpose of a matrix A is obtained using either A.transpose() or A.T.

b. The inverse of a matrix A is obtained using either A.inverse() or A^-1.

c. The determinant of A is A.det().

d. A basis for the null space Nul(A) is found with A.right_kernel().

e. Pull out a column of A using, for instance, A.column(0), which returns the vector that is the first column of A.

f. The command A.matrix_from_columns([0,1,2]) returns the matrix formed by the first three columns of A.

Eigenvectors and eigenvalues

a. The eigenvalues of a matrix A can be found with A.eigenvalues(). The number of times that an eigenvalue appears in the list equals its multiplicity.

b. The eigenvectors of a matrix having rational entries can be found with A.eigenvectors_right().

c. If A can be diagonalized as $A = PDP^{-1}$, then

   D, P = A.right_eigenmatrix()

   provides the matrices D and P.

d. The characteristic polynomial of A is A.charpoly('x') and its factored form A.fcp('x').

Matrix factorizations

a. The LU factorization of a matrix

   P, L, U = A.LU()

   gives matrices so that $PA = LU$.

b. A singular value decomposition is obtained with

   U, Sigma, V = A.SVD()

   It's important to note that the matrix must be defined using RDF. For instance, A = matrix(RDF, 3,2,[1,0,-1,1,1,1]).

c. The QR factorization of A is A.QR() provided that A is defined using RDF.

Index

RGB color model, 179
YC_bC_r color model, 179

augmented matrix, 15

back substitution, 11
basic variable, 18
basis, 160
basis, standard, 162

characteristic equation, 244
characteristic polynomial, 245
chrominance, 179
coefficient matrix, 15
coefficient of determination, 400
column space, 220
consistent system, 35

decoupled system, 11
determinant, 203
diagonalizable, 259
dimension, 219
discrete dynamical system, 117
Discrete Fourier Transform, 185
dot product, 336

eigenspace, 246
eigenvalue, 230
eigenvalue, dominant, 324
eigenvector, 230

free variable, 18

Gaussian elimination, 10
Gram matrix, 459
Gram-Schmidt, 384

inconsistent system, 35

invertible, 146

linear combination, 51
linear equation, 8
linear system, 8
linearly dependent, 98
linearly independent, 98
lower triangular matrix, 150
luminance, 179

Markov chain, 294
matrix transformation, 109
matrix, addition, 63
matrix, elementary, 152
matrix, identity, 64
matrix, inverse, 146
matrix, rank, 222
matrix, scalar multiplication, 63
matrix, shape, 63
matrix, square, 146
matrix-vector multiplication, 65
Moore-Penrose psuedoinverse, 476
multiplicity, 248

normal equation, 398
null space, 222

orthogonal, 341
orthogonal complement, 356
orthogonal diagonalization, 419
orthogonal matrix, 378
orthogonal projection, 372
orthogonal set, 368
othonormal set, 370

parametric description, 18
partial pivoting, 315

pivot position, 34
positive matrix, 296
power method, 326
principal components, 447
probability vector, 292

quadratic form, 433

R squared, 400
rank, 222
reduced row echelon form, 17
reduced row echelon matrix, 17
round off error, 312
row equivalent, 15
row space, 467

scalar multiplication, 46
similarity, 263
solution, 8
solution space, 8

span, 81
state vector, 117
stationary vector, 295
steady-state vector, 295
stochastic matrix, 292
subspace, 217
symmetric matrix, 419

transition function, 117
transpose, 358
triangular system, 11

unit vector, 369
upper triangular matrix, 150

vector, 45
vector addition, 46

weights, 51

Colophon

This book was authored in PreTeXt.

Made in the USA
Monee, IL
15 July 2024

61834223R00284